The Fibonacci Resonance

The Fibonacci Resonance

and other new Golden Ratio discoveries

*Maths, music, archaeology, architecture, art,
quasicrystals, metamaterials,
Lucas numbers, Ori32*

by

Clive N. Menhinick

Ori32 geometry & cryptochromatology (CC) series:

Book 1

OnPerson

Published by
OnPerson International Limited, Poynton, Cheshire, England.

British Library Cataloguing-in-Publication Data
A catalogue record for this book is available from the British Library.
Includes bibliographic references (pp.487–576) and index.
1. Fibonacci numbers.   2. Golden section.
3. Recurrent sequences (Mathematics).

512.7'2–dc23.   MSC2010: 11B39   Fibonacci and Lucas numbers.

BISAC subjects:

MAT022000	MATHEMATICS/Number Theory,
SOC003000	SOCIAL SCIENCE/Archaeology,
ART015100	ART/History/Modern (late 19th Century to 1945),
MUS007000	MUSIC/Instruction & Study/Composition,
SCI016000	SCIENCE/Physics/Crystallography.

Typeset using LaTeX in Adobe® Utopia, Nimbus, and Fourier fonts. The author has asserted his moral rights.   The publisher cannot warrant that online resources referred to in this book will remain available.   Printed on acid-free paper: ∞. *Reasonable efforts have been made to trace owners of copyright in respect of material reproduced in this book, but should any have been inadvertently overlooked the publisher will be pleased to hear from them.*

Disclaimer: All information in this book is provided in good faith including for example but not limited to data regarding the bio-compatibility of quasicrystals. However it is the responsibility of the reader to check before making any use of such information. Neither author nor publisher shall be held liable for any consequences arising from the application of information presented in this text.

Cover picture credits   [*Refs. p.487–*], (*Figs. p.577–*).

Tiwanaku Sun Gate by Mhwater, Public Domain (Fig. 5).

A Sunday afternoon on the Island of la Grande Jatte by Georges Seurat,
      PD [823] (extracts from Fig. 64, and analysis Fig. 66).

Combined image showing 10-fold symmetry (Fig. 217), comprising
      —diffraction pattern by Materialscientist, CC-BY-SA-3.0 Licence [632, 144],
      —Samarkand tiling by Patrickkringgenberg, CC-BY-SA-3.0 Licence [731, 144].

Sunflower by US Department of Agriculture, PD (Fig. 55, detail).

Penrose tiling by Solarflare100, CC-BY-3.0 Licence [847, 143] (Fig. 187, detail).

Treble clef by Gringer, PD [277] (Fig. 81, detail).

To the Vedic Rishis *(Seers).*

योगस्थः कुरु कर्माणि

Yogasthaḥ kuru karmāṇi

'Established in Being, perform action.' [1]

---

[1] Veda Vyāsa [921].

Contents

IV TO PELL AND BEYOND

V ϕ IN SCIENCE

VI APPENDICES

Preface

Worldwide interest in Fibonacci numbers and the Golden Ratio[2] *is surging—that is, if* millions *of web references are anything to go by.*

Fibonacci and ϕ, the Golden Rectangle, and the Golden Spiral: these are still, even after thousands of years, very much *hot topics*. Together, these four have acquired an iconic status to rival gold itself. Over the centuries, they have exercised the minds of some of the world's greatest thinkers—mathematicians, architects, artists, botanists, educators, engineers, physicists, philosophers, and psychologists.

But while there are many popular works covering the basics and also many advanced resources, there remains something of a gap in between. This book seeks to fill that gap in a single volume. While it brings together in one place knowledge already familiar to the adepts, it also adds a fair amount of new material. The mathematics included is to high-school level, and straightforward examples are worked in detail, step by step. This is a recreational maths and science book, written to be enjoyed, but one still furnished with over 800 bibliographic references and a detailed index.

Black holes cosmologist Stephen Hawking once declared that he preferred to see things in terms of geometry rather than equations[3]. Invisibility physicist John Pendry says he 'thinks in geometry'—that he always likes to have pictures in his mind.[4] And so too, apparently, did Isaac Newton and Leonardo da Vinci who both consistently illustrated their writings. So, on good recommendation, *The Fibonacci Resonance* is lavishly illustrated (with over 233 images and diagrams —mostly in colour). The emphasis throughout is on *visualization*. Our goal is to *see* what the equations are saying.

---

[2]   Also called phi (ϕ), the Golden Number, Golden Section, Divine Proportion, ...
[3]   Kristine Larsen [523].
[4]   Anjana Ahuja [12].

In 618 pages, we celebrate Fibonacci and ϕ with three books in one:

1. The history of the Fibonacci numbers and the Golden Ratio. We review their beginnings in the ancient world—focussing on the people who made the key discoveries in nature and mathematics. We analyse the Golden Spiral (and its spiral family), and then move on to Fibonacci and ϕ examples in music (Bartók, Debussy, Xenakis); art (Boulanger, Seurat, Toulouse-Lautrec, Mondrian, Steinlen); and architecture (Le Corbusier and his 'Modulor'). Further, we visit the Bohemian haunts of *fin-de-siècle* Paris in order to understand how it became 'the capital of ϕ'.

2. The announcement of a new discovery. It is often said that as the Fibonacci numbers grow, the ratios of successive Fibonacci numbers progress as 'better and better approximations to the Golden Ratio'. Could it be possible to go beyond this approximation and pin down the exact relationship by which this growth progresses? Ideally such a scheme would pull together Fibonacci numbers, their companions the Lucas numbers, Golden Spirals, and the Golden Ratio—and yet retain a beautiful abacus bead simplicity of visualization...
The Fibonacci Resonance demonstrates this.

Further, we see how this visual way of thinking about growth may be generalized and extended to Pell numbers and to other such sequences. We also touch on links to the perfect numbers and record-breaking Mersenne primes, and we review the 'Ori32' angle geometry that originally led to this discovery (via a Golden partitioning of the circle). A new concept is introduced—that of *the companion axes offset angle.* For the Fibonacci and Lucas numbers, this is $150.505°$.

3. A review of Fibonacci and ϕ in modern science and maths. We look at how the same mathematical patterns found in pine cones and sunflowers also appear in scientific experiments studying the structure of proteins, magnetic cactuses, and superconductor flux bundles. The fascinating (and at times dramatic) story of the ϕ-based Penrose tiles and the discovery of quasicrystals is told. This began with Albrecht Dürer and Johannes Kepler (in the 16th and 17th centuries) and led to:

- a recent crisis in crystallography—
    only resolved by redefining the crystal!
- a Nobel prize, and
- exciting applications in 21st century light-based technologies

Clive Menhinick,   Poynton, Cheshire, UK   *28th October 2015.*

Acknowledgements

Thank you: To my family for their enduring patience and support.

To Benedict Heal for his strong encouragement over a long period.

To Dr. Sabah Merad for her heroic checking of chaotic early drafts of both the photographic and the initial Ori32 work.

To both Benedict and Sabah for their detailed readings of the final drafts of this book, and for their comments and suggestions.

To computing consultant and author Derek Andrews for recommending the use of the LaTeX typesetting system. To His Excellency Roberto Calzadilla, Bolivia's Ambassador to London, for inviting the author to the Bolivian Embassy to discuss the ancient Tiwanaku civilisation, for giving his valuable comments and suggestions, and for introducing the author to expert Tiwanaku researcher El Ing. Jorge Emilio Molina Rivera. To computing consultant and author Ian Graham for his advice on writing and his contribution to the copy-editing of the final version. To architect George A Hall for relating his experience of using Le Corbusier's Modulor system. To Trudy Hope of Wordsmithery for proof-reading services. To music scholar Roy Howat for his expert comments and suggestions regarding the sections on Bartók and Debussy. To computing consultant Dr. Hubert Matthews for timely help with Eclipse (software for building software) during the photographic phase of the overall project. To Professor Ronan McGrath, Head of School, Physical

Sciences, Liverpool University for his reading of the final drafts of the quasicrystal and related chapters and for his comments and suggestions and for inviting the author to lunch with Professor Danny Shechtman. To my wife, Laurence Menhinick for her detailed reading of the final draft and her comments and suggestions. To Susie Metcalfe at The Writers' Bureau for advice on style. To Trevor Palin of Palin Images for publicity photographs. To Professor Sir John Pendry, Chair in Theoretical Solid State Physics, Imperial College London, for his time discussing metamaterials after his Edwards Lecture 2014, and for informing the author of the metamaterial satellite antenna work of Kymeta. To David Renshaw, of the Royal Astronomical Society for his comments on the final drafts. To Hilary Richardson at the Mondrian/Holtzman Trust c/o HCR International for her expert comments and suggestions. To Tiwanacologist engineer Jorge Emilio Molina Rivera for checking the Tiwanaku Sun Gate and Moon Gate analyses. He used (respectively): Escalante, JF (1993) *Arquitectura Prehispánica en los Andes Bolivianos*, CIMA, La Paz, Bolivia; and Posnansky, A (1945) *Tihuanacu: Cuna del Hombre Americano*, Augustin, NY, USA; and he concluded that: 'The ϕ dimensions found on both Tiwanaku Gates are correct and objects of future and urgent investigations [this being] such an important cultural topic for mankind.' To nuclear physicist David Roberts, formerly of the Institute of Physics and the Safety and Reliability Association (retired) for his comments on the final drafts. To Nobel Laureate Professor Dan Shechtman for his time in discussing his quasicrystal discovery and clarifying the crucial distinction between TEM and X-ray techniques. To Sanskrit scholar Dr. Valerie J Roebuck for her cautionary advice regarding ancient Indian sources. To the staff of: University of Manchester libraries— main, John Rylands, Art & Archaeology, Joule, Kantorowich, and Lenegan; the Tate Library, London; and the Special Collections departments at Liverpool and Leeds University libraries, and The National Library of Scotland. And to all well-wishing friends and colleagues.

The author accepts full responsibility for any errors that may remain in this book, and requests notification—mailto: found_fr@ori32.com

Part I

BACKGROUND

Chapter 1

Fibonacci numbers and the Golden Ratio

Welcome to this book about Fibonacci numbers and the Golden Ratio. It presents a number of new discoveries, in particular, one that relates Fibonacci numbers, Lucas numbers, the Golden Ratio, and Golden Spirals—*the Fibonacci Resonance*. But first, we shall trace a little history and see how the Golden Ratio and the Fibonacci numbers cross paths—then eventually weave together into one thread. Towards the end of the book, we shall consider the importance of these numbers in the latest science and technology. Also, for a deeper understanding, we shall (from time to time) draw comparisons with related numbers such as the Silver Ratio.

What are Fibonacci numbers? F_n

The Fibonacci numbers are a deceptively simple set of numbers obtained by repeated addition. Starting with 'seeds' 0 and 1, we add the two preceding numbers to get the next: 0+1=1, 1+1=2, 1+2=3, and so on:

$$0, 1, 1, 2, 3, 5, 8, 13, 21, 34, 55, 89, 144, 233, 377, 610, 987, \ldots$$

Sequences such as this never end. The 'recurrence' rule is written

$$F_n = F_{n-1} + F_{n-2} \quad \text{with } F_0 = 0, \quad F_1 = 1. \tag{1.1}$$

As we shall see, the Fibonacci numbers are found in many places in nature—including sunflowers, pineapples, and pine cones.

The Golden Ratio 'phi' ϕ

As for the Golden Ratio, (also known as the Divine Proportion, the Golden Mean, the Golden number, and many other similar names,[1]) this emerged quite independently. In Fig. 1 the ABC line at the top illustrates the 'equal ratios' principle of the Golden Section—that the greater is to the smaller (AB/BC), as the whole is to the greater (AC/AB). This special ratio we call 'phi' where $\phi = 1.618\ldots$ There are many constructions leading to this division—here we have some of the more notable.[2] To the left we see the equilateral triangle mid-points relation—which was only very recently discovered by artist and geometer George Odom. Geometer H S M Coxeter published this work in Odom's name, saying just: 'Behold!' [710].[3] To the right of that is the classic construction whereby a square is extended to form a Golden Rectangle (also p.475). Below these is another triangular construction which relies on width being twice height giving a hypotenuse of $\sqrt{5}\times$ height—the swept arcs then deliver the Golden Section. And at the bottom, we have a pentagon with crossed chords each dividing the other in the Golden Section. Lastly we see how the Golden Spiral exhibits the Golden Section 'at every turn' (and we shall prove this shortly). Now, taken apart, the ratios AB/BC and AC/AB could each have arbitrary values; but once set equal to each other ('in proportion'), they give rise to *two* exact situations. The first is the famous one, where BC extends AB (as shown at the top), and the second—resulting from the following algebraic formulation—is that AC 'goes negative'. We show this in the diagram by placing C (call it C') *to the left of* A. To see this, we start by letting $AB = 1$, and $AC = x$. From this it follows that

$$\frac{AB}{BC} = \frac{1}{x-1} \quad = \quad \frac{AC}{AB} = \frac{x}{1}. \qquad (1.2)$$

Rearranging (1.2) gives us the equation

$$x^2 - x - 1 = 0. \qquad (1.3)$$

We solve this using the quadratic formula and obtain two roots,[4]

$$x = \left(1 \pm \sqrt{5}\right)/2. \qquad (1.4)$$

---

[1]   Grimaldi also mentions Sacred Ratio, and attributes Sacred Chapter to Leonardo da Vinci, and Divine Section to Kepler [274]. We also have Golden Proportion, Continuous Proportion, Mean of Phidias, Mean and Extreme Ratio...

[2]   See also for example: Walser [934], Posamentier and Lehmann [762].

[3]   Numbers in square brackets—see the References section starting on page 487.

[4]   If $ax^2 + bx + c = 0$, then $x = \left(-b \pm \sqrt{b^2 - 4ac}\right)/2a$. Here $a = 1$, $b = -1$, $c = -1$.

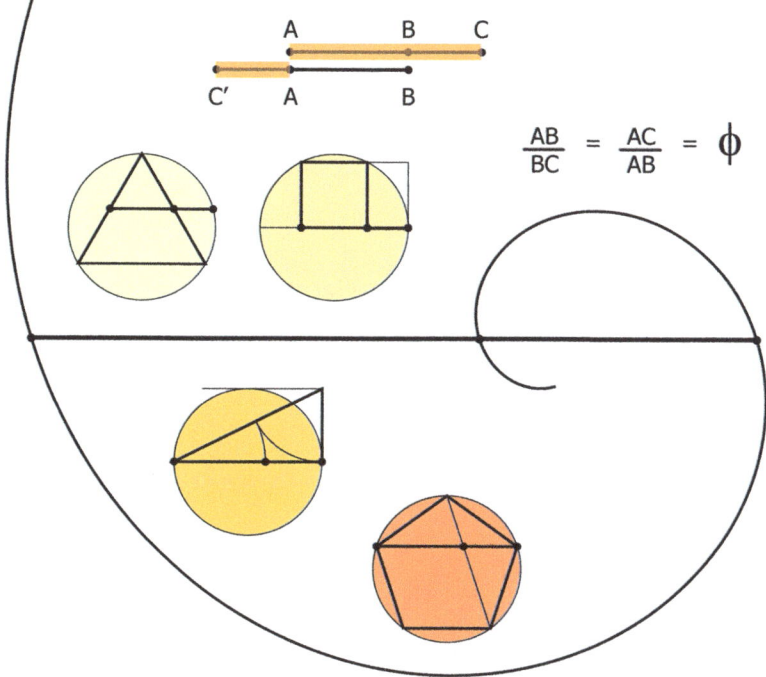

Figure 1: **The Golden Section**
—an ever abundant source of mathematical delight.

We could just take the positive result in (1.4) and identify it with the Golden Ratio ϕ and be done. However, as we shall be delving rather deeper, it will assist us to keep both roots in mind and to think of them on an equal footing—as α and β. We see it is the negative root β that provides the alternative geometric configuration AC', resulting in the ratio AC'/AB.

$$\alpha = \frac{1+\sqrt{5}}{2} \qquad = +\phi^{+1} \qquad = +1.618\ldots \qquad (1.5)$$

$$\beta = \frac{1-\sqrt{5}}{2} \qquad = -\phi^{-1} \qquad = -0.618\ldots \qquad (1.6)$$

3

The simple rule for converting α to β is the same as that for β to α—namely: 'take the reciprocal and negate'. Sometimes, β is described as the 'conjugate' of α, as α and β belong together. Accordingly, it is interesting to see how these roots combine and compare [275] :

$$\alpha + \beta = 1 \qquad \alpha^2 + \beta^2 = 3 \qquad \alpha\beta = -1 \qquad \alpha^2 = \alpha + 1$$

$$\alpha - \beta = \sqrt{5} \qquad \alpha^2 - \beta^2 = \sqrt{5} \qquad 1 - 2\beta = \sqrt{5} \qquad \beta^2 = \beta + 1.$$

Conics

In Fig. 2 we further witness the complementarity of α and β—in terms of conics. We see how equation (1.3) (rearranged as $x^2 = x+1$) can be shown as a straight line intersecting a parabola; then similarly, we see how equation (1.2) (rearranged as $1/x = x-1$) can be shown as a straight line intersecting a hyperbola. Both the lines have unity gradient. There also exists a 'Golden Ellipse' with major and minor axes in Golden proportion which has interesting properties [399].

A different angle

When the applications of the Fibonacci numbers and the Golden Ratio are considered, certain questions recur, such as:

- Are there obvious (or debatable) examples of F_n and ϕ being used by a particular artist, musician, or architect?

- Were these uses intentional, or the result of a some intuitive process (which somehow produced mathematical structure)?

- Did the resulting work *benefit* from the use of F_n and ϕ—that is, was any aesthetic, functional, or musical value added?

By answering these kinds of questions piecemeal, it is possible to build up a rather disjointed overall view of the Fibonacci numbers and the Golden Ratio. However, in this book, we shall approach the subject differently. We shall focus on the mathematics first and foremost, striving to find consistent themes and to gain an in-depth understanding and then ask:

- Do the Fibonacci numbers and the Golden Ratio provide excellent tools—available to artists, musicians, architects, engineers, and scientists... *should they choose to use them*?

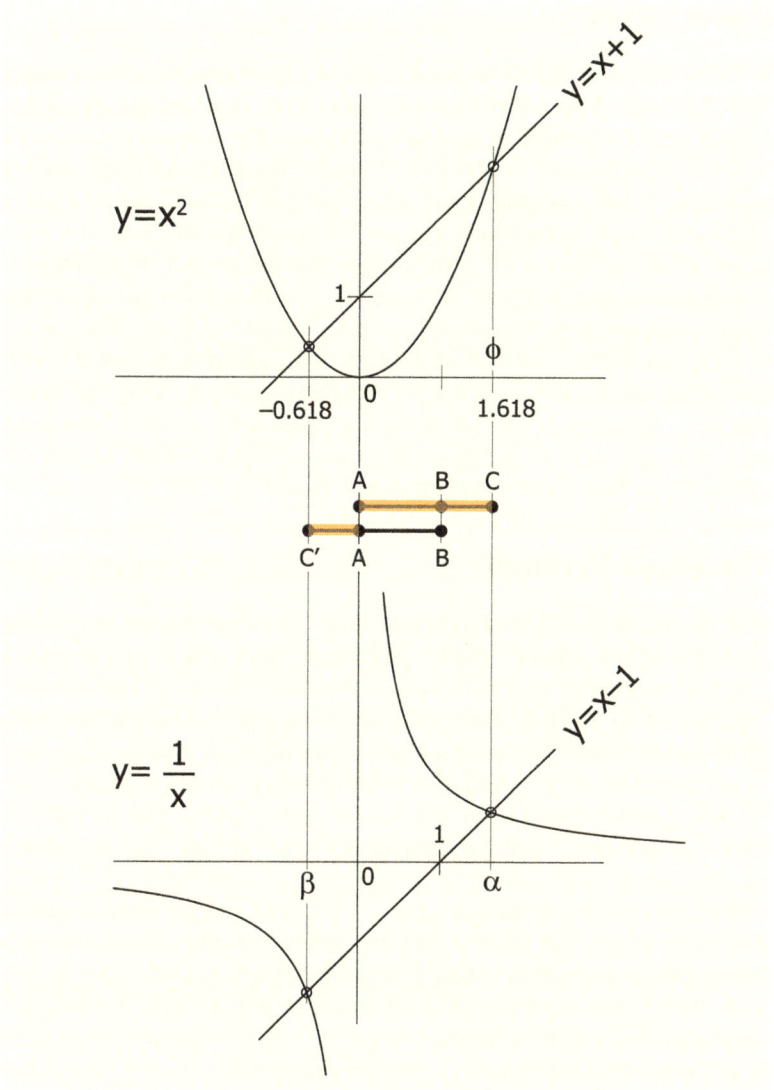

Figure 2: **The Golden Section and Conics.**

We shall still consider particular examples (ones which *suggest* Golden proportions), but then primarily to illustrate the principles involved. It will be a bonus if any of our analysis appears to fit well, but this will not be our main concern. We are here for the mathematics.

Historical fact

In order to appreciate the extent of the togetherness of the Golden Ratio and the Fibonacci numbers, we shall now review some key figures and discoveries of the past. But, from the start (especially), we find that much is not as it seems. Experts disagree—although having said that, there is a good consensus that the Platonic solids are not Plato's, Fibonacci's numbers are not Fibonacci's, and Binet's formula is not Binet's. As a result, the phrase 'historical fact' may begin to sound like a contradiction in terms... Further, as maths historian George Gheverghese Joseph points out, despite much academic effort to date, the history of mathematics is still regarded as being a very European affair [452]. Western historians prefer to think of mathematics as beginning (for the most part), in Greece.[5] They have been astonishingly slow to acknowledge the mathematicians of ancient India, Egypt, Mesopotamia, China, and Islam.

The Great Pyramid ϕ? π?—(undated)

The pyramids at Giza, Egypt are the last surviving of the Seven Wonders of the Ancient World. We know the Great Pyramid (Fig. 3, left) was built with approximately 2.3 million blocks of stone mostly weighing 2 to 15 tons each, but opinions about when it was built differ widely. Also, this pyramid has been said to have had ϕ, $\sqrt{\phi}$, and or π included in its design.[6] Mathematics lecturer Roger Herz-Fischler has studied 11 such design theories [347]. But let's review (just) the numbers, using authoritative data. William Flinders Petrie made an extensive and meticulous survey at Giza from 1880 to 1882 and published his findings in 1883. From his precise measurements and calculations, Petrie wrote (of the Great Pyramid): 'The mean base being 9068.8 ± .5 inches, this yields a height of 5776.0 ± 7.0 inches' [746, 747].[7] We represent this pyramid in Fig. 4 (top left) with the height and base dimensions marked h and b. One theory is that if the base perimeter ($4b$) is taken to be the circumference of a circle, then its radius will be equal to the pyramid height. To check, we calculate $h = 4b/(2\pi) = 5773.4$ inches, and even if we account for the uncertainty in b (± 0.5 inch)—h remains well within Petrie's height range.

---

[5] Joseph considers the West as being 'Europe and her cultural dependencies'—but he excepts Córdoba and Toledo, the two great centres of Arab learning and culture that received their scientific and mathematical knowledge via Baghdad [453].

[6] In passing, $14/11=1.2727...$ provides a very good approximation to $\sqrt{\phi}=1.2720...$

[7] Petrie deduced this height as the original peak was gone. (1 inch = 2.54 cm.)

Figure 3: **The Giza Pyramids.** Image by Robster1983 [785].

Another theory is that the height is given by Kepler's triangle—a right triangle with sides in the ratios $1 : \sqrt{\phi} : \phi$, here, $(b/2) : h : s$ [353, 348]. To check this, we calculate $h = \sqrt{\phi} \cdot (b/2) = 5767.8$ inches. In a later survey (1925), J H Cole found the mean base to be 0.6 inches wider than had Petrie [121, 748]. Fig. 4 shows how this affects π- and ϕ-theory heights. Although Cole did not infer a height, Herz-Fischler, using slope data, estimates the original height at 146.6 m (= 5771.7 inches) [350, 351, 352]. Herz-Fischler also discusses how well several out of 9 other theories fit, but concludes that questions such as: 'Was there a design principle?' and if so: 'What was it?' are currently undecidable [354]. Even at this scaling, all these theoretical and inferred heights (remarkably) still fall within only a few inches of each other (approximately 0.1% of height). So, based on numbers alone, both π and ϕ claims are credible. And *if* the designers had deliberately put either of the π or ϕ theories into practice, then they would have got the other 'for free' (had they been aware of it or not).

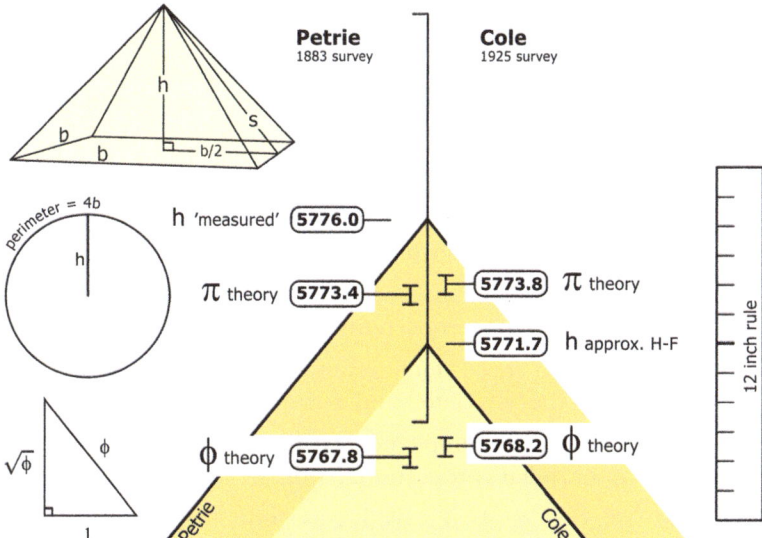

Figure 4: **Top of the Great Pyramid—heights in inches.**

The Gate of the Sun ϕ—(undated)

High in the Andean Altiplano (at 3,850 metres), not far from the south shore of Lake Titicaca in Bolivia, is the fenced archaeological site of Tiwanaku. It is about 64 km west of La Paz and roughly 1000 km from Nazca in neighbouring Peru. One of the best known monuments at the site is the megalithic (great stone) 'Gate of the Sun' (Fig. 5). It is approximately 4 metres wide × 3 metres high; and before it broke in two, the single carved andesite block would have weighed about 10 tonnes [922]. But, just as with the pyramids, there is no agreement about when the Gateway was created, and again opinions differ wildly. It is however, extremely interesting from our point of view, as its proportions *indicate* that its designer(s) had a comprehensive knowledge of the Golden Ratio. The original location of the gate is not now known as it was moved to its present position by General Sucre in 1825 (during the Wars of Independence) [797]. However, it is possible that it came from the nearby Puma Punku ancient megalithic complex [768]. The piece appears unfinished overall; however, the carving of the frieze (in the very hard rock) is of the highest quality. The side we shall study is well preserved, whereas the other side shows considerable weathering (consistent with it lying flat for a very long period).

The Gate of the Moon ϕ—(undated)

As a complementary partner to the Gate of the Sun, there is also a 'Gate of the Moon' (Fig. 6). While maintaining the pillars and lintel shape along with a number of proportions, it beautifully incorporates several mirrored concepts. These two monuments are joined both in their similarities and in their complementary differences:

- In frontal view, the Gate of the Sun exhibits a landscape format, while the Moon Gate is in portrait format.
- Both doorways are close to $1:\sqrt{5}$ in proportion.
- Both gates appear to have been designed *overall* according to Golden proportions.
- Each gate has a $1:\phi^3$ long rectangle element. This shows in the width of the Sun Gate—and in the height of the Moon Gate.
- The gates perfectly complement each other in their 'pillar : door : pillar' proportional widths…
    - Sun: $\phi : 1 : \phi$ (with ϕ as stone, 1 as air), and
    - Moon: $1 : \phi : 1$ (1 stone, ϕ air)—Figs. 8 and 9, p.11.

Figure 5: **The Gate of the Sun (Tiwanaku, Bolivia).**
Also known as Puerta del Sol.   Image by Mhwater [644].

Figure 6: **The Gate of the Moon.**
Also known as Puerta de la Luna.   Image by Daniel Maciel [605, 140].

These mutual relationships are exceptionally significant as each gate 'corroborates the story' of the other—providing very high confidence in this ϕ-based interpretation of the geometry. In Fig. 8, with door width as 1, the door height is $\phi + \phi^{-1} = \sqrt{5}$, and the whole width is $\phi^2 + \phi = \phi^3$. And in Fig. 9 with column width as 1, the whole height is ϕ^3. (The ancient Egyptian Temple of Dendur—now in the Metropolitan Museum, New York—has a basically similar gate. In that, the foot-gap-foot is close to 1:ϕ:1, but relative door height is ϕ^3—so the doorway is 1:ϕ^2; and door + 1 pillar is 1:ϕ. We shall return to ancient Egypt in Book 2 of this series.) We may further check these correspondences by asking: 'What needs adding to each gate to make it square ($\phi^3 \times \phi^3$)?' Well, the landscape Sun Gate would need a $1 \times \phi^3$ top lintel copy, still horizontal, but underground: $(1 + \sqrt{5}) + 1 = \phi^3$ (Fig. 7, centre). Whereas, the portrait Moon Gate would also need the height of its top lintel adding—this time as a vertical bar width, to give total width $(1 + \phi + 1) + \phi^{-1} = \phi^3$ (Fig. 7, right).

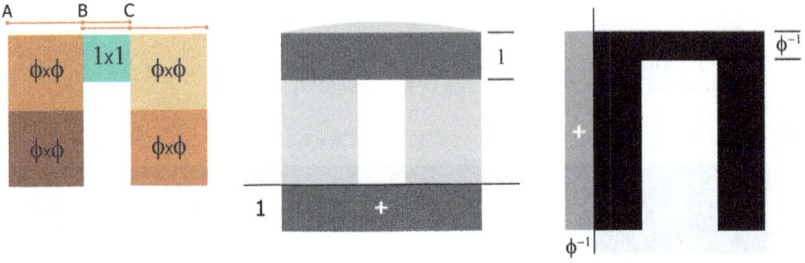

Figure 7: **Simplest partitioning (left), and 'completing the squares'.** (Different scales: the Sun Gate is bigger.)

The design of the Gate of the Sun demonstrates a 4-stage Golden Proportional growth sequence $\langle\ \phi^0,\ \phi^1,\ \phi^2,\ \phi^3\ \rangle$. Also the factor 2 is incorporated as 2ϕ. (As these gates have such rich mathematical content, we shall revisit them several times as we venture deeper into our study of Fibonacci and ϕ.)

Ancient India F_n—c.1000 BC and before

(c. *circa* 'about'.) A search for the first appearance of the Fibonacci sequence takes us back thousands of years to ancient India. There, the study of rhythm and metre (prosody) was as old as the Vedas—Sanskrit works embodying knowledge, especially that of spiritual development.

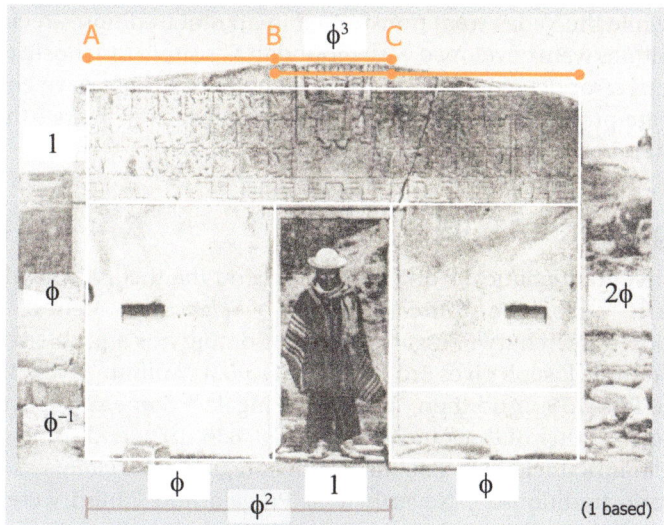

Figure 8: **The Gate of the Sun—'$\phi : 1 : \phi$'.** Photograph courtesy of the Secret Museum of Mankind [815].

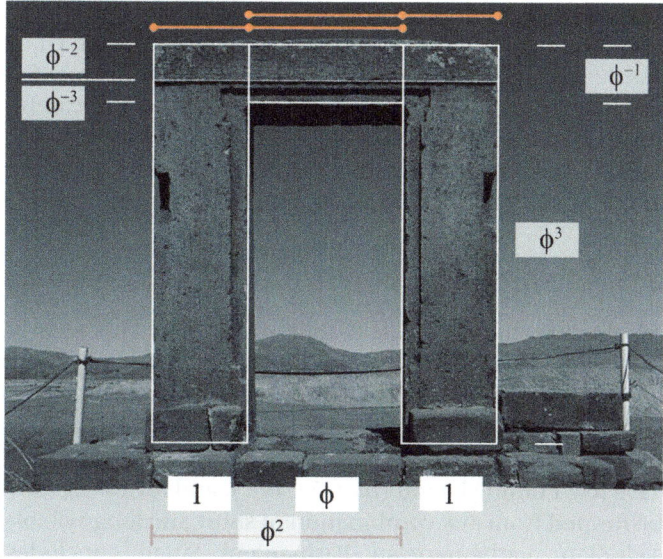

Figure 9: **The Gate of the Moon—'$1 : \phi : 1$'.** (This image is perspective corrected.) Original photograph Fig. 6, page 9 by Maciel [605, 140].

And while the Vedas were transmitted in an oral tradition—recitation techniques were developed with redundancies similar to those used in modern error detection and correction methods. These have ensured accurate preservation of content across thousands of years—thereby maintaining the purity of the knowledge.

Pingala and Pāṇini F_n—450 BC?

The first mathematical work that we have on the theory of rhythm in Sanskrit verse, the *Chandaḥśāstram—(Science of Metres)*,[8] was composed by Āchārya Pingala Nāga, but dating it is a problem [191]. For example Joseph gives 3rd century BC [455], Wolfram gives '450 BC or 200 BC' [751]. And then, 'Who was Pingala?' Some say he was the younger brother of the famous Pāṇini [10, 838, 192], and some say he was Pāṇini's uncle [929, 838], and others regard such claims as being 'folkloric traditions'.[9] Nevertheless, Pingala and Pāṇini were both intellectual giants, and they are linked in our story. Pāṇini is famous for what we now might call 'Sanskrit 2.0'—his great work on Sanskrit grammar, the *Āṣṭādhyāyī*. This took the Vedic spoken language of the time, discarded all its irregularities and rigorously formalized the essence that remained into the fixed, (almost mathematical) Classical Sanskrit. He did this in *8 Chapters* (the name of the work), which he further structured into 8 × 4 = 32 quarter chapters— perhaps 32 was an important number to him. The name 'Sanskrit' itself means polished or perfected, reflecting its careful and systematic construction. It is because of Sanskrit's highly logical structure that the language has been proposed for knowledge representation and manipulation in machine intelligence [84].

Short and long syllables

Pāṇini's shortest grammar rule simply states 'a, ā', ('uh' as in *up* and 'aaah')—the short and long 'A' vowel syllables. He is requiring that these be properly distinguished. In both Vedic and Classical Sanskrit verse, this difference between light and heavy syllables is funda-mental to the construction of the metre. By considering those metres that have fixed timing length (*mātrā vṛttas*), and reviewing how these may be formed from any combination of short and long syllables that fits... Indian mathematician Parmanand Singh concludes that

---

[8] Sometimes *Chandaḥśāstram* is called *Chandaḥ Sūtra*.

[9] Dr. Valerie J Roebuck—(translator of *The Dhammapada* and *Upanishads*, Penguin Classics, 2010 and 2004)—personal correspondence, quoted with kind permission.

1 ◡	2 ◡	3 ◡	4 ◡	5 ◡
◡	—	◡ —	— —	◡ — —
	◡ ◡	— ◡	◡ ◡ —	— ◡ —
		◡ ◡ ◡	◡ — ◡	◡ ◡ ◡ —
			— ◡ ◡	— — ◡
			◡ ◡ ◡ ◡	◡ ◡ — ◡
				◡ — ◡ ◡
				— ◡ ◡ ◡
				◡ ◡ ◡ ◡ ◡
1 way	2 ways	3 ways	5 ways	8 ways

Figure 10:   **Short and long syllables.** Adapted from Singh with permission [837].

knowledge of the Fibonacci sequence was an essential component of ancient Indian prosody [837, 500, 501, 454]. In Fig. 10, for each fixed overall duration, we list all the possible sequences of shorts and longs from which it may be composed.[10] The standard notation for shorts and longs is   ◡ = 1   and   — = 2,   so — = ◡◡ and (for example) 5 shorts (of ◡ = 1 each) can be replaced by 1 short (◡), and 2 longs (of — = 2 each). By making different combinations of these, we form the total durations: 1, 2, 3, 4, and 5, and for each of these we count the number of ways it can be built; and we then recognize these counts as Fibonacci numbers.   Whether or not Piṅgala originated all the knowledge in his *Science of Metres*, we do not know [193]. But, given the fact that the Vedas are built in a set of sacred metres, it seems (to the present author at least) more likely that he was consolidating and reporting more ancient knowledge. Notwithstanding this, in his work, Piṅgala demonstrates a remarkable understanding of combinatorics (and possibly the Fibonacci numbers).   He also includes both the binary number system and the first record of Pascal's triangle, which he called 'Mount Meru, the mountain of cadence' [836, 751].[11]

---

[10]   In Sanskrit, short syllables are called *laghu* (light), and long called *guru* (weighty).

[11]   Pascal's triangle and the Fibonacci numbers are strongly related.   In fact, the Fibonacci numbers starting with $F_1 = 1$ are obtained by summing shallow diagonals of Pascal's triangle [372].

Pythagoras ϕ—c.569 to 475 BC

The emblem of the ancient Greek Pythagorean mathematical and mystical school was the 5-pointed star, whose geometry depends on ϕ (as we saw in Fig. 1—the crossed chords example). They called it *pentalpha* as it could be formed by successively rotating and super-imposing 5 copies of the capital letter 'A' (Alpha). Perhaps Pythagoras (of Samos) and his followers specifically identified the Golden Ratio, perhaps not. Either way, they certainly had the use of it as a consequence of their geometrical constructions. But then again, the 5-pointed star itself had a vast prior history. We know that Pythagoras had learnt much from (among others) the Egyptians and the Babylonians. Each of these studied mathematics and used this star. The Babylonians in their turn had inherited mathematical knowledge from the Sumerians. One Babylonian clay tablet shows $\sqrt{2}$ accurate to 6 decimal places[12] and another shows a list of 'Pythagorean' triples.[13] Each of these date from between 1800 BC and 1600 BC. Excavations of the ancient Sumerian city of Uruk have produced pentagrams from the 4th millennium BC [572]. The earliest Egyptian pentagram we know of was found on a jar from Naqadah near Thebes—dating from about 3100 BC [563].

Theætetus: Platonic solids ϕ—c.380 BC

It was around the time when the academy of (Greek philosopher) Plato was at its height that we first hear of the five Platonic solids (Fig. 11), and these are now mostly associated with Theætetus.[14] Two of these regular polyhedrons—the dodecahedron and the icosa-hedron—are proportioned according to the Golden Ratio. The do-decahedron has 12 faces that are regular pentagons and the icosa-hedron has 20 faces that are equilateral triangles [965]. Each of the Platonic solids has a 'dual'—a related partner. To create the dual, simply swap faces for vertices and vice versa. This reversible process leaves the number of edges unchanged. The dodecahedron and the icosahedron are duals; so too are the cube and the octahedron. But what about the tetrahedron? Given that there are only 5 Platonic solids, we just 'ran out'.

---

[12]   Yale University, Yale Babylonian Collection—YBC 7289. (We discuss the simple yet powerful method they used on p.74.)

[13]   Columbia University—Plimpton 322, cuneiform tablet, Larsa, Iraq.

[14]   Heath [328] and Livio [571]. Though Scottish Neolithic people had rudimentary models of these 1,000 years before Plato [459, 328, 571].

Figure 11: **The 5 Platonic solids with triple Golden Rectangle.**
Image created in *Blender* [787].

The resolution here is to imagine linking the centre points of the four faces of the tetrahedron—and *voilà*, we have another tetrahedron (upside down)—the tetrahedron is its own dual. Now, could it be easy to understand how the Golden Ratio determines the structures of the two higher Platonic solids—that is, how ϕ determines the geometry of both the icosahedron and the dodecahedron? On the lower left of Fig. 11, we first consider three Golden Rectangles set at right angles to each other, with coincident centres. Next right we see the icosahedron, with its vertices corresponding one-to-one to the corners of the Golden Rectangles. Moving right again, we see how the vertices of the icosahedron become the pentagonal faces of the dodecahedron, its dual. By linking the ends of opposite edges of the dodecahedron we get rectangles with sides in the ratio $1{:}\phi^2$. Looking again at the dodecahedron, at each vertex we see pentagons meeting—three meet in each case. This configuration may be described using the 'Schläfli symbol' $\{p, q\}$, here as $\{5, 3\}$— 'pentagons three'.

Similarly, the icosahedron—the dual of the dodecahedron having vertices and faces swapped—is denoted {3, 5}—'triangles five'. So again, we see the same pair of (adjacent) Fibonacci numbers.

Plato's Republic (maybe) ϕ—380 BC

At around the time of Piṅgala's work, Plato wrote his dialogues, one of which is his great work *The Republic* [752]. In it, he introduces the idea of 'the Divided Line'. And while some philosophers (including Pierre Grimes [276]) regard the division as being according to the Golden Section, others such as Yuri Balashov argue this is not evidenced by the text [26]. Balashov points out that Plato's description is 'notoriously incomplete'. Nevertheless, through this debate, we do start to see the Golden Ratio flickering into view.

The Epidaurus Theatre F_n—c.340 BC

This ancient Greek theatre (Fig. 12) was built by architect Polykleitos the Younger, and when considered complete with its stage, it has been found to have exceptionally good acoustics—amplifying actors' voices while reducing crowd noise [574]. Originally it had 34 rows of seats; then an upper tier of 21 rows was added around the 2nd century BC [811]. These numbers: 21, 34, and total 55, are all Fibonacci, with $55/34$ and $34/21$ being good approximations to ϕ. The implied association with ϕ is further confirmed by the extensions of the directions of the aisles, which reveal two Golden Triangles (isosceles triangles with apex 36°) with slightly overlapping sharp vertices [813, 259, 945]. In Fig. 13, the 12 inner sectors are $(2\pi)/20$ radians each, and the 22 outer are $(2\pi)/40$ each. Hence all are subdivisions of the 'pentagram angle' of $(2\pi)/5$ radians—that is, 72°.

Figure 12: **Epidaurus theatre 55 rows (34 + 21).** By Ronny Siegel [832, 143].

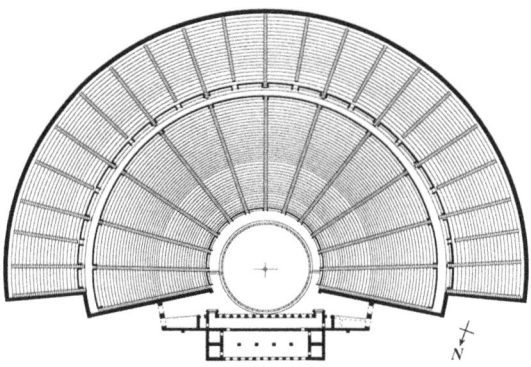

Figure 13: **Epidaurus theatre plan.**
Image: The Ancient Theatre Archive [369].

Euclid's Elements ϕ—300 BC

Euclid's *Elements* is a set of 13 books covering geometry and early number theory. It has been vastly influential—all the more so since the invention of the printing press. In it, Euclid ('of Alexandria'—with its then good local library) devotes the 13th book to a comprehensive study of 'the so-called Platonic figures' [327]. However, our focus here is in Book VI where we find Euclid's description of 'the Mean and Extreme ratio'—now called the Golden Ratio, Golden mean, or Golden Section. It appears as Definition 3 and Proposition 30, and it describes the special division of a line such that the ratio of the length of the whole to the greater is the same as the ratio of the greater to the smaller (exactly as detailed on page 2) [326]. As we shall soon see though, while this ratio evaluates to near 1.6180, many years would pass before the decimal calculation was made. Initially the Golden Ratio existed solely as a geometric concept.

Virahāṅka F_n—c.500 to 700

As discussed by Singh [836], the work of Indian mathematician Āchārya Virahāṅka was studied in detail by Velankar [923]. Virahāṅka extended Piṅgala's work on prosody, and he emerges as the first to state the 'Fibonacci' recurrence relation explicitly—again counting the numbers of possible variations of (*mātrā vṛttas*) rhythms.

Al'Khwārizmī and Abu Kamil ϕ—c.790 to c.930

Figure 14: **Abu Ja'far Muḥammad ibn Mūsā al-Khwārizmī (780–850).**

Al'Khwārizmī was one of the greatest Arab Islamic mathematicians (Fig. 14). The term 'algorithm' is derived from his name, and the word 'algebra' comes from the title of his treatise on algebra, the *Hisab al-jabr w'al-muqabala* [690]. He also wrote the first Arabic treatise on the use of the Indian numbering system, with its ten digits and place value innovations [690]—this was hundreds of years before Fibonacci wrote his *Liber Abaci*, the Western equivalent that we shall discuss shortly. One of the problems that Al'Khwārizmī gives concerns dividing a line that happens to include the Golden Ratio, but he does not mention it as such [697]. Al'Khwārizmī's work was acknowledged, praised, and developed by Abu Kamil Shuja, an Egyptian mathematician, whose texts in turn were used by Fibonacci in compiling his *Liber Abaci* [689]. From these, Fibonacci had a very clear idea of ϕ and in his own book he gives the division of a line length ten as $\sqrt{125} - 5 : 15 - \sqrt{125}$, which today we might write as $10\phi^{-1} : 10\phi^{-2}$ (because $\phi^{-1} + \phi^{-2} = 1$, and $\phi^{-1}/\phi^{-2} = \phi$).

Gopāla, Hemachandra F_n—1130 to 1150

Further building on the *mātrā vṛttas* prosody work of Virahāṅka, Indian mathematicians Gopāla and Hemachandra each wrote down the number sequence: 1, 2, 3, 5, 8, 13, 21, ... some decades before Fibonacci was born [923].

Fibonacci's book—*Liber Abaci* F_n—1202

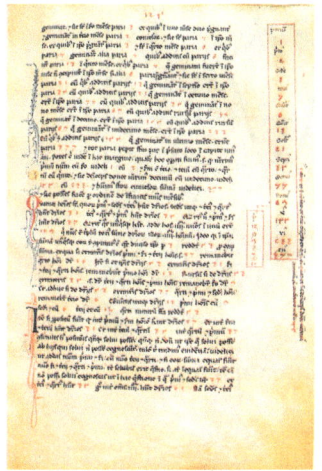

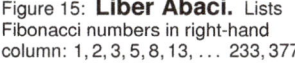

Figure 15: **Liber Abaci.** Lists Fibonacci numbers in right-hand column: 1, 2, 3, 5, 8, 13, ... 233, 377.

Figure 16: **Fibonacci (c.1170–1250).**

The Fibonacci numbers were introduced to the Western world in 1202 by Leonardo da Pisa, (also known as *Fibonacci*) in his book, *Liber Abaci—The Book of Calculation* [221] (Figs. 15 and 16).[15] Fibonacci's knowledge of the Indian system of numerals and calculation was gained during his youth, in the port of Bugia, North Africa (now Bejaïa, Algeria), where his father was clerk or dragoman[16] to Pisan merchants [172, 695, 893]. But although we have now adopted the peculiar habit of calling these numerals 'Arabic'—Fibonacci was crystal clear in his Latin—describing them as *Novem figuræ Indorum*.[17] For the Western world, the book was revolutionary in its ambition to introduce (and demonstrate how to use) the Indian '9 digits plus a zero sign' number system (based on place value) for commercial calculations. In the book, as an example, Fibonacci famously included a problem about the growth of an idealized rabbit population, to which the solution is (what we now call) the Fibonacci sequence. We shall return to Fibonacci's rabbits briefly later.

---

[15]   The Italian city of Pisa is these days mainly known for its leaning tower. Fibonacci was also known as Leonardo of Pisa, Leonardo Pisano (Bigollo), Leonardo Bonacci, Leonardo Fibonacci... There is now a statue of Fibonacci in the Camposanto, Pisa.

[16]   Interpreter & translator between Arabic, Persian, Turkish, & European languages.

[17]   Translation: 'new *Indian* figures (numerals, numbers, digits, ...)'.

The Darb-i-Imam shrine, Iran ϕ—1453

Figure 17: **The Darb-i-Imam shrine, Iran.**
Image: AAAS/Science [587].

Perhaps it is not obvious what the Islamic design in Fig. 17 has to do with ϕ. And in the absence of any evidence of numerical calculation, it may be that the ϕ element here is purely geometric—just as we suspected with the Pythagoreans' *pentalpha* 5-pointed star. Nonetheless, this composite design works on quite a different level compared with the pentagram. Indeed, it has taken over 500 years to discover the depth and the sophistication of the pattern's relationship to the Golden Ratio. We shall review this recent research later, in the 'ϕ Science' part, but in the meantime, we should consider what a great achievement this piece represents. For a start, the angles used are those of the decagon (with its ϕ-based geometry)—angles rarely found in ornamental pattern designs for reasons of symmetry (which again, we shall discuss later). And further, this tiling is based on a two-tier hierarchy that has the potential for infinite subdivision in non-repeating patterns. Elements of this kind of pattern are seen in Islamic architecture in: India, Iran, Iraq, and Turkey (mostly built from 1200CE onwards [579])—and the Darb-i-Imam shrine is a special example of how they may all come together to give an advanced design with local 5- and 10-fold symmetries. The result also has a remarkable mathematical consistency. When mapped to a quite recently discovered relevant mathematical tiling, only 11 defects were found across a scheme of 3700 tiles [766]. Further, the faults present were not problems with the overall system; rather they were point errors (presumably made by the builders or repairers)—isolated and locally correctable [581].

Leonardo da Vinci, Luca Pacioli ϕ—c.1497

Figure 18: **Leonardo da Vinci (1452–1519).** Self portrait c.1512 [542].

Figure 19: **Luca Pacioli (1445–1517),** possibly accompanied by Albrecht Dürer (right). (Top left: rhombicuboctahedron.)

Polymath genius Leonardo da Vinci and his friend Luca Pacioli— mathematician and Franciscan friar (Figs. 18 and 19) together produced a 3-manuscript work entitled *De Divina Proportione*. Leonardo provided 60 illustrations for this study in proportion [704]. It claims to disclose a 'secret science' and is based on the works of Euclid, Plato, and Vitruvius (his *Canon*) [716, 77]. In 1509, it was published in printed form, and its title appears to be the first instance of the Golden Ratio being referred to as 'the Divine Proportion'.

Michael Mästlin F_n and ϕ—1597

In 1597, leading German astronomer Michael Mästlin, at the University of Tübingen, wrote a letter to his former student Johannes Kepler, giving the Golden Ratio correct to 7 places of decimals—the first known example of its decimal calculation [697]. Now given that the Golden Ratio was well understood by Fibonacci, and that (as mentioned on p.18) he gave the Golden Section of 10 as $\sqrt{125} - 5 : 15 - \sqrt{125}$, and that methods for finding square roots were fairly well known; it does seem unlikely that nobody had made the calculation earlier. But perhaps the most interesting thing is that Mästlin sent his result to Johannes Kepler. We shall meet Kepler over and over again in our story, even up to the present day where his work has contributed to our understanding of the 5- and 10-fold symmetries found in recently discovered crystal structures.

Johannes Kepler

F_n and ϕ—1611

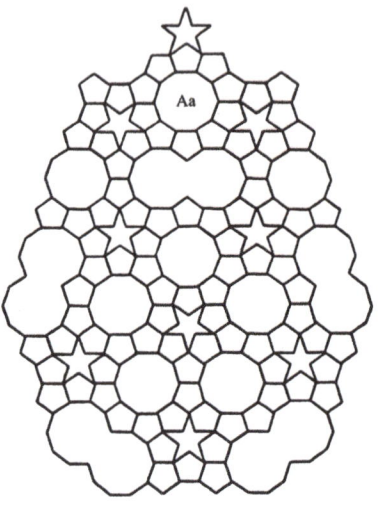

Figure 20: **Johannes Kepler (1571–1630).** Unknown artist, reproduced with kind permission of Kremsmünster Observatory [512].

Figure 21: **The Kepler Aa tiling.** This is from his *Harmonices Mundi* 1619. Extra lines in the top 'monster' have been removed, not being part of the Aa pattern as such [468].

In 1611, German mathematician and astronomer Johannes Kepler (Fig. 20) wrote an essay entitled *Strena Seu De Nive Sexangula*, (*The Six Cornered Snowflake*) [467]. In a preamble to his main discussion, Kepler muses on how the geometry of the two higher Platonic solids depends on the Golden Ratio. He then goes on to consider the sequence: 1, 2, 3, 5, 8, 13, and 21, and he compares one-to-the-next growth ratios with ϕ, apparently without having read *Liber Abaci* [221, 570].[18] He notes that as the numbers get bigger, the match to ϕ is 'more and more perfect':

$$\frac{1}{1} = 1, \qquad \frac{2}{1} = 2, \qquad \frac{3}{2} = 1.5, \qquad \frac{5}{3} = 1.\overline{6}, \qquad \frac{8}{5} = 1.6,$$

$$\frac{13}{8} = 1.625, \qquad \frac{21}{13} = 1.615, \qquad \frac{34}{21} = 1.619, \qquad \dots \qquad \phi = 1.618033989\dots$$

---

[18] Kepler refers to ϕ as 'the Divine Proportion' [467].

However, he also makes the very key point that no matter how big they get, it is impossible to represent the exact Golden Ratio as the ratio of two whole numbers. We would now regard this as being a comparison between a rational number (F_{n+1}/F_n) and an irrational number (ϕ). This important result also appears as an annotation of Pacioli's *Euclid* [697, 704]. It was also noted in a 1560 book by German reckoning master Simon Jacob [806], and again discovered independently by Albert Girard (who published in 1634) [697]. Kepler regarded the Golden Ratio as 'a precious jewel'—'one of the two great treasures of geometry' (the other being Pythagoras' theorem) [565]. Kepler is also famous for his fascination with the number five [5]. He found the number five in the cycles of leaves around plants, in the seeds of an apple, also in the five Platonic solids, and in geometry—in non-periodic tilings. Trying to tile the plane just using pentagons though, will of course leave gaps. But Kepler showed these could be filled using three more shapes: 5-point stars, decagons, and 'monsters' (two decagons overlapped). His book *Harmonices Mundi* (1619) shows an unusual arrangement of polygons he called 'Aa' (Fig. 21). Before Kepler, German artist and mathematician Albrecht Dürer (perhaps appearing to the right in Fig. 19, p.21) published some experimental combinations of pentagons in his *Unterweysung* [190, 600]. As an aside, just earlier we mentioned conics (p.4). The first major work on conics was by Apollonius of Perga (c.262–190 BC) who gave us the terms: parabola, ellipse, and hyperbola. But it was Kepler who first applied this pure mathematics to a scientific problem—that of modelling the paths of the planets [900, 700].

Albert Girard F_n and ϕ—1634

French engineer, lutenist, and mathematician Albert Girard (1595–1632) was first to interpret negative quantities as geometrically 'moving backwards' (we used this idea when we showed negative root β as AC' in Fig. 1 on p.3) [696]. He also studied the Fibonacci sequence and defined it as the recurrence relation (1.1) (p.1). He, like Kepler, knew the ratios of successive Fibonacci numbers converged to the Golden Ratio [696]. Girard also observed that an isosceles triangle with sides F_n, F_n, and F_{n+1} has an obtuse apex angle very close to those of a regular pentagon [596].[19]

---

[19] The larger the Fibonacci numbers, the closer this angle becomes. In the limit, the ratio of half the long side to a shorter side will tend to $\phi/2$. And as this ratio is the cosine of both acute angles, these tend to $36°$, and hence the obtuse angle will tend to $108°$.

Giovanni Cassini F_n—1680

Figure 22: **Cassini-Huygens mission.**
Image courtesy NASA/JPL-Caltech [669].

Figure 23: **Giovanni Cassini (1625–1712).**

Almost a contemporary of Kepler, Giovanni Domenico Cassini was another great astronomer (Fig. 23). He is remembered today (Fig. 22) particularly for discovering 4 moons of Saturn,[20] and he is now sometimes referred to as Cassini I, being the first of a famous family of astronomers [691]. In fact, Kepler's last 5 years overlapped with Cassini's first 5, and Cassini shared Kepler's interest in the Fibonacci numbers, which resulted in his (re-)discovery of the quadratic identity which now bears his name [489]

$$F_{n+1}F_{n-1} - F_n^2 = (-1)^n. \tag{1.7}$$

It is remarkable that the square of any Fibonacci number differs by exactly 1 from the product of its neighbours—and that this difference alternates completely regularly, between one more and one less. This identity (1.7) was also later independently discovered by Scottish geometer Robert Simson. However, as noted by Josef Brandmüller, Kepler had been well aware of this relationship (in 1619), describing it in (biological) detail in his book *Harmonices Mundi* [80].

Robert Simson F_n and ϕ—1753

Although Kepler and Girard both observed the connection between the Fibonacci numbers and ϕ, it was only in 1753 that Robert Simson at the University of Glasgow showed that by dividing successive pairs of Fibonacci numbers, the resulting ratios would tend towards $(1 + \sqrt{5})/2$ [968, 288, 83]. We explore such ϕ 'convergents' on p.422.

---

[20] Including the mysterious Iapetus—with its separately coloured white and black hemispheres and its unique 'walnut' equatorial ridge (discovered in 2005, Fig. 22) [43].

'Binet's Formula' (De Moivre, Bernoulli, and Euler)

F_n and ϕ— 1718 onwards

Figure 24: **Abraham de Moivre (1667–1754)** by Faber after Highmore.

Figure 25: **Leonhard Euler (1707–1783).** Handmann, Kunstmuseum, Basel.

$$F_n = \frac{\alpha^n - \beta^n}{\sqrt{5}} = \frac{\phi^n - (-\phi)^{-n}}{\sqrt{5}} = \frac{(1 + \sqrt{5})^n - (1 - \sqrt{5})^n}{2^n\sqrt{5}}. \qquad (1.8)$$

It is now generally agreed that 'Binet's formula for the Fibonacci numbers' was known long before Binet rediscovered it in 1843 [943]. It appears to have been known to Abraham de Moivre (Fig. 24) in 1718 [654]. (And in passing, we note that it was De Moivre who coined the term 'recurrence' for describing sequences such as the Fibonacci [49].) Daniel Bernoulli discovered the 'Binet' formula and proved it in 1726 [23, 54]. Euler (Fig. 25) presented the formula to the Petersburg Academy in 1763, and in 1767, he published it in a paper. This paper is now referred to simply by its Eneström number, E326 [206] [203, 207]. When we first see (1.8), we may find it incredible that it can reliably produce integers—let alone Fibonacci numbers (given its radicals and powers). In the meantime, we shall simply state it, and then as we get into the material, we shall find it increasingly useful. Later, because of its importance to us, we too shall 'rediscover' it (in detail) in Appendices E and F, from page 439 onwards.

Lamé and his theorem F_n—1844

Figure 26: **Gabriel Lamé (1795–1870).**

Gabriel Léon Jean Baptiste Lamé (Fig. 26) was one of the 72 names (all male French mathematicians and scientists) chosen by Gustave Eiffel to be commemorated on his tower in Paris.[21] The list also included Joseph-Louis Lagrange, René Just Haüy, and Joseph Fourier—each to whom we shall refer later in this book (albeit briefly). Lamé used the—soon to be named—Fibonacci numbers in his theorem which gives the 'running time' for the Euclidean algorithm, (which is a way to find the greatest common divisor of two integers).[22] His work on differential geometry, curvilinear coordinates, and Fermat's last theorem are considered particularly important [701]. He also (independently) rediscovered the Binet formula—(1.8) on p.25 [507]. Sometimes the Fibonacci sequence is referred to as the Lamé sequence. They are identical.

---

[21] Omitting (for example) mathematician Sophie Germain, whose work on elasticity had been important in the design of the tower [269].

[22] For $n \geq 1$, let u and v be integers with $u > v > 0$ such that Euclid's algorithm applied to u and v requires exactly n division steps, and such that u is as small as possible satisfying these conditions. Then $u = F_{n+2}$ and $v = F_{n+1}$ [597, 955, 502].

Lucas and Lucas numbers F_n, L_n, and ϕ—1876

Figure 27: **Édouard Lucas (1842–1891).** Family archive [144].

Édouard Lucas (Fig. 27) studied both the Fibonacci numbers and the numbers we now call Lucas numbers; and it was Lucas who gave the Fibonacci numbers their name [594]. In France he is described as a *'normalien'*—that is, one who has been a student at the élite *École normale supérieure* in Paris, well known for only accepting exceptional candidates. The Lucas numbers are often referred to as the 'companion sequence' of the Fibonacci numbers. This reflects the very close relationship between the two sequences. In (1.9) we see they share the same formula (recurrence relation), but instead of starting with seeds of 0 and 1 (as do the Fibonacci numbers), the Lucas numbers start with 2 and 1.

$$L_n = L_{n-1} + L_{n-2} \quad \text{with} \quad L_0 = 2, \ L_1 = 1. \tag{1.9}$$

Hence, the Lucas number sequence begins:

2, 1, 3, 4, 7, 11, 18, 29, 47, 76, 123, 199, 322, 521, 843, 1364, 2207, . . .

and it continues indefinitely.

Just earlier we saw the Binet formula for the Fibonacci numbers (p.25), and here is the equivalent relationship for the Lucas numbers,

$$L_n = \alpha^n + \beta^n = \phi^n + (-\phi)^{-n} = \frac{(1+\sqrt{5})^n + (1-\sqrt{5})^n}{2^n}. \quad (1.10)$$

Lucas went on to use his significant Fibonacci results to prove that Mersenne number $M_{127} = 2^{127} - 1$ is prime [499, 702].

$$M_{127} = 170141183460469231731687303715884105727.$$

(We shall be discussing Mersenne numbers later.) Lucas also regarded recreational mathematics as being extremely important—he invented the *Tower of Hanoi Puzzle* and wrote a 4-volume work, *Récréations mathématiques* [599].

Felix Klein: The icosahedron ϕ—1884

German mathematician Felix Klein had a major impact bringing together different areas of mathematics.   His 1872 'Erlangen programme' synthesized different geometries in terms of symmetry. He then went on to identify the ϕ-based icosahedron (page 15) as a 'master' object that related no less than five fundamental areas of mathematics [478, 479, 861].[23]

Mark Barr: Phi notation ϕ—c.1909

Although ϕ is now in widespread use to denote the Golden Ratio, this association is a relatively recent development.   American mathematician Mark Barr is credited with introducing it around 1909 [124]. He used the beginning of the name of Phidias, the Greek sculptor, painter, and architect, who (some say) used the Golden Ratio in his architecture—for example when designing the Parthenon in Athens. However, the actual facts about whether or not the Golden Ratio appears in ancient architecture, human anatomy, perception, music, literature, and art are increasingly admixed with myth and misconception—while writers such as Roger Herz-Fischler [341]–[363], Ron Knott [480]–[497], and Mario Livio [563]–[572] continue to work to set the record straight.

---

[23] The five areas being: geometry, Galois theory, group theory, the theory of invariants, and differential equations.

Growth

A major theme of this book is growth. We shall be interested in:

- the step-by-step growth of the Fibonacci and Lucas numbers
- crystal-like growth in plants, flowers and fruits
- step growth by the Golden Ratio—the integer powers of ϕ
- the steady growth of spirals—especially the Golden Spiral
- a remarkable, newly discovered kind of crystal growth
- (even) what happens *in-between* the whole numbers of the Fibonacci sequence—smooth growth that includes a loop-back

So, as we near the end of this chapter, let's consider the Fibonacci and Lucas growth sequences in more detail, in particular how these relate to roots α and β. But where to start? Some authors begin: $0, 1, 1, 2, 3, \ldots$, while others start: $1, 1, \ldots$, and still others start: $1, 2 \ldots$ In this book, however, we shall prefer not to be constrained by such starting positions—ideally thinking of such sequences as extending seamlessly in both directions indefinitely—if that is possible.

Negative-index Fibonacci (and Lucas) numbers

At the outset, (when we started with seeds 0 and 1), we saw how the Fibonacci recurrence $F_{n+2} = F_{n+1} + F_n$ works forwards (with increasing index n) to generate the positive Fibonacci number sequence. Now, let's rearrange this formula to 'work backwards' making n increasingly negative (still keeping $F_1=1$ and $F_0=0$):

$$F_n = F_{n+2} - F_{n+1}$$

$$
\begin{aligned}
F_{-1} &= F_1 - F_0 & &= 1 - 0 & &= +1 \\
F_{-2} &= F_0 - F_{-1} & &= 0 - 1 & &= -1 \\
F_{-3} &= F_{-1} - F_{-2} & &= 1 - (-1) & &= +2 \\
F_{-4} &= F_{-2} - F_{-3} & &= -1 - 2 & &= -3 \\
F_{-5} &= F_{-3} - F_{-4} & &= 2 - (-3) & &= +5 & & \ldots
\end{aligned}
$$

The numbers we generate are very familiar—they are just the Fibonacci numbers over again, but with alternating signs—therefore we may write

$$F_{-n} = (-1)^{n+1} F_n. \qquad [482]$$

This is an important standard identity, and also, the negative-index Lucas numbers behave similarly. Backwards from $F_0 = 2$ we have: $+2, -1, +3, -4, +7, \ldots$ giving the complementary Lucas relation as

$$L_{-n} = (-1)^n L_n. \qquad [483]$$

Fibonacci and Lucas growth rates

In Fig. 28, starting top right with $F_8 = 21$ and working back leftwards, if we consider smaller and smaller Fibonacci numbers, we see their growth rate begin to fluctuate (zigzag). Then as we go through zero into negative indices, the F_n numbers start to thrash between '+' and '−', with ever wider swings (as we saw in the last section). This might give the impression that the wobble near zero worked itself up into a wild 'bridge-breaking' oscillation of undefined growth. Yet as Fig. 28 shows, quite the opposite is the case. Said thrashing arises from true negative growth from step to step. (Whereas usually, when we talk about 'negative growth', we mean *a negative percentage*—that is, growth by a positive factor less than 1.) In the diagram we see how quickly the actual growth rates (i.e. F_n/F_{n-1} and L_n/L_{n-1})—settle to within a fraction of a percent in just 8 units either side of $n = 0$.

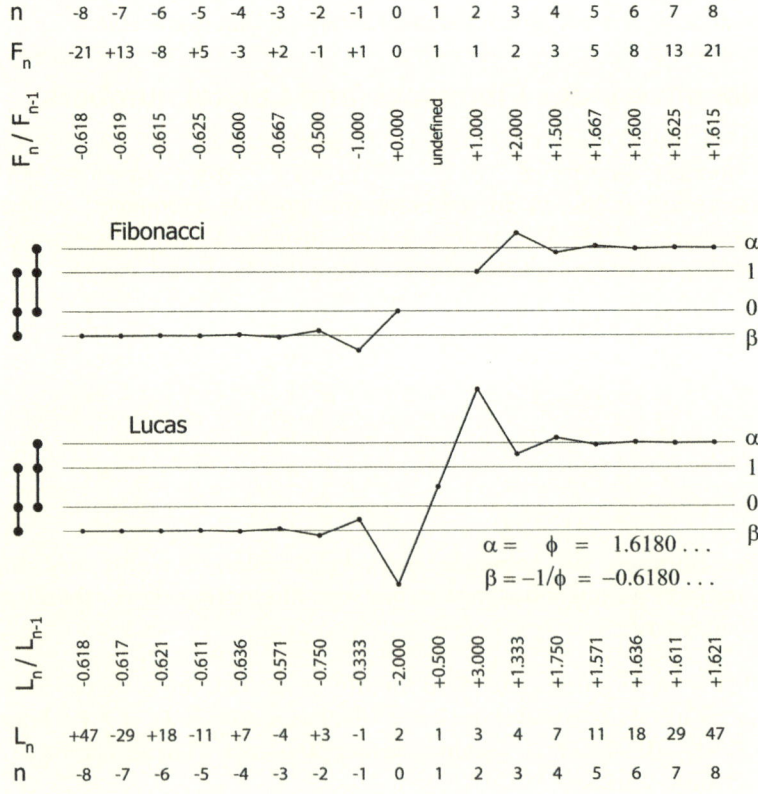

n	-8	-7	-6	-5	-4	-3	-2	-1	0	1	2	3	4	5	6	7	8
F_n	-21	+13	-8	+5	-3	+2	-1	+1	0	1	1	2	3	5	8	13	21
F_n/F_{n-1}	-0.618	-0.619	-0.615	-0.625	-0.600	-0.667	-0.500	-1.000	+0.000	undefined	+1.000	+2.000	+1.500	+1.667	+1.600	+1.625	+1.615

$$\alpha = \phi = 1.6180\ldots$$
$$\beta = -1/\phi = -0.6180\ldots$$

L_n/L_{n-1}	-0.618	-0.617	-0.621	-0.611	-0.636	-0.571	-0.750	-0.333	-2.000	+0.500	+3.000	+1.333	+1.750	+1.571	+1.636	+1.611	+1.621
L_n	+47	-29	+18	-11	+7	-4	+3	-1	2	1	3	4	7	11	18	29	47
n	-8	-7	-6	-5	-4	-3	-2	-1	0	1	2	3	4	5	6	7	8

Figure 28: **Fibonacci and Lucas step-by-step growth rates.**

Relating Fibonacci recurrence and Golden Ratio

We know that the growth rate for the Fibonacci numbers (from one to the next), approximates to the Golden Ratio, with the same being true for the Lucas numbers. On page 22 we saw how these ratios alternate between too small and too big. We also know that as the numbers get bigger, the approximation gets better—the variances become arbitrarily small. We might therefore begin to think in terms of some underlying growth rate, with the Fibonacci numbers and Lucas numbers then being the result of 'adjustments' away from this steady growth. So let's look again at the Fibonacci recurrence formula, but this time at just its *operation*—'add preceding two to get the next', Eqn. (1.11). We do this without reference to the Fibonacci numbers themselves, and ask: 'What numbers could satisfy this recurrence, if we also require that growth be perfectly steady?'[24]   (To distance ourselves from the Fibonacci numbers, we shall use u_n instead of F_n.)

$$u_n = u_{n-1} + u_{n-2}. \tag{1.11}$$

To explore (1.11), we first assume that steady growth is possible—that these u_n members may scale consistently from one to the next by a single growth factor—let's call it x. In Equation (1.11) therefore, we let $u_n = x^n$, and hence $u_{n-1} = x^{n-1}$ and $u_{n-2} = x^{n-2}$ which gives us

$$x^n = x^{n-1} + x^{n-2}.$$

Dividing through by x^{n-2} and rearranging we get

$$x^2 - x - 1 = 0. \tag{1.12}$$

This (1.12) is called the 'characteristic' quadratic equation for the Fibonacci recurrence, and it is the very same quadratic that defines both the Golden Ratio ($\alpha = \phi$) and its conjugate ($\beta = -\phi^{-1}$) —Eqn. (1.3), p.2. Here is the deep link between the Golden Ratio and the Fibonacci recurrence. So, we have found that using this approach we get two possible steady growth rates, $\alpha = \phi$ and $\beta = -\phi^{-1}$, where each suggests a geometric sequence. For α, we have (along with its ϕ equivalent version):

$$\ldots, \quad \alpha^{-3}, \quad \alpha^{-2}, \quad \alpha^{-1}, \quad \alpha^0, \quad \alpha^1, \quad \alpha^2, \quad \alpha^3, \quad \ldots$$
$$\ldots, \quad \phi^{-3}, \quad \phi^{-2}, \quad \phi^{-1}, \quad \phi^0, \quad \phi^1, \quad \phi^2, \quad \phi^3, \quad \ldots$$

---

[24]   Richard Dunlap (a physicist specializing in atmospheric science) gave a similar treatment to this in his *The Golden Ratio and Fibonacci numbers*, © 1997 World Scientific Publishing [189]. This adaptation is made with kind permission.

where each member is ϕ times its predecessor, and each is also the sum of the two preceding members. These properties will also survive the scaling of all members by a constant—e.g. $k\alpha^{-3}$, $k\alpha^{-2}$, $k\alpha^{-1}$, ...

And similarly for $\beta = -\phi^{-1}$ we have:

$$\ldots, \quad \beta^{-3}, \quad \beta^{-2}, \quad \beta^{-1}, \quad \beta^{0}, \quad \beta^{1}, \quad \beta^{2}, \quad \beta^{3}, \quad \ldots$$
$$\ldots, \quad -\phi^{3}, \quad +\phi^{2}, \quad -\phi^{1}, \quad +\phi^{0}, \quad -\phi^{-1}, \quad +\phi^{-2}, \quad -\phi^{-3}, \quad \ldots$$

where again each member is the sum of its two predecessors, and mathematically we say there is a steady step-by-step growth of $\beta = -\phi^{-1} = -0.618\ldots$ But, as we saw in the previous section, true steady negative growth rates can produce unexpected effects. Here we see both: shrinking—as $|\beta| < 1$, and sign flipping—as $\beta < 0$. In Fig. 29 we use the Binet formulæ (1.8) and (1.10) from pages 25 and 28 to look at how powers of α and β combine to make $\sqrt{5}$-scaled Fibonacci numbers and non-scaled Lucas numbers. We saw numeric value identities such as $\alpha^{2} + \beta^{2} = 3$ summarized on page 4. Again, the top row α terms provide a steady 'backbone' growth of α—the Golden Ratio—for each step. In contrast, the β terms alternately add and subtract as 'adjustments' that get us to the scaled Fibonacci numbers and non-scaled Lucas numbers. We shall return again and again to these ideas of steady 'underlying growth' (the α terms) and of 'adjustment' by numeric values with alternating sign (the β terms).

Backbone:	α^{-2}	α^{-1}	1	α^{1}	α^{2}	...	α^{n}
− adj.	β^{-2}	β^{-1}	1	β^{1}	β^{2}	...	β^{n}
differences	$-\sqrt{5}$	$+\sqrt{5}$	0	$\sqrt{5}$	$\sqrt{5}$	...	$\alpha^{n} - \beta^{n}$
= scaled F_n	$\sqrt{5}F_{-2}$	$\sqrt{5}F_{-1}$	$\sqrt{5}F_{0}$	$\sqrt{5}F_{1}$	$\sqrt{5}F_{2}$	...	$\sqrt{5}F_{n}$.

Backbone:	α^{-2}	α^{-1}	1	α^{1}	α^{2}	...	α^{n}
+ adj.	β^{-2}	β^{-1}	1	β^{1}	β^{2}	...	β^{n}
sums	$+3$	-1	2	1	3	...	$\alpha^{n} + \beta^{n}$
= L_n	L_{-2}	L_{-1}	L_{0}	L_{1}	L_{2}	...	L_{n}.

Figure 29: **Binet's F_n and L_n formulæ for several n values.**

Constructing Kepler's $\sqrt{\phi}$ triangle

In Fig. 30 (left) we derive Kepler's 'Root Golden' triangle (p.7) from a Golden Rectangle.[25] By Pythagoras, $(1)^2 + \left(\sqrt{\phi}\right)^2 = (\phi)^2$, gives the familiar $1 + \phi = \phi^2$. On the right, we swing back arc LB to create an ABC Golden Section and a spiral sequence of Kepler's triangles— their hypotenuses decreasing by $\sqrt{\phi}$. By noting point S, where the perpendicular from B intersects with AL, we see how the spiralling-in continues indefinitely. (We shall return to this special spiral later).

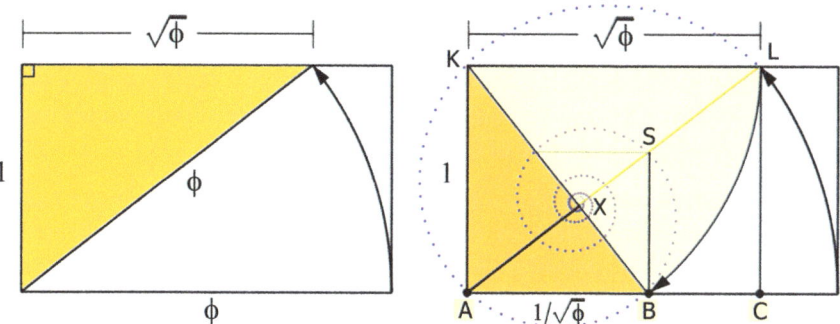

Figure 30: **Kepler's triangles:** ($\triangle$KAL), $\triangle$LKX, $\triangle$KAX, $\triangle$ABX, $\triangle$BSX, ...
Golden Section ABC: $\triangle$LKX, $\triangle$KAX, and $\triangle$ABX are all similar, hence AB$=1/\sqrt{\phi}$.

Golden Ratio v. Golden Section

While these terms are often employed interchangeably, it will help us to associate 'Golden Ratio' with ϕ (1.618...), and 'Golden Section' with $1/\phi$ (0.618...), and to think of the latter as the 'cutting' of a length in Golden Proportion—to give lengths 61.8% and 38.2% of the original.

The new circle constant $\qquad \tau = (2\pi)$

The Greek letter *tau* has lately been promoted as being a 'new and improved' circle constant [728, 321]. It surely makes eminent sense. However, as τ-evangelist Michael Hartl himself now acknowledges, τ is widely used in academic science and engineering to denote the Golden Section [322].[26] Having said that though, there does exist a temporary compromise, *which we shall therefore adopt*, viz. that we treat (2π) as a single symbol that denotes the angle of one full turn in radians. So, when we see $(2\pi)/10$, we shall resist the urge to write $\pi/5$.

---

[25] We study this construction a little further in Appendix D on page 431.
[26] τ was chosen as an abbreviation for *'tome'*—the Greek verb 'to cut' [952].

Resources

As interest in Fibonacci and ϕ has grown, high-quality resources have become increasingly available.[27]

The Fibonacci Quarterly (FQ). This journal is published by the Fibonacci Association, which was founded by Verner E Hoggatt Jnr. and Brother Alfred Brousseau. Its focus is on: 'Fibonacci numbers and related mathematics, emphasizing new results, research proposals, challenging problems, and new proofs of old ideas'. The Association also provides a website, and, in even years, it holds an international conference. The FQ's current editor is Curtis Cooper—well known for finding record-breaking Mersenne primes [987]. The journal is (at time of writing) provided online, and a large block of back issues has been made available free of charge [220].

Ron Knott's website. Since 1996, now retired computing lecturer Ron Knott has maintained an extensive and award winning dedicated website which provides a wealth of authoritative information, formulæ, examples, puzzles, and links. In this book, we shall refer a number of times to various of the identities which he lists [480].

Neil Sloane's online encyclopedia. In 1964, Neil J A Sloane, a graduate maths student at AT&T, started his now legendary 'Online Encyclopedia of Integer Sequences' (OEIS). Sloane's initiative was greatly welcomed by the mathematical community; and on top of the considerable work by Sloane himself, plus that of collaborator Simon Plouffe (from early 1990s), contributions were submitted from all over the world. A milestone of 200,000 sequences was reached on December 1, 1990. The OEIS press release picks three example sequences: the Fibonacci numbers Sloane A000045, the pentagonal numbers A000326,[28] and the 'Lazy Caterer's' sequence A000124 [711, 839].

Melvyn Bragg's radio programme. In 2007 the BBC devoted an edition of its *In Our Time* radio series to the Fibonacci sequence. This 45-minute programme may be downloaded as an mp3 file. It gives a good overview, and it highlights the way the trajectories of the Golden Ratio and the Fibonacci numbers draw together into a single study. It is chaired by Melvyn Bragg, who is joined by expert guests: Marcus du Sautoy, Jackie Stedall, and Ron Knott [79].

---

[27] See also, for example, Roger Herz-Fischler [343], and Prof. R C Archibald's notes in Hambidge's *Greek Vase* [17].

[28] All positive numbers can be expressed using 5 pentagonal numbers [961].

Chapter 2

Spirals

The famous mathematical amalgam of the Golden Ratio and the logarithmic spiral is the Golden Spiral. And in order to understand the Golden Spiral properly, we shall review the history of the logarithmic spiral, and as a contrasting introduction to that, we shall take a quick look at the Archimedes spiral. But, ahead of any of these, we had best select a suitable coordinate system.

Polar coordinates

If we wish to define spirals in the simplest mathematical formulæ, then polar coordinates are a natural choice. The polar convention is to denote the length of a radial vector by r and the angle of turning by θ ('theta'), with positive increasing θ being defined as anticlockwise (Fig. 31). Such coordinates are far better suited to our purpose than say, cartesian (x, y) coordinates.

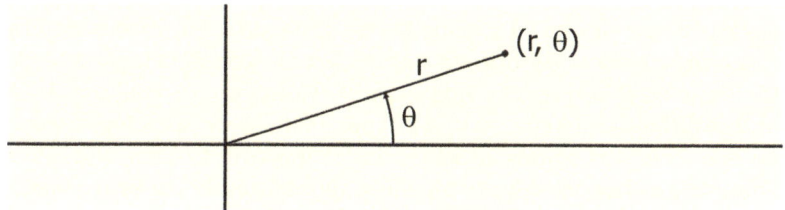

Figure 31: **Polar coordinates.**

Types of spiral—from the simplest

Conon of Samos (c.280–220 BC) and Archimedes were good friends, and although Conon is credited by Pappus with discovering the spiral of Archimedes, this has been questioned—and in any case, it was Archimedes who did most work on this spiral [692] (Fig. 32). This is the very simplest of spirals. The spoke length (radial vector) is directly proportional to the amount of turning. While not an exact analogy, 'the free end of a taut string unwinding from a fixed rod,' gives the idea.[1]

$$r \; = \; k\theta.$$

That is, for every turn, the spoke length increases by a constant amount—it increases arithmetically. There are many other types of spiral too: conical, Cotes, epispiral, Euler (Cornu, clothoid), Fermat, hyperbolic, lituus, logarithmic, Poinsot, rational, Theodorus, ... plus a couple of common approximations to the Golden Spiral.

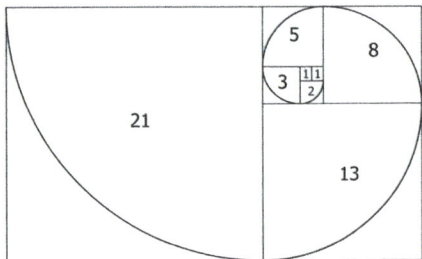

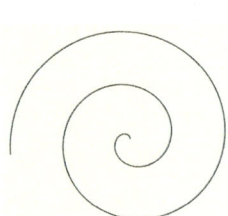

Figure 32:
Archimedes spiral.

Figure 33: **Whirling squares spiral.**

Approximating the Golden Spiral

Fig. 33 above shows one such approximation. What it lacks in smooth growth, it makes up for by being quick and easy to construct—in the successive squares of a whirling Fibonacci squares format, quarter-circle arcs are drawn. A slightly better approximation is obtained by repeating the exercise but starting with a Golden Rectangle and (in the same way) adding whirling squares. However, we shall not dwell on the differences between spiral types. Rather, we shall focus solely on the family of spirals which includes the true Golden and Silver Spirals.[2] Further, in order to highlight spiral formulæ (and also make their exponents more readable), we shall show them in larger type.

---

[1]   This rod-based unwinding produces a curve known as 'the involute of a circle'.

[2]   Silver Spirals are based on the Silver Ratio $(1 + \sqrt{2})$—more on these shortly.

René Descartes

Figure 34: **René Descartes (1596–1650).**
Frans Hals, Musée du Louvre.

In 1638, French philosopher and mathematician René Descartes (Fig. 34) wrote to his friend Marin Mersenne about his discovery of the logarithmic spiral while he was studying dynamics (Fig. 35). In this spiral the radial vector increases geometrically. Mersenne in turn corresponded with Pierre de Fermat and John Pell [703].

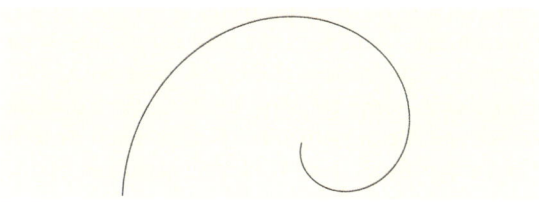

Figure 35: **Logarithmic spiral.**

37

Understanding the growth spiral

The growth spiral is also known as the logarithmic spiral and this curve has a natural simplicity. However, expressions such as

$$r = ae^{\theta \cdot \cot(b)} \tag{2.1}$$

tend to obscure this. In fact, the rule for its construction is just 'growth rate per turn is constant'. Let's consider what this means in practice for the spoke length. If we start with a radial vector (spoke) of 1, then: [3]

- after one turn, we shall have $1 \times$ *growth rate*
- after 2 turns, $(1 \times$ *growth rate*$) \times$ *growth rate* $=$ *(growth rate)*^{2}
- after 3 turns, $1 \times$ *(growth rate)*^{3}, and so on...

Hence,

$$spoke\ length = 1 \times (growth\ rate\ per\ turn)^{(number\ of\ turns)}. \tag{2.2}$$

And further, to find the lengths in between full turns, we just raise the growth rate to a fractional power—the fraction of a turn, $\theta/(2\pi)$. Now

$$spoke\ length = 1 \times (growth\ rate\ per\ turn)^{\frac{\theta}{(2\pi)}}. \tag{2.3}$$

As a consequence of the above, turning by equal amounts (starting anywhere on the curve) will produce the same growth. In (2.1) we observe an overall scaling factor of a—but its effect may not be quite as we expect. We know that if we scale a logarithmic spiral by its per-turn growth factor, the new spiral will fit on top of the existing one exactly (by definition). And we may also mimic the scaling of a log spiral by rotating it. It follows therefore, that we may replace *any non-zero scaling* with some rotation between zero and one full turn. Also in (2.1), we see the growth term is expressed in terms of e, the base of natural logarithms, and it further involves $\cot(b)$. We shall return to such refinements shortly.

---

[3] The '$1\times$' in this introduction is retained as a placeholder for the start length.

Jakob Bernoulli

Figure 36: **Jakob Bernoulli (1654–1705).**

Jakob Bernoulli (Fig. 36) was a prominent member of a family of mathematicians. He was fascinated by the logarithmic spiral—so much so that (in 1692) he called it *spira mirabilis* (marvellous spiral). The name 'logarithmic spiral' is also due to him [17]. He even asked for one such to be engraved on his headstone (after Archimedes), along with the inscription: *eadem mutata resurgo*, (I shall arise the same though changed) [949]. He was referring not just to those properties of the logarithmic spiral that we have just discussed, but also to its other self-referral qualities, such as it being its own evolute and involute (which he proved) [688].

Equiangular—true log spiral

As Descartes noted, it is a defining property of the logarithmic spiral that its radial vectors (spokes) always cross the curve at the same angle [17]. Let's consider the simplest possible Golden Spiral, starting it with a unit length spoke, and growing it by ϕ^4 per full turn of (2π) radians— here, $\theta/(2\pi)$ will give us the fraction of a turn represented by θ

$$r = (\phi^4)^{\frac{\theta}{(2\pi)}} = \phi^{\frac{4\theta}{(2\pi)}}. \tag{2.4}$$

Now, can we express this as a power of e, the base of natural logarithms? (We shall prefer this later.) Let

$$e^{b\theta} = \phi^{\frac{4\theta}{(2\pi)}}.$$

Then taking (natural) logs and noting that $\ln(e^x) = x$,

$$b\theta = \frac{4\theta}{(2\pi)}\ln(\phi), \qquad \text{hence,}$$

$$b = \frac{4\ln(\phi)}{(2\pi)}. \tag{2.5}$$

In Fig. 37 we see how the radials cross the spiral—always at the same angle (marked σ). We then look at how the spoke length increases as a result of a tiny amount of turning $\delta\theta$. This will give us a way to calculate the angle σ.

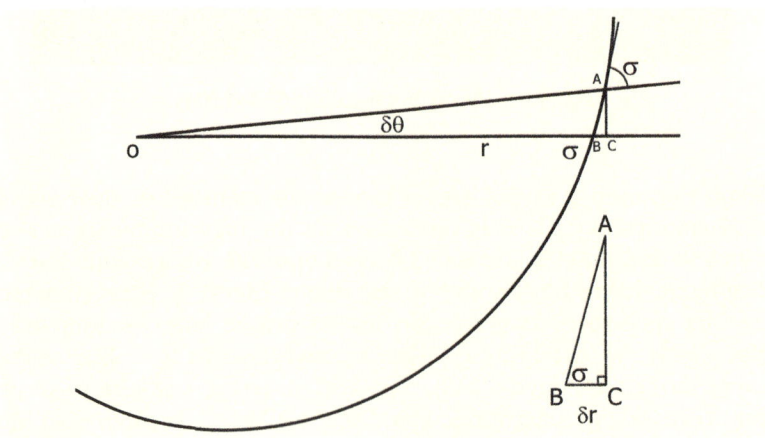

Figure 37: **'Equiangular'**.

40

The initial spoke length here is $OB = r$. With a turn of $\delta\theta$ the spoke becomes OA. In the diagram, we may initially think of AC as a shallow arc centred on O. But, as we reduce $\delta\theta$ towards zero, AC will tend towards a straight line perpendicular to OB. If we make $\delta\theta$ small enough,[4] we may regard ABC as a right triangle.

$$BC = \delta r, \quad \delta\theta \to 0 \quad \Longrightarrow \quad \begin{cases} \angle ABC & \to \sigma \\[2mm] AC & \to r\delta\theta \\[2mm] \cot(\sigma) = \dfrac{BC}{AC} = \dfrac{\delta r}{r\delta\theta} & \to \dfrac{1}{r}\dfrac{dr}{d\theta} \end{cases} \tag{2.6}$$

then as $r = e^{b\theta}$,

$$\frac{dr}{d\theta} = be^{b\theta} = br. \tag{2.7}$$

Applying (2.7) in (2.6) we get

$$\cot(\sigma) = \frac{1}{r}\cdot br = b, \qquad \sigma = \operatorname{arccot}(b) \tag{2.8}$$

and we already have b in (2.5), so

$$\sigma = \operatorname{arccot}\left(\frac{4\ln(\phi)}{(2\pi)}\right) = 72.9676089\ldots^{\circ} \tag{2.9}$$

The Silver Spiral

We shall come back to the Silver Spiral in more detail later (in relation to Pell numbers). But in the meantime, we should just note that it is another logarithmic spiral—differing from the Golden Spiral only in its growth rate.[5] For the Golden Spiral, the growth is the Golden Ratio ϕ per quadrant; for the Silver Spiral, it is the Silver Ratio $\delta = 1+\sqrt{2}$ per quadrant—so (2.9) becomes

$$\sigma = \operatorname{arccot}\left(\frac{4\ln(\delta)}{(2\pi)}\right) = 60.703223\ldots^{\circ}$$

---

4 That is, small enough that second-order terms may be ignored.
5 In passing we note that the logarithmic spiral was also studied by Evangelista Torricelli (who *rectified* it), John Wallis, Christopher Wren, Isaac Newton, Guido Grandi, and James Clerk Maxwell [17].

The Golden Spiral and its Golden Section

The Golden Spiral is a logarithmic spiral that grows at ϕ per quarter turn—normally anticlockwise. But here in Fig. 38, we grow clockwise to match Fig. 1 on page 3. If we start at B with radial vector 1, and we turn clockwise for two quadrants, the spiral will grow by the Golden Ratio twice—that is, to $(1 \times \phi) \times \phi = \phi^2$ at C. Similarly in the next half turn it will grow ϕ^2 more, reaching $\phi^2 \times \phi \times \phi = \phi^4$ at A. So, $AB = \phi^4 - 1$, and $BC = \phi^2 + 1$.

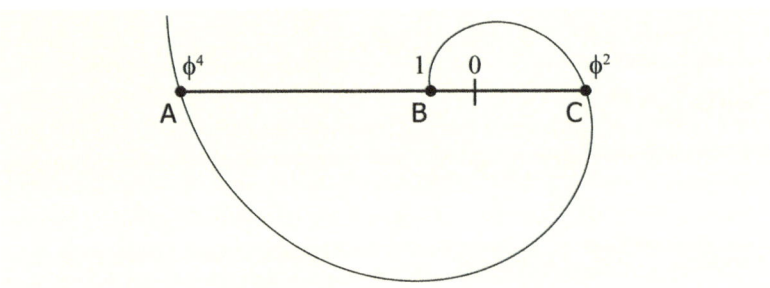

Figure 38: **The Golden Spiral Golden Section.**

To convince ourselves that the ratio AB/BC in Fig. 38 is indeed the Golden Ratio ϕ, we first factorize the expression for AB

$$AB = \phi^4 - 1,$$

using the identity $(a^2 - b^2) = (a + b)(a - b)$. This gives

$$AB = (\phi^2 + 1)(\phi^2 - 1). \tag{2.10}$$

We now consider BC,

$$BC = \phi^2 + 1. \tag{2.11}$$

And using (2.11) to substitute in (2.10) we get

$$AB = BC \cdot (\phi^2 - 1)$$

and from (1.3) p.2 we have $\phi^2 = \phi + 1$, so

$$AB = BC \cdot \phi.$$

True Golden Spiral 'with walls'   17.0 32 39 11°

Using equiangular and Golden Section approaches from pages 40 and 42, we are now in a position to check how the true Golden Spiral may be 'walled' (with vertical and horizontal tangents as in Fig. 39). The tangent points (P, Q, R, S, T, U) on this true curve do not correspond to the corners of squares (as they did in the approximation, p.36). Instead, the lines from the spiral centre 'O' to these tangent points make a shallow angle with the axes of $90 - 72.9676089 = 17.0323911°$.

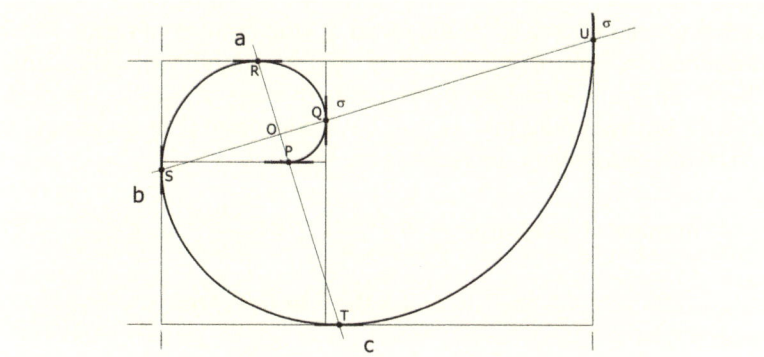

Figure 39: **Analysing the proportions of the true Golden Spiral.**

The structure in Fig. 39 grows by repeatedly adding a square to a Golden Rectangle. We now imagine the vertical and horizontal lines as walls—'containing' the true Golden Spiral. And because this spiral is equiangular, its radials will always cross the walls at the same angle (σ from Eqn. (2.9), p.41).

If we set up our spiral dimensions so that OP (centrally located) is just the right length so that P has a y coordinate of $P_y = -1$, then because the Golden Spiral grows by ϕ each quarter turn, the equivalent for Q will be $Q_x = \phi$. If we continue round, we shall have $R_y = \phi^2$, $S_x = -\phi^3$, $T_y = -\phi^4$, $U_x = \phi^5$. Then by taking differences, we shall find a, b, and c:

$$a = Q_x - S_x \;=\phi +\phi^3 \;= \phi\,(1+\phi^2)$$

$$b = R_y - T_y \;=\phi^2 +\phi^4 \;= \phi^2(1+\phi^2)$$

$$c = U_x - S_x \;=\phi^5 +\phi^3 \;= \phi^3(1+\phi^2).$$

In these expressions we see that $c = \phi b$ and $b = \phi a$, confirming that the ratios $b : a$ and $c : b$ are Golden.

Working in quadrants

Because the Golden Spiral grows by ϕ for each quarter turn (quadrant), it will help us to take a moment to familiarize ourselves with how this right-angle may show up in formulæ. For example,

$$\theta = n \cdot \frac{(2\pi)}{4} \qquad \text{should be read as: 'Theta equals } n \text{ quadrants.'}$$

Also we saw an example just earlier (Eqn. (2.4) on page 40) of the angle θ radians effectively being converted into quadrants. This was so that we could *raise ϕ^4 to the power of* (the number of turns) which simplified to *raise ϕ to the power of* (the number of quadrants). There, we approached from the point of view of a full turn of ϕ^4 growth, but we could just as well have converted θ to quadrants to start with—either way, the result is the same, viz.

$$\text{number of quadrants} = \theta \bigg/ \frac{(2\pi)}{4} = \frac{4\theta}{(2\pi)} = 4 \cdot \frac{\theta}{(2\pi)}$$

$$= 4 \times \text{the number of turns.}$$

Visualizing the interplay between discrete and continuous—the role of spirals

For thousands of years, mathematicians have been fascinated with the relationship between the discrete ('counting') and the continuous ('measuring'). With the Fibonacci numbers, we have a discrete sequence $\langle F_n \rangle$, and with increasing n, the ratio of one number to its predecessor tends towards a common growth factor, ϕ. And as we saw (on page 30 at the end of Chapter 1), these ratios alternate greater than ϕ and less than ϕ, ever closer to ϕ, but never quite reaching it. Now, let us think of the underlying growth as a continuous function—say a logarithmic spiral. We shall choose the Golden Spiral for this, because it grows at exactly the Golden Ratio per quadrant (by definition). We may then regard sequence members as being 'variations from' or 'adjustments to' this reference. It will be these differences that then constitute the relationship between the discrete and the continuous that we are interested in here. In order to visualize these differences, we must devise a scheme capable of comparing number points with a spiral. At first sight this might seem like comparing chalk and cheese, but let's see how we might achieve it.

'Placing' Fibonacci numbers—in a polar layout

If we place Fibonacci numbers on quadrant axes (i.e. 90° apart) at a distance of F_n from the origin, then *provided we choose a suitable scaling* for the Golden Spiral, we shall expect the number points to be close to the spiral as it crosses the axes. As we have seen, polar coordinates are ideal for this kind of work. In this format, the Golden Spiral will grow at exactly ϕ per quarter turn, and (keeping close, increasingly close) the Fibonacci numbers will grow at approximately ϕ per quarter turn. Accordingly, (while remembering to read $n \cdot (2\pi)/4$ simply as 'n quadrants'), we use Eqn. (2.12)

$$(r, \theta) = \left(F_n, \ n \cdot \frac{(2\pi)}{4} \right) \qquad (2.12)$$

to place the successive F_n number points 'spiral fashion' anticlockwise around the origin, as in Fig. 40.

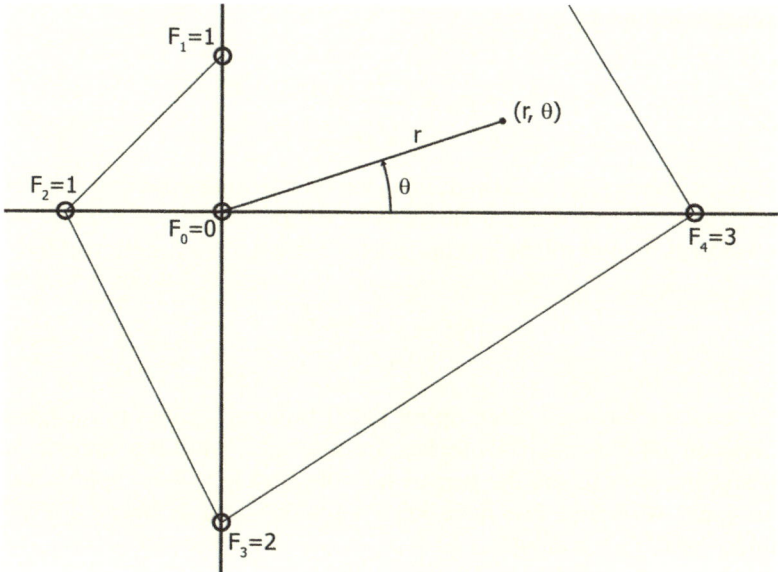

Figure 40: **Polar-placed Fibonacci numbers.**

Scaling the Golden Spiral

In equation (2.4) (page 40), we considered the simplest case—that of a Golden Spiral having unit spoke length (radial vector) when θ=0. As we shall soon see though, in this simple form it is not yet suited to our purpose of comparing with the placed Fibonacci numbers. To resolve this, we need to look at Binet's formula for a moment—(1.8) on p.25—

$$F_n \;=\; \frac{\alpha^n - \beta^n}{\sqrt{5}} \;=\; \frac{\phi^n - (-\phi)^{-n}}{\sqrt{5}}. \tag{2.13}$$

We know that for large n, the α^n term is dominant (the same as saying ϕ^n is dominant). Conversely, the β^n term (with increasing n) forever numerically dances either side of zero while rapidly shrinking and all but disappearing. So, if we fit the curve to the α^n term only, then our spiral will cut between all the infinity of number points that result from such 'beta-adjustments'. We shall better appreciate the extreme delicacy of this positioning as we construct the diagrams—and realize that our Golden Spiral must pass arbitrarily close to all the number points, but never touch any of them. Now, in order to evolve (2.13) into the required spiral, we first state it (with the polar Fibonacci number replaced by a polar radial r_n, while ignoring the β^n adjustment terms) as

$$r_n = \frac{1}{\sqrt{5}}\phi^n.$$

Also, for every step increment of n, we turn by one quadrant, so just as before: 'If the number of turns is $\theta/(2\pi)$, then the number of quadrants turned will be $4\times$ this (p.44).' Hence r as a function of θ is

$$r(\theta) = \frac{1}{\sqrt{5}}\phi^{\frac{4\theta}{(2\pi)}}. \tag{2.14}$$

So, now we have a Golden Spiral (2.14) precisely scaled to navigate between all the polar Fibonacci numbers.[6] With the benefit of hindsight, we can see the reason why the simple Golden Spiral (2.4) on page 40 would not have been suitable—it just lacked the all important $1/\sqrt{5}$ scaling.

---

[6] It is a little like a slalom ski course—except here, the flags zigzag either side of a long, slowly curved piste, getting ever closer to it.

Adjusting spoke length to get to F_n

We are now ready to combine the plots of the polar Fibonacci numbers and the chosen Golden Spiral (Fig. 41). In this configuration, the differences—the adjustments—are set to occur along the quadrant axes. These are $(d_0, d_1, d_2, \dots)$ and they get us from the spiral radial vector (spoke) length, to the polar Fibonacci number. Adjustments along the horizontal $(d_0, d_2, d_4, \dots)$ are on zero and half turns, have even indices, and are *inwards* towards the polar origin. Adjustments along vertical quadrant axes $(d_1, d_3, \dots)$ are on the quarter turns between and are *outwards*. The in-out pattern reflects the way the convergents of ϕ alternate greater than, less than ϕ.

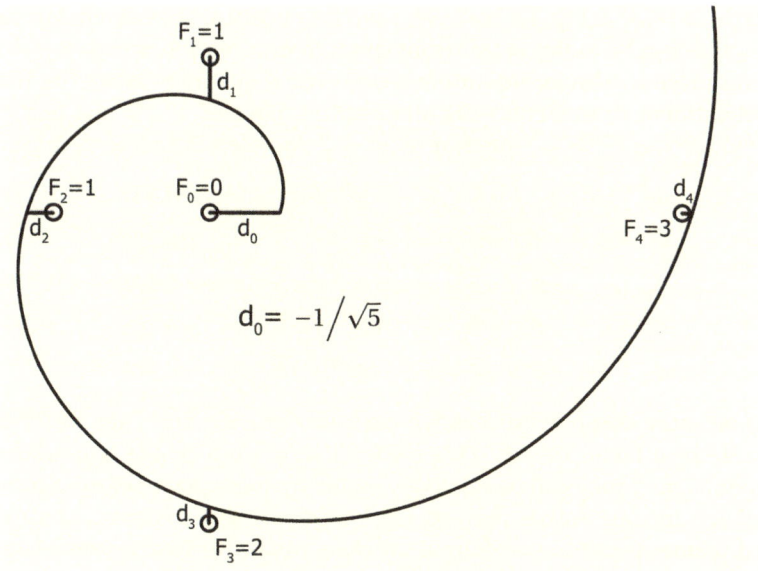

Figure 41: **Polar Fibonacci numbers near a chosen Golden Spiral.**

We now have a regular 'backbone growth' (the spiral) from which we then make smaller and smaller radial adjustments. Our spiral is supplying the $\alpha^n/\sqrt{5}$ part of the Fibonacci numbers; but the focus now moves to the $\beta^n/\sqrt{5}$ terms which each get us from the spiral to the relevant polar Fibonacci number point. Throughout this book, we shall consistently refer to such distances as adjustments.

(Adjustment d_n) = (Fibonacci number F_n) − (spiral spoke r_n)

$$= F_n - \frac{1}{\sqrt{5}}\alpha^n.$$

We substitute for F_n with the Binet form (2.13)

$$d_n = \frac{\alpha^n - \beta^n}{\sqrt{5}} - \frac{1}{\sqrt{5}}\alpha^n$$

$$= \frac{-\beta^n}{\sqrt{5}} \qquad\qquad = \frac{-(-\phi)^{-n}}{\sqrt{5}}.$$

Let's evaluate the first few of these adjustments, and observe their sign and size in Fig. 41, p.47. We note that positive adjustments mean increasing the radial vector from spiral to number point (e.g. top d_1), and these will be for odd indices on vertical quadrant axes. The first adjustment d_0 must take the initial spiral spoke of $1/\sqrt{5}$ and reduce it to zero (as $F_0 = 0$). Therefore $d_0 = -1/\sqrt{5}$. Here is a table of values:

$$d_0 \;=\; -1\big/\sqrt{5} \;=\; -\phi^0\big/\sqrt{5} \;=\; -0.447$$

$$d_1 \;=\; -\beta\big/\sqrt{5} \;=\; +\phi^{-1}\big/\sqrt{5} \;=\; +0.276$$

$$d_2 \;=\; -\beta^2\big/\sqrt{5} \;=\; -\phi^{-2}\big/\sqrt{5} \;=\; -0.171$$

$$d_3 \;=\; -\beta^3\big/\sqrt{5} \;=\; +\phi^{-3}\big/\sqrt{5} \;=\; +0.106$$

$$d_4 \;=\; -\beta^4\big/\sqrt{5} \;=\; -\phi^{-4}\big/\sqrt{5} \;=\; -0.065$$

If we now compare successive adjustments, we find they differ in scale by a factor of $-\phi$. This means that if we pick *any* two adjustments, then they will be relatively scaled by some power of this factor. Which further means that no matter which pair is chosen, the two adjustments can never share a common measure. That is, there is no unit of distance with which *both* adjustments can be measured without remainder. Some might even say that the relationship between the adjustments here is 'maximally incommensurate'.[7] Nevertheless, the overall result—that we find the integer number points are at irrational distances from the smoothly growing spiral, and that these distances share no common measure—is hardly a great surprise. (Yet our efforts here are far from wasted—the reason for introducing this approach will become very clear later.)

---

[7]   In Appendix C, *Continued fractions*, we explore why ϕ is sometimes called 'the most irrational number'—identity (C.4) on page 422. However, it is the transcendental numbers (those real but not algebraic) that have the greater claim to being called 'the most irrational'.

Chapter 3

In nature

The chambered nautilus shell—Golden or not?

Is the shell of a chambered nautilus a good example of a logarithmic spiral? Yes, it is. But is it a good example of a *Golden Spiral* (as many would have us believe)? Let's check this once and for all, in Fig. 42. We overlay a true Golden Spiral (the one from page 43) to see if it matches the spiral edge of the nautilus shell, and we see that the Golden Spiral grows much faster than the nautilus shell. So now we know, the chambered nautilus may have the shape of a growth spiral, but the growth is *not* in the form of a Golden Spiral. However, some nautilus shells and snails do have spirals that grow at approximately half the Golden Spiral rate.

Figure 42: **Chambered nautilus shell—very beautiful, but not Golden.** Shell by Chris 73 [115, 144].

The world's first Golden Spiral nautilus

George Hart (co-founder of MoMath New York, research professor, geometer, and sculptor) noticed this discrepancy too, but uniquely, he took the next step by asking: 'What would a nautilus look like if it actually was a Golden Spiral?'   He then imagined, designed and fabricated an example using a two-colour 3D (3-dimension) printer (Fig. 43) [320].   The weird look of the over-large opening clearly demonstrates that the growth rate per turn of the Golden Spiral is far too rapid to match that of the natural nautilus.

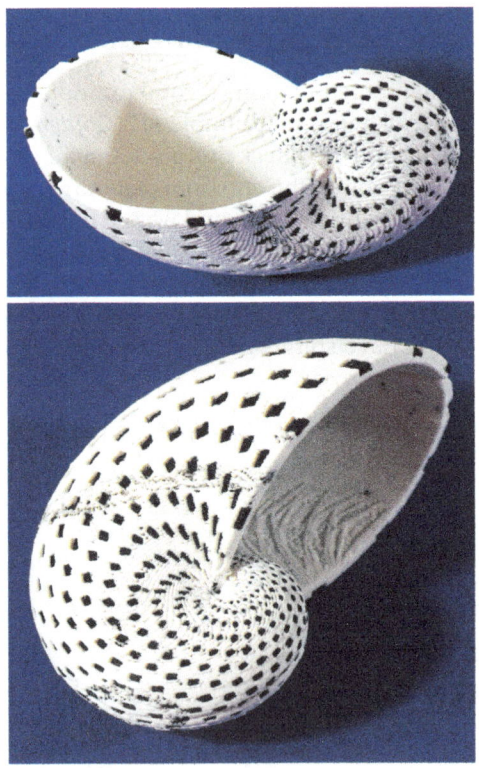

Figure 43: **The Golden Ratio nautilus.** Images courtesy of and © George Hart, http://georgehart.com who has also made a short video which explains the model in full.

Petal counts

One might expect that the subject of counting petals on flowers would be one of the simpler, more well defined topics that we might cover. However it is not.

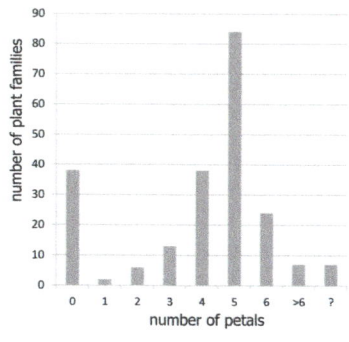

Figure 44: **Petal counting—** *Prunus Cerasifera* has 5. Image by Cb89 [103, 144].

Figure 45: **Table** data source: Nishiyama 2012. Data reproduced with permission of International Journal of Pure and Applied Mathematics [681].

It is certainly possible to map Fibonacci numbers (especially the number 5) to the numbers of petals of named flowers [484]. But does this give us the 'whole picture'? Indeed, just what could or should the whole picture be? Are we interested in the number of species having n petals, or the number of families, or even the number of individual plants worldwide or in a particular region? In Fig. 44 we see a flower with five petals—generally held to be the most common number.[1] And in Fig. 45 we plot the data found by researcher Yutaka Nishiyama [681]—the '?' column is for 'unknown number of petals'.[2] It is sometimes said that having 4 petals is evidence of membership of the Lucas number sequence, but this does not then correspond with the relative lack of 7-petal flowers. Also, 6 is neither a Fibonacci nor a Lucas number. The best we can say is that flower petal numbers often match to Fibonacci numbers. However the arrangements of leaves and seed heads *do* follow much more of a 'law' (but still not 100% reliably). As a result of evolutionary processes (for example maximizing light capture), patterns of growth may include Fibonacci or Lucas numbers of spirals. This phenomenon is called 'phyllotaxis'.

---

[1] Kepler noted that trees and bushes tend to produce 5-petal flowers, and that apples and pears (which are tree fruits) exhibit 5- and 10-fold symmetry [467].

[2] Nishiyama used Makino's extensive (1989) *New Illustrated Flora of Japan.*

Phyllotaxis (leaf arrangement)

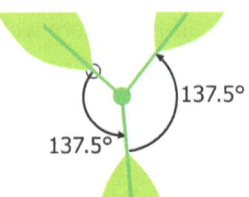

Figure 46:

'Ideal angle' between leaves

Interest in general phyllotaxis has recently mushroomed, leading to an immense multi-disciplinary literature. As an example, mathematician, scientist, and phyllotaxis specialist Irving Adler (1913-2012) has written many analytical articles about phyllotaxis, and his papers have recently been compiled into a book [3].[3] As Adler documents, both the ancient Egyptians and Greeks observed the regular spacings of leaves. Leonardo da Vinci noted that these were often in a spiral arrangement [3], and also that: 'Nature has so placed the leaves of the latest shoots of many plants that the sixth leaf is always above the first, and so on in succession, if the rule is not (accidentally) interfered with' [541]. Johannes Kepler conjectured on the involvement of Fibonacci numbers [467], Genevan naturalist Charles Bonnet observed the effect [74], and the same was confirmed in the first detailed study—made by German botanist Karl Schimper in 1830 [805]. Kepler's remarkable insight led him however, to the incorrect conclusion that the Fibonacci sequence itself was part of the mechanism of plant growth [5]. The Bravais brothers (1837) built on Schimper's work by 'unwrapping the stem', thus allowing them to represent the leaf distribution as a point lattice [81, 6]. Looking back, we can now appreciate how perfectly suited the Bravais brothers were to their task—one being a botanist and the other a crystallographer [415]. They also first introduced the concept of 'living crystals' and that of the 'ideal angle' (of divergence)—being 360° divided by the square of the Golden Ratio—i.e. 137.5° (Fig. 46, looking down on plant with stalk centre). Their work prepared the way for modern studies. However, progress in understanding has by no means been steady. For example in 1882 physiologist anatomist Julius von Sachs published his very influential *Textbook on botany*, in which he completely rejected the mathematical approach, dismissing it all as 'nothing but a game of numbers', and also saying that the spirals seen were a matter of 'irreducible subjectivity' [795, 7].

---

[3] These papers were previously published in the *Journal of Theoretical Biology.*

The Golden or Fibonacci Angle   137.5°

In the mention of the 'ideal angle' above, it may seem strange that the reciprocal of the *square* of ϕ was referred to. But this is just a one-step calculation to get the result

$$\text{Golden Angle} = 360 / \phi^2 = 137.50776405...°$$

This irrational value is normally rounded to one decimal place. We also have:

$360/\phi$	$= 222.5$	$= 1 \times 360 - 137.5°$ or
$360 \times \phi$	$= 582.5$	$= 2 \times 360 - 137.5°$ or even
$360 \times \phi^2$	$= 942.5$	$= 3 \times 360 - 137.5°$.

All these calculations end up giving us the same amount of change in direction (ignoring sign)—that is, we still arrive at the same divergence angle of 137.5°. (The range of ϕ powers with this property is from -2 to $+2$, missing out ϕ^0.) In the cactus shown in Fig. 47, the smallest, youngest leaf (pointing up towards us) is marked A, and the next older leaf B. We see that the angle A to B equals 137.5°—which is the same as the angle B to C, and so on.

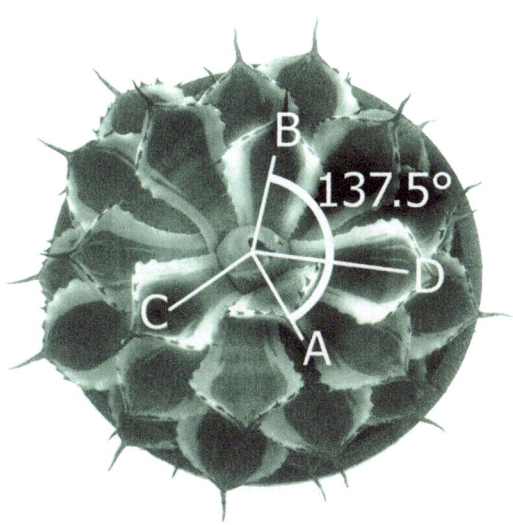

Figure 47: **Leaf arrangement—example cactus.**

Conspicuous spirals

Figure 48: **Pineapple 'crystals'** from Costa Rica.

Spirals are found throughout the natural world—they remind us how often growth occurs as a cyclic process. In particular we are very familiar with the spirals seen: of leaves on stems, florets in flower heads, the seeds of a sunflower (Fig. 55, p.60), and scales of pine cones, pineapples, and Romanesco broccoli (e.g. Fig. 49 with 8 shallow and 13 steep spirals). In Fig. 48 on the right, we see 3 sets of helical spirals. The shallowest-sloped set runs from top left to bottom right, and there are 8 of these counting around the girth. Then from top right to bottom left there are 13 steeper spirals, and finally there is the steepest set of 21—almost vertical. Such conspicuous spirals are called 'parastichies'. The counts: of 8, 13, and 21, are successive Fibonacci numbers. As Adler points out, the study of these repeated patterns in many ways resembles the study of *crystals*—which echoes the Bravais brothers' consideration (mentioned earlier) of botanical examples such as the pineapple being *'living crystals'* [4].

Figure 49: **Fibonacci (8,13) Romanesco broccoli** from United Kingdom.

Pine cones

The Scots pine (*Pinus sylvestris*) is native to the United Kingdom [224], and in Fig. 50 we see a mature open pine cone (female) with two views of its base. This example shows the very common parastichy pair $(5, 8)$—but other pine cones may show pairs of $(2, 3)$ or $(3, 5)$ [418]. Conifers have been very successful worldwide—the earliest known trees date back 300 million years [750].

Figure 50: **Fibonacci $(5, 8)$ pine cone.**

In past studies of phyllotaxis, one of the driving questions has been: 'Why do the Fibonacci numbers appear as spiral counts *in plants*?' However, exactly the same effect has recently been discovered in non-botanical systems too—so its scope has significantly widened.

Vogel's better sunflower formula

Much has been written on the mathematical modelling of plants and trees. In their *The Algorithmic Beauty of Plants*, Prusinkiewicz & Lindenmayer [769], discuss Vogel's formula [930], which places sunflower floret number n at

$$\begin{cases} r &= scale \times \sqrt{n} \\ \theta &= n \times 137.508°, \quad n = 0, 1, 2, 3, \ldots \end{cases} \quad (3.1)$$

We see the effect of this very simple rule and the way it regularly packs the florets in Figs. 51 and 52 and (modelling a different stepping angle) in Figs. 53 and 54.

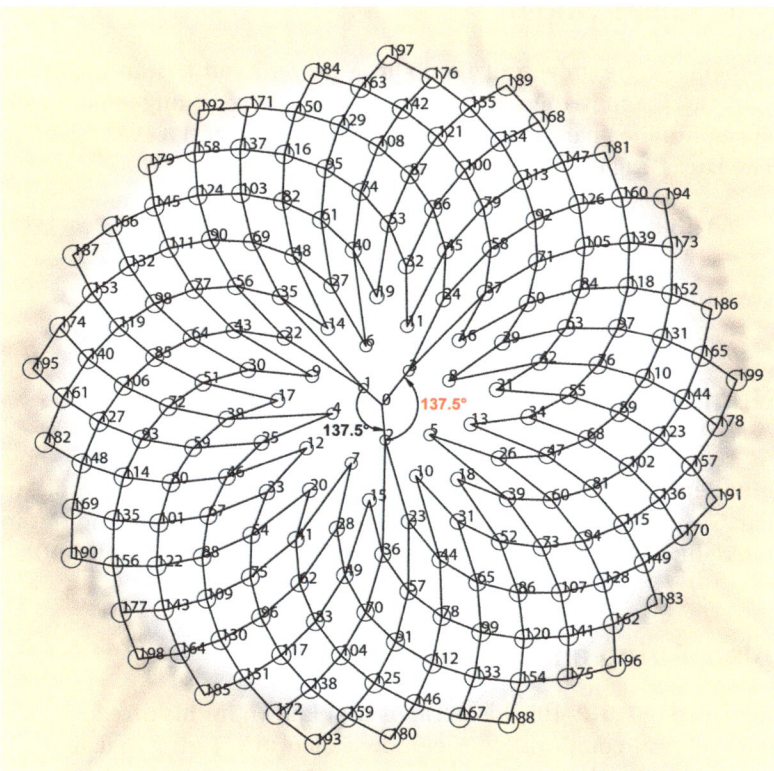

Figure 51: **Fibonacci (21, 34) sunflower head with numbered florets** from Vogel's formula. At the centre we see the initial turns of 137.5° from floret 1 to floret 2 and from floret 2 to floret 3. Such rotations are repeated round and round, up to a final count of 199. Here we see the conspicuous steep and shallow spirals—the parastichies—34 steep and 21 shallow (these being consecutive Fibonacci numbers)—and all this from equation (3.1) ! In Fig. 52 below, we look closer to find a yet steeper set of spirals, and also F_n number grouping.

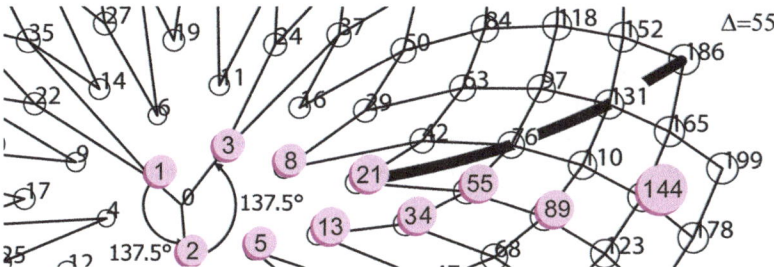

Figure 52: **Fibonacci (21, 34, 55) detail.** As with the pineapple, we see a very steep parastichy; this one links florets: 21, 76, 131, and 186. The constant difference of 55 gives us the total number of these steep spirals—and 55 follows 21 and 34 in the Fibonacci sequence.

The 'Lucas Angle'   99.5°

Sometimes the growth of florets in a flower head has an imperfect start, and it may then settle into an alternative arrangement[4] with Lucas number parastichies rather than Fibonacci (Figs. 53 and 54). The Lucas Angle is

$$360 \cdot \left(3 + \frac{1}{\phi}\right)^{-1} = \frac{360}{3 + \phi^{-1}} = \frac{360}{2 + \phi} = \frac{360}{1 + \phi^2} = 99.50155°. \quad [417]$$

We return to phyllotaxis later (in 'ϕ Science' research, p.289).

Number spirals

As computer scientist John Williamson notes, it is interesting how numbers group when positioned in Vogel spirals—for example in Fig. 52 (p.57), the Fibonacci numbers group in the centre and out to the right; and in Fig. 54 (p.59), the Lucas numbers do the same [981].

Alan Turing and sunflowers

Alan Turing (1912–1954) is perhaps best known for his invention of a hypothetical computing device now called the 'Turing machine'. It clarified the idea of 'definite process' in solving mathematical problems and was the precursor to modern stored-program computing. Turing will always be remembered too for his key role in cracking the Enigma code [381]. As a schoolboy, he had a keen interest in nature and was familiar with Scottish biologist and mathematician D'Arcy Thompson's master work, *On growth and form* [893, 380, 370]. The mathematical aspects of natural growth became a lifelong interest, leading (for example) to his work on the patterns of stripes and spots on animals. Up until the end of his life, he worked on theories about the appearance of the Fibonacci numbers in nature. His goal was to explain emergent phenomena such as phyllotaxis in terms of maths, physics, and chemistry [905].

---

[4]   H S M Coxeter wrote: 'It should be frankly admitted that in some plants the numbers do not belong to the sequence of [Fibonacci numbers] but to the sequence of [Lucas numbers] or even to the still more anomalous sequences $3, 1, 4, 5, 9, \ldots$ or $5, 2, 7, 9, 16, \ldots$ Thus we must face the fact that phyllotaxis is really not a universal law but only a fascinatingly prevalent tendency.'   © 1961 Wiley, quoted with permission [139].

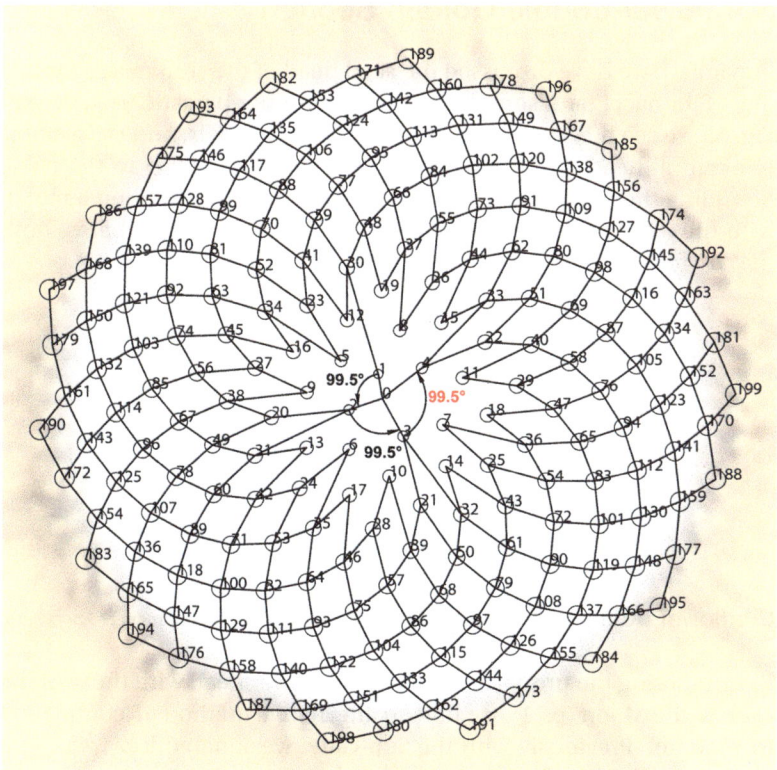

Figure 53: **Lucas (18, 29) sunflower head with numbered florets.** (Vogel's formula—but with the Lucas divergence angle.) At the centre we see the initial turns of 99.5°. And again we repeat the turning and growing up to a final count of 199. This time we see 18 shallow and 29 steep parastichies; these counts are consecutive Lucas numbers. And as with the Fibonacci head, in Fig. 54 below we discover a yet steeper spiral.

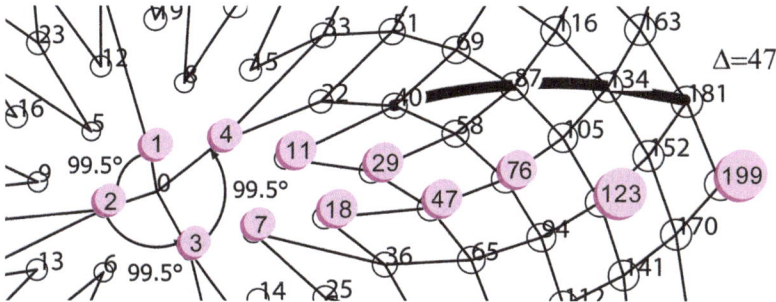

Figure 54: **Lucas (18, 29, 47) detail.** Here the steepest spiral links florets: 40, 87, 134, and 181. The constant difference this time is 47 which again gives the total number of these parastichies. Also, 47 follows on from 18 and 29 in the Lucas number sequence.

Sunflower double Golden Section

Previously, we have discussed the seed head of the sunflower in terms of its Fibonacci (and sometimes Lucas) number parastichies. Here in Fig. 55 we see the seed head complete with its petals, and we see how the central head disk diameter is (closely) related by ϕ to the circle bounding the petal tips. When we look at it this way, we find a match with the ABC and $C'AB$ Golden Sections in Fig. 1 (p.3)—that is, both the $\alpha = \phi$ and $\beta = -\phi^{-1}$ cases.

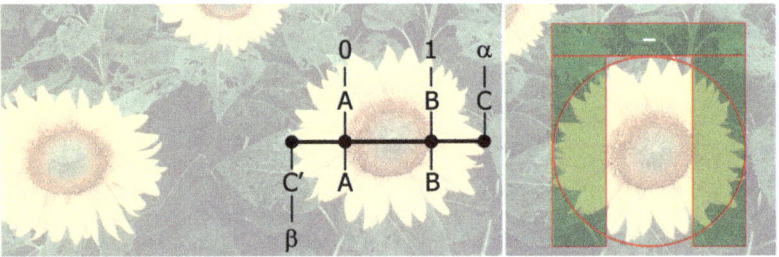

Figure 55:
Sunflower both-roots Golden Section. Flowers: US Department of Agriculture.

By comparing the proportions of the sunflower head with those of the Gate of the Moon (p.11), we find an alternative method of 'completing the square'. Previously with the Sun Gate, we squared it by *adding* a copy of its top lintel; here we square the Moon Gate by *removing* its top lintel—again a perfectly complementary operation.

Music, art, and architecture:
The importance of Paris

For the rest of Part I, we shall explore the appearance of ϕ in music, art, and architecture. But, rather than cover a wide range of examples (and then necessarily briefly)—we shall instead focus on a few—and analyse those in more detail. For music, we shall concentrate on the works of Bartók, Debussy, and Xenakis, with mentions of Satie and Ravel. For art and architecture, we shall consider works of Boulanger, Seurat, Toulouse-Lautrec, Mondrian, Steinlen, and Le Corbusier. Although we shall aim for date order, the need to develop concepts will sometimes take priority. In reviewing these examples, it will become apparent how (from the late-1800s) Paris became the *de facto* centre for the study and application of Fibonacci and ϕ.

Chapter 4

Music—Bartók

Much has been written discussing whether ϕ (and or the Fibonacci numbers) have or have not been used in particular pieces of music or architecture. But in contrast, as mentioned earlier, our emphasis will instead be in exploring the *relationships* that make Fibonacci numbers and Golden Ratio such excellent 'engineering tools'—tools which are freely available to musicians, architects, and artists alike. The subtlety and power of such relationships are not generally appreciated—so a composer might for example include a lone cymbal clash at the Golden Section point in a piece (61.8% of the way through), and commentators will then list that composer as one who has 'managed to incorporate the Golden Section into his or her work'.[1] However, as we shall see both here and in the *Binet spirals* chapter, some composers demonstrate a deep understanding of the mathematics and make good use of recurrence relations in structuring their work. Earlier, we touched on the importance of rhythm and metre in Sanskrit—time was structured using long and short syllable durations. In music we add:

- different 'colours' of musical instruments

- many more note lengths

- the pitch of notes (frequency)

- dynamic range (loud to soft)

- parallel 'threads' of harmony

---

[1] (Illustration by exaggeration.)

A composer builds his or her structures at a number of levels. There are units for the lengths of notes, for groupings of notes, phrases, and so on up to the full length of a piece. Much (but not all) music uses nominally standardized pitch frequencies for notes, along with agreed note-length conventions, and widely adopted written forms. A very common approach to choosing note lengths is based on the factor 2—a duration may be halved and halved, again and again (or doubled similarly). Bars[2] may be grouped as: 2, 4, 8, 16, and so on. Nevertheless, composers (especially modern ones) have sought different ways to go beyond such conventions, and to do this, many have apparently applied Fibonacci and ϕ: Johann Sebastian Bach, Béla Bartók, Claude Debussy, Jean-Claude Éloy ('El-wah'), Ernst Křenek, Per Nørgård, Erik Satie, and Franz Schubert [497, 509]; also Gustav Mahler, Leoš Janáček, Paul Hindemith, Iannis Xenakis, Karlheinz Stockhausen, Luigi Nono [477].[3] In the first of two related articles in the Fibonacci Quarterly, Edward Lowman makes a key observation regarding composers' exploration of time and its musical organization—he notes their discovery that Fibonacci proportions (in rhythms and duration structures) *offer an alternative to both old techniques and randomness* [577].

Bartók: Sunflowers and pine cones

In his second article, Lowman writes that the great Hungarian composer Béla Bartók (Fig. 56) had a deep interest in nature, the Golden Ratio, and the Fibonacci sequence [578]. Bartók's particular fondness for sunflowers and pine cones is very widely reported— these being natural embodiments of, and 'publishers' of, Fibonacci numbers [638, 535]. Bartók was also an able mathematician [782] as well as a musician, and he was very much influenced by Debussy (whose works are sometimes analysed from a mathematical point of view). As a result, it appears that Bartók was drawn to consider the musical possibilities of the Fibonacci sequence for his own compositions. Lowman discusses Hungarian musicologist Ernő Lendvai's analysis of one of Bartók's best known pieces—his 1936, four movement *Music for String Instruments, Percussion, and Celesta*— often referred to as just 'MFSPC' [40]. Ernő Lendvai was a pianist, music professor, broadcast sound engineer, and a controversial pioneer in the study of Bartók.

---

2   Measures based on rhythm.
3   With the statutory qualification: 'consciously or otherwise... '

Figure 56: **Béla Bartók (1881–1945).**

He recalled that as early as the 1940s, he felt that he had intuitively discovered key elements of Bartók's compositional techniques—as if 'struck by spiritual lightning'. It was not until 1955 that he published these findings—in *Bartók's Style* [534]. This book was closely followed by a collaborative work led by Bence Szabolcsi of the Budapest Conservatoire, which included input from (Hungarian composer) Zoltán Kodály: *Bartók his life and works* [880]. The unfortunate controversy (still rumbling on today) arose in equal parts from Lendvai's deep insights and also (in the view of a number of serious analysts) that various of his conclusions were based on flawed arithmetic. For example Lendvai concluded that Bartók overtly structured the first movement of MFSPC in Fibonacci number proportions (Fig. 57)—and did so in a way that strongly implied an overall structure of 89 measures, indeed Lendvai shows 89 in his diagram [537]. However the piece actually finishes after 88 measures.

It is for reasons such as this, that various musicologists, including Ruth Tatlow and Paul Griffiths for example, regard Lendvai's work as 'dubious' [568]. However, a simple explanation of how a pure Fibonacci structure (initially envisioned) could end up distorted could just be that 'the composition required it'. Bartók chose never to clarify; he was extremely reticent, secretive even, about his methods. He said: 'Let my music speak for itself; I lay no claim to any explanation of my works.' [540]. Lowman states (after Lendvai) that in the first movement of MFSPC ('MFSPC I'), the dynamic climax, *(fortissimo possibile),* occurs after a total of 55 measures, (effectively at the Golden Section of the first movement); and leading up to this point, there are 34 measures played with mutes, followed by 21 without. In Fig. 57 (lower left) we see how a phrase of 21 measures may be followed by one of 13, and how a parallel phrase of 34 measures will end at the same time.

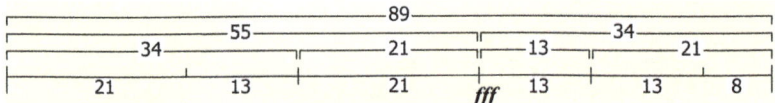

Figure 57: **Bartók: Simplified structure of the first movement of** *Music for String Instruments, Percussion, and Celesta* **[40].** Diagram by Edward A Lowman [578] (after Lendvai [537]), © 1971 *The Fibonacci Quarterly*, reproduced with kind permission.

Indeed, the Fibonacci numbers give enormous flexibility in constructing such same-length groups in music[4]—threads may go off, subdivide differently, work with or against each other, but *still arrive at the end of a phrase together.* These whole-number-based, exact-fitting, coterminal properties are crucial, and they provide self-consistent and orderly structure. Another source of order is the similarity and repetition typically found in Fibonacci-based structures in both the short to medium and the long ranges—and in both numbers and proportions. Even in this very compact example, we see the sequence: 21, 13, 21, appearing twice; and 13 following 21 in three instances, while 21/13 closely approximates the Golden Ratio.[5]

---

[4] As do (similarly) the powers of ϕ. Le Corbusier realized how this special property could be systematically applied to architecture and to packaging. (We discuss this in Chapter 8.)

[5] The ratio $21/13 = 1.615\ldots$ With such 'convergents of ϕ' (p.22 and p.422) we do not need very large number pairs to get excellent approximations to ϕ.

> In his 1973 article, *The Fibonacci series in twentieth-century music,* music theorist Jonathan Kramer went straight to the heart of the matter— saying of the Fibonacci sequence:
>
> **'It avoids periodicity and regularity, yet it is well ordered ... '** [510]

In a more recent and very detailed analysis, pianist and music scholar Roy Howat examines the arguments for and against the exact use of Fibonacci numbers by Bartók. From this, the picture emerges that at the very least Bartók did use Fibonacci structures as basic starting points in various of his compositions, and much of his original structure remains. As regards the lack of documentary evidence from Bartók, Howat concludes that he probably had a policy of disposing of any working papers that might give insight into his methods [397]. However his supposed eradication efforts were not 100% effective, and Howat discovered an example of Bartók's notes analysing a Turkish folk song, explicitly revealing his application of Lucas numbers: 3, 4, 7, 11, and 18   [35].   We shall return later to the important question of written evidence, but meantime, we shall just note that (for example) no one is surprised if a great chef keeps a recipe secret.   And again, Debussy had very clear views about deconstruction: 'The beauty of a work of art is something that will always remain mysterious... At all costs let us preserve this element of magic peculiar to music.'—he stresses: 'at all costs' [86, 942]. And later too, in his article *Taste*, Debussy asks that in-depth analysis be avoided—so that: *'one can never find out exactly how it is done'* [166].

So, to summarize this analysis of the MFSPC I, we find Bartók does (broadly) structure according to ϕ/Fibonacci. But set against this, Howat contrasts the tonal structure which he finds symmetric about bar 44 (out of 88)—exactly half way. So, in addition to $1 : \phi$, we see the ratio $1 : 2$ [393].   Yet in case we think '1, 2, and ϕ' might be Bartók-specific, we should note that Howat also found these in his study of Debussy [382]. Also we shall find this theme continued in art and architecture. At the end of this Part, in the *Binet spirals* chapter, we shall be meeting Lendvai, Bartók, and Debussy again. For now, we move on to see how *Belle Époque* Paris became a hub of artistic and (Golden) mathematical activity—that is, *the* hub. And to set the jubilant scene... we have a contemporary poster by Henri Boulanger, Fig. 58. The geometric overlay reveals its Golden design basis—3 landscape Golden Rectangles, stacked to give the ratio $3:\phi$.

Figure 58: **Henri Gray—pseudonym of Henri Boulanger (1858–1924),** **_Théâtre de l'Opéra, Bal Masqué,_ (Paris) 1899.** Colour lithograph, proof print before addition of text announcing the event, 127×97 cm. Golden Rectangles analysis added.

The vertical Golden Sections of each rectangle create a single key vertical in the design—marking the left of the central columns. This poster well captures the ebullience of _Belle Époque/fin de siècle_ Paris. (The _Belle Époque_ is often taken as the period 1871–1914, though some commentators shorten this.)

Chapter 5

Paris—capital of ϕ

But first... Zeising, Fechner, and Hambidge

In mathematics, agreement is expected: for example consensus about what constitutes a correct proof. Conversely, in art it is expected that every viewer will have their own unique response and opinion—and that such opinions may differ greatly. So it is not surprising that the simple recipe of 'mathematics plus art' will reliably provoke debate. In 1854, German psychologist Adolf Zeising published his *New theory of the proportions of the human body*, in which he proposed that the Golden Ratio was the 'underlying secret' of all natural and artistic form [1000, 718]. Art history researcher and lecturer Marcus Frings argues convincingly that prior to Zeising, ϕ was *not* applied in art and architecture—save in the addition of Platonic solids as ornament —a practice specifically approved of by Luca Pacioli (p.21). Pacioli related the story of Phidias giving his statue of Ceres in Rome an icosahedron [230, 717].[1] Zeising's work (among other things) prompted philosopher, physicist, and experimental psychologist Gustav Fechner to conduct a series of experiments, the results of which he published in 1876 [211]. In one famous experiment, Fechner asked individuals which of ten rectangle shapes they liked best, and which they liked least. Fechner was surprised to find his measurements confirmed an overall preference by his test group for the Golden Rectangle aspect ratio $(1 : \phi)$. But, although Fechner's work in Germany was tremendously influential,[2, 3] it was Paris (as mentioned earlier) which was destined to become the *de facto* centre for the

---

[1] But, we have already discussed pre-Zeising works—from Bolivia and Egypt (p.10).

[2] It launched the new field of study 'experimental psycho-aesthetics'—which seeks to understand what factors are involved in aesthetic appreciation.

[3] It also led to the use of ϕ by certain groups of German artists shortly after [355].

study and application of the Golden Ratio in the arts, music, and architecture, for at least the first half of the twentieth century. Next though, we discuss Jay Hambidge and several mathematicians.

Dynamic symmetry

Jay Hambidge (1867–1924) was a Canadian-born American artist [296, 300]. In his works he wrote of the distinction between (what he called) static and dynamic symmetry, and he promoted the latter. For Hambidge, static symmetry was the simple 'unit-based' type—for example the 1:1 left-right symmetry of the idealized human body shape, or the 6-fold symmetry of the snowflake [297]. In geometric patterns it would include those based on squares or equilateral triangles. In contrast, he defined dynamic symmetry as that based on irrational ratios.[4] Now, $1{:}\phi$ comes under this heading, and Hambidge also included ratios such as $1{:}\sqrt{2}$, $1{:}\sqrt{3}$, and $1{:}\sqrt{5}$. Hambidge preferred to think of these as being the aspect ratios of rectangles—hence the term 'dynamic rectangle' (Fig. 59).

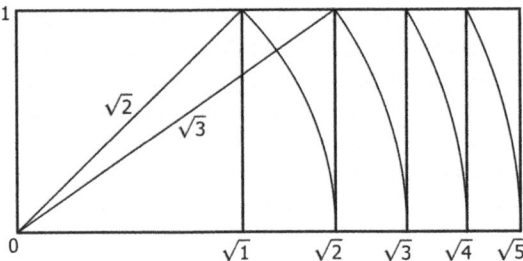

Figure 59: **Hambidge root rectangles** redrawn from *Dynamic symmetry: the Greek vase* (1920) [300].

Hambidge associates dynamic symmetry with life and movement [299]. His studies overall encompassed Greek and Egyptian art, human and plant form (including phyllotaxis), and the Platonic solids. By considering the proportions of available statues, he concluded that the Greeks got their knowledge of dynamic symmetry from the Egyptians in around the 6th century BC [298]. The Golden Section is an example of symmetry where as we know, *proportion* (rather than length) is mirrored—where the ratio of 'greater to the

---

[4] One way to think of this distinction is that 'static' aspect ratios will have continued fractions that stop, and that continued fractions for 'dynamic' ratios will 'run on'. (We explore continued fractions in Appendix C on page 421).

lesser' is the same as the 'whole to greater'. The resulting asymmetry of lengths gives a 'dynamic quality' to the Golden Section which changes according to 'greater first' or 'lesser first', when set in a particular context and reading direction—our ABC and $C'AB$ again.

(In his 1956 book, Czech-born artist and architect Wolfgang von Wersin identified 12 rectangles as his 'orthogons', which he claimed were used historically by artists, architects, and calligraphers to guide their designs. He gave each one a new name, calling the Golden Rectangle 1:ϕ the 'auron', and the 1:$\sqrt{2}$ rectangle the 'diagon', and the ϕ:2 the 'biauron' [973]. He also showed a Renaissance drawing of 7 rectangles (said to be) known to, and revered by, ancient architects —6 static and one $\sqrt{2}$ [972].)

Mathematicians

Jay Hambidge's ideas were put into practice by artist Maynard Dixon who was also informed by his friend, mathematician Albert Barrows [291, 292]. Earlier (p.16) we saw an architectural work of Polykleitos (the Younger)—the Theatre at Epidaurus. His father was mathematician Polykleitos (the Elder)[5] [775]. We also recall how Leonardo da Vinci partnered with Luca Pacioli to produce their book *De Divina Proportione* [704]. And in more modern times artist M C Escher (1898–1972) used input from mathematicians George Pólya, H S M Coxeter, and Roger Penrose [803, 311]. Such relationships have brought non-trivial mathematics into art, music, and architecture over the centuries and around the world.

Paris—how it became the capital of ϕ

In the same year that Fechner published the results of his aesthetic experiments—1876—Édouard Lucas in Paris presented his landmark mathematical studies of the Fibonacci numbers and related sequences [211, 592]. Shortly after this, Parisian ϕ-evangelists such as Charles Henry (librarian, colour theorist, friend of Lucas, and 'mathematician to the Symbolists'); and Joséphin Péladan, (Martinist and Rosicrucian), spread the knowledge. As we shall see, such was the level of interest that it was probably rather difficult for any (even slightly interested) intellectual living in Paris *not* to become thor-

---

[5] Polykleitos the Elder devised a system of mathematical proportion—his *Canon*— which he used to create statues with idealized human form. This ambitious exercise was repeated by Vitruvius and Leonardo da Vinci, by Adolf Zeising, and (to an extent, for a different purpose) by Le Corbusier with his 'Modulor Man'.

oughly acquainted with and captivated by the Golden Ratio—especially in the case of those involved in esoteric groups such as Theosophy (Piet Mondrian, Alexander Scriabin, Léonce Rosenberg), Martinism (Joséphin Péladan), and Rosicrucianism (Claude Debussy, Erik Satie).

The Table in Fig. 60 gathers several examples of Parisian artists, composers, and architects, (and those linked to Paris), and their relationships with mathematicians. Self-proclaimed High Priest, ('*Sâr'*) Joséphin Péladan promoted ϕ mathematics from a mystical point of view. Charles Henry gave lectures at the Sorbonne which included discussion of the Golden Section where the audience included Parisian divisionist artist Georges Seurat. Seurat reportedly applied this knowledge in his work [335]. (But, our purpose in this chapter is not to decide which artist did or did not employ Fibonacci and ϕ in their work. Rather it is to appreciate what a crucible of artistic, philosophical, and mathematical activity Paris became.)

Maurice Princet (1875–1973) was a Parisian actuary who became known as 'the mathematician of cubism'—because of his role in its genesis and development. 'His' cubists first congregated in the artistic hub *Le Bateau-Lavoir* in Montmartre [647] (map page 143, near 'L'). But, following the departure of his wife to André Derain in 1907, Princet turned his attention to the nascent *Section d'Or* group in Puteaux. Yet, while their name translates simply as 'Golden Section', they were not concerned with exhibiting specifically ϕ-based art [174]. Instead (they announced that) by so naming their salon, they were claiming continuity with the great artistic traditions of the past [170]. (Some viewed the setting up of this group as 'a homage to Seurat'.) The name *Section d'Or* was also chosen—according to musician and Dadaïste Gabrielle Buffet—to give association with philosophic and scientific research [360]. This Puteaux group took a wider view of cubism than did the Montmartre group (which included Picasso and Braque).

After the start of World War I in 1914, many of the Montmartre group moved to Montparnasse [648]. The geometry here is interesting: Haussmann's new Paris respected the historic axis which (from west-east) tilts northwards by 26° when going west. This axis continues through Puteaux/Courbevoie/La Défense—(in the map on page 143, top left, marked 'N'). Thus because Haussmann centred his plan on *Le Palais du Louvre* (marked 'A'), the radial axis out through Montparnasse ('H') makes close to a right angle with the *Axe Historique*. So, the two groups (Puteaux and Montparnasse) were not so much diametrically opposed—rather they were orthogonal.

Artist / Composer / Architect	Mathematics from	
Georges Seurat	Charles Henry.	[335]
Paul Signac	Charles Henry.	[685]
Henri de Toulouse-Lautrec	Formal education, & maybe Signac, & *Chat Noir* [a] regulars?	[216]
The Symbolists and	Charles Henry,	[385]
Claude Debussy and	Joséphin Péladan.	[386]
Erik Satie		
Paul Sérusier via Jan Verkade	Didier Lenz at Beuron.	[356, 567]
Théophile Steinlen	Formal education (Lausanne), also *Chat Noir* regulars?	[379]
The Cubists (esp. Picasso) and *Section d'Or* (Puteaux[b] group)	Maurice Princet (who introduced the work of Henri Poincaré).	[645]
Gino Severini	Raoul Bricard.	[358]
Juan Gris	Formal education: maths, physics, & engineering.	[671]
Jacques Lipchitz	Juan Gris & other cubists.	[357]
Amédée Ozenfant	Juan Gris.	[359, 361]
Piet Mondrian	Mathieu Schoenmaekers.	[975]
Le Corbusier	Architectural training & Matila Ghyka.	[564, 362, 260]
Salvadore Dalí	Matila Ghyka, René Thom.	[575, 261] [159]
Iannis Xenakis	Formal education: engineering, maths, & physics.	[34, 635]

Figure 60: **Artists' sources of mathematical knowledge.**

---

[a] We visit the acclaimed *Chat Noir* (Montmartre) cabaret in a later chapter.

[b] Puteaux/La Défense (Paris West) now hosts '*La Grande Arche*'—reputedly a monument to the 4th dimension (suggesting a 3D projection of the hypercube).

(Romanian Prince) Matila Ghyka (1881–1965) was a mathematician, philosopher, and diplomat. Much of his education took place in Paris, and he became well known for his books on the Golden Number:

- (1927) *Esthétique des proportions dans la nature et dans les arts* (Aesthetics of proportions in nature and in the arts) [260]
- (1931) *Le Nombre d'Or; rites et rythmes pythagoriciens dans le développement de la civilisation occidentale* (The Golden Number, Pythagorean rites and rhythms in the development of Western civilization) [261]
- (1946) *The geometry of art and life* [262]

Surrealist Salvador Dalí was so impressed by Ghyka's work that he met him and adopted him as mathematical mentor (a relationship similar to that between Leonardo da Vinci and Luca Pacioli) [575]. Ghyka's work also had a turnaround effect on the previously negative attitudes of Parisian architect Le Corbusier [564]. Roger Herz-Fischler notes that having read the Ghyka's 1927 work, Le Corbusier in 1928 redrew a 1926 plan to show his prior use of the Golden Ratio [362].

The Golden and Silver Ratios are 'Pisot numbers'

Like Édouard Lucas, p.27, Charles Pisot (1910–1984) was another *normalien*—having studied at the prestigious *École normale supérieure* in Paris. His 1938 thesis popularized the numbers now known by his name [964].[6] Both Golden and Silver Ratios are Pisot numbers, respectively having the forms $(1 + \sqrt{5})/2$ with conjugate $(1 - \sqrt{5})/2$; and $1 + \sqrt{2}$ with conjugate $1 - \sqrt{2}$. Taking a Pisot number to higher powers quickly produces near-integers. Higher powers of ϕ always produce near-Lucas numbers—for example $\phi^{15} = 1364.0007\ldots$ ($\approx L_{15}$, p.477). It is easy to see why this should be from the Binet formula $L_n = \alpha^n + \beta^n$ (p.28) where with $n >> 0$, the α^n term provides the near-integer and the β^n term makes an exact tiny adjustment to leave an integer result.

Summary

The aim in this chapter has been to introduce certain key characters. Later, in the *Binet spirals* chapter, we shall return to Paris and attempt to analyse some of the inter-relationships and events along a time line.

---

[6] Pisot numbers are also known as Pisot-Vijayaraghavan or PV numbers. They were discovered by Axel Thue (1912) [897], and again by G H Hardy (1919) [303] and later studied (in 1941) by Tirukkannapuram Vijayaraghavan [928]. The smallest Pisot number is the plastic number Ψ (p.94, footnote 5). (Definition—p.457.)

Chapter 6

Art—Seurat, Toulouse-Lautrec

The Golden Section of 2

In the *Bartók* chapter we noted his use of 1, ϕ, and 2 in creating proportioned structures in his music (with close approximations to ϕ arising from the use of Fibonacci and Lucas numbers). Musing on this, we might then wonder: 'Could the Golden Section of (length) 2 have some role to play in music, art, and architecture?'

One source of aesthetic pleasure in a design can be the repetition of similar shapes that echo and resonate with each other; and as we shall see, the 'Golden' design approach makes this remarkably easy to accomplish. In Golden-based schemes we shall find not only the Golden Rectangle echoed, but also rectangles proportioned according to the Golden Section of 2, along with visually very similar (and mathematically related) rectangles. Just as with ϕ and π in the Great Pyramid ('use one, get one free'), these additional ratios appear automatically in many Golden layouts. They are a gift back from the geometry, regardless of whether the designer is aware of them or not. To understand the mechanism, we shall first note that $2/\phi$ is within 3% of $\sqrt{\phi}$, and hence $2/\phi$ could be said to be an 'artistic approximation' to $\sqrt{\phi}$, Root Golden, [RG]. Or put another way,

> The Golden Section of (length) 2
> is close to $\sqrt{\phi}$

73

Now, back in Chapter 1, we mentioned that the Babylonians had a method to calculate $\sqrt{2}$ accurately (p.14) [171]. They knew that if x_1 is an approximation to $\sqrt{2}$, and x_1 is less than $\sqrt{2}$ then $2/x_1$ will be greater. Therefore a better approximation is to be found between these, and the mid-value is chosen: $x_2 = \frac{1}{2}\left(x_1 + 2/x_1\right)$. So, what if we apply this logic to derive better approximations to $\sqrt{\phi}$? Starting with low-valued $\rho_1 = 2/\phi$, we shall find a high-valued approximation by dividing this into ϕ. This gives $\rho_1' = \phi/(2/\phi)$, where ρ_1' is high by about 3%—Eqn. (6.1).

$$\rho_1 = \frac{2}{\phi}, \qquad \rho_1' = \frac{\phi}{(2/\phi)} = \frac{\phi^2}{2}. \qquad (6.1)$$

And as with $\sqrt{2}$, we get a better approximation by taking the average— Eqn. (6.2)—noting that $1/\phi = \phi - 1$, and $\phi^2 = \phi + 1$

$$\rho_2 = \frac{1}{2}\left(\rho_1 + \rho_1'\right) = \frac{1}{2}\left(\frac{2}{\phi} + \frac{\phi^2}{2}\right) = \frac{5\phi - 3}{4}. \qquad (6.2)$$

With this ρ_2, we have $\sqrt{\phi}$ to within 0.05%.[1] However, for art, our interest will be back with ρ_1 and ρ_1'—which from now on, we shall refer to as [RG−] and [RG+] as in Fig. 61.[2] So to recap, [RG−] and [RG+] are first-order approximants of $\sqrt{\phi}$, Root Golden, [RG]. They form a complementary pair in the Babylonian square root method.

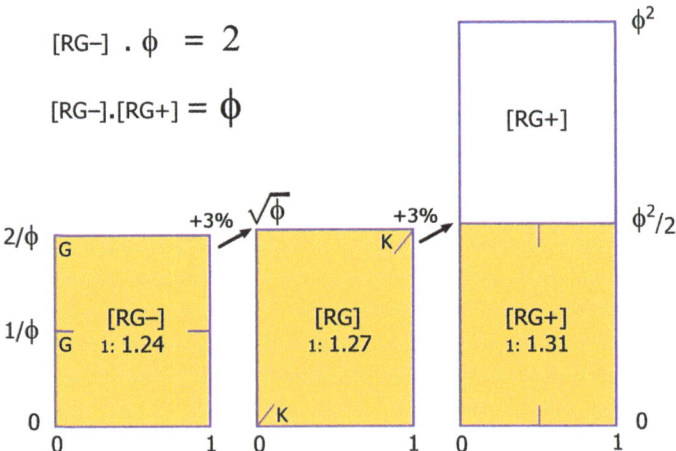

Figure 61: **Root Golden Rectangle [RG] & approximants [RG−], [RG+].**

---

[1] Then from ρ_2, $\rho_3 = (11\phi + 34)/\left[8\,(5\phi - 3)\right]$—being $\sqrt{\phi}$ to within 0.00001%. And in passing, we here spot one Lucas number: 11; then 4 Fibonacci: 3, 5, 8, and 34.

[2] The [RG−] divides *horizontally* as 2 stacked landscape Golden Rectangles 'G'— (Boulanger stacked 3, p.66). The [RG] divides *diagonally* into 2 Kepler's triangles 'K'. And the [RG+] divides *vertically* as 2 adjacent portrait 1:ϕ^2 rectangles.

Georges Seurat

Seurat and divisionism

Georges-Pierre Seurat (1859–1891), the 'father of neo-impressionism', was a scientific artist who (with Paul Signac) followed the colour theories of: Charles Henry; 'colour wheel' Michel Eugène Chevreul; '*Grammaire*' Charles Blanc; and others [385, 119]. Under white light, paint pigment reflects back its characteristic colour and absorbs the others. So (typically), when paints are premixed, more colour is *subtracted*, and less light is returned. For example, red + green = muddy brown. However, in 'divisionism' (sometimes called 'point-illism'), colours are dabbed next to each other. This became Seurat's signature technique. From a suitable distance, the marks blur together and colours *add in the eye*. Now, red + green = 'yellow'.[3] Seurat's theory-based motivation was to produce brighter, purer colour, and he named the method 'chromoluminarism'.[4] Paul Signac also applied divisionism, often using short strokes of colour.

Seurat: *A Sunday afternoon on the Island of la Grande Jatte* (1884–1886)

This is Seurat's best known masterpiece and also his largest canvas— measuring near to 3 × 2 metres. It reveals Seurat's interest in ancient Egyptian art (Fig. 64) [823]; the central figure (holding the copper-coloured parasol) is said to be based on the 'Lady Tuya'—a 1300 BC statuette displayed in the Louvre [337, 46]. This painting is often paired with Seurat's earlier (pre-divisionist, 1884) *Bathers at Asnières* (Fig. 65) showing the other side of the Seine from la Grande Jatte.[5] However, here we shall compare the *Grande Jatte* painting with Seurat's later (1888) *Parade* (Circus Sideshow). Now, given his extremely scientific approach to art, it seems only logical that Seurat would use a clear mathematical basis for his design. Nevertheless this painting has defied proper geometric analysis for over 120 years. Perhaps the reason for this, is that in 1889 when Seurat set about

---

[3] But we now know that a good yellow pigment will (on its own) easily outdo such a combination in terms of luminosity.

[4] These days, we rely on exactly this optical mixing effect for our computer displays, mobile phones, and television screens. The dots we use are much much smaller— organized in a dense array of millions of tiny 'pixels'. Typically, each pixel has 3 dot light sources: a red, a green, and a blue—each with its own responsive brightness control. The ability to change perceived pixel colour and brightness quickly enables video display.

[5] In the Paris map on page 143, see X (*Asnières*) and Y (*la Grande Jatte*) top left.

making his final change to this work—that of adding the border—he decided to re-stretch the painting onto a slightly larger frame first. And in the process of doing so, (by intention or accident), he *removed the relationship between the edges and the original design structure.* It was as if he had slid across locking catches on a briefcase—such a small distance was moved—yet quite sufficient to seal away his secrets. Nevertheless, once the horizontal and vertical offsets are taken into account, then Seurat's Golden layout shines forth simply and precisely. He partitions his 3×2 geometry area (relative units) as a large square of side 2, starting at the left edge; with (to the right, and adjacent), 2 unit squares stacked. He then cuts all 3 squares in what we call $C'AB$ Golden Section (viz. with AB extended left—p.3, here Fig. 62). The key verticals now comprise: the right edge of large square (green), 2 Golden Sections (red), and we add an overall centre line (yellow)—to make just 4 in all. We also note (Fig. 63) how the horizontal centre line in the large square crosses the Golden Section to form 2 types of rectangle—the related shapes [RG+] portrait and [RG−] landscape. They are not exactly similar as would be similar triangles, rather they are artistically similar, with relative characters of short versus long (as we saw on p.74).

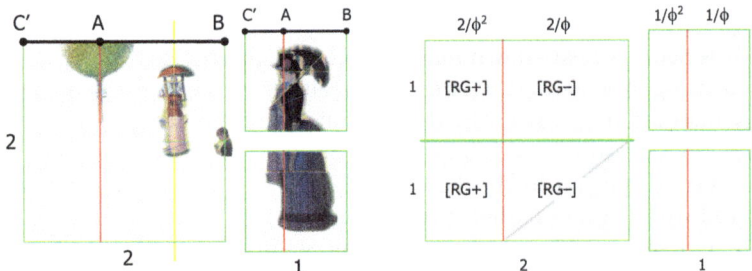

Figure 62: *La Grande Jatte* : $C'AB$ s.　　Figure 63: **[RG] approximants.**

A full analysis is shown in Fig. 66 (p.78). For horizontals—we have the 50% of height centre line (green), plus 4 more horizontals—these being at: 10%, 40%, and 80% (yellow), and 67% ($\approx 2/3$, blue). So, the red Golden Sections are irrational;[6] but all the other divisions are in simple whole number ratios. To the left, the purple line extending the cane locates the centre of an arc.[7] Apparently Seurat has noticed that the left Golden Section (red) is almost at quarter-width position (overall).

---

[6]　Irrationality is *a concept*—in art and design, sufficiently close ϕ convergents may be used—we discuss such practicalities later as they relate to recent crystal discoveries.

[7]　And its reflection in the red Golden Section locates the head of the black dog.

Figure 64: **Georges Seurat (1859–1891)** *Un dimanche après-midi à l'île de la Grande Jatte,* **[A Sunday afternoon on the Island of la Grande Jatte.] 1884–1886.** (Paris: River Seine, Neuilly-sur-Seine.) Oil on canvas, 207.6 × 308 cm, Art Institute, Chicago [823]. The contained 2×3 format echoes that of the *Bathers*.

Figure 65: **Georges Seurat (1859–1891)** *Une baignade à Asnières,* **[Bathers at Asnières.] 1884 + some dots added 1887.** (Paris: River Seine, Asnières-sur-Seine.) Oil on canvas, 201 × 300 cm, London: National Gallery [822]. The 2×3 format is partitioned with a full square to the left (tints added), and half-square right. The vertical extent of the water closely matches the Golden Section line at horizontal level 'B' from the 'ABC' shown far left. The semicircular area is cleared from the tints to reveal the original 2:1 horizontal division (e.g. Seurat's own vertical line in the boy's cheek, and in the water above).

Figure 66: *La Grande Jatte* analysis.

To test this, he sweeps out an arc and finds an overlap at the 'Lady Tuya' central woman. He then echoes this overlap on the right as the (cyan) square containing the tree trunks overlapping the right Golden Section, locating the gentleman's profile. The central black dog is aligned on a diagonal of the lower [RG−] rectangle (from Fig. 63).

Seurat: *Parade du Cirque* (1888)

Fig. 67 shows the later Seurat masterpiece, *La Parade*—Circus Sideshow (or *La Parade du Cirque*) [824]. This was painted just a few years before Seurat's untimely death in 1891, age 31. In it, we shall see the geometry Seurat had developed in *La Grande Jatte,* but here with more vertical irrational sections, and more horizontal rational divisions—Seurat even includes a basic Fibonacci sequence. The advanced design may owe much to Seurat's 1886 meeting with mathematician Charles Henry (p.69) [339].

Figure 67: **Georges Seurat (1859–1891) Circus Sideshow—*La Parade du Cirque* 1888.** Oil on canvas, 99.7×149.9 cm, Metropolitan Museum, New York [824].

The central figure in this piece has various similarities with that in *La Grande Jatte* (Fig. 68). The body is square to us. The hips and left elbow are similar.

Figure 68: ***Grande Jatte* v. *Parade* comparisons.** Centre-line characters and top-right '2+3' verticals (trees and 'piano keys') with (nearer) characters in profile, facing left.

In each there is a horizontal line to the right of the jacket bottom and a white area above that (with similar over their left shoulders). Each has a conically tapering hat. Then there are their left arm positions, with each left hand holding a thin rod. Each has a 'little standing character with hat' to their right. In *La Grande Jatte* the parasol stick tilts left— in *Parade* (Fig. 67) the stick (wittily) floats up and away, tilting right, with its gas flare top now a parasol blown inside out—by the gusting winds of change perhaps? To our right, the *Parade* ringmaster with his slanting cane echoes the couple on the right in *La Grande Jatte*. The *Parade* central figure is certainly an odd character—he presents as a beguiling mix of Pied Piper, Glen Miller-esque dance band leader, and (cryptically understated) giant squid (with head + hat = stabilizing fins, Figs. 68 and 69).

Circus fun

The 'Piper' beckons his audience to follow—ostensibly to attend and enjoy the close-by Circus Corvi which was set up near *place de la Nation* in spring 1887 [824]. However, this is only a surface reading— and one completely at odds with the lugubrious mood of the piece. On a deeper level, (*in this author's opinion*) it would appear that Seurat is warning of the looming dystopia of the 2nd Industrial Revolution—with its (relatively untested) science-based advances and its mass production. Perhaps Seurat's trombonist is in the business of persuading workers to follow him into his new factories, each to become a single-function component in some mechanical process. Behind him are 3 ('and a bit') musicians—portrayed as emotionless gingerbread cut-outs, equispaced as if behind the non-stop conveyor belt of a sorting- or assembly-line.[8]

The trombonist—a metaphoric squid?

The giant squid, like the related octopus, is a highly developed predator—and as such, it has long been used as a metaphor for voracious and evil financial institutions. Consequently, the gas pipe across the top of *Parade* may also give the idea of an ocean surface, with the gas flares now splashes or bio-luminous jellyfish—'We are all underwater, maybe in the subconscious, maybe in a dream.' In the murky (possibly kilometre) depths of the ocean, the giant squid lurks looking for prey, its outsized eyes positioned halfway along its body. Indeed, the 'colossal' squid holds the record for the largest eye of any animal—27 cm across (dinner plate size). And although such squids typically operate in darkness, they have evolved their own light sources (one glowing strip on the side of each eye) [890].

Figure 69: **Giant squid**—found in 1877 at Trinity Bay, Newfoundland. The 2 extra-long feeding tentacles with their club ends are clearly shown to the left [908].

---

[8] Did Seurat have a prophetic dream, or was he already aware of the 1880s emergent developments of electric conveyor belts and single-station, single-function working (*à la* Chicago meat industry)—advances which Henry Ford would very soon combine into mass production assembly lines [223]—the kind of mechanisation which German expressionist Fritz Lang would later present in his 1927 futuristic film, *Metropolis*?

Back with the trombone, the 2 long straight tubes of the instrument's slide (which can be rapidly extended) vividly recall the 2 super-long feeding tentacles these squids have—2 'arms' that can shoot forward in an instant to grab prey [789]. All along these tentacles are powerful suckers (with extras on the club-shaped ends). Seurat shows these suckers (giant size, circular) towering directly behind the end-of-row musician—the latest catch maybe? (In addition, the 2 zeroes in the bright panel suggest yet more suckers.) When, out of the gloom, we notice the glow of that area (right of the trombonist)—we correctly guess we are now in range. The tentacles can lash out 10 metres to seize prey and draw it back towards the 8 legs (with even more suckers), thence towards the feeding beak [790]. Seurat's (apparent) squid idea may have had its origins in the work of French author Jules Verne. From 1865 to 1870, Verne lived in the villa *La Solitude* in the medieval coastal town of Le Crotoy, (Picardy, northern France), and there he wrote his *Twenty thousand leagues under the sea* [926]. Towards the end of the story, Captain Nemo's submarine—*The Nautilus*—is attacked by a gigantic octopus.[9] Both Seurat and Henri de Toulouse-Lautrec also (later) stayed in Le Crotoy. (Nautilus, octopus, and squid, are all cephalopods.)

'*Look into my eyes...* '

The mood of *Parade* is subdued, sad, and sombre[10]—the central figure stares directly at us, offering no 'bright and happy outcome'. What then could be the secret of his recruiting style? Maybe he is using some kind of 'eye-fixation induction' method? Hypnosis was extremely topical in Paris during the 1880s. Renowned neurologist Dr. Jean-Martin Charcot and his pupil Georges Gilles de la Tourette[11] were famously performing pioneering hypnosis experiments and demonstrations at the Parisian Salpêtrière Hospital—sometimes to packed audiences (Fig. 70).[12] Using hypnotism therapeutically, Charcot quickly gained an international reputation as an expert in 'female hysteria'.

---

[9] *Poulpe* in French, variously translated as octopus or giant squid. Also, in 1885 and 1886, cephalopod expert Dr. William Evans Hoyle published his octopus and squid report (from the 4-year *Challenger Expedition*) [988].

[10] Seurat developed a language of line directions, based on the theories of Humbert de Superville [896], where (for example) the downward-pointing trombone sounds a note of melancholy—compare this with the raised horn being played in *La Grande Jatte*.

[11] The Tourette of *Tourette's syndrome*—*TS*.

[12] Sigmund Freud framed a copy of this André Brouillet painting (a lithograph by Eugène Pirodon) and kept it as 'the picture over the couch' in his Vienna consulting rooms for more than 50 years [88, 664].

Figure 70: **André Brouillet (1857–1914)** *Une leçon clinique à la Salpêtrière* **1887 (detail).** Hypnosis demonstration by Dr. Jean-Martin Charcot [88].

In 1885, Sigmund Freud, having been awarded a 6-month travelling grant, beat a path to Charcot's door to learn all he could [664]. A teenaged Jane Avril was treated by Charcot—that is, the Jane Avril who later achieved great box-office success as the sophisticated yet 'Crazy Jane', dancing solo *cancan* while being immortalized by Toulouse-Lautrec in his exceptional poster designs and portraits of her [73, 138].[13] However, amongst the public there was a widespread fear that crimes were being committed using non-medical hypnosis —that women were being 'mesmerized' and taken advantage of—that men were being 'programmed' to do things against their will (or better judgement), being transformed into 'automatons' [69, 70]. One way or another, this trombonist wants us to dance to *his* tune.

Masterclass in proportion + simple Fibonacci

To appreciate the carefully considered geometric scaffold of *Parade*, we must first note that this is a painting within a painting. For a start, the tube at the top—with its row of 9 gas-jet flare lights—spans the full width. This, while keeping the existing background, separates the top as a kind of decorative border—almost floral.[14] This long tube with its row of flames also echoes a long squid tentacle with its row of dangerous suckers *with sharps*. Gas lighting was still a relatively

---

[13] E.g. p.88. Jane was also nicknamed *La Mélinite*—after a powerful explosive [138].

[14] Some have compared this top border with those seen in Egyptian art.

recent innovation, and in 1875—little more than a decade before Seurat started work on *Parade*, Charles Garnier's opulent new *Opéra de Paris* was constructed.[15] It boasted no less than 28 miles of gas piping and 960 gas jets [743].[16] Again, in addition to the top 'gas' border, each side has a narrow border. This leaves a Golden Rectangle area in which all the geometry takes place; the narrow border along the bottom is *within* this geometry area.[17] Also we note (in the main area) the very clear horizontal and vertical line divisions between the many rectangular elements. As we shall see, this work demonstrates 8 rational and 5 irrational ratios (8 and 5 of course being Fibonacci numbers), then 6 Fibonacci numbers, themselves using 29 graphic elements.[18] The rational proportions comprise 1:1, 1:2, 1:3, up to 1:8, (mostly landscape rectangles). The irrationals are 4 Golden proportions: $\sqrt{\phi}$, ϕ, ϕ^2, and ϕ^3, plus one Silver, $\delta = 1 + \sqrt{2}$ (these being mostly portrait). The Fibonacci numbers: $1, 2, 3, 5, 8$, and 13, are also simply (and once seen, very obviously) spelt out as steps in the Fibonacci recurrence.

Parade analysis—the rationals   R

To analyse, in Fig. 71 we take the overall Golden Rectangle as having width 'w', and note that the Golden Section (red vertical 'G') exactly coincides with the right edge of the trombonist's plinth. This major section leaves a square to the left, (demonstrating the 1:1 proportion) and a secondary Golden Rectangle, (now in portrait), to the right. Next we observe how the performer is placed centre-canvas, halving the width and thereby showing the 1:2 ratio. Further, we mark a rectangle across the top area, full width (w), with height 1/3rd of this. We then do likewise at the bottom with a long 1:4 rectangle. These 1:3 and 1:4 now locate the 'conveyor belt' between. And as this may herald the start of a pattern, we guess there may be a long 1:5 rectangle included too.

---

[15] Here, Gaston Leroux set his 1910 *The Phantom of the Opera*—a novel inspired by George du Maurier's extremely popular 1894 novel *Trilby*. In *Trilby* du Maurier directly confronts the problem of unethical hypnosis (noted on p.83), warning his readers. But on a lighter note, from the first London Haymarket theatre performance of the stage version (in 1895), the felt hat worn by the heroine has been known as a 'trilby'—and it is typically brown. Perhaps Toulouse-Lautrec is making ambiguous references to du Maurier, brown hats, and hypnotic puppetry in his 1899 *Jane Avril* poster (p.88).

[16] These were controlled with a gas table of 88 stopcocks—a head from which the hundreds of 'octopus tentacle' pipes radiated out.

[17] First observed in 1935 by sculptor, painter, teacher, and writer, André Lhote [557].

[18] 29 is a Lucas number; 6 is 'perfect'. We discuss perfect numbers later (p.280).

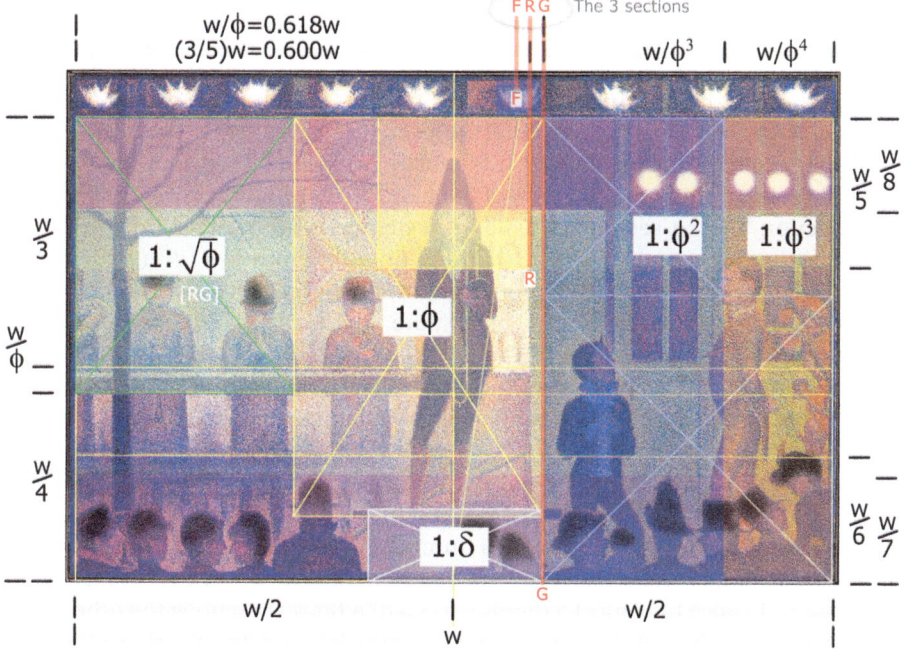

Figure 71: **Parade analysis.**
ϕ **powers increase left to right:** $\sqrt{\phi},\ \ \phi,\ \ \phi^2,\ \ \phi^3$.

This we find across the top, and similarly, we find a further 1:6 long rectangle along the bottom then 1:7, and even 1:8. The 1:5 at the top is especially interesting as, if we think of it as 5 squares lined up across the width then the performer's portrait will be framed in the centre (3rd) square (shown); and this square will have its right edge at 3/5=0.6 from the left horizontally (rational section 'R')—which matches the primary light-to-dark subdivision in the upper half of the geometry area. Compared to the Golden Section 'G', this clearly exposes the difference between the true $1/\phi = 0.6180...$ and the (Fibonacci numbers) $3/5 = 0.6$ convergent. Meantime, we note that the w/3 does not fit as well as the other rectangles. Given the accuracy in those other cases, then perhaps this was a misdirection to foil analysts. Many castles have staircases which include a 'trick step' (or 'stumble step') to wrong-foot intruders—Seurat was extremely secretive about his methods (*even more so* than Bartók) [267].

85

Parade analysis—Golden and Silver Ratios   G

As for the irrationals, Seurat starts top left with a $1:\sqrt{\phi}$ rectangle (pale green)—as too will Mondrian, some 40 years later. Seurat scales this rectangle to meet the top of the $1:4$ rectangle, and in the gap to the right between this (green) and the primary vertical section ('G'), he fits a Golden Rectangle which is mostly bright-lit. This then has its lower edge along the junction of the plinth and its floorboard. Right of section 'G', the secondary Golden Rectangle is vertically subdivided into long rectangles: one $1:\phi^2$ (blue tinted), and one $1:\phi^3$ (golden tinted). Within these, in the upper right, the ticket office windows are arranged 'piano keyboard black-note style' in 2+3=5 Fibonacci grouping.[19] On the vertical 'keys', 2 (electric?) lights label the $1:\phi^2$ rectangle, and 3 lights label the $1:\phi^3$ rectangle. At the bottom, the plinth and board together form a Silver Rectangle 'with ears'.

Parade analysis—Fibonacci numbers   F

Seurat also presents 6 Fibonacci numbers in recurrent sequence. Our clue for this is near the top—the 'parasol stick' under one of the gas flares 'F' (also hinting at a flaming torch or a flower) sections the other flares into Fibonacci-count groups (5 to the left and 3 to the right). In the analysis we see how all 3 such section dividers group together here: 'G'—Golden Section (irrational), 'R'—3/5 (rational), and this 'F'—the Fibonacci flare counts 5 and 3. Now we see the pattern— that of 1+2=3 musicians—the 'one' being the taller (but ignoring the edge of the mystery tuba player); 2+3=5 black-note piano keys; and the 5+3=8 'either side' division at the top of flares (or flowers/splashes/ jellyfish/parasols). Finally (along the bottom) we have the silhou- etted audience in left and right groups: 5+8=13. Counting the elements used, there are: 1+2+2+3+3+5+5+8 = 29, and 29 is Lucas number L_7. Now, if we follow the numbers, starting with tall musician left as 1, then moving right to 2 other musicians, then right to 2 piano keys, right to 3 keys, up to 3 flares, left to 5 flares, down to 5 spectators and right to 8 more—we trace an 'e' shape—a kind of spiral.

Henri de Toulouse-Lautrec

Henri de Toulouse-Lautrec was a leading post-impressionist French painter, printmaker, and draughtsman. Inherited weaknesses and injuries to his legs as a teenager resulted in his developing an adult

---

[19]   We compared these with (top-right) trees in *La Grande Jatte* (Fig. 68, p.80).

Figure 72: **Henri de Toulouse-Lautrec (1864–1901) Ambassadeurs, Aristide Bruant dans son Cabaret (1892).** Colour lithograph 150 × 100 cm. Again in 1892, Henri used this same design (mirrored) for his Aristide Bruant poster *El Dorado.*

torso but with very stunted legs. In response, Henri immersed himself in art, and in Montmartre—in all it had to offer. He was influenced by the works of Degas, Manet, and Van Gogh, and especially by the outlined, flat colour style of *Ukiyo-e* Japanese prints (e.g. by Hokusai & Hiroshige)—as were so many other artists during the printmaking craze of *fin de siècle* Paris. Fig. 72 shows his use of just 4 measures (1, 2, ϕ, and 2/ϕ) to achieve a perfectly balanced structure using 'all irrational' components, contained in a 3×2 format (3ϕ×2ϕ)—one which spirals out: 1 bottom left (2 × 2ϕ), 2 right, 3 above. The bottom-left, top-left, and top-right rectangles are all Golden; the remaining 3 are [RG+]. The primary Golden Section is shown ABC.

Figure 73: **Henri de Toulouse-Lautrec (1864–1901) Jane Avril (1899).**
Colour lithograph 56 × 38 cm. Analysis & inset added. Washington: National Gallery of Art.

This (7 years later) lithograph Fig. 73 is interesting, but not because of its Golden geometry—here (maybe) a single *ABC* serves to scale the lesser arc. Instead the design shows Toulouse-Lautrec's developed interest in geometric layout—this example being both wildly ambitious and triumphantly successful. It is based on a doubled *vesica piscis* (interlocked circles) with arcs: 2 major and 1 minor. If we view the work at 45° (Jane's elbows level), then who could be the man suggested in Jane's brown hat—a Charcot/Svengali to Jane's Trilby maybe?[20] Or (noticing his jaw-dropped look of astonished concern), Henri himself, hand-holding, supporting—saving her from being swept away in the swirling undercurrents of Belle-Époque night-life?

---

[20] Trilby footnote 15 on p.84. Also, compare Jane with the (mirrored) inset from p.83.

Chapter 7

Art—Mondrian, gnomons, and megaliths

Piet Mondrian

Piet Mondrian (1872–1944) is the most celebrated member of a group of Dutch artists who assembled around a journal called *De Stijl* and who shared a utopian idea [68]. Mondrian had joined the Netherlands Theosophical Society in 1909 [759] and in the following years he had been developing a new philosophy of art based on theosophic principles of universal harmony. Then in 1917, Mondrian was contacted by artist and writer Theo van Doesburg (1883–1931), and along with artists Anthony Kok, J J P Oud, Bart van der Leck, and Vilmos Huszar—Mondrian and van Doesburg founded the *De Stijl* journal. In a series of articles spanning the first 12 issues, Mondrian published his theories—which he called *Niewe Beelding* or 'neo-plasticism' ('a new kind of abstract art'). Mondrian referred to all his work from 1920 onwards as being 'neo-plastic', and although a few other artists used this designation in their titles, the idea was Mondrian's alone.

The *De Stijl* group included painters, architects, and furniture designers, but because Mondrian moved to Paris in 1919, he never met many of them [760]. He did write further articles for the journal until 1922, but in 1925 he split with van Doesburg and left the group. Van Doesburg's death in 1931 signalled the end of the group, and in January 1932, the last issue of *De Stijl* was published [761]. Although *De Stijl* art was preceded by and informed by cubism, it featured extreme geometric abstraction—finding beauty through radical simplification. It may be true that some of Mondrian's abstractions had their distant origins in trees and landscapes, but Mondrian went on to exclude manifest nature from his work completely. Mondrian maintained the highest goals in that he endeavoured to extract and record the hidden, underlying workings of the universe in line and colour. As he developed his now instantly recognisable (but not instantly understandable) visual language, he rejected all representation of perceived nature (green was forbidden, flowers were left long behind) [983]. In New York, towards the end of his life, Mondrian is quoted as saying: 'I don't want pictures, I want to find things out.' [374, 918]. He tried to paint universal spiritual truths—the precursors of nature as we know it.

We shall spend some time with Mondrian because the example we shall discuss is *suggestive* of a great deal of ϕ-based mathematics— basic Golden principles and growth sequences. Mondrian wrestled with the dynamic balance of conflicting forces and (as we shall see) may have hinted at resonant systems. But the question of whether Mondrian deliberately based his work on geometric theory is still debated. Did he intend, calculate and measure out every last last detail? Or did he (as many in the art world now hold) paint by eye and intuition alone? Most likely, the truth lies somewhere between.

Van Doesburg and Mondrian together experimented with horizontal and vertical lines in a quest for purity and universality. Mondrian (especially) sought to balance the particular with the universal [781]. The *De Stijl* group believed that once presented, such pure art could change the world and usher mankind into an ideal society. This undertaking often resulted in works that included 'more or less' Golden and (perhaps) other ϕ-based rectangles. Unusually however, their canvases omitted an outer border—this being done on a matter of principle. Mondrian was a theosophist,[1] dedicated to the removal of boundaries; he wanted his work to *extend out into its surroundings*. It may seem strange to us to see pictures with a forced absence of framing—so much so, that at first sight, they may appear to have been badly cropped. Indeed, some companies that sell art prints paint in the 'missing' heavy outline, or supply an imposing black frame (to satisfy customer expectation).

---

[1] In one of a number of definitions, Theosophy co-founder Helena Petrovna Blavatsky states: 'Theosophy is, then, the archaic *Wisdom-Religion*, the esoteric doctrine once known in every ancient country having claims to civilization.' [59]. Agnita Feis (van Doesburg's first wife) and Janus de Winter were also theosophists [179].

While analysing an example of Mondrian's work, we need to distinguish clearly between a possibly underlying mathematical framework and the finished piece itself. After all, what separates a work of art from an exercise in pure geometry is that art is checked, adjusted, and finished by eye rather than by taking measurements. We also need to try and keep an open mind about whether Mondrian consciously applied any mathematical structure, or whether his geometry somehow emerged (with unknown measurements) from deep within him. It is true that when 41 years old, Mondrian claimed (in a letter) not to use any calculation (translation from the Dutch quoted with kind permission of Reaktion Books Ltd.) :

'I construct lines and colour combinations on a flat surface, it is with the aim of portraying *universal beauty* as consciously as possible. ... I believe that it is possible by means of horizontal and vertical lines, constructed *consciously* but not *calculatingly*, guided by a higher intuition and brought to harmony and rhythm—I believe that these fundamental aesthetic shapes—where necessary supplemented by lines in other directions or curved lines, make it possible to arrive at a work of art which is as strong as it is true.'

Mondrian, Paris, 1914 [60].

© 2014 Mondrian/Holtzman Trust c/o HCR International.

But the 1930 work we shall study was painted when Mondrian was nearly 60. Also, as we mentioned earlier, intellectual life in Paris was (by then) saturated with discussions of the Golden Section and experiments in its application. Still further, we must acknowledge that we have hit upon a tricky area. While we are *very reassured* that our surgeons and pilots are trained to employ established procedures —we tend to expect quite the opposite from artists and composers. For example, in his article *Art and Mathesis: Mondrian's structures,* Anthony Hill starts out by asserting that there is a consensus that Mondrian used neither measurement nor calculation in his work. He then goes on to say that: 'It follows from this' that if 'mathematical proportions' are found, then Mondrian must have created them 'unwittingly and/or unconsciously.' [368]. Later we shall discuss how musicologist László Somfai published remarkably similar arguments about Bartók's methods.[2]

---

[2] On page 138.

In yet another case, Arthur I Miller notes how Picasso swung between admission and denial of his use of mathematics and trigonometry [646]. Nevertheless, there is also one perennial objection (to suggestions of intentionality and calculation)—namely that particular artists and composers left no written evidence to explain how they did what they did.[3] One possible reason for this—at least in some cases—could be that the artist or composer received their knowledge through membership of an esoteric group. In these cases, they may have been bound by oath not to discuss such matters with the uninitiated. Then again artists might argue that (like jokes), works of art will lose *possibly forever* their impact, their gestalt, their ability to engage and detain, their effectiveness—once they are sequentially explained and deconstructed. We have already mentioned Debussy's views on technical analysis (page 65). But regardless of all this, and given the choice, how would any creative person prefer to be remembered:

- as a great artist and accomplished technician,   *or*
- as an intuitive genius ?

In his book *The painter's secret geometry*, Charles Bouleau analyses a very spare lozenge shaped canvas by Mondrian: *Tableau I: Lozenge with four lines and gray* (1926) [656]. He shows how the artist is making a characteristically reduced, but nevertheless obvious, statement using ϕ. Mondrian arranges that the unseen diagonals of both a larger 45° rotated square (the lozenge canvas itself) and a smaller upright square (the outer edges of his four black bars design) intersect each other at exactly their Golden Section points [76]. The fact that *Tableau I: Lozenge with four lines and gray* was completed in 1926 is also interesting as it shows Mondrian already elegantly applying the Golden Ratio (a year at least) before Matila Ghyka started publishing (in Paris) his very influential works on the subject (p.72).

Composition in Red, Blue, and Yellow (1930)

The canvas we shall study here is one of Mondrian's best known neo-plastic works—his 1930 *Composition with Red, Blue, and Yellow* (Fig. 74). This piece was chosen by the Pompidou Centre for the front cover of their 2010–2011 *Mondrian / De Stijl Album de l'exposition*: a 60-page guide to accompany their adjacently running Mondrian and

---

[3] Should we then, require artists and composers to supply 'adequate supporting documentation' in a standardized format?

Figure 74:

Piet Mondrian (1872-1944)
Composition with Red, Blue, and Yellow,
1930. Oil on canvas, 46 x 46 cm. Kunsthaus Zürich.
© 2014 Mondrian/Holtzman Trust c/o HCR International.

De Stijl exhibitions [758].[4] On a quick first reading, its black horizontals and verticals might appear to be very stylized and unsubtle outlines for some nearly-squares, and a vaguely-Golden-rectangle. People do talk of Mondrian's grid lines as if they were intended as blatant barriers between the rectangular spaces, coloured or otherwise. But as we shall see, Mondrian had actually developed a visual language and using this he was usually saying very different things. We recall that the *De Stijl* principles included the abolition of boundaries—so what if we consider first just rectangular areas of

---

[4] It is interesting to compare this layout with Vilmos Huszár's 1917 design for the first cover of the *De Stijl* journal [400]. That has a 'nearly square' in the bottom left, long rectangle along the bottom, and to the left, suggested bars and rectangles stacking above the 'square'. These elements are much easier to see when Huszár's design is viewed as a negative image. This technique also reveals some classic *De Stijl* long black bars.

colour themselves (with the black bars absent)—as in several of Mondrian's earlier works? Architect and author Richard Padovan[5] addresses this with a quote from an article by Mondrian himself in the *De Stijl* journal [721]. Mondrian says: 'By means of rectangularity, colour is delimited without being enclosed.' [655]. Mondrian's black bars are not (usually) walls between areas, they are primarily abstracted forms that 'sit above' a layer of un-walled colour below [722]. They are abstractions, but abstractions of what?[6] Well, at one level, *all* things that have (in their essence) strong horizontal and vertical components:

- skyscrapers in a city
- pines by a lake
- people standing—when giving lessons, artist Cecile Elstein would say: 'We stand vertical on a horizontal.' [202][7]
- (and post-Mondrian) the obelisk/mysterious black panel in Stanley Kubrick's film—*2001 A Space Odyssey* [517][8]

In 1938 Mondrian made a train journey from Paris to London with the artist Winifred Nicholson. As they sped through the Somme landscape, Mondrian delightedly remarked: 'Isn't it wonderful!' Yet as Winifred soon realized, his joy was not with the landscape itself, but rather with the rhythm of the telegraph poles passing by [852]. In his layered visual language, rounded poles become flat black silhouettes interposing between him and the flat horizon. The horizon 'layer' retains its integrity—the poles do not chop it into pieces. But, in his quest for universality, Mondrian reaches to a far higher level of abstraction—where vertical and horizontal together symbolize *all dualities*. Examples of these could be masculine/feminine, hot/cold, good/bad, active/passive, self/non-self, rational/irrational, and so on. In his visual neo-plastic language, a horizontal/vertical crossing point marks the resolution of duality—transcendent one-ness.

---

[5] Richard Padovan has written several books on proportion and has a Fibonacci-like sequence named after him: $P_n = P_{n-2} + P_{n-3}$, with seeds $P_3 = P_2 = P_1 = 1$, A000931. Gérard Cordonnier who first studied it in 1924 (age 17) called its limiting growth the 'radiant number'. Dom Hans van der Laan (in 1928) called it the 'plastic number' ($\Psi = 1.324718\ldots$) and he based his new system of proportion in architecture upon it [630]. Ψ is the unique real solution of $x^3 = x + 1$.

[6] The meaning of the term 'abstraction' in art has changed during the last 100 years or so. It used to be about finding the essence of forms and motion in nature; but now it is usually understood as making no reference to the real world [336]. It is interesting to consider just where Mondrian fits in the spectrum between these two meanings—as he appears to present concepts (such as growth, branching, duality, and asymmetry).

[7] Quoted with kind permission.

[8] In which, Kubrick positions the *Intermission* 61.8% through the film [925].

Mondrian: Branching

Now, a key design goal for Mondrian was irreducibility, so if something appears (i.e. 'remains') in one of his paintings, then there will certainly be a very strong reason for it—take for example, the short black horizontal in the upper left (Fig. 74). This is at one with all things that branch (including living things), but at the same time it relates to none of these in particular. As we mentioned, Mondrian was no longer concerned with representation. The branch here is the 'result' of an experiment in his laboratory; it is not about (say) wooden trees; it represents nothing other than itself. Having said that, and looking back—trees *were* an important earlier subject for Mondrian (1911–1913)—for example *The Grey Tree 1911* [431], *Flowering Appletree 1912* [435], *Composition Trees 2, 1912–1913* [439], and many more: [430, 432, 433, 434, 436, 437, 438, 440, 441].

As we have noted, Mondrian's ambition was (through art) to go beyond nature to discover a deep holistic and mathematical order in the Universe—'the laws behind the laws'.[9] So, considering further sources of abstraction, (if we so wish) we are free to associate the branching growth of a vast array of examples from rivers to railway systems and even to conceptual systems with 'A or B?' decision branching, still remembering Mondrian's goal of being all-encompassing. And while simplicity risks ridicule, that upper-left peg (visually) does make the most basic possible statement about the nature of branching itself. Then in the bottom right, we see this branching form echoed again and again. Visually, the branch to the right balances the original branch to the left (from the vertical). But from a mathematical point of view, we are seeing the growth that occurs by the repeated application of simple rules.[10]

Leonardo da Vinci considered the branching of trees, trying to relate sums of cross sections of branches with the cross section of the trunk [652]. But Mondrian's terse visual distillation here is far more fundamental, and it comes a decade after (*normaliens*) Gaston Julia and Pierre Fatou (separately) did their pioneering work in Paris on iterated functions.[11] Their interest was in growth behaviour with repeated application of a function—taking the output of one stage as

---

[9] An idea similar to that of the 'implicate order' later proposed by David Bohm [72].

[10] This downward branching may also be considered in terms of complementarity and balance—in the same way that a tree grows upwards and branches out while its roots grow downwards and branch out too.

[11] Édouard Lucas and Charles Pisot (p.72), Gaston Julia (1893–1978)—'*Julia sets*' and Pierre Fatou (1878–1929)—'*Fatou sets*', all studied at the *École normale supérieure* in Paris (p.27) [524, 698, 694].

input for the next and studying when results remain bounded and when (and how fast) they 'blow up' (shoot off to infinity).[12] Yet Mondrian's minimalist statement of repeated branching predates Benoît Mandelbrot's work on fractals in nature (from 1967 onwards) [622, 623]—with *its* trees and ferns, and the work of Aristid Lindenmayer (a colleague of Mandelbrot) with his 'L-Systems'—a powerful recursive approach used to model organic growth, fractals, and string substitution sequences [770].[13] A current example of the scientific application of fractal branching mathematics is in the study of blood vessels [340].

Mondrian: Backbone diagonal

Let us now look at the major structuring of Mondrian's design by considering diagonals.[14] In Fig. 75 by drawing the diagonal from bottom left to top right, we find it coincides with the crossing point of the right edge of the vertical bar and the lower edge of the horizontal bar. To understand this in simply symmetric terms, we may now imagine the red area R extending *under* the horizontal bar to make an exact red square—and in just the same way we may think of the blue area B extending to the right, tucking under the vertical bar to form an underlying blue square. But going back to seeing 'as is', we note a neat complementarity between the blue being squeezed horizontally and the red vertically. The 'pivot point', where the major diagonal and both relevant bar edges all cross, is obviously rather important here—we shall call it 'P'. This pivot, when taken with the bottom-left corner of the canvas, provides a reference square; and from now on, we shall regard this as our unit square—the basis for all the geometry that we discuss. For convenience we might also refer to it as 'the blue square'. As we shall see again and again in this example, Mondrian's famous black bars are in most cases *not* about partitioning rectangular areas of colour or white. Typically, just one edge of a black bar serves this purpose and, as we have noted, the bars themselves stand vertically (or repose horizontally) as extreme abstractions of form—they then 'sit over' rectangles and squares on a lower layer.

---

[12] When published, Julia's work was greatly and widely appreciated (not just in mathematical circles), and he was awarded the Grand Prix of the Academy of Sciences.

[13] We examine substitution sequences later on—in the '-*Onics*' chapter, page 383.

[14] There is a popular legend that an argument over diagonals was the direct cause of the break up of Mondrian and van Doesburg [66]. However, art historian Susanne Deicher finds very little evidence that this was the real reason [169].

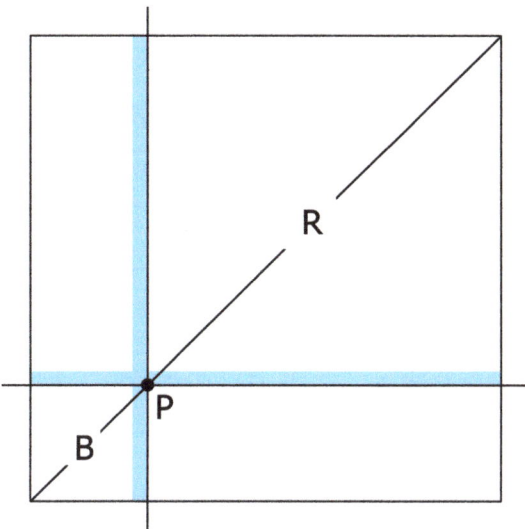

Figure 75: **Backbone diagonal and pivot point P.**

This layer logic may seem strange initially, but it will become familiar as we see how Mondrian has quietly applied it to achieve a ϕ-based structural scaffolding— 'hidden in plain sight'. However, we should also note that this layering is conceptual and does not determine the way a painting is built up in practice. In the same way that Mondrian's bars serve as extreme abstractions of form, he intends *the primary colours* red, blue, and yellow, to serve as an abstraction of *all* the colours of the 'real world' [720]. Is it not ironic then, that these uncompromising mathematical expressions have been so fully absorbed back into that real world—as fashionable Mondrian-style phone cases, mouse-mats, handbags, shoes, raincoats, *haute couture*,[15] swimwear, and even as art prints 'complete' with heavy black frames?

---

[15] The 1965 Yves Saint Laurent *Mondrian Collection* [999].

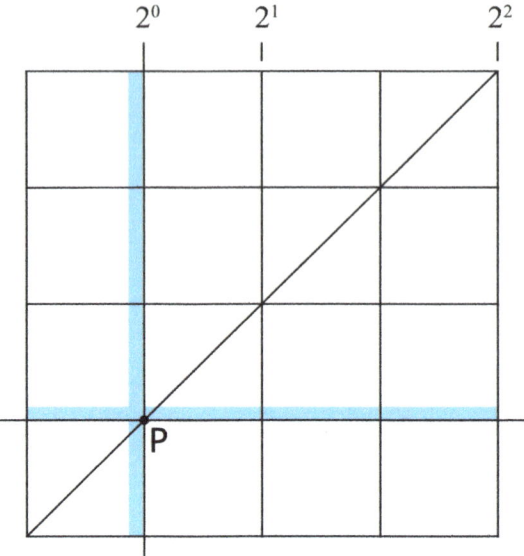

Figure 76: **Backbone diagonal, pivot point, and 4×4 grid.**

2 schemes—4×4 grid and Golden

In Fig. 76 we discover that the position of the pivot P is at exactly a quarter of the width, and a quarter of the height. This therefore suggests a 4 × 4 grid pattern based on blue filling one square (which is then overlapped by the main vertical bar), and the red filling 9 squares (then overlapped by the main horizontal bar). If we look at examples of Mondrian's preparatory studies and unfinished works—for example in the *Catalogue Raisonnée*, we see that grid lines such as these were the first elements applied to a new canvas [427, 428]. But back in the original painting on page 93—what about the white rectangle above the blue square—surely it has nothing to do with the grid? It would appear that Mondrian has based this on a Golden Rectangle with overlaid black bars right and bottom. Also, along the bottom, to the right of the blue square, the long white rectangle appears to be based on the ratio $1 : \phi^2$.

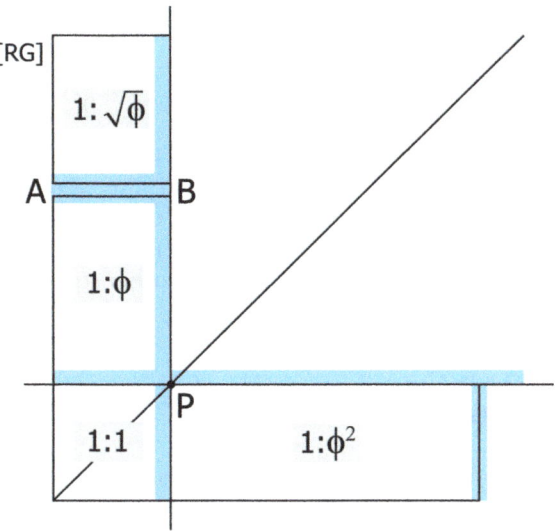

Figure 77: **Locating the Gold. Again the ϕ powers increase: $\sqrt{\phi}$, ϕ, ϕ^2** —reminding us of Seurat's *Parade* (p.85)

Root Golden too...

Encouraged by our ϕ and ϕ^2 Golden finds, we draw them in Fig. 77 and further discover that a Root Golden Rectangle will fit comfortably in the top left, leaving a gap between which is more or less centred underneath the thick black horizontal bar AB branching to the left. It is interesting that Mondrian has used a $\sqrt{\phi}$ rectangle in the top left—just as Seurat did in *Parade* (page 86). When compared to the familiar Golden Rectangle it still might seem exotic and unusual. However, as we saw back on page 33, its geometric construction from a Golden Rectangle is quite trivial: we merely swing the ϕ side of a '1, ϕ' Golden Rectangle across to meet the other ϕ side at $\sqrt{\phi}$.[16] However, this Golden-based interpretation does very much depend on just what is happening under that thick horizontal bar in the upper left (AB)—'behind the scenes' as it were.

---

[16] Also, for 'artistic $\sqrt{\phi}$', two [RG+] rectangles fit exactly into the 1:ϕ^2 (Fig. 61, p.74).

Gap width

In passing, it is interesting to ask: 'How big is this vertical gap between the Golden and the Root Golden Rectangles?' We see this will be

$$4 - 1 - \phi - \sqrt{\phi} \;=\; 0.10995 \qquad \approx \; \frac{1}{9}.$$

So in words, this gap (from the point of view of paint on canvas), is an excellent 'one-ninth'. The difference from the true one-ninth (i.e. 0.11111...) is a mere one-thousandth of the '1' blue height reference. This 1 : 9 ratio—based on our (blue) '1' reference—recalls the ratio between the grid areas of the 1 blue corner square and the 9 red squares.

Scientific theories

(Soviet crystallographer) Aleksandr Kitaigorodskii [309] once said: 'A first-rate theory predicts; a second-rate theory forbids; and a third-rate theory explains after the event.' [827]. But so far here, we have been very much attempting to explain. So instead, let us use our analysis to venture a prediction. Namely that (given some way of 'seeing' into the thick width AB bar in Fig. 77) we might notice an original thin black bar (the gap between the Golden and Root Golden Rectangle), and further, we would see how this has been widened (above and below) into the thick bar of the finished work.

Putting theory to test

Now, if we do (say) go back to, and carefully examine the photo on the front of the Pompidou *Album de l'exposition* [758] then *we do indeed find a central darker black bar in just the right place, and we see this has been widened each side in a slightly less dark black.*[17] It is almost as if Mondrian wanted this to be discovered (like a secret diary)—as if eventually, he wanted the full extent of his geometry to be appreciated —otherwise he could have just as easily applied a sufficiently opaque black to hide his working.

---

[17] This was checked using a digital photographic scan of the catalogue picture. Gamma, levels and hue were all adjusted, allowing the two-stage painting to be seen very clearly.

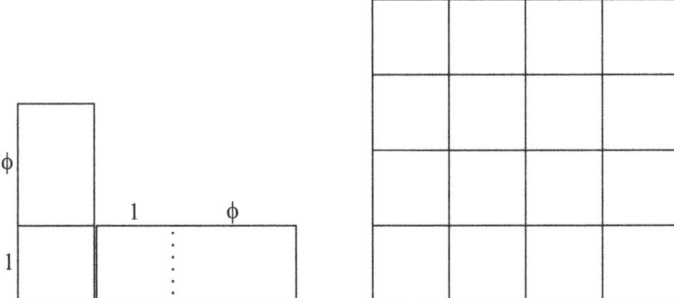

Figure 78: **Mondrian's two schemes.**

Unifying principle

But surely this is not the full story. This is only the manifest logic of the canvas itself—the gross elements, there for all to see. As such, (apart from the white panels going to the edge of the canvas), it is far from clear how this design might: 'extend into its surroundings'—an oft-stated goal of Mondrian. What we now seek is (ideally) a single geometric unifying principle that would hold the work together as a whole—a principle able to reconcile both the grid and Golden forms. In Fig. 78 we visually review the currently conflicted state of affairs. At the bottom left we have the reference square, and above it is a Golden Rectangle. Now, because $1 + \phi = \phi^2$, then taken together, the square and Golden Rectangle form a ϕ^2 rectangle; and it is this shape that is repeated (on its side), along the bottom to the right of the reference square. (On the left of Fig. 78, the Golden blocks are slightly separated to show the $1 + \phi = \phi^2$ groupings.) So, one ϕ^2 rectangle shape echoes the other in an asymmetric harmony which had been much favoured in *De Stijl*, with vertical considered active and horizontal passive. Together, these echo the long vertical and horizontal black bars. Then on the right of Fig. 78, we see the simple (but then puzzling) 4×4 'quarter-chessboard' grid. Had we seen a Fibonacci number for its side, such as 3, 5 or 8, then we should have immediately spotted a link to the Golden Ratio. But 4?

The 'Golden' forms to the left are the polar opposites of the integer grid to the right. Could it ever be possible to reconcile them? Here we have here a perfect example of just the kind of duality that Mondrian is abstracting with his opposing vertical and horizontal bars. The key to solving this riddle is to recall Mondrian's ambition to abolish barriers, along with his intention that his work should:

> 'Extend in that virtual space outside—beyond the canvas and into the surroundings.'

[723, 724]

Mondrian: Scheme unification by gnomon

In summary, can Mondrian resolve the unresolvable? The best clue we have is that the crossing of vertical and horizontal takes place at point P, which both horizontally and vertically divides the 4 into $1 + 3$ (Fig. 75, page 97). Aha! The hidden unifying principle is now revealed to us. Here is 'mustard to mix oil and vinegar,' namely...
the Lucas number sequence:

$$\langle\ (2),\ 1,\ 3,\ 4,\ 7,\ \ldots\ \rangle.$$

As we know, the Lucas numbers are generated using exactly the same recurrence relation as the Fibonacci numbers, (add the preceding two to get the next). Also, odd-index Lucas numbers (such as this one, $L_3 = 4$) have a property that—we guess—would have greatly delighted Mondrian: that they result from the difference of two powers of ϕ. As we mentioned just earlier, a persistent goal for Mondrian was to do work that extends out into its surroundings—and as we shall soon see, a layout based on an odd-index Lucas number could offer him just that.

Earlier we met the Lucas Binet form—(1.10) on page 28–that is,

$$L_n = \alpha^n + \beta^n \ = \ \phi^n + \left(\frac{-1}{\phi}\right)^n.$$

So for Lucas number $L_3 = 4$

$$L_3 = \alpha^3 + \beta^3 \ = \ \phi^3 + \left(\frac{-1}{\phi}\right)^3.$$

That is,

$$4 \ = \ \phi^3 - \frac{1}{\phi^3}.$$

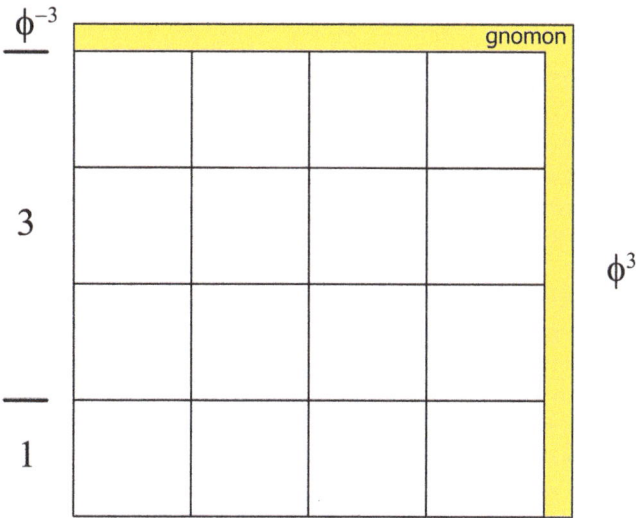

Figure 79:   **4 × 4 grid and ϕ^{-3} gnomon.**

Or, as we view the component heights on the left side of Fig. 79,

$$1 + 3 + \frac{1}{\phi^3} = \phi^3.$$

Here we see how an added ϕ^{-3} element unifies the rational and irrational schemes, while extending the design out 'virtually'—beyond the canvas—and disclosing a simple reciprocal relation ($\phi^{-3} : \phi^3$). Also this same logic must apply both horizontally and vertically to the square canvas. So, in Fig. 79 we see these extensions combine into a 'half-turn-rotated-L' shape (of width ϕ^{-3}). This kind of growth is very common and it was studied in both ancient India and ancient Greece. 'There are certain things which suffer no alteration save in magnitude when they grow.' said Aristotle (as quoted by Lawlor [526]).

The shape that is added to effect the growth is called a 'gnomon' [951].[18] This provides the perfect bridge between the integer-based 4×4 square and the Golden-based $\phi^3 \times \phi^3$ virtual square.

The bigger picture

Now, with the addition of this gnomon, the complete design 'makes sense'. In Fig. 80, considering the left side, and taking the bottom-left blue square as having side one, the overall composition square—the 'bigger picture'—will have side $1 + \phi + \phi = \phi^2 + \phi = \phi^3$. In the bottom-right corner, there is substantial potential for the detailed matching of geometric relationships. However, we shall just note two important ones. First, the upper edge of the short horizontal black bar, when compared with the end of the ϕ^2 rectangle, divides the height in the Golden Section. Second, the width of the yellow rectangle (as seen, not imagined) is very close to $1/(2\phi)$. This is significant because in the final scheme (including the gnomon), the width of the red area is 2ϕ—exposing a further reciprocal arrangement.[19] In the upper left, we see how easy it would have been for Mondrian to construct his [RG] rectangle—just by swinging across an arc radius ϕ to give a Kepler triangle construction (p.33).

While most probably it is just a coincidence, in a later version of this painting, his *Composition No. II / Composition I / Composition en rouge bleu et jaune, 1930* (B219) [447]—if we follow the 'Lucas' logic through—then Mondrian might possibly have chosen the size for this canvas with the intention of evidencing his virtual gnomon design. That is, by fixing the 4-unit side at 51cm, the ϕ^{-3} gnomon width becomes almost exactly 3 cm—with the resulting arithmetic of 50 + the units $(1 + 3 = 4)$. We find that the rule of thumb for this set-up is that every 17 cm added to the canvas side dimension increases the gnomon width by 1 cm giving 18 cm—here we have $3 \times 17 = 51$ and $3 \times 18 = 54$, difference 3.

The 18th camel

Some readers may find this kind of situation very familiar—where an outcome is made possible by the addition of something that is not 'used up' and that remains separate at the end.

---

[18] Also: Glossary, page 454.

[19] Mondrian was always exceptionally neat; his studio was the complete opposite of (say) that of Francis Bacon.

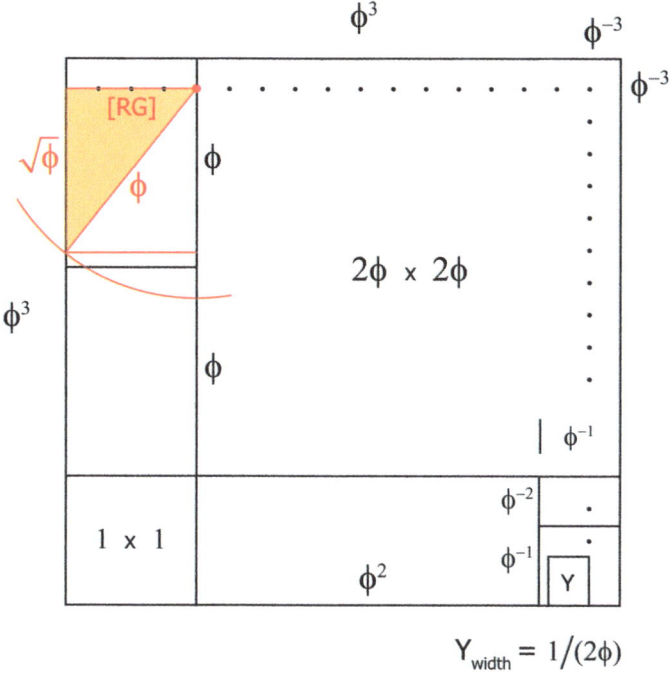

Figure 80: **Mondrian: (virtual) extended Golden structure.**

Chemists will think of catalysis. Arab readers may recall the ancient parable of the 18th camel. In the proportional formula (which we just found) 17 cm + 1 cm = 18 cm, the 'cm' (centimetres) might also stand for camels... The old old story goes as follows—a camel owner dies and leaves his 17 camels to his 3 sons in the proportions one half, one third, and one ninth. The deadlock of this impossible division is then resolved by a wise neighbour who lends the brothers one of his own camels to make 18. The total can now be divided to give 9 + 6 + 2 = 17 camels—with one left over, which is taken back again by the neighbour. The wisdom here concerns 'win-win' negotiation—where all parties may benefit if they can only find 'the 18th camel'. In Mondrian's case the 18th camel is the virtual gnomon which 'makes sense' of the design, reconciling and integrating the rational and the irrational into a singular wholeness.

Mondrian: 'Chord' relationships

Mondrian loved jazz—and jazz in particular inspired his later 'Boogie-Woogie' series of paintings [725]. But well before these, there was ample music in his canvases. Discussing Mondrian's blocks of primary colour being shown together, art critic Pierre Sterckx made the comparison between these and musical chords [870]. This is surely a very insightful observation, as the pitches (i.e. frequencies) of the notes in a chord grow one to the next according to a mathematical relationship; and in Mondrian's (apparent) geometry we find a storehouse of such growth sequences. Furthermore, these are often composed of three elements. To appreciate Sterckx's point, we had best check the musical structure of a simple chord first—for example, the major triad. And as soon as we do (Fig. 81), we discover this particular chord is characterized by the (Lucas number) intervals of $4 + 3 = 7$ semitones. Fig. 82 lists some 'chord' growth sequences that now become evident.

Mondrian's sustained focus

If this type of design had been a mere passing interest to Mondrian, then we would probably be right to discount any deep mathematical intentionality outright, but we must keep in mind that he spent many years, day by day, narrowly focused on these works. We also know that (early on) Mondrian was significantly influenced by mathematician and theosophist Dr. Mathieu Schoenmaekers with his views that mathematics was: 'the only purity', and that: 'A work of art must always have a mathematical foundation.' [975].[20] In the same way that Mondrian 'locked on' to trees as a subject, he moved on to be captivated by the essence of the design we are currently studying, and he produced a number of (often very similar) variations: [442, 443] [444, 445, 446, 447, 448, 449, 450, 451]. Mondrian's intense concentration with its laser-like focus, his neatness, and his repetitive activity—all these have led some to associate him with AS/HFA—Asperger's Syndrome/High Functioning Autism [201, 473].

---

[20] When they met in 1915, Schoenmaekers had just published *Plastic Mathematics* (from which Mondrian repurposed certain vocabulary for his neo-plasticism [977]). Van Doesburg wrote a year later that Mondrian (and Jakob van Domselaer) were obsessed by the theories of Dr. Schoenmaekers—ideas of a higher reality expressing itself through dualities [976], though Mondrian would later credit Blavatsky [64]. However, seeing as Schoenmaekers had been a member of the Theosophical Society as early as 1905 [727] (but soon left, in favour of his own 'Christosophy'), it seems likely Mondrian was referring back to the original source rather than denying Schoenmaekers' influence.

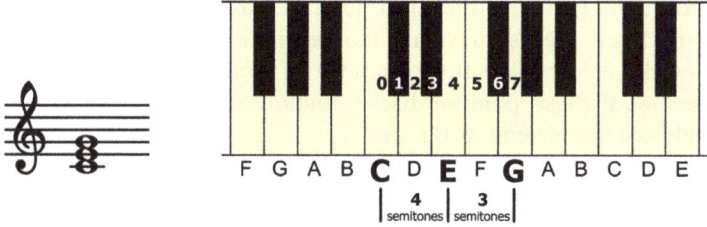

Figure 81: **The major triad chord—C major, 4 + 3 = 7 semitones.**
Major Chord image by Gringer [277]. Original keyboard image by Tobias R [898, 142].
Semitone annotation was added.

$\langle 2^0, 2^1, 2^2 \rangle$ — The factor 2 grid divisions—octaves (Fig. 76, p.98).

$\langle 1, 3, 9 \rangle$ — Factor 3 rational growth, with the bottom-left blue as 1 square, the 3 grid squares directly above it (and the bottom 3 to the right), then the remaining 9 unit squares (red) centre/top right (Fig. 76, p.98).

$\langle 1, \phi, \phi^2 \rangle$ — Simple ϕ irrational growth: 1—blue unit, ϕ— Golden Rectangle above it, and ϕ^2—rectangle to the right of the blue (Fig. 78, p.101).

$\langle \dots \phi^{-3}, \phi^{-2}, \phi^{-1}, \phi^0, \phi^1, \phi^2, \phi^3 \dots \rangle$
Diverse inter-relationships (Fig. 80, p.105).

$\langle 1/\phi^3, 1, \phi^3 \rangle$ — The reciprocal relation between gnomon width and outer square side, with blue side as 1 (Fig. 80, p.105).

$\langle 1/(2\phi), 1, 2\phi \rangle$ — Similarly—the reciprocal relation of red and yellow widths—again with blue side as 1 (Fig. 78, p.101).

$\langle \phi^2, 2\phi, \phi^3 \rangle$ — An 'artistic $\sqrt{\phi}$' growth sequence, bottom to top, growing [RG−] $=2/\phi$ then [RG+] $=\phi^2/2$ (p.105, p.74).

$\langle \sqrt{\phi}, \phi, \phi^2 \rangle$ — The squaring sequence of the sides of the white ϕ-based rectangles (Fig. 77, p.99).

$\langle \phi^{\frac{1}{2}}, \phi^1, \phi^2 \rangle$ — Same as above, but here seen as yet another neat reciprocal arrangement—this time in powers of ϕ.

Figure 82: **Mondrian: some 'mathematical chords'.** (The angle bracket notation is used to emphasize the sequential aspect of the relationships.)

To add a yet further dimension, Jonah Lehrer (of *Psychology Today*) reports Semir Zeki's view that the geometric art of Mondrian (and Malevich) is very like the geometry of lines sensed by the visual cortex, as if these painters had somehow got inside the brain and could: 'see how seeing works' [531].

Mondrian's asymmetric balance

As we have seen, Mondrian's work was primarily a spiritual quest, one of reconciling dualities and resolving paradoxes—all in order to realize transcendent harmony. Not long before his death, Mondrian concluded:

> 'The great struggle for artists is the annihilation of static equilibrium in their paintings through continuous oppositions (contrasts) among the means of expression. It is always natural for human beings to seek static balance. This balance of course is necessary to existence in time. But vitality in the continual succession of time always destroys this balance.'

Mondrian, New York, 1942 [659].

Static balance

Let's deal with the 'necessary' static balance first and look back to consider Fig. 75 on page 97—with its red and blue 'squares' either side of the pivot point *P*. We rearrange these here in Fig. 83 to see how static asymmetric balance works. The blue colour is visually heavier than the red, and as the blue area is smaller, it balances. In more concrete terms, we might imagine a blue pot full of paint, and a red carton empty.

Dynamic balance

So much for the static, but how could it be possible to achieve dynamism in an asymmetrically balanced, purely geometric work of art—that is, without any help say, from leaping horses or crashing waves? To understand this, we need to delve into just how we see things—the mechanics of perception. As Monika Krishan notes, the expression: 'Representation is explanation.' (now in frequent use by cognitive scientists) was coined by cognitive psychologist Michael Leyton in his 1992 book *Symmetry, Causality, Mind* [513, 554].

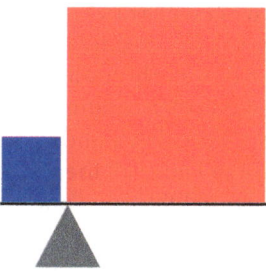

Figure 83: **Static asymmetric balance.**

Leyton's theory is essentially that when we see something, we automatically and without effort construct our own much simplified version, and then we memorize, and in various ways account for, the differences between the real object and our simplified idealized model. He says that it is in the initial simplification process that we employ innate concepts of symmetry. For example when we see a rectangle, we first think of a square, and then we consider and memorize how it has been distorted to become what it is. As another example, if we see a dented tin, we may imagine a perfect cylinder, and then guess what kind of energy was exerted to cause its deformation [552]. In both of these examples, elements of time and history are implied—a narrative about the before, during and after [551]. However, (as Leyton points out), if we consider the Platonic solids (p.15), we find they do not immediately suggest deformation from some other shape—they seem to be as they ever were [550].[21] Leyton draws the far reaching conclusion that the whole purpose of cognitive systems is to generate and process causal explanations [553, 555]. He changes the emphasis from: 'Who/what/where?' to: 'Why and how and when?', and further, from matching to remembered finished real objects to comparison with abstract ideals.

---

[21] Although an experienced geometer might look at a dodecahedron and thereby recall the icosahedron, its dual. But this would not be in some 'before-then-after' sense—rather the forms would each assert their mutual duality, their underlying 'points-for-faces' equivalence and alternate, oscillate, resonate. Looked at this way, we might see the (later) Mondrian lozenge shaped canvas as the 'dual' of his square canvas (swapping corners for sides and vice versa).

Mondrian's dynamic stillness and resonance

While the idea of dynamic equilibrium is well defined in physics and chemistry; the same cannot be said for art. However, as Mondrian specifically intended to address the inner workings of the universe, we may well achieve some understanding of his work with a simple physics experiment. By sliding back and forth in an oblong bathtub at just the right rate, one can cause all the water to move as one mass; it suddenly starts to slosh wildly from one end to the other—with major risk of spillage. This is resonance—one of the themes of this book. When the water has heaped at one end as high as it will go—it is still and silent, just for a special moment, before it starts to swish back, only to heap at the other end. In the Mondrian example, this heaping would correspond to the compressed blue corner—compared with the extended red panel.[22] Such 'special moments' were of overwhelming interest to Mondrian—these occur on the cusp of change, as a direction begins to reverse, as night becomes day and day becomes night [411]. In his own working, these were the moments where he broke his own rules, destroying the old to make way for the new—in order to evolve [410, 63]. The Greeks represented such moments with their god Kairos—a god of time, but not the tick-tock time of Chronos who transports us from day to day. Kairos rather was about the potentiality of the moment. A vast history of causal events lead up to a moment, and the consequences of what happens in that moment then convolve and radiate outwards forever after. Above all Kairos is about fleeting opportunity—opportunity only available to those quick enough to take advantage [679].[23]

With Leyton's insight and guidance, we can now appreciate how a few static (and very geometric) straight bars painted on a canvas can evoke in us impressions of energy and movement. For him, the viewer is not a passive recorder as a simple camera would be. As a design is being perceived, it is at the same time being analysed for history and causality—associations are being made and preferences decided upon. In this Mondrian example, we find its dynamism is generated by a succession of implied duals. In Fig. 84, the *asymmetry* in *A* is immediately suggestive of the possible *symmetry* in *B*, then these together suggest the possibility of the *opposite asymmetry* seen in *C*.

---

[22] John Milner notes Mondrian's achievement of: 'movement in stillness' [650].

[23] To show this discriminating effect, Kairos is invariably depicted as having a quiff of hair at the front of his head while being bald at the back. As he speeds by (with wheels on his feet or wings on his ankles), an agile young man can react quickly. He reaches out and grasps the hair—he seizes the opportunity. The old man reaches out to the head too, but... more slowly... He misses the tuft, and his hand slides off the bald part.

Now from the static image we have the suggestion of a movement from lower left to upper right *ABC*. And again, this suggests the contrasting possibility of a movement in the opposite direction *CBA*.

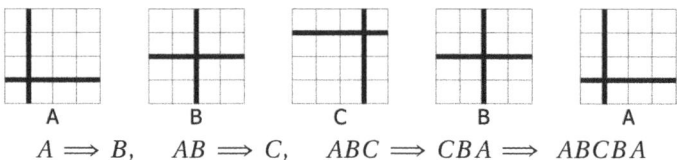

$$A \Longrightarrow B, \quad AB \Longrightarrow C, \quad ABC \Longrightarrow CBA \Longrightarrow ABCBA$$

Figure 84: **Successively implied dualities.**

Finally, this single cycle suggests we might be seeing a snapshot of a balanced process *ABCBABCBABCBA...* a 'steady state' that could continue forever. Now we have an exactly complementary kind of dynamic equilibrium. The active nature of the mind has boot-strapped us up from a special moment to an eternity (Fig. 85).

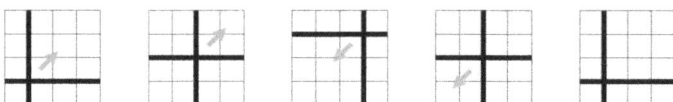

Figure 85: **Mondrian: resonance—with moments of stillness.**

Another Leyton measure is that of energy—think for example of a bow and arrow. When the string of the bow is in tension, but straight and still, it is in its lowest energy state. Energy is transferred to it as we pull the string back, and the elasticity of the bow stores this energy [552]. When we release the string, it flies back and forth oscillating. In the same way, looking at Fig. 84, our ideas of symmetry would suggest that the symmetric centred lines in B constitute the lowest energy resting state, and that we must *do work* to force them into position A. Then when we release them, they will shoot back through the centre and oscillate to and fro, in just the same way as did the bow string.

Mondrian, boxing and colours

In the Western conventions of colour, red and blue (the primaries near opposite ends of the visible light spectrum) are often used to identify contenders—be they in the politics of the United Kingdom or the United States, in military exercises or in international boxing. In fact, the subject of boxing seems surprisingly pertinent to us here; a boxing match is typically fought in a square shaped 'ring' which has opposite corners marked red and blue, and with the other two corners white. So, could boxing have been an influence on Mondrian? Did he go to any matches? (It seems unlikely.) Boxing was, we find, very popular in Paris during his career. In 1920, the *Fédération Internationale de Boxe Olympique* was founded in Paris (as the first such international organization). Also, Mondrian's agent/art dealer (and fellow theosophist) Léonce Rosenberg boxed [65]. Even Pablo Picasso, whose works Mondrian '*greatly* admired' [62], enjoyed watching the matches [755] and boxed occasionally himself [187]. Picasso portrayed boxers in ink (c.1893), crayon (c.1896), and oil (1912) [27]. In a closely fought match, one boxer might fight the other into one corner, only to find he is then fought back into the opposite corner, and this give and take may repeat over and over—which is again just the dynamic that we see in the resonance diagram Fig. 85, p.111.

Further, art historians John Milner and Mark Taylor both note that in Theosophy, red is considered as sensual, and blue as spiritual [651, 888, 56]. But either way, if we accept that the colours red and blue are paired in opposition as a dual, then we might guess that the yellow symbolizes the resolution of their conflict. And as we do this, we realize that Mondrian's three colour scheme exactly echoes the function of his long black bars—the horizontal and vertical being duals and their crossing point being the non-dual resolution.

Confirmation?

Should any lingering doubt remain whether this painting really is based on Lucas numbers $\langle 1, 3, 4, 7 \rangle$ then we might also count the panels: 4 white, plus 3 coloured, total 7 (again reflecting the 4+3 semitone major triad musical chord structure).[24] The way that Mondrian appears to have achieved so many conflicting goals with one canvas—integer and irrational division in the same piece, the 'dynamic stillness' (with its stored energy) that suspends the viewer in

---

[24] Addendum: Theosophy has much to say about 4+3=7,
[for example, see Blavatsky (1888) *The Secret Doctrine*, vol.2, pp.590–598].

a *special moment*, a design that virtually extends out into its surroundings, the presentation of the opposition of all dualities and even their resolution (not once but twice, in 'line' *and* colour)— surely, these all taken together, mark it out as a true masterpiece.

Mondrian: Theory v. intuition

How then can we reconcile all this geometry with Mondrian's earlier claim of using only intuition—a claim now so widely accepted in the art world? In the view of Mondrian authority Carel Blotkamp, his denial of theory was a shrewd response to criticism that his work was over-cerebral [61].[25] Also we have already mentioned the 'intuitive genius' consideration (p.92). But given Mondrian's possible association with AS/HFA (p.106), perhaps his geometry could have been the workings of some savant ability of which truly, he was not consciously aware. Or again, could it be yet another case of Mondrian breaking his own rules, which he did systematically and as a matter of principle [410]?

Research on Golden Section aesthetics

We saw earlier (p.67) how the science of experimental psycho-aesthetics owed its beginning to ϕ through the work of Fechner (1876). In the period from then until 1930, there followed 6 more ϕ studies, and then up to 1991 there were a further 27. In a considered and comprehensive review, psychologist Christopher Green discusses all 34 studies and summarizes them in a Table [271, 272]. His motivation is to answer what has been seen as the key question: 'Do there exist measurable aesthetic effects associated with the Golden Ratio?' But his conclusion is, that the substantial effort to date has only provided conflicting data and conclusions. So, instead of offering a 'yes' or a 'no' to the question, he persuasively floats the idea that *there could be some fragile effect*, but this has not yet been satisfactorily confirmed owing to the debunking zeal of some of the researchers. We note that many (not all) of the experiments performed focused on subjects' shape preferences for plain (i.e. empty) rectangles. This kind of simplification is made in order to standardize experiments and therefore allow valid comparisons between them.

---

[25]   And again we compare Mondrian's situation with that of Bartók [396, 783, 883].

However, from a ϕ point of view, it does not give subjects much of a chance to compare both 'mean and extreme ratios' visually—as they might (for example) in a painting where a prominent spire provides a Golden division in the design. But let's step back for a moment. Doesn't this single-instance approach recall more the lone cymbal clash at the Golden Section moment? By contrast, in the next chapter we shall review the 'all-singing, all-dancing' systems-engineering approach of Le Corbusier.

ϕ—The Gate of the Sun: F_n, and L_n too

Looking back at the original picture of the Gate of the Sun (Fig. 5 on page 9), we see at the top 4 rows—a bottom 'meander' row plus the 3 rows above it. This may remind us of how we have just seen Mondrian using the $1 + 3 = 4$ Lucas number sequence as the basis for his design, which (we propose) he regarded as extending beyond his canvas, with a virtual ϕ^{-3} width gnomon. Could it be that something similar was incorporated into the Gate of the Sun? To find out, we scale up by 4× so that each of the 4 rows now has height one unit (Fig. 87). As a result of this, we are initially surprised to find near integers, and near half-integers (Figs. 86 and 88).[26]

ϕ-based dimension	decimal value	near to	difference relative to height (13)
$4\phi^3$	16.94	**17.0**	−0.43%
4	4.00	**4.0**	0.00%
4ϕ	6.47	**6.5**	−0.21%
$4/\phi$	2.47	**2.5**	−0.21%
8ϕ	12.94	**13.0**	−0.43%

Figure 86: **From ϕ-based to number-based (nearest halves).**

---

[26] That is, before we realize that starting with 8 and multiplying by ϕ again and again will produce near-Fibonacci numbers. Hence by starting with 4, we shall get half F_n's.

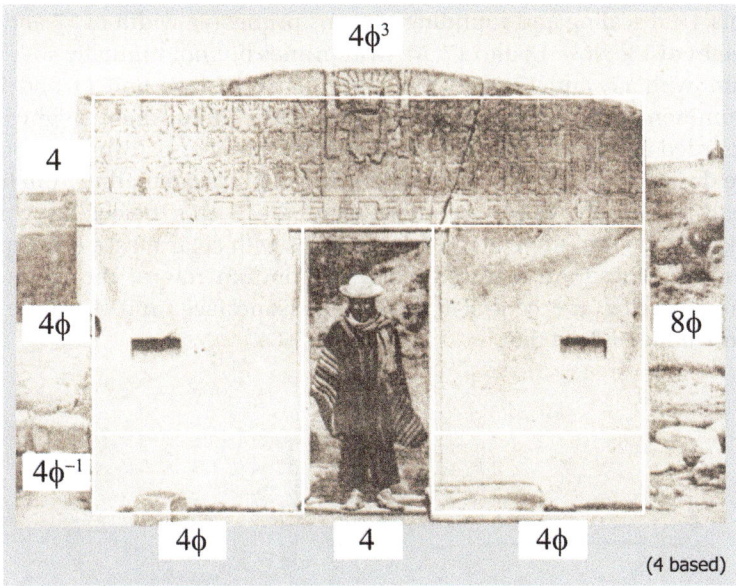

Figure 87: **The Gate of the Sun $4 \times -\phi$ measures based on single row height as one unit.** Original photograph: Secret Museum of Mankind [815].

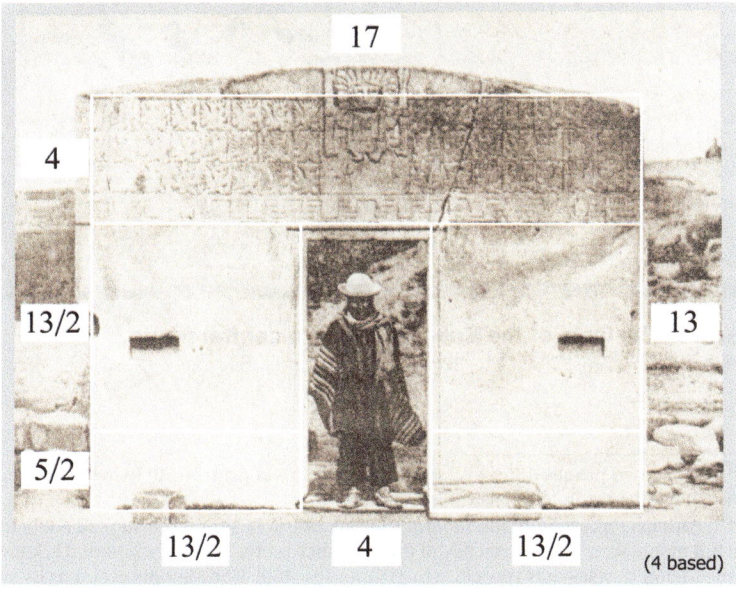

Figure 88: **The Gate of the Sun $4 \times$ —associated numeric values.**

115

This 4× rescaling and rounding gives us primes—a width of 17 and a height of 13. Now, 13 and 17 are twin primes but not mutually so—13 pairs with *11*, and 17 pairs with *19*. [27]   So can we find 11 and 19 prominent somewhere on the monument? (If so, these might be regarded as a kind of confirmation that we have correctly understood the design.) And yes, the central figure has 19 rays emanating from his head—these are shown separate from the 5 rays below his chin (Fig. 89).[28] Also, there are 11 frontal faces with solar masks ('chasqui icons'). These are featured along the bottom row of the frieze—directly below the central figure there is one face, and there are 5 more either side of that.

Figure 89: **The Gate of the Sun—zoom onto central figure.**
Image by Arthur Posnansky [764, 765].

---

[27]   When two primes differ by 2, they are called 'twin primes'. All twin primes apart from (3,5) differ by one from a multiple of 6 [971].

[28]   Perhaps representing the moon's 19 years of travel 'round the sun's head' before the full moon occurs on a given day of the year once more. This 19 year period is known as the Metonic cycle—it is very close to 235 months [460]. The Metonic cycle has its own 'Golden Numbers'—these are nothing to do with the Golden Ratio, and are assigned to years in the cycle to assist in the calculation of Christian and Runic calendars.

Aymaran Great Master, shaman, and Harvard visiting assistant professor 'Willka' (Dr. Álvaro Rodrigo Zarate Huayta) associates these 11 icons with the 11 calendar positions in the calendar wall outside the Kalasasaya compound (10 positions have pillars, one position marks a lost pillar) [982, 980]. But what if we add to the '4' dimension in the same way that we did in the Mondrian analysis—again adding ϕ^{-3} to the height? We recall that

$$4 + \phi^{-3} = \phi^3.$$

So, with the height now at ϕ^3, the 'bigger picture' Sun Gate lintel can now fit 4 ϕ^3 squares across its width. In Fig. 90 (and more clearly in Fig. 91), we see how the chasqui icons align into the bottom left of the left-of-centre $\phi^3 \times \phi^3$ square (in 4 × 4 format). This leaves—at the top and the right—the now familiar shape of a ϕ^{-3} width gnomon. And symmetric about the centre line, we have the same arrangement mirrored on the right-hand side of the frieze. In Fig. 91 we see how the 4x4 square aligns with the left edge and bottom of the $\phi^3 \times \phi^3$ square. This means the gnomon is top and right. We note how the horizontal and vertical 'eyelines' from the central figure correspond with the edges of the gnomons (Figs. 90 and 91)—in particular how the line through the tops of the ϕ^3 squares passes through (radiates from?) the eyes of the central character.

Work in progress

From the point of view of the stone carving, (and integration into a wall), this monument is unfinished. We note also the way that the top does not fit the bottom. It would appear that the simplest mathematical design for width 17—allowing 1 unit width in the centre, with 8 units either side—was overruled in favour of a larger central section at the top (for a bigger central character). This has caused the square icons to have their width squeezed, and the '8 column, 4 row' block on the left to be shunted further leftwards. This in turn has caused an overhang when compared with the integral left pillar below it. To the right, the column of final icon 'squares' stacked at the edge is almost down to half width. We can only guess at what difficult decisions were taken during the work, leading to this state of affairs. 'Would the greater respect arise from a larger central figure or from a perfected geometry?' But once the gateway was built into a wall (as surely intended), perhaps with an extended frieze, the expanded design width would have all but blended in.

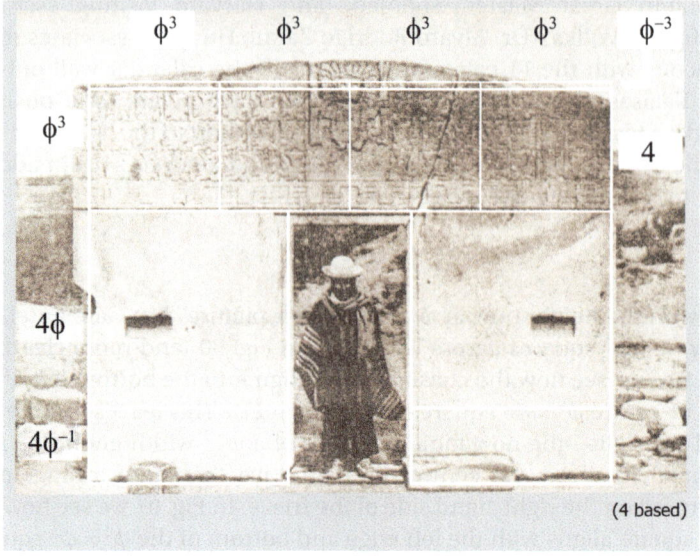

Figure 90: **The Gate of the Sun—four ϕ^3 squares.**
Original photograph: Secret Museum of Mankind [815].

Figure 91: **The Gate of the Sun—central figure with gnomons.**

The Gate of the Moon: F_n, and L_n too

Let's now apply the same logic that we used on the Gate of the Sun in order to envisage this companion gate in terms of Fibonacci and Lucas numbers too. As we try multipliers for the ϕ dimensions (just as we did in Fig. 86, p.114), then we find the first that will give all 'very-near-integer' results is ×8. Now, by using 8 as the basis (Fig. 92), a side 29 square is formed immediately below the lintel cross-piece. This width 29 of the Gate of the Moon is interesting for two reasons. First, it demonstrates 8+13+8=29—that is, $F_6 + F_7 + F_6 = L_7$ —which itself is an example of the general form $L_n = F_n + 2F_{n-1}$ [487]. And second, we note that the prime number 29 is close to the number of days in a lunar month.[29]

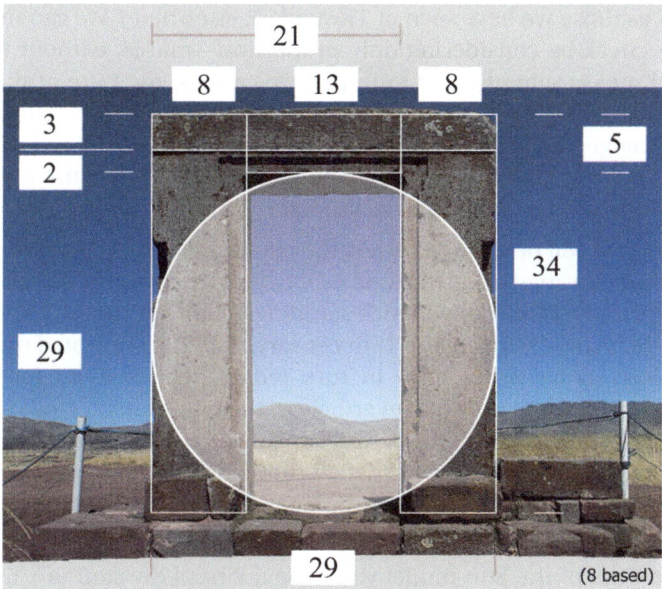

Figure 92: **The Gate of the Moon. 29: 'moon prime'.**
Original photograph by Daniel Maciel [605, 140].

Disclaimer—our overriding interest is in the 'Golden' mathematics suggested by these examples. Finding and considering closely matching (mathematically consistent) proportional relationships is

---

[29] In fact, it is 1.8% short of the average lunar month (that is, the 'synodic' month— new moon to new moon) which is 29.530589 days [205]. The prime number 29 occurs in both the Lucas and the Pell number sequences (as L_7 and P_5)—for Moon Gate/Pell see p.273.

our goal here. In contrast, to test these models rigorously, for example against the artefacts themselves, using precise measurements—say to decide if an original design was ϕ- or integer-based, (e.g. using F_n)—well, that is quite another matter, and one that is outside the scope of this book.[30]

Stonehenge $\qquad$ ϕ? $\sqrt{5}$?

Megalithic monuments and structures are found all around the globe —in Bolivia, Cambodia, Easter Island, Egypt, England, Ethiopia, France, Greece, Guatemala, India, Iran, Iraq, Ireland, Israel, Italy, Japan, Korea, Lebanon, Malta, Mexico, Peru, Russia, Scotland, Spain, Sudan, Tonga, Turkey... So, might we discover ϕ-based proportions (such as those we have seen at Tiwanaku), elsewhere? We can make a quick check by considering only proportion—that is, without taking actual measurements. For example, we can use the Gate of the Sun model as an overlay (from Fig. 8, page 11), and apply it to a picture of Stonehenge (Wiltshire, England)—Fig. 93. And when we do this (Fig. 94), we immediately find an encouraging fit (given the irregular shapes of the stones) for part of the Gate of the Sun design.

However, it will be good to review the known history of this site before we draw too many conclusions. Drawings and paintings from the 16th century show many stones standing [114]. But the site deteriorated, particularly during the 18th century, and this prompted modern rebuilds. Such works have in turn led to debate about just how 'authentic' the site now is. Nevertheless, the outer sarsen circle stones we are using for our check do appear in the old drawings and paintings—giving some confidence they might still now be in their original positions. Archaeologist Julian Richards described this group as: 'the best preserved section of surviving lintels' [779].

In the Gate of the Sun model we made a virtual division of $\sqrt{5}$ into $\phi + \phi^{-1}$ to aid ϕ-based analysis. In Fig. 94 we use the doorway (marked '1') and the ϕ-square and rectangle to its left. The fit shown is of course completely dependent on the earth-level around the bases. However, (considered together with the Sun and Moon gates), it may point to the possibility of other megalithic instances of ϕ geometry. It does seem likely that these stones were originally evenly spaced, centre to centre. (But with such irregularly shaped stones we do not have simple heights and widths.)

---

[30] Addendum: shortly before this book went to press, the Sun and Moon Gate dimensions were checked by Tiwanaku expert El Ing. Jorge Emilio Molina Rivera (p.xiv).

Figure 93: **Stonehenge, Wiltshire, England.** By Gareth Wiscombe [984, 140].

Figure 94: **Relating proportions from the *Gate of the Sun* (Bolivia)**

Figure 95: **$\sqrt{5}$ door proportion overlaid onto pillar+lintel heights.**

However, the ratio of height of pillar+lintel to the pillar width does appear to be fairly constant—in the same proportion as the Sun Gate doorway—viz. $1 : \sqrt{5}$ (Fig. 95). In this configuration, if the pillar widths were standardized as one unit (relative) with the 'pillar (left edge) -to-next pillar (left edge)' distance as ϕ (as in Fig. 94), then the resulting (idealized) gap would be $\phi - 1 = (1/\phi)$—that is, in simple Golden Ratio.

Figure 96: **Eagle headed being—presenting the twin primes 17 and 19.**
Assyrian c.865–860 BC, from Nimrud, North West Palace, Room G, panel 27. V&A London.

Fig. 96 above shows another instance of the twin primes 17 and 19 from the ancient world (p.116). Could we relate these to Golden/ Fibonacci geometry? Let's think of 17 and 19 as the sides of a rectangle, but doubled so as to provide a Fibonacci context: 34×38 (blue tint area). We see that by adding half the short side to the long we get 34×(38+17)=34×55, which is 'φ-convergent' Golden. This half-square addition recalls the simplest *ABC* Golden construction (Fig. 269, p.475). Here, by Pythagoras, *r*=38.01... (a factor of 1.0003 bigger than 38). We also see that two near-[RG–] rectangles are formed: the ratio of their sides of 21 and 17 provides a φ-convergent approximation to the Golden Section of 2 (i.e. 2/φ)—in this case, to within 0.07% (p.73).

Chapter 8

Architecture—Le Corbusier

Figure 97: **Le Corbusier (1887–1965) & Albert Einstein (1879–1955).**
Princeton, New Jersey, USA, 1946. Einstein immediately appreciated the value of Le Corbusier's ϕ-based system. © FLC/ADAGP: Paris and DACS, London 2015.

Charles-Édouard Jeanneret-Gris was born in Switzerland in 1887, and at age 30, he 'rebranded' himself as architect *Le Corbusier*[1] (Fig. 97, p.123). This was in part a result of his meeting Amédée Ozenfant in 1917—together they founded the Purist movement. Later, in 1930, Le Corbusier took French nationality. One of his most famous buildings is the Villa Savoye (Fig. 98)—a Purist masterpiece, 'a machine for living (in)'—which he constructed with his cousin Pierre Jeanneret. He designed it while he was still consolidating his ideas about the Golden Ratio—ahead of his work on 'the Modulor' (which we shall discuss shortly). Looking at the villa, a quick check for proportions upon which Le Corbusier may have based this work suggests those in Fig. 99. In the diagram, the long rectangle represents the left facing elevation in Fig. 98, with f: floor thickness and h: height of window row [126]. The final design keeps very close to these proportions, but not slavishly so.[2] The front elevation appears to be soundly based on the Golden Ratio—with an audaciously long $(1:3\phi)$ horizontal rectangle.[3] But the vertical dimensions within this (to size and position the window strip) include division according to the Silver Ratio $\delta = 1 + \sqrt{2}$. That long panoramic window facing out from a bright white background is perhaps remembered in Stanley Kubrick's *2001 A Space Odyssey*—in the design of the large tadpole-shaped white spacecraft *Discovery One* [517]. And if we think of this window strip as 'the eyes' of the building, then the white above will become the forehead. And given the somewhat unusual height of this (the $\sqrt{2}$ part of the Silver Ratio), the building takes on an air of 'high-brow' contemplation.

Le Corbusier began his seminal ϕ-based work *the Modulor* by describing the naturally continuous quality of sound [127]. He noted that in order to make music—particularly music that can be written down and replayed—sound must be: 'divided into sections and measured' [129]. He then observed that the basic tools then available for making music—standard scales, timing conventions, etc. had no equivalent in architecture. Music seemed to him a more mature discipline. He went on to note that in the past... the Parthenon, Indian temples, and cathedrals were all built according to measures such as the foot and the cubit—measures of the human body—a source of dimensional harmony and beauty [130].

---

[1] A registered trademark, now held by *Fondation Le Corbusier,* Paris.

[2] Le Corbusier used his 'regulating lines' as guides only [341].

[3] This shape also appears vertically in the Gray poster p.66—to the left of the pillar/section line—a stack of 3 portrait Golden Rectangles.

Figure 98: **Le Corbusier (1887–1965) Villa Savoye, Poissy, France (1931)** [125]. © FLC/ADAGP: Paris and DACS, London 2014.

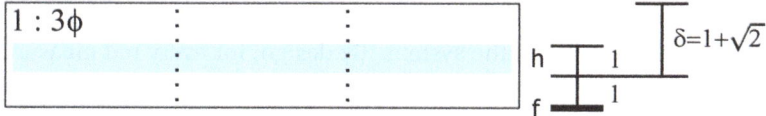

Figure 99: **Both Golden and Silver Ratios—ϕ and δ.**

Le Corbusier then noted the disconnect between these and the more recent metric measure, (originally based upon the circumference of the Earth).[4] His goal became to devise a universal system of measurement that would fulfil the needs not just of architects but of product designers and packagers too—a self-consistent and efficient system, based on human dimension. Just earlier, we discussed grouping in music: ... *threads may go off, subdivide differently, work with or against each other, but then still arrive back together at the end of a phrase.* Now, just these same properties are highly desirable in architecture. Spatial dimensions must be filled exactly in a variety of ways (to include windows, doors, roof, and so on).

---

[4] As an aside: while the conversion of miles to kilometres is often approximated using the Fibonacci number ratio 8/5 (a convergent of ϕ), actually the conversion is nearer to ϕ than that (being 1.609...). So to estimate the conversion of 21 miles to kilometres, take the next Fibonacci number... 34 (km)—and for 'in betweens' the Lucas numbers will work in the same way.

In a regular periodic design, such as a factory, aesthetics are often traded in favour of simplicity of build and lowest cost. Alternatively, a superior ϕ-based design can provide the opportunity to maintain a harmonious repetition of proportion while reducing periodicity. Though without some new standardization that would support such ϕ-based architecture, every build would be a one-off—with costly custom-sized components. So here we have Le Corbusier's grand vision: a universal system of measure—in fact, a standard sequence of lengths—where one member grows to the next by ϕ. This system (begun in 1943 and published in 1948, 1954, and 1955), he called: 'The Modulor'.

The Modulor: red and blue series

In order to move from theory to practice (in both imperial and metric units), he devised a scale of measurements he called his 'red series' [134]. For this sequence he chose his starting reference as six feet: 'the height of an English policeman' [133]. And to complement the red series, he added a second sequence, his 'blue series' that provided measures in-between those of the red series (Figs. 100 and 101). His goal was that building components (and packaging) would be manufactured to fit the system. By design, for every red measurement, there is a blue twice its size. For Le Corbusier, this arose from what he regarded as 'his' three fundamental measures: 1, 2, and ϕ [132]. So, the Modulor has 3 essential elements:

- red series—a geometric progression which grows by ϕ
- blue series—the same, but offset by a factor of 2
- a single (human) reference dimension—to set the scale

We recall seeing the same basics of: 1, 2, and ϕ in the works of both Bartók and Mondrian (p.65 & p.107). We also saw how when Golden Rectangles are composed as the framework of an artistic design, Root Golden approximant rectangles [RG−] and [RG+] may appear whether intended or not—for example in Toulouse-Lautrec's *Aristide Bruant* poster (p.87). So, in retrospect, we can appreciate how well Le Corbusier's intuition for an in-between sequence coincides with the mathematics. If we take the geometric mean of ϕ and ϕ^2—we find $\sqrt{\phi \cdot \phi^2} = \phi^{3/2} = 2.058$—a number quite near to 2. It therefore follows that the blue series closely approximates the mid-growth points between successive members of the red series and vice versa. In the ϕ Table in Fig. 101 we see how the red series begins: 1, ϕ; and the blue begins: $2/\phi$—which is the ratio [RG−], then 2.

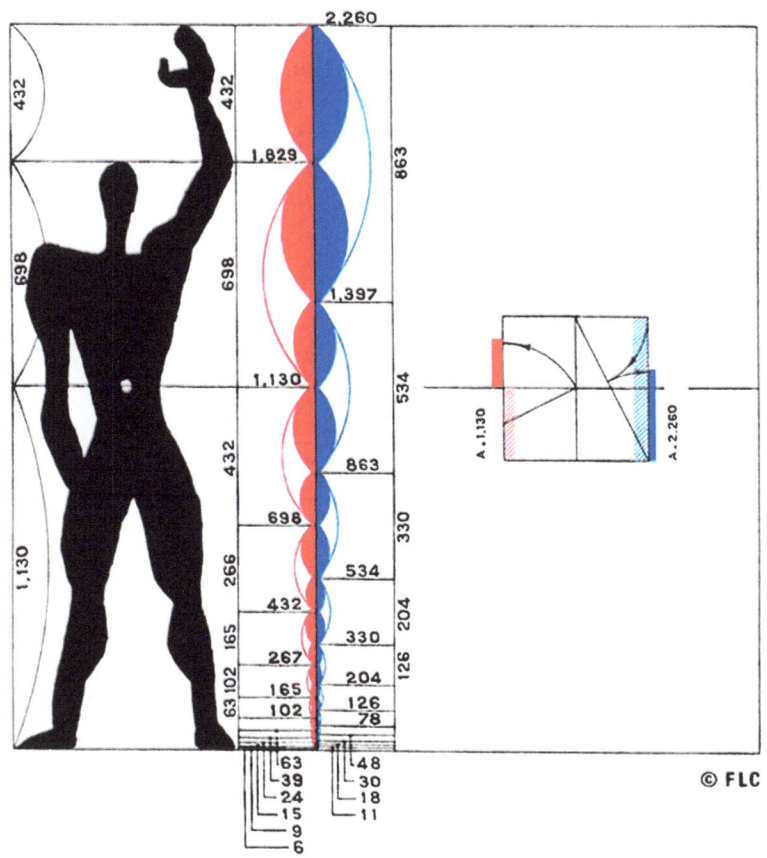

Figure 100: **Le Corbusier's Modulor Man.**
© FLC/ADAGP: Paris and DACS, London 2014.
On the right are two Golden Ratio constructions (red: c.f. p.475; and blue: c.f. p.3).

Red and blue series [134]. (Each blue doubles a red value, ±1 mm.) Higher values progress,

red series:	...,	1130,	1829,	2959,	4788,	7747,	12535,	20282, ...
blue series:	...,	2260,	3658,	5918,	9576,	15494,	25069,	40563, ...

Red	1		ϕ	ϕ^2	ϕ^3	ϕ^4	...
Blue	...	$2/\phi$	2	2ϕ	$2\phi^2$	$2\phi^3$	

Figure 101: **Modulor red & blue series: relative values in terms of ϕ.**

Looking back to the Gate of the Sun (p.11) we see the ϕ^2 dimension at the bottom (left and centre), the 2ϕ on the right side, and ϕ^3 on the top. This we now recognize as a closely approximated Root Golden growth sequence.[5] We also saw this same growth sequence earlier in Kepler's triangle[6] (Fig. 4, p.7), but now if we pair Kepler triangles 'back to back', we realize the Root Golden growth (just about) continues—becoming the base itself—as '2' in our proportional units (Fig. 102), the last step being × [RG−].

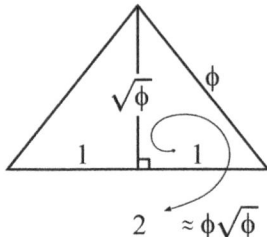

Figure 102: **Kepler pair—Root Golden growth:** $\langle \phi^{\frac{0}{2}}, \phi^{\frac{1}{2}}, \phi^{\frac{2}{2}}, \approx \phi^{\frac{3}{2}} \rangle$.

In Fig. 103 we 'try out' the Modulor concept by re-designing a bookshelf. This will be our version of Le Corbusier's 'panel game' where he experimented with many combinations of Modulor rectangles [135]. To understand the theory we shall first sketch a shape in relative units (ϕ and Fibonacci), but then for actual dimensions we shall scale up to 'real world' Modulor dimensions. We shall again see (much as we did in the Bartók example), that harmony results from:

- dense repetition of the same (Golden) proportion
- many different combinations of dimensions exactly fitting into a chosen width or a height
- similarities between the multiple fitted combinations

Such qualities prompted Einstein to describe the Modulor as: 'a scale of proportions which makes the bad difficult and the good easy' [128]. Applying the Modulor in even a very basic design will fairly effort-

---

[5] The error in the growth $\phi^2 : 2\phi$ is exactly compensated by the error in that of $2\phi : \phi^3$ so the overall (two step), growth is ϕ. Also from the ancient world, the theatre at Dodona, Epirus, Northern Greece has been proposed as including an example of a $\sqrt{\phi}$ approximant, $19/15 = 1.2666...$ (this being within 0.5%) [814]—compare $14/11$, p.6, fn.6.

[6] Kepler's triangle, Herz-Fischler notes, was rediscovered by Mascheroni (1797), Wiegand (1847), and Röber (1855) [344]. Kepler himself learnt of it from a music professor—Magirus [348].

lessly generate a symphony of Golden Sections. In Fig. 103 to the left we see the original bookshelf, and to the right, a Modulor design for comparison. Shelves harmoniously make Golden Rectangles. Three shelves butt onto the left vertical divider; 2 each mark the Golden Section of the shape the other side (be it square or Golden Rectangle). But this visual harmony is different from that of musical notes, where euphony depends on the ratios of small numbers (3/2, 5/4, etc.). Here the ratios are nearly all irrational —excepting the factor of 2 (as 1+1) on the right. Not only is the Modulor design visually more interesting, but also its stages of ϕ scaling make it far more functional. The design now acknowledges that we have mostly medium sized books, a good number of small books, and rather fewer outsize ones.

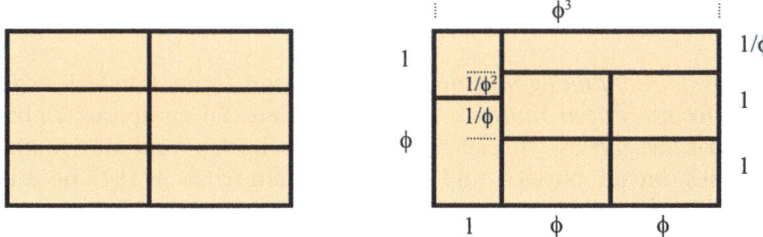

Figure 103: **Bookshelves: simple, then using the Modulor principle.**

In Fig. 104 we first see how the shape may be implemented (to a good approximation) using Fibonacci numbers, and also (as a practical example) by using the Modulor 'blue series'.

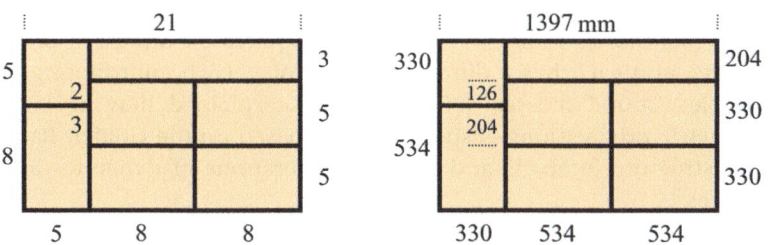

Figure 104: **Fibonacci and Modulor (blue series millimetres, p.127).**

In some situations, a practical problem arises with the Modulor—that is, where packaging thickness is taken into account, or gaps are required between elements. In the above shelving example, we meet this as soon as we take into account the thickness of the shelves and sides. As another example, a set of Modulor dimensioned building bricks would all fit together in many ways exactly, but without leaving room for any mortar. To add mortar as well, but still 'have everything fit' and to retain proportions—this would call for a little calculation and re-dimensioning.[7] Such problems are usually relatively easy to address. (And for a no-gap, generalized, precision Modulor approach, see page 486.) Although the Modulor system was not widely adopted, it still has strong supporters today—including Liverpool architect George A Hall, who says: 'In practice, Modulor dimensions are useful up to person size, but above that, the measures initially become too coarse to be used in architecture. However, still bigger Modulor dimensions may be applied in town planning.' [295].

Xenakis—honorary Parisian

'I have found, among my papers,' said Goethe, *'[one] in which I call architecture frozen music.'* [195]. Experimental composer Iannis Xenakis deserves a special mention. His background was mathematics, music, physics, and technology, and when in 1947 he was exiled from Greece, he moved to Paris and was hired by Le Corbusier as an engineer. Le Corbusier had taken French nationality and so too (later) did Xenakis. In addition to his architectural work, Xenakis studied music and began to apply mathematics to composition—he also received mentoring from Messiaen, during which he met Karlheinz Stockhausen. Based in Paris, Xenakis worked with Le Corbusier for close to 12 years after which he devoted all his time to music [34]. Xenakis was one of the first composers to go on record confirming his use of ϕ and Fibonacci [475]—for example in his 1953 score *Metastaseis B* [992].[8] The piece requires an orchestra of 61 players, and each has a different part to play, each contributing to complex 'sound masses' [858]. Xenakis explained how he used geometric progressions—especially ones based on the Golden Ratio —to structure intervals and to set the durations of dynamics and timbres.[9]

---

[7] In the shelving design, we could simply re-design on the basis that the theoretical lines—as in Fig. 104—are taken as centre lines, locating the centre of the thickness of each shelf and vertical section.

[8] 'After + a state of standstills = dialectic transformation', (Xenakis notes).

[9] For us, this is a refreshing change from those artists and composers obsessed with keeping their methods secret—but were Debussy alive at the time, he might well have regarded Xenakis' open approach and his detailed disclosures as rather 'letting the side down' (p.65, [86, 942]).

Figure 105: **Iannis Xenakis (1922–2001)—Philips building, Expo58, 1958—frozen music?** Photograph by Wouter Hagens [290, 144].

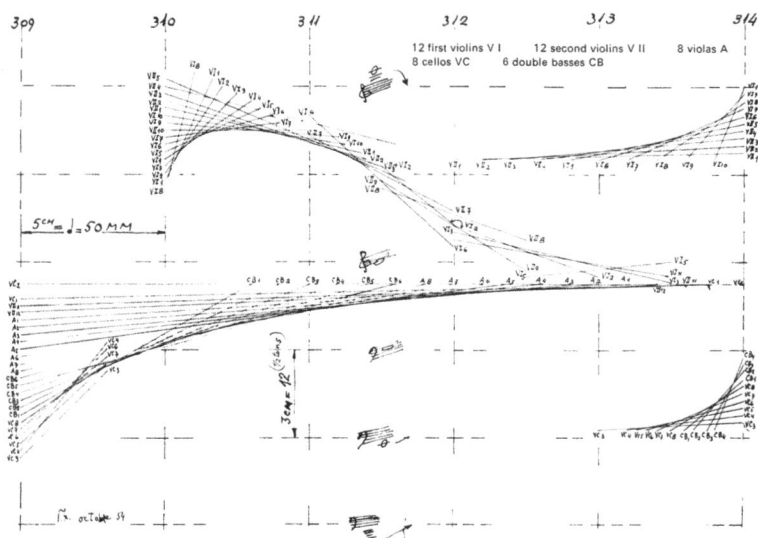

Figure 106: **Xenakis—Metastaseis 1953. String glissandi, bars 309–314.** From *Formalized Music* by Iannis Xenakis, © 1971 Pendragon Press, NY, reproduced with kind permission [993].

Xenakis drew an analogy between these concepts and those he applied in his architectural designs for the façades of the *Convent de La Tourette* near Lyon. This late modernist building incorporates key Le Corbusier architectural styles including the ϕ-based *Modulor* [136]. He also notes how the sound shapes of the *glissandi* that he systematically used in *Metastaseis*—Fig. 106, the hyperbolic paraboloids (saddle-curves), [993]—later became transformed into the exterior architecture of the Philips Pavillion that he created for the 1958 Brussels Exposition (Fig. 105) [992].[10] His design combined ruled surfaces in hyperbolic paraboloid forms to give an ultra-modern tent-like mathematical effect—and he created and practically engineered all this before the days of 3D-modelling software [117].

Ernst Neufert

German architect Ernst Neufert (1900–1986) was an assistant to Walter Gropius. Marcus Frings (p.67) shows Neufert to be another important ϕ evangelist [229]. In his 1936 *Bauentwurfslehre* ('Architects' Data'), the Golden Ratio plays a key role. This technical handbook is copiously illustrated and has since gone through many editions, putting on weight all the way. Over the years, it has found a worldwide audience and is still very much in use today. Modern editions retain discussions of the Golden Section:

- in the introduction to measurement, with human dimensions being regarded as: 'the universal standard' (referencing the work of Zeising, later endorsed by researcher E Moessel) [675][11]
- in a summary of Le Corbusier's Modulor which includes various diagrams and the red and blue series dimensions [676][12]
- in the context of a stalls theatre auditorium design, recommending that the proscenium height to width ratio should be Golden [677]

Also, in his 1943 *Bauordnungslehre* (teaching manual), Neufert describes systems of proportion and gives special attention to the Golden Ratio [678].

---

[10] Though Joseph Clarke plays down the interpretation of a *direct* correspondence between the curved web graph at the end of *Metastaseis* and the Philips Pavillion's structure [118].

[11] Page 67.

[12] Page 127.

Chapter 9

Binet spirals, Debussy, and a black cat

In order to gain a yet deeper understanding of the Binet formula, we shall consider non-integer n—that is, by asking: 'What happens in between the integers (between each of: $0, 1, 1, 2, 3,$ and so on)?' And finally we ask if Bartók, Debussy, and Ravel had use of this knowledge.

Between the Fibonacci numbers

This section (up to Root Golden Spiral) *requires complex numbers. The important thing here though, is the shape of the spiral obtained.*

As pointed out by Ron Knott, it is interesting to apply the Binet formula to study the gaps between the Fibonacci numbers [481]. Normally we feed an integer n into the Binet formula, and it gives us back an integer—a Fibonacci number F_n—that is

$$F_n \quad = \quad \frac{\alpha^n - \beta^n}{\alpha - \beta} \quad = \quad \frac{\phi^n - (-\phi)^{-n}}{\sqrt{5}}, \qquad n, F_n \text{ in } \mathbb{Z}.$$

But what if we feed in a non-integer, say $n = 0.5$? We can immediately see that α^n will evaluate simply as a square root of a positive number ($\alpha^{0.5} = \sqrt{\alpha} = \sqrt{\phi}$). However, as β is negative, the β^n term takes us into another realm entirely...

$$\beta^{0.5} \quad = \quad \left(-\phi^{-1}\right)^{0.5} \quad = \quad \left(\frac{-1}{\phi}\right)^{0.5} \quad = \quad (-1)^{0.5} \cdot \frac{1}{\phi^{0.5}} \quad = \quad \sqrt{-1} \cdot \frac{1}{\sqrt{\phi}}.$$

Taking the $1/\sqrt{\phi}$ first, if we compare $1/\phi^1$ with $1/\phi^0 = 1$ then the ratio between these is ϕ, and growth-wise $1/\sqrt{\phi}$ is half way between these two. In comparison, with $\sqrt{-1}$, we have a number that growth-wise is half way between +1 and −1. A way to think of this is to imagine standing on the origin of a number line (at zero) holding out a bar length one, reaching to +1 on the number line. We now turn to the left, swinging the bar round until we are facing in the opposite direction, and the bar now reaches to −1. So far so good. We are thinking of 'multiply by −1' as a rotation of half a turn. It should now be logical to think of the square root of minus one as the situation that was half way round that half turn we made—in plan view, one quarter turn anticlockwise. The convention is to say the bar then reaches to an imaginary number called 'i', with the bar now at right angles to the real number line. The correspondence between multiplication and turning is exact, so that $i \times i = -1$, quarter turn plus quarter turn = the half turn (from +1 to −1), and $i^4 = 1$. As a result, in this now two-dimensional representation (called an 'Argand diagram'[1]), we can see that arbitrary non-integer powers will have associated angles of turning, and we have already seen that the growth aspect can be separated out. So let's rewrite the Binet formula to emphasize that the power we supply may be non-integer (call it 'w'), and the result that we get may be a complex number—that is, it may contain both real and imaginary parts (call it 'z')

$$z \;=\; \frac{\alpha^w - \beta^w}{\alpha - \beta} \;=\; \frac{\phi^w - (-\phi)^{-w}}{\sqrt{5}}, \qquad w \text{ in } \mathbb{R},\; z \text{ in } \mathbb{C}. \quad (9.1)$$

The interesting spiral behaviour of this formula is easy to see in the Argand diagram Fig. 107.[2] Also, to see how the curve passes through 1, and then loops back on itself twice at 1 to take in all of F_{-1}, F_1 and F_2; we zoom in for a closer look in Fig. 108.

'Regions of Dominance'—a drama in 3 acts

To understand this Binet formula (9.1) better, we shall track the terms α^w and β^w and see how significance may pass from one to the other. Specifically, when w is large and negative, α^w is small, and β^w is large (in absolute magnitude).

---

[1] This was independently discovered by Wessel [708], Buée, Mourey, Warren, and Gauss [98].

[2] The Argand diagrams in this section were all created using Oracle Java® [412] and Dr. Jon S Squire's *Complex* class [859].

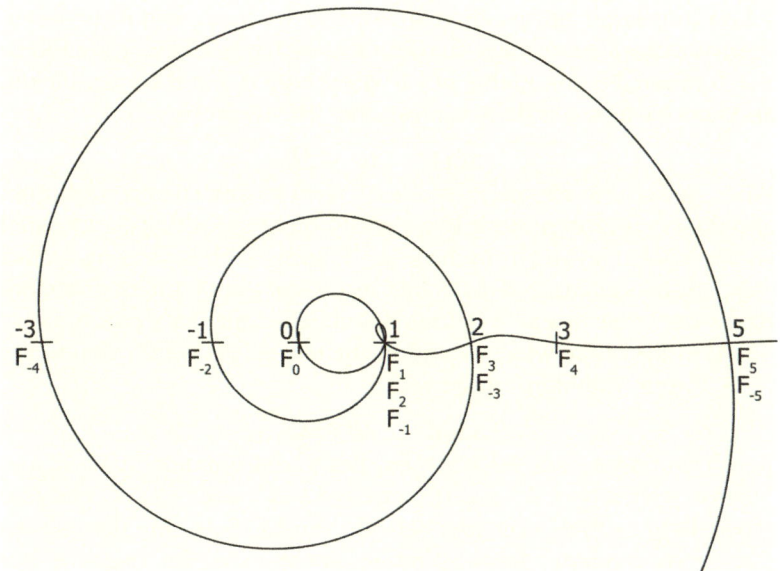

Figure 107: **In between the Fibonacci numbers–the Binet spiral** (Argand diagram). As w increases successive points are plotted. That is, w increases along the curve itself—not along any axis—such plots are called 'parametric'. We note the points where w becomes integer: -5.0, -4.0, and so on. We start on the right at F_{-5} and follow the curve up and back (anticlockwise) to F_{-4} (left), then to F_{-3} and continue on.

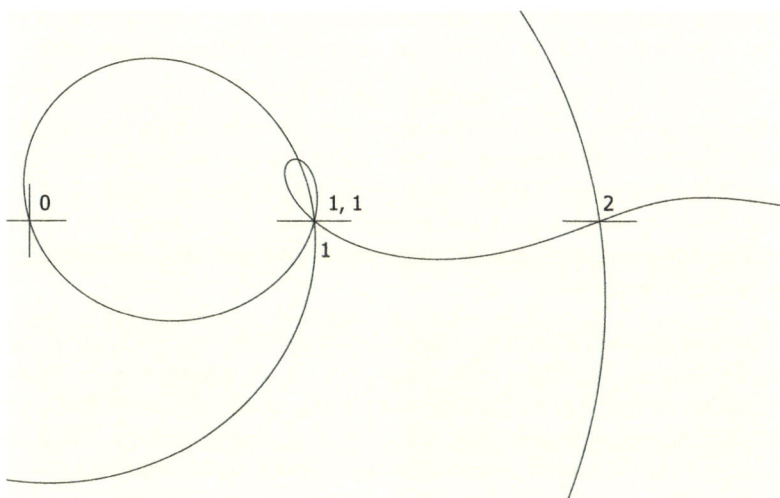

Figure 108: **Complex Binet—zoom view.** Starting bottom left: As the curve spirals up and inwards it passes through 1 (as F_{-1}) then 0. But next it loops back twice more through 1, thereby correctly placing all three '+1' entries in the Fibonacci sequence $\langle \dots 1, 0, 1, 1, 2, \dots \rangle$.

When w is large and positive, the opposite is true, and in between, there is a changeover stage. In summary, we have 3 distinct situations: $w < 0$ where β is dominant, $w \approx 0$ where both α and β are significant, and $w > 0$ where α is dominant—rather like a 3 act play.

Act I $\quad w < 0$

When power w is very negative the ϕ^w term makes a tiny contribution and the curve is asymptotic to a logarithmic spiral with its large radial vector (spoke) arising from the $(-\phi)^{-w}$ term. With increasing w, this spiral shrinks at a rate of ϕ per half turn—*not* per quadrant, therefore this is not what we call a Golden Spiral. The imaginary part of every point on the curve which is not on the real axis is wholly due to the $(-\phi)^{-w}$ term with non-integer w.

Act II $\quad w \approx 0$

In its traditional form, the 3 act play has a confrontation in Act II, and here we see the rival α^w and β^w terms—i.e. ϕ^w and $(-\phi)^{-w}$, vying for recognition as their strengths become similar. It is here that we see the curve changing from spiral to linear. As the negative but increasing w approaches zero and passes through it, the spiral continues to shrink away—both its real and complex parts tending towards zero. But now, the (real, positive) ϕ^w term is beginning to become significant—rapidly growing and taking over. As it does so, the smaller and smaller coils of the spiral are seen to be 'dragged out' to the right. This is because the new influence is only real, leaving any existing spiral-based imaginary value as is. It is as if an outer arm of the spiral were held fixed, while the centre is teased out to the right.

Act III $\quad w > 0$

Here the positive term ϕ^w dominates—it gives growth out along the real axis only (horizontal to the right)—with no imaginary (i.e. vertical) component. The curve quickly approaches a straight line as the residual influence of the complex term fades away. In other words, the curve here is asymptotic to positive real axis: Act III brings a peaceful resolution.

When we zoom out (and also mark in asymptotes), we see each of the three 'Acts' in Fig. 109—with the (Act I) limiting log spiral for large negative p, the (Act II) 'pulling out' changeover region and the (Act III) limiting straight line—the real axis—for large positive p. We also note the *ABC* Golden Section for comparison with those shown on page 3. This sectioning closely follows the logic we used on page 43. For more about the complex Binet curve, see de Bruijn [90] and Wunderlich [990].

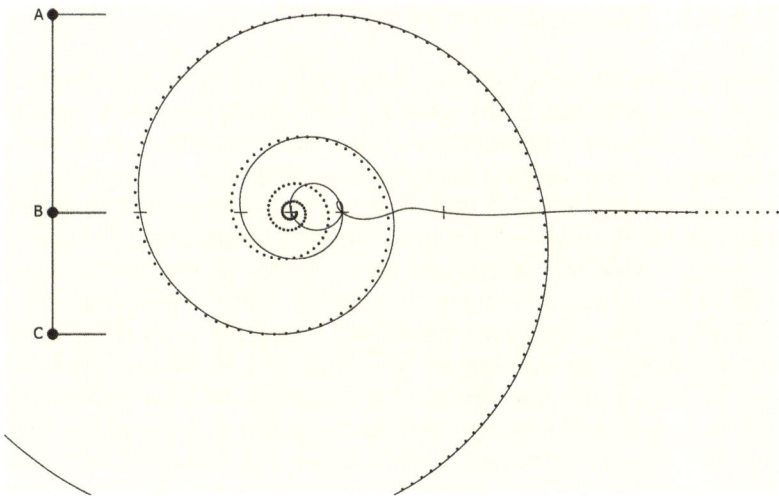

Figure 109: **Complex Binet asymptotes (dotted).**

The Root Golden Spiral

In 'Act I' we met the underlying logarithmic spiral arising from the $(-\phi)^{-p}$ term. For increasingly negative p, this (asymptote) spiral grows at a rate of ϕ per half turn (shown dotted in Fig. 109) We met this dotted spiral briefly at the end of the first chapter (in relation to Kepler's triangle, p.33). Then in the *Spirals* chapter we saw the polar formula for the Golden Spiral—Eqn. (2.4), page 40), which is

$$r = (\phi^4)^{\frac{\theta}{(2\pi)}} = \phi^{\frac{4\theta}{(2\pi)}}.$$

So, as the Golden Spiral's radial vector grows at ϕ^4 per turn—our new curve, which grows at ϕ^2 per turn, will have the formula

$$r = (\phi^2)^{\frac{\theta}{(2\pi)}} = \phi^{\frac{1}{2} \cdot \frac{4\theta}{(2\pi)}}.$$

That is,

$$\boxed{r = \left(\sqrt{\phi}\right)^{\frac{4\theta}{(2\pi)}}} \qquad (9.2)$$

Therefore, to speak of its relationship to the Golden Spiral, let's call this curve (9.2) the 'Root Golden Spiral', as it grows at $\sqrt{\phi}$ per quadrant. In the following sections we shall see how this curve has become important in the analysis of some composers' work.

Bartók: The question of intent

Much as a result of the initial work by Ernő Lendvai (whom we met in Chapter 4, page 62), many writers have discussed Bartók's *'possible use'* of Fibonacci numbers and the Golden Ratio. One such, Hungarian musicologist László Somfai, is often misrepresented as having written that he: 'found no evidence of the use of the Golden Ratio in Bartók's work'—even after 3 decades of effort with 3600 pages of notes. What Somfai actually argues is, (given the lack of documentary evidence of numerical planning), that Bartók must have created his mathematically based works through *intuition*—that he unconsciously created detailed structures in the Golden Proportion. Somfai does admit that Ernő Lendvai found Golden proportions in Bartók's work (and makes no argument against this understanding), however, he regards the lack of relevant pen and paper calculations as 'solid evidence' against *deliberate planning* [848]. It is of course impossible for Somfai to prove this kind of negative. A lack of something may be suggestive (even strongly so [185]), but it is not 'solid evidence' as such. Somfai's is (what has been called) 'an argument from ignorance'—that is, from a lack of evidence to the contrary. As we discussed in Chapter 4 (page 64), Bartók refused to describe his compositional methods, saying:

> **'Let my music speak for itself ... '** [540]

And so it does: in the next section, we find that the first note of the slow 3rd movement of *Music for String Instruments, Percussion, and Celesta* (MFSPC), speaks on behalf of Bartók providing us with a master key to understanding his advanced use of Fibonacci and the Golden Ratio.

Xylophone palindrome

It is fairly well known now that in the opening of the 3rd movement of MFSPC, Bartók trumpets his use of the Fibonacci sequence—with a xylophone solo. This phrase is normally reported as having an explicitly written accelerated and decelerated repeat of a single note—in a Fibonacci rhythmic palindrome: 1, 1, 2, 3, 5, 8, 5, 3, 2, 1, 1.

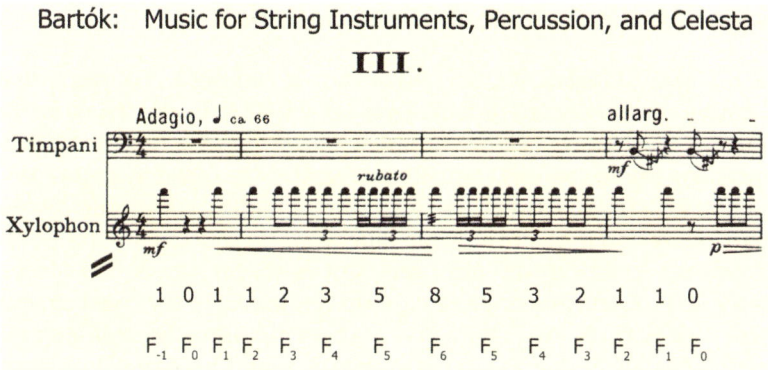

Figure 110: **Béla Bartók (1881–1945), MFSPC 1936. Xylophone opening of 3rd movement [40].** Reproduced by permission of Universal Edition (London) Ltd. All Rights Reserved.

What has been missed so far though, is that with the very opening note (ahead of the palindrome), Bartók is announcing his interest in the *negative-index* Fibonacci numbers—which we first met on page 29. In Fig. 110, starting at the very beginning, we find a '1' (quarter) note and then a two beat rest—which provides a suitably slow '0'— consistent with it being an earlier start to the acceleration. This now gives us: '1 0 1 1 2 3 5 8 5 3 2 1 1 0'. Bartók has rather overtly pushed the boundary into negative-index Fibonacci numbers—starting $F_{-1} = 1$ (with exactly the 1 0 1 1 2 sequence we saw in Fig. 108 on page 135). (And although the idea that Bartók deliberately used a negative index might sound 'rather unlikely' at this stage—as we move on to consider the spiral insights of both Lendvai and Howat, and further, to discover a particular cascade form used by Bartók, things will fall into place.) At the centre of the palindrome, (the climax of the acceleration), we find number 32 again—here as a quarter note divided into eight. We guess it might be quite difficult to play those central eight 1/32nd notes perfectly accurately, and... Bartók relents. He invites the percussionist to 'rob' him of his intended true Fibonacci sequence—by allowing flexibility even from the group of five notes (F_5). We wonder if that *rubato* (permission to adjust the rhythm) was in the original score, or was negotiated at first playing.[3]

---

[3]   To try playing this phrase, set a metronome to 66 beats per minute (or use the seconds tick of a clock), and for each beat, think one word at a time, while evenly tapping once for each syllable: 'I, (wait), (wait), I, am, Bar-tók, com-po-ser, ex-tra-ord-in-aire, su-per-ca-li-fra-gil-ist-ic, ex-tra-ord-in-aire, com-po-ser, Bar-tók, am, I.' (For the full word, 'supercalifragilistic...', see Zimmer [1002].)

Bluebeard's Castle: A Binet palindrome?

In his essay *Bluebeard and Erwartung: a Notebook*, Pulitzer prize winning composer Ned Rorem discusses the palindrome structure he finds in Bartók's psychological opera *Bluebeard's Castle* [36]. It has only two singing voices, and Rorem tracks a gradual and complementary change between them. Young Judith starts out being: 'passionate and protective', while Bluebeard is: 'reticent and rational'. Bluebeard then grows stronger and stronger as Judith fades away [791]. Does this underlying form remind us of the Binet drama Acts I–III with *its* two characters—where the initially strong spiral term fades, while the real term overtakes it and becomes dominant (p.136)? But, bearing initials BB and being born in 1881, we should not be surprised that Bartók was fond of palindromes. We just noted the: '0 1 1 2 3 5 8 5 3 2 1 1 0' xylophone solo, and to this we could add (for example) his five movement *Fourth Quartet* [33, 38]. Somfai finds that this string quartet was originally written as four movements but that movement IV was added in later to create an ABCBA 'arch' symmetry [849]. Bartók again used the arch form in his fifth String Quartet [39, 111] and in a number of other pieces too. Such structures are also described as being 'mirror-symmetric', or 'in bridge form', and these might be regarded as further examples of the 1:2 proportion that Howat noted in the MFSPC I (p.65) [393].

The Lendvai Spiral

Now we return to the subject of spirals. For Bartók, the spirals of nature were echoed in his own career. Hungarian musicologist Bence Szabolcsi recalls Bartók describing his artistic development as a spiral form: '[dealing] with the same problems on an ever rising level, with correspondingly rising success' [882]. And it was Szabolcsi's colleague Ernő Lendvai who introduced spirals into the proportional analysis of Bartók's works. Lendvai's original spiral (Fig. 111) appeared in the 1968 (i.e. 2nd edition) *Bartók sa vie et son oeuvre*, 'Bartók his life and work'. This was written by Bence Szabolcsi with four collaborators —two of whom were Lendvai and Zoltán Kodály [881].[4] In Lendvai's description of how he identified the Golden Sections in his musical analysis, he notes the same two possibilities that we did at the start of this book (Fig. 1 p.3).

---

[4] The said spiral was not present in the 1956 first edition [880]. Both 1st and 2nd editions were in French, and in 1971 Lendvai published a summary of his theories in English, again showing the same spiral [539].

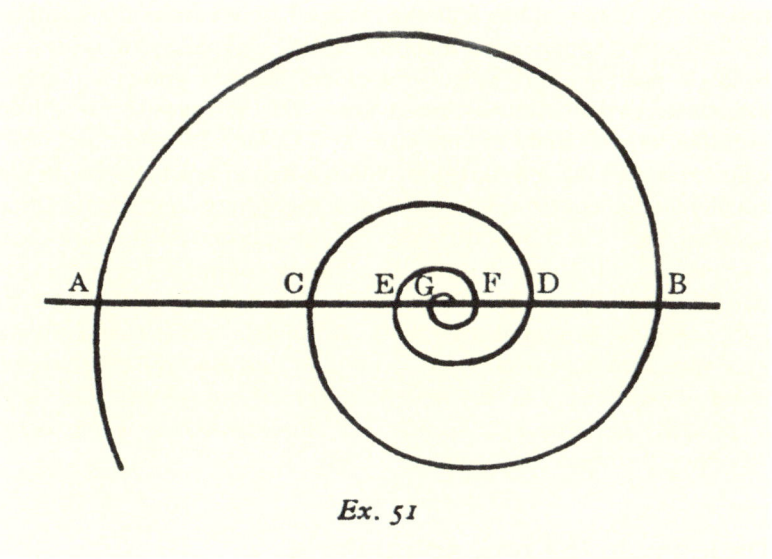

Ex. 51

Figure 111: **Lendvai's original spiral (Root Golden) [881, 539].**
Copyright Hawkes & Son (London) Ltd. Reproduced by permission of
Boosey & Hawkes Music Publishers Ltd.

The first Golden division (our *ABC*—where the smaller section is added) he calls 'positive' and the complementary division (our *C'AB* where the smaller section precedes the original unit) he calls 'negative' [538, 476]. As we shall see, it was Lendvai's noting of the relative positions of the positive and negative sections that helped guide him towards the spiral form. Lendvai's chapter on the Fibonacci numbers ends with a small section about phyllotaxis which he briefly interrupts to consider the *cephalopod nautilus* [536]. He notes Jules Verne's interest in this chambered shell mollusc and how Verne had named his fictional submarine after it.[5] With remarkable insight, he then drew a *Root* Golden Spiral—Fig. 111. He did *not* draw the standard Golden Spiral with its unjustified popular association with the nautilus (which we reviewed back on page 49). He points out that with this (Root Golden) spiral, any diagonal through the centre will always produce Golden Sections. If we call the centre *O* in Fig. 111, then

$$\frac{OA}{OB} = \frac{OB}{OC} = \frac{OC}{OD} = \ldots = \phi.$$

---

[5] Which we too noted in the Seurat section, page 82.

Indeed the Root Golden Spiral gives a far better fit to the nautilus shell's shape (though still not good). Lendvai's goal was to represent Bartók's positive and negative Golden Section points as spiral intersections with the horizontal axis. This is because the music example he was analysing climaxes at a Golden Section point, and when approaching the climax the Golden Ratio phrasing sections get smaller, and in contrast, they are increasingly drawn out after it. Once Lendvai had this qualitative fit, he then proposed a numerical comparison with the structures of his Bartók examples: the *Sonata for two pianos and percussion* [41] and the *MFSPC fugue—1st movement* [40], complete with all the ϕ ratios. And while (with these) Lendvai may appear to have been 'onto something', Roy Howat [395] finds far more convincing fits in the *MFSPC 3rd movement* [40] and (especially) in Debussy's *La Mer 3rd movement* [164] where exact Fibonacci numbers are in evidence [383].

Debussy in the new spiral Paris

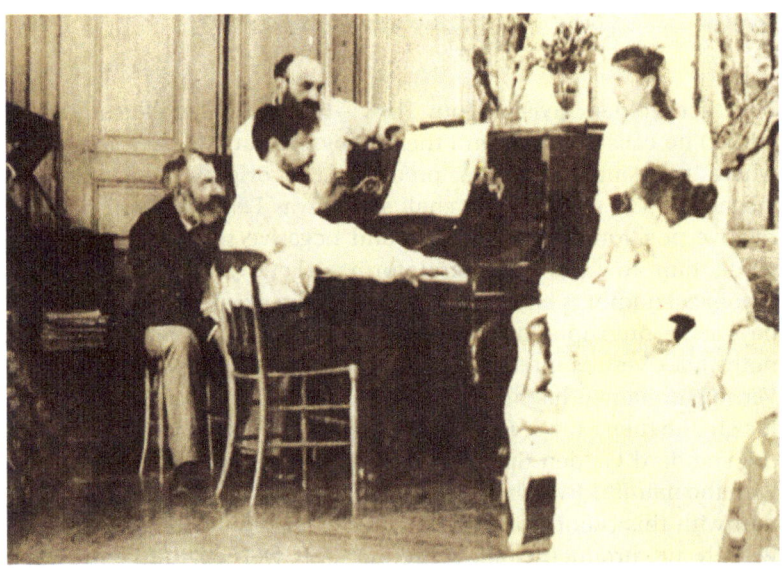

Figure 112: **Claude Debussy (1862–1918).** Image from *Mestres da Música*, Planeta De Agostini—Debussy at the piano *en famille* with the Chaussons at their Luzancy house, summer of 1893 [686].

Claude Debussy (Fig. 112) was a tremendously influential composer. He grew up in Baron Haussmann's newly renovated Paris, with its city

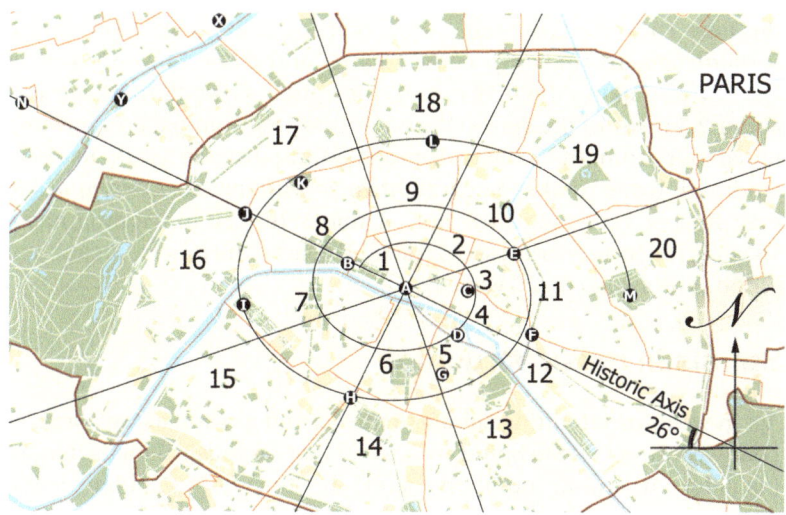

A:	*Le Louvre*	H:	*Tour (Maine-)Montparnasse*
B:	*Place de la Concorde*	I:	*Tour Eiffel*
C:	*Centre Pompidou*	J:	*Arc de Triomphe*
D:	*Notre Dame*	K:	*Parc Monceau*
E:	*Place de la République*	L:	*Sacré-Cœur (Montmartre)*
F:	*Place de la Bastille*	M:	*Père Lachaise Cemetery*
G:	*Panthéon-Sorbonne*	N:	*La Grande Arche, La Défense*
X:	*Asnières-sur-Seine*	Y:	*L'île de la Grande Jatte*

Figure 113: **Paris: Numbered *arrondissements municipales*—spiralling out around the Louvre, with several landmarks.**
Based on a plan by The Promenader [767, 144].

limits expanded in 1860 by Napoleon III and its administrative districts set out in a spiral format—sometimes called *'l'escargot'* (the snail) [388]. In the analysed plan Fig. 113 we see from upper left to lower right, the 'historic axis' that tilts at 26° and upon which Haussmann's whole design rests.[6] The spiral comprises 20 numbered *arrondissements* (municipal administrative areas) each subdivided into 4 quarters. The spiral curve (approximately) fitted and overlaid in Fig. 113 is logarithmic, but with a 22% compression in the north-south direction. Centred on the Louvre, the spiral is 'born' close to the Place de la Concorde, and then turns and grows, finally ending up in

---

[6] Notre Dame is said to be oriented on this same 26° offset.

the Père Lachaise Cemetery—as if charting a lifetime.[7] The first 8 *arrondissements* make up central Paris, and these have $8 \times 4 = 32$ '*quartiers*', making the first turn of the spiral. Overall, there are just about 2.5 turns of the spiral. We mentioned back on page 70 how the radial out from the Louvre through Montparnasse (with its 1973 landmark Tower) makes a near right angle with the *Axe Historique*. We may further note that now buried under the 210m Montparnasse Tower is the former address 26 rue du Départ—the precise location of Mondrian's studio for 25 years [978].

Spirals in music structure

Again we return to fitting spirals to music, and as soon as we consider Lendvai's Root Golden Spiral—we recall that (a) musical measures are (usually) in whole numbers, and (b) the Root Golden Spiral dimensions are based on the 'very irrational' ϕ. So how could the Root Golden Spiral distort in order for us to have whole number intervals between the intersection points of spiral and horizontal axis? Well, (you are ahead of me...)—we have exactly this already in the form of the Binet spiral for the negative-index Fibonacci numbers. Fig. 114 shows the Argand diagram of the spiral, vertically flipped to match with the Lendvai spiral diagram Fig. 111 (and with related 'spiral-proportional' music analysis since). It is a zoomed out view of the spiral part of Fig. 107 on page 135, showing F_0 centre, then: $F_{-1} = +1$, $F_{-2} = -1$, $F_{-3} = +2$, and so on up to $F_{-10} = -55$. Now, it was Lendvai's 'spiral' insight that set us on the road to here. But, as we start to wonder (that is, doubt) if Debussy and or Bartók might have been familiar with complex Binet spirals, we realize that although such spirals are very beautiful, they themselves are not necessary. All we need are the crossing points on the real axis, and these points are *simply* the negative-index Fibonacci numbers—the ones Bartók pointed us to in his xylophone opening that we studied earlier. Starting at F_{-1} and taking more negative indices, the numbers are: +1, -1, +2, -3, +5, -8, +13, -21, +34, -55, ..., as shown in the 'volcano' Fig. 115. We now see that this simple layout is ready-to-use *as is*.[8]

---

[7] This cemetery holds the remains of many famous people: Sarah Bernhardt, Edith Piaf, Oscar Wilde, Frédéric Chopin—and (mentioned in this book), Jane Avril, Joseph Fourier, Sophie Germain, Georges Haussmann, Georges-Pierre Seurat, and Paul Signac.

[8] We just add 55 to each number to get the ideal 'Golden Proportional' bar numbers: ⟨ 0, 34, 47, 52, 54, 55, 56, 57, 60, 68, 89 ⟩ as in the offset diagram on p.475. These compare well with the corresponding bar numbers of Roy Howat's MFSPC III spiral [395].

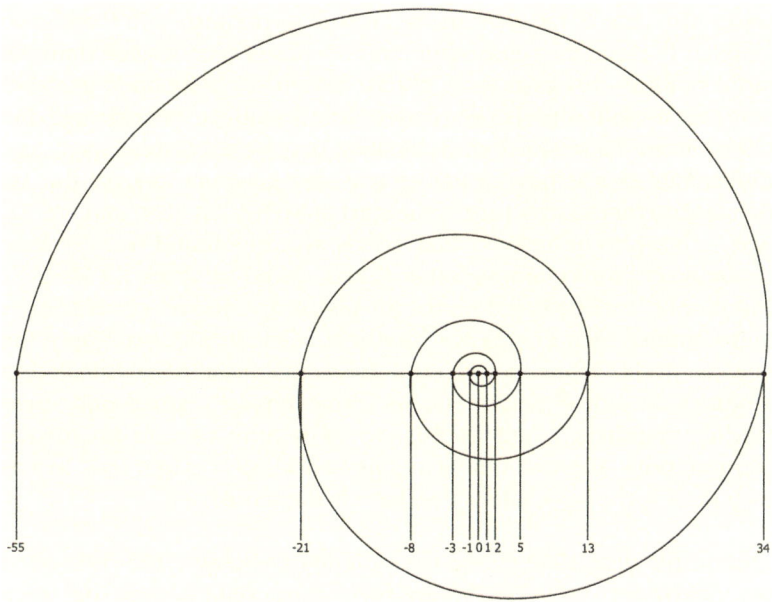

Figure 114: **Binet Spiral: Argand diagram for negative-index F_n.**

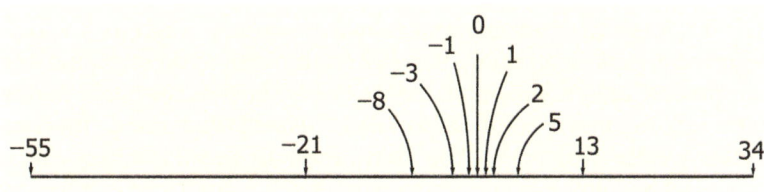

Figure 115: **A 'volcano' of 11 negative-index Fibonacci numbers F_0 to F_{-10}—a way of showing their layout along the (real) number line.**

It is immediately applicable to musical composition—complete with whole unit divisions and overall Golden Section climax. However, if used 'raw' without further subdivision, these Golden-based phrase lengths scale one to the next not by (close to) ϕ but by (close to) ϕ^2. This would result in a slower start, then an accelerated more dramatic build to climax (compared with a scheme where successive phrase lengths were in Golden Ratio). Also, these phrases are perfectly balanced: 'positive' long to short accelerating before the climax and 'negative' short to long, decelerating after. This is the same idea as

145

using *ABC* and *C'AB* sections, but with higher contrast in the phrase lengths. Further, the reintroduction of ϕ-based sections could not be more natural—because each phrase length is a Fibonacci number that may be split into its two predecessor members. For example, the phrase from -55 to -21 (that is, 34 bars) may be subdivided as 21 and 13—indeed this is how both Lendvai and Lowman analyse the 1st theme/2nd theme structure of the start of MFSPC III [394] and Fig. 57, p.64.　　It seems quite likely that spirals, whirlpools and vortices were on Debussy's mind (among other things) when he wrote *La Mer*, but whether he needed spiral maths for his compositional structuring is quite another matter.　As we have just seen, in order to apply the negative-index-Fibonacci layout, the actual minimum calculation required is '+ or −' arithmetic on some typically small and quite familiar numbers.　And as this would require next to no mental computation—there would be no busy workings or paper trail. Just as with Bartók, the key evidence we have is the score itself.

Around the year 1900, composers Claude Debussy (1862–1918) and Maurice Ravel (1875–1937) were for a short time friends, but then (with the help of journalists) they fell out—though this did not diminish their admiration for each other's work. Debussy was also a very great influence on Bartók. Kodály had brought Debussy's music from Paris back to Hungary and in 1907 he introduced these works to Bartók.　Although Debussy and Bartók never met, there is a widely reported anecdote revealing Bartók's strong wish to do so. Critic and composer Virgil Thomson related that when in Paris (in 1909), Bartók declined to meet Saint-Saëns and instead suggested meeting Debussy.　He was told: 'You don't want to meet Debussy, he is antisocial and will just be rude to you; do you want to be insulted by Debussy?' 'Yes,' replied Bartók.　In tracing the influences on Debussy, Howat notes his early involvement with the Symbolists.　This group had twin interests in number; one was the practical—the measuring of sizes and shapes (e.g. in metres or inches), and the other numero-logical—where numbers have intrinsic meanings, associations, and power [391].　In particular Howat proposes the pivotal influence of Charles Henry (mathematician and assistant, then librarian at the Sorbonne) through the publication of his *Introduction à une esthétique scientifique* in 1885 [334]. Howat finds that after this date, Debussy's work included the use of both symmetry and Golden Proportion in 'a thoroughly organized way'—particularly in *Spleen* [384, 165].

Henry was a kind of 'mathematician in residence' for the Paris Symbolists [385]. He had a very wide knowledge of the mathematical aspects of art and music. Howat also mentions Debussy's involvement in esotericism from 1885 onwards [386].

Skip-Fibonacci: Make Lucas

In the Binet Spiral layout (with its negative-index Fibonacci number axis crossing points), we next consider the sizes of the *gaps* between the number points, and see how these 'skip' positions in the Fibonacci sequence. In Fig. 115 the gaps on the left are: 34, 13, 5, 2, and 1 (odd-index F_n), and to the right (even-index): 1, 3, 8, 21. As a result, we find that if we add two adjacent gaps, we produce a Lucas number (in fact, the 'mid-index' Lucas number)—according to the identity $F_{n+2} + F_n = L_{n+1}$ [488]. For example in $34 + 13 = 47$—that is $F_9 + F_7 = L_8$—we see that odd-index Fibonacci numbers give even-index Lucas numbers. So, in a musical composition based on the (Binet spiral) negative-index Fibonacci numbers—Lucas numbers will appear quite naturally without further refinement. In Fig. 116 we see the even-indexed gap combinations played out, taking successive pairs of gaps: $F_2 + F_4 = 1+3 = 4$, $F_4 + F_6 = 3+8 = 11$, and so on. This generates a cascade of skip-Lucas numbers. Each new interval starts half way through the previous one. Interestingly, in the score for the MFSPC I, Bartók has just such a cascade shape in bars 65 to 68—but just the shape itself—the Fibonacci and Lucas numbers themselves have no further role here.

This 'skip-and-combine' property is remarkable as it allows Lucas-based structures to be incorporated into Fibonacci-based ones seamlessly. So to summarize, by using this technique, the already very flexible Fibonacci tool that is 'non-periodic, but highly ordered', becomes even more flexible with the admission of Lucas numbers along with their own possible partitions. For example, the 'F-skip interval' of 5+13=18 can be taken as an 'L interval' 18. And this may then be 'Lucas partitioned' as 11+7, or 7+4+7, or 3+4+4+7, and so on. Now, the appearance of Lucas numbers has the potential to bring a new ratio onto the composer's 'palette'.

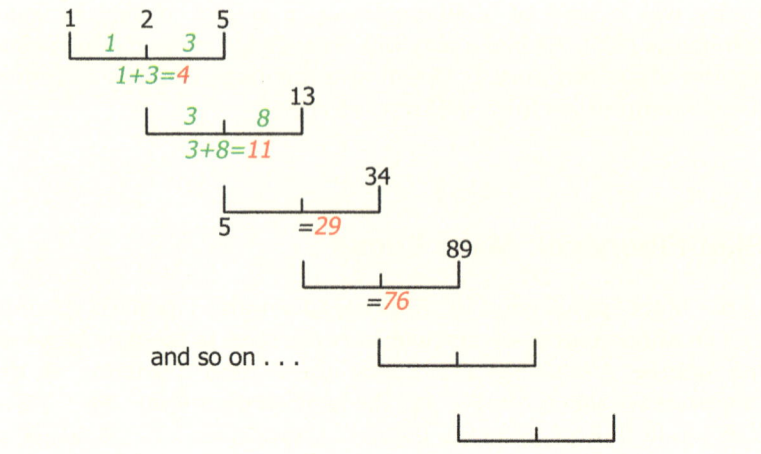

Figure 116: **Skip-Fibonacci Lucas cascade.**

This is because, as we shall see later, successive Lucas numbers are greater than Fibonacci numbers by a factor increasingly close to $\sqrt{5}$. So, for the example we have just seen, (and in round figures), the 18 might be regarded as as being Fibonacci number $F_6 = 8$, grown by $\sqrt{5}$. This effectively adds another engineering technique to the 'toolbox'.

A green Modulor series?

Using this, we might make a simple extension to the Modulor (p.127)—viz. a 'green series'. In this, each member would be $\sqrt{5}\times$ greater than a red series member. So that in any circumstance where two red series measurements (with indices that differ by 2) appear adjacent ('skip Fibonacci'), then one green dimension will fit instead ('make Lucas'). For example, if we take the red series 7747 mm, skip one (12535 mm) and add the next (20282 mm); this gives (green) 28029 mm—which we confirm is $\sqrt{5}\times$ 12535 mm. This new section may then be exactly subdivided using smaller green series members, for example in Golden Section using the two next smaller greens: 17323 + 10706 mm. This approach introduces a different (yet contained) scaling, still keeping within the overall Modulor 'system' and retaining the ϕ growth for each (green) member to its next. Such a local and coordinated scale change might be likened to a key change in music.

Ravel's $\sqrt{5}$

Roy Howat, while studying Ravel's application of Fibonacci numbers and the Golden Ratio, discovered an example of the $\sqrt{5}$ scaling in the *Alborada del gracioso* [390]. The *Alborada* is the fourth movement of Ravel's 1905 *Miroirs* suite for solo piano [774].

What did they know, and when... ?

Following on from the 'Paris' Table (p. 71), if we try to trace a Golden path linking Lucas, Henry, Seurat, Debussy, Ravel, and Bartók, then soon we meet Erik Satie too. Here are several dates and events, with some (clearly marked) conjectures about how they are linked.

?— Conjectures are shown in italics, with enclosing question marks —?

1876	Gustav **Fechner**'s Golden Rectangle experimental results are published in Germany [211].	p.67
1876–1877	Édouard **Lucas** (in Paris) publishes his key works on Fibonacci numbers and Lucas numbers— confirming him as pre-eminent authority [592, 595].	p.27
1879–1882	**Lucas** and Charles **Henry** work together in Paris and Florence [167].[9]	
1881	*Le Chat Noir* artistically élite nightclub opened in Montmartre by Rodolphe Salis [181, 924].	p.153
	?—a clandestine 'den of mathematics' for top artists and composers, cunningly disguised as a cabaret—?	
	Bartók is born in Hungary.	p.140
1884	**Seurat** regularly attends **Henry**'s lecture series at the Sorbonne [186, 373]	
1884–1886	**Seurat** works on his painting *A Sunday afternoon on the Island of la Grande Jatte* structuring it in horizontal Golden and vertical rational proportions.	p.77
1885	*Le Chat Noir* moves to bigger premises.	p.153
	Debussy leaves Paris to spend 2 years in Rome (after winning the Prix de Rome), but keeps up to date with the Parisian intellectual life by letter [387].	
	Henry publishes pamphlet on aesthetics (in Paris) [334]. This and others include discussions of the Golden Section.	p.146

---

[9] On a book: *The complete works of Fermat*. Henry was an expert in the history of mathematics, and was interested in how maths and science related to art.

?— **Henry***'s several publications alert* **Debussy** *to the Fibonacci numbers and* ϕ *(Howat [385]) —?*

1885–1887	**Debussy** composes *Spleen* [165] marking a new style of writing, organized on symmetry and ϕ [384]. **Debussy**'s 'letters from Rome' also reveal his interest in the esoteric (Rosicrucianism) [386].	p.146
1886	**Seurat** meets **Henry** [339].	
	Hoyle publishes his octopus and squid report [988]	p.82
1887–1888	**Seurat** paints *'Parade'—(La Parade du Cirque)* demonstrating proportions: $1, 2, 3, 4, 5, 6, 7, 8, \delta,$ $\sqrt{\phi}, \phi, \phi^2, \phi^3,$ and the Fibonacci numbers: $1, 2, 3, 5, 8, 13$—using Golden geometry from **Henry** [895, 335].	p.79
1891	**Seurat** dies in Paris.	p.79
1892	*'Gymnopédiste'* Erik **Satie** (sometime Rosicrucian and friend of **Debussy** and **Ravel**) publishes *Trois Sonneries de la Rose+Croix* [799] and uses Golden Proportion [263, 264].	
	Toulouse-Lautrec creates poster *Ambassadeurs, Aristide Bruant Dans Son Cabaret* with its 3 Golden Rectangles and 3 [RG+] rectangles.	p.87
	?— **Debussy***, by experimenting with Fibonacci layouts (or hearing sufficient from* **Henry***), discovers the negative-index Fibonacci numbers—?*	p.145
1896	**Steinlen** creates poster for *Le Chat Noir* tour.	p.153
1897	*Le Chat Noir* cabaret closes following death of Salis.	p.153
1900	International Congress of Mathematicians (ICM) is held in Paris. We discuss 'Hilbert's 10th' in Appendix G (p.450).	
1903	**Debussy** in a letter to his publisher, owns up to missing a bar out of the *Estampes* manuscript, but says: 'It is necessary for *the divine number.*' [546].	
1903–1905	**Debussy** writes *La Mer* [164].	p.146
	?— and also discusses Fibonacci and Golden Section with his friend Maurice **Ravel** *—?*	
1904–1905	**Ravel** writes *Miroirs* [774] and uses Fibonacci [389].	p.149
1907	**Bartók** is introduced to works of **Debussy** by Kodály and is strongly influenced by these.	p.146
	?— finds Fibonacci numbers in certain pieces —?	

1909	**Bartók** visits Paris but does not meet **Debussy**. **Bartók** was a capable mathematician interested in nature, Fibonacci and Lucas numbers.	p.146
	?— As such, he would have been aware of the Binet formula. Bartók is therefore able to apply his deeper mathematical understanding —?	
1911	Rue Ravignan in Montmartre (the *Bateau-Lavoir* was at No. 13) renamed as 'place Émile Goudeau'.	p.70 p.152
1911–1917	**Bartók** composes *Bluebeard's Castle* [36].	p.140
	?— in overall 'Binet form' —?	
1913	**Debussy** makes his plea against deconstruction: 'Let us preserve this … magic peculiar to music.' [86, 942].	p.65
1918	**Debussy** dies in Paris.	p.142
1926	**Mondrian's** *Tableau I: Lozenge with four lines and gray* demonstrates his expert use of ϕ [656].	p.92
1927	**Bartók** composes *String Quartet no. 4* [38].	p.140
	Ghyka publishes *Esthétique des proportions* [260].	p.72
	Le Corbusier reads **Ghyka** [564, 342].	p.72
1929	**Mondrian** begins a series which includes *Composition with Red, Blue, and Yellow, 1930* (with its Lucas numbers and ϕ powers) [442, 446].	p.93
1931	**Le Corbusier** completes the *Villa Savoye* [125].	p.125
1934	**Bartók** composes *String Quartet no. 5* [39].	p.140
1936	**Bartók** composes *MFSPC* [40]. Roy Howat guesses, though (as he says), this can only be conjecture[10]	p.62
	*?— that **Bartók** could have modelled his MFSPC 3rd movement spiral form [395] on that used by **Debussy** in 'La Mer 3rd movement: Dialogue du vent et de la mer' [383] —?*	
1937	**Bartók** composes *Sonata for two pianos and percussion* [41].	p.142
1945	**Le Corbusier** begins work on 'The Modulor' [131].	p.127
1947	**Le Corbusier** hires **Xenakis** [34].	p.130
1954	**Xenakis** composes *Metastasis* [992].	p.130

---

[10] Personal communication with author quoted with kind permission.

Le Chat Noir

Debussy and Satie frequented Rodolphe Salis' *Chat Noir* cabaret in Montmartre, and for a period, Satie was employed playing piano there [544]. The great success of the *Chat Noir* followed some inspired decisions by Salis. His 'Black Cat' name recalled Poe's short story [754] and hinted at night-life, mystery, and independence. Salis also hired a friend of Aristide Bruant and Adolphe Willette—Swiss artist Théophile Steinlen (who understood cats)—to design his now iconic logo Fig. 117 [379].[11] Salis chose Louis XIII furnishings—eccentric bric-a-brac and heavy furniture; he also promoted paintings by young artists. For his opening night on 18.11.1881 (5×1, 3×8), he invited Émile Goudeau (*'goût d'eau'*), founder and president of a substantial Latin Quarter literary club—*'les Hydropathes'*. With their cat-like avoidance of water, they preferred wine, beer, and absinthe. And Goudeau *brought them all with him—the whole club.* They paraded from the Left Bank over the Seine on up to Montmartre, their new home ('G' to 'L' in p.143 map). Salis fostered chaotic 'open stage' spontaneity, parody, disrespect, and derision—insulting his clients and introducing acts with mock politeness—acts he paid in drinks alone. His rude waiters (uniformed à la *Académie Française*) served bocks of beer [181].[12] Also, Henri Rivière's shadow plays (from 1886) were a key attraction, running 11 years. And in parallel, the *Chat Noir* journal ran for over 10 years—a 4-page tabloid with articles and satirical drawings [924]. But let's now analyse Steinlen's poster, which seems to be based on a central (landscape) Golden Rectangle, with ϕ^2 rectangles above and below it. The ϕ^2 rectangle at the top contains the moon/aura behind the cat's head. And because the width is ϕ then the upper area of the poster (light background) makes a $\phi \times \phi$ square. The cat then sits on a ϕ^2 rectangle at the bottom of the design, (reminiscent of Mondrian's ϕ^2 rectangle, Fig. 77 on p.99, which came decades later). Perhaps also, the curl of the tail is based on a Golden Spiral (green overlaid),[13] with its centreline used to align nearby text. Horizontally, the design is further subdivided by a vertical Golden Section line which separates the cat from the lettering. This now produces a 'Golden chequerboard' of 3 Golden Rectangles and 3 squares. Overall vertically, we see both positive and negative Golden Sections (to use Lendvai's terms)—that is, the *ABC* and *C'AB* which we met at the very beginning (page 3). Also, in these, we associated *C*

---

11 Théophile Steinlen was accepted into the artists' community in Montmartre.

12 Under Louis XIII (1601–1643), Richelieu founded the *Académie Française* in 1635.

13 This is the page 3 Golden Spiral, rotated 180°.

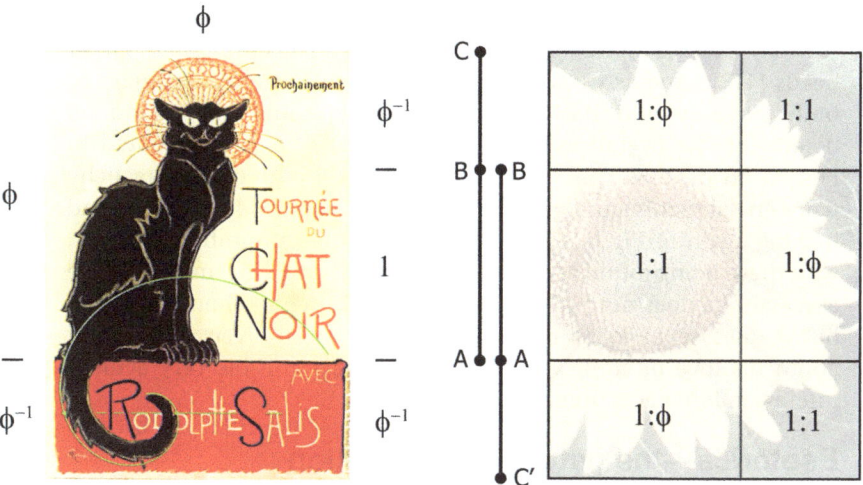

Figure 117: **Théophile Steinlen (1859–1923)**
***Chat Noir* (tour) poster, 1896**—spiral added.
Colour lithograph on paper, 140.8 × 100 cm.
Bibliothèque nationale de France [869].

(Sunflower for geometric
comparison, from p.60.)

with the root α, and C' with its conjugate β—and here the extent of the height is from C' up to C. Hence, this height represents the distance between the roots, $(\alpha - \beta) = \sqrt{5}$. We also remember seeing this both-roots Golden Sectioning of the sunflower on page 60.

Bartók's sunflowers and pine cones

Lendvai reports Bartók writing: 'We follow nature in composition.' He also recalls Bartók's love of sunflowers and pine cones, and his conviction that folk music should be considered alongside these—as an equally natural phenomenon [37]. Now that we have studied both negative-index Fibonacci numbers (pages 29 and 145) and phyllotactic parastichies (page 52)… we are fully equipped to consider possible underlying reasons why numbers such as 5 and 8, 55 and 89 (or 88) were chosen in the structures that we have been looking at—be they considered simply in terms of a row of negative-index numbers (blips on a line), or instead, in their full spiral splendour.   Back in Chapter 3, (for simplicity) we looked at an idealized model of the sunflower head that showed Fibonacci sets of: 21, 34, and 55 parastichies (Fig. 52 on page 57). In the real world however, sunflower parastichy pairs range from (13,21) up to the (89,144) pair [416], and the pair (55, 89) is common in larger heads [419]. And as we see in Fig. 118 (which relates to examples of both Debussy and Bartók), the Fibonacci numbers 55 and 89 determine the overall musical structure. Further, as illustrated in Fig. 50 on page 56, (5, 8) is an example of a pine cone parastichy pair—in fact, this configuration is very common [418, 496]. And looking again at Fig. 118, we have the Fibonacci numbers 5 and 8, with these being found at the core of the structure. Now making these associations does not prove that Bartók had such a scheme in mind, but it does seem to be more than coincidence—given all we hear about his love of sunflowers and pine cones, and his very evident interest in Fibonacci numbers.

Esoterica—the nature of completion

By refusing to discuss his methods though, Bartók invites conjecture. For a start, could it be that the 88–89 discrepancy (which we saw on page 62), is deliberate? In his 2009 talk *Symmetry, reality's riddle* [801], mathematician Marcus du Sautoy recalls being taken by his Japanese colleague Nobushige Kurokawa up to the temples of Nikkō. There they discussed the fact that, in an otherwise regular row of columns, one column had been set upside down. Professor Kurokawa explained this by quoting a 14th century Japanese essayist (Yoshida Kenkō) who wrote that: 'Leaving something incomplete makes it interesting, and gives one the feeling that there is room for growth.' [465, 464]. And then again, what about the identification of the sunflower and the pine cone in the Debussy/Bartók Binet spiral scheme? Let's try some free association… Might the sunflower represent the sun itself? And does the sun in turn symbolize *the* self? For many aspirants, self-realization is the *completion* of a human being, wholeness, perfection, the Philosopher's Stone [456]. But completion may well not be simple and final—in nature it is typically cyclic and evolutionary. Endings double as next beginnings in a round of rebirth and renewal. On pollination, both sunflowers and pine cones become seed carriers. And in the spiral in Fig. 118, if we do associate the central (5, 8) with a pine cone, and we muse on those over-arching spiral curves… the resulting picture (in our mind's eye) might then remind us of the pine-cone-shaped *pineal body* at the core of the brain [268]). This tiny gland enjoys vast esoteric importance—it was at one time considered by Descartes to be the very 'seat of the soul' [573]. We find it located at about the Golden Section point (front-to-back) in the brain, as shown.

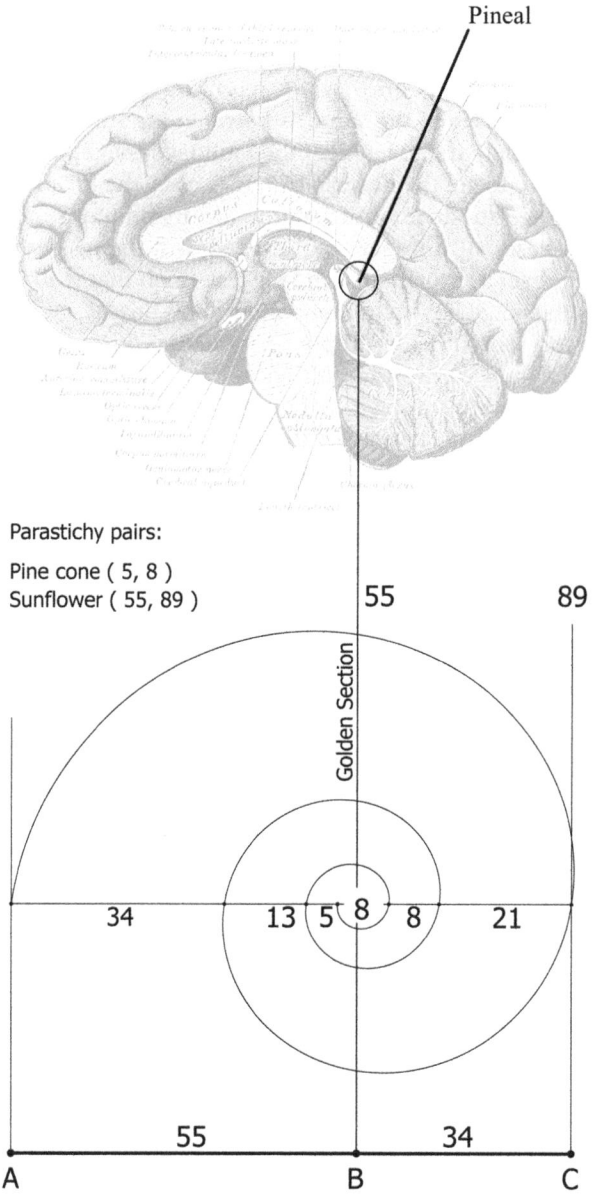

Figure 118: **Bartók's work was criticized as being over-cerebral** [396, 883, 783]. The spiral element here is a simplified version of a diagram by Roy Howat (relating to Bartók [395] but also relevant to Debussy [383]), adapted and shown with kind permission. (This permission to show the spiral should not be construed as endorsement of the overall composition.)

155

Conclusion: ϕ's potential is (relatively) untapped

With *ABC* on page 3 we saw how the Golden Ratio arises from adding two lengths to get a third, while requiring the proportion of the greater to the lesser be the same as between the sum and the greater. This is the underlying essence of the Fibonacci sequence: adding the preceding two to get the next. In the *ABC* case, the (irrational) ϕ ratio is exact, stage to stage, and lengths are incommensurate. In the Fibonacci case, all numbers are integer, and the 'ϕ' ratio (stage to stage) is forever approximate. The biggest 'myth that won't go away' is that the first stage has standalone magic and beauty. Accordingly we may be presented with empty rectangles with sides in Golden Ratio—and then we wonder: 'If so many people regard this as the pinnacle of aesthetic appeal—why can *I* not see it?' But by now, we know the real magic comes in the combination of ϕ-based shapes (e.g. rectangles)—with their almost unreasonable composability. Charles Henry apparently understood this power of ϕ, and (through his lectures, writings, and discussions) he helped Parisian artists and musicians apply this knowledge in their work. When Le Corbusier 'got it', it was something of an epiphany for him; he wrote excitedly of: 'an infinity of combinations . . . unity within diversity, an inestimable boon, the miracle of numbers', and he published pages and pages of reconfigured panels to prove the point [135]. We conclude that a ϕ-based layout has major aesthetic potential because of:

- how neatly it all fits together—*closely followed by*
- how the same (not unpleasant) proportions are harmoniously echoed over and over again—*and*
- how these together generate great orderliness—*and*
- how the (potentially boring) 'simple repetition of equal lengths' is avoided

Surely it is not that a particular ratio gives 'exquisite delight' as is. If this were the case, we should know for certain by now. Rather, it is that this very special ratio delivers all the above visual (and architectural, and engineering) benefits. Further, the fact that ϕ can do this is not a matter of opinion—it is instead an eternal mathematical truth. When Le Corbusier showed the ϕ-based Modulor to Einstein, Einstein instantly 'got it', and remarked on its power to: 'make the bad difficult, and the good easy' [128]. Inspired by Le Corbusier, architect Richard Padovan also readily 'got it' himself, but he notes how, at a 1957 meeting of architects in London (the RIBA) a motion that: 'Systems of Proportion make good design easier and bad design more difficult.' was defeated by 60 votes to 48 [719].

Part II

LEADING TO A DISCOVERY

Chapter 10

Introducing Ori32 geometry

(This relates to 32 orientations and so is pronounced 'orree-thirty-two.')

In their case study, Brillhart and Morton speak of: *'the standard and time-honoured practice in mathematics to erase all hint of the development of a subject or proof… '* [and note that:] *'Very seldom in textbooks or research papers is there a hint of the original questions that motivated the researchers.'* [85].[1]

They then break with such tradition, and we shall follow their example —by setting background contexts and disclosing the detailed thinking.

Rock surfaces?

This book begins a series whose first titles will focus on geometric analysis. Further books will then look back over what has been a long-term personal project. It started in 2005, originally as a study in light—photographic research. In the later books, we shall explore a sample of the many interesting images that together led on to the geometry. It was in 2005 that I became aware of various state-of-the-art developments in DSLR (digital single-lens reflex) camera technology. In particular, one leap forward was a vast increase in colour resolution ('bit depth'). I saw this almost universally unsung aspect of the advanced sensor design as opening up a new frontier for research in natural science. Over a seven-year period I developed digital-photographic techniques and custom software to explore this new territory. The first subjects were rock surfaces (photo p.166), but the scope widened rapidly to take in many kinds of natural subjects. During that work, it was very surprising to find that many of the

---

results *appeared* to depend on an angle-based geometry which (as will be discussed shortly) I came to call 'Ori32'. It was the study of this geometry that provided a clear path to the discovery of the Fibonacci Resonance, which adds a new understanding of the interrelationships that unify the Fibonacci numbers, Lucas numbers, Golden Spirals, and the Golden Ratio.

Cryptochromatology (CC): finding 32 orientations

In the analysis of over ten thousand photo images, certain (supposedly natural) motifs showed up again and again, and these recognisable shapes seemed (for no apparent reason) to favour orientations based on thirty-seconds of a turn (that is, $360°/32$ $= 11.25°$)—relative to the horizontal or to some reference edge. This then raised the question as to whether the effect was:

- objective—some physical attribute (e.g. of a rock), or
- interactive—some quality that stimulated a psychological human-perception effect, or
- independent—completely imagined—little or nothing to do with the object and everything to do with the observer, or
- something else?

The results and possible interpretations will be discussed in future books. Now, because such work involves revealing patterns otherwise hidden in subtle colour variations, we shall name the new field 'cryptochromatology' (informally: 'CC-studies')—which shall include 'cryptochromatography' for the image processing methods, and 'cryptochromatoscope' for any suitable imaging device—thence 'cryptochromatoscopy' for 'scope use.[2]

Geometry

But whether real or imagined, the angle relationships based on a thirty-second of a turn became increasingly absorbing. In the photo (Fig. 119), we clearly see the Ori32 angles—but we do not know if (Serbian genius) Nikola Tesla had any particular interest in them. (Tesla invented the 3-phase AC electrical power distribution now standard across the globe.) In this chapter we make the briefest introduction to Ori32—a minimal appreciation sufficient to carry us forward towards the *Penny drops* discovery chapter.

---

[2]   *kryptós,* Greek (κρυπτός)—'secret or hidden'; *khrōma,* Greek (χρῶμα)—'colour'; *-logía,* Greek (λογία)—'study of' (art, science);   *-graphía,* Greek (γραφία)—'writing'; *-skopeîn,* Greek (σκοπεῖν)—'to look' or 'see'.

Figure 119: **Nikola Tesla (1856–1943).**

First we note that, in general, much simple geometry is about positions and lines and lengths; and angles then appear only as by-products. However, in Ori32 geometry, the angles are first class, and lengths may not even be specified. The main interest is in simple shapes— how they relate and fit with each other. Now, while pen and paper (and especially the backs of envelopes) were ideal for developing such ideas—the available standard angle notations did present a serious handicap. Surely notation is meant to help not hinder. The choice was between awkward degree values, measured to the quarter (for example $213.75°$) and sundry multiples of $\pi/16$ radians—(e.g. $19\pi/16$)— and both these complicate and obscure the basic '32nd-of-a-turn' simplicity.

Navigation

There does already exist a 1/32nd-turn unit, traditionally used in navigation, and known as a 'point'. This results from dividing the north, south, east, west compass into 32 directions—a system which goes all the way back to the ancient Polynesian voyagers [894]. However the scope for confusion in a geometry with two kinds of 'point' does not bear thinking about. The mantra says: *'Introduce new terminology or notation only when it is clearly needed.'* But in this case, I did sense a clear need. Reluctantly therefore, I coined the name 'MIK' (pronounced 'Mick' with plural the same—2 MIK).[3] Also, I found that noting the number of MIK in a square box helped greatly. (This box style is already used to mark temperatures on some weather forecasts.)[4] To take an example, in Ori32, $\boxed{14}$ (14 MIK) simply denotes 14/32nds of a turn anticlockwise.[5] Hence a right-angle, being a quarter turn, will be 8 MIK, written $\boxed{8}$.

A name for the geometry

The name Ori32 is chosen simply to reflect the concern with orientation and the even division of a full turn by 32. The form naturally suggests other subdivisions, for example Ori8 and Ori64.

The four Fibons

(Fibon is pronounced just as in 'Fibonacci'—that is, *'Fibbon'*.) In a triangle, the internal angles sum to 180° (in plane Euclidean geometry). Measured in MIK this total will be $\boxed{16}$. Therefore it follows that there are only 4 right triangles in this geometry (not counting flips)—the smallest angle must be: $\boxed{1}$, $\boxed{2}$, $\boxed{3}$, or $\boxed{4}$. In Ori32 these four right triangles are called 'Fibons', and they are distinguished according to their smallest vertex angle—for example the Fibon1 has the smallest vertex $\boxed{1}$ (Fig. 120). Fibon2 is half of a Silver Rectangle—so it has legs in the Silver Ratio, $1 : (1 + \sqrt{2})$. Fibon3 is the 'Fibonacci 3 5 8', and Fibon4 is the 45° isosceles triangle found in every school geometry set.

---

[3] A sound chosen to be suggestive of a vertex tapering to a point.

[4] A (largish) boxed number is very easy to render in the LaTeX typesetting system, for example with just \fbox{3} (but for more compact rendering, see p.482). Where boxes are not available, square brackets may offer a workable alternative.

[5] We use the maths-standard turning direction, not its seafaring/geographic/bearing opposite.

Fibon1

has vertices ⬜1, ⬜7, and ⬜8. It is a thin wedge.

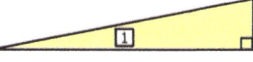

Fibon2

has vertices ⬜2, ⬜6, and ⬜8, and is otherwise known as the Silver triangle.

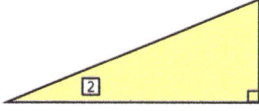

Fibon3

is interesting as its vertices are ⬜3, ⬜5, and ⬜8—a Fibonacci sequence.

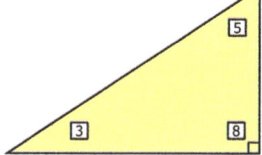

Fibon4

is isosceles, with vertices ⬜4, ⬜4, and ⬜8. It is half of a square.

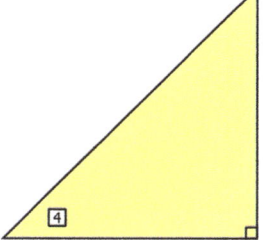

Figure 120: **Ori32—the four Fibons.**

A near-Golden section

As an introductory example of working in Ori32, we shall now look at a 'near-Golden' construction which is followed by a numerically dimensioned version named *The Nine and* ϕ. As we were saying, Ori32 is mainly about shapes, angular relationships, and proportions, and here we have a construction that approximates the Golden Ratio to better than one fifth of a percent (Fig. 121).

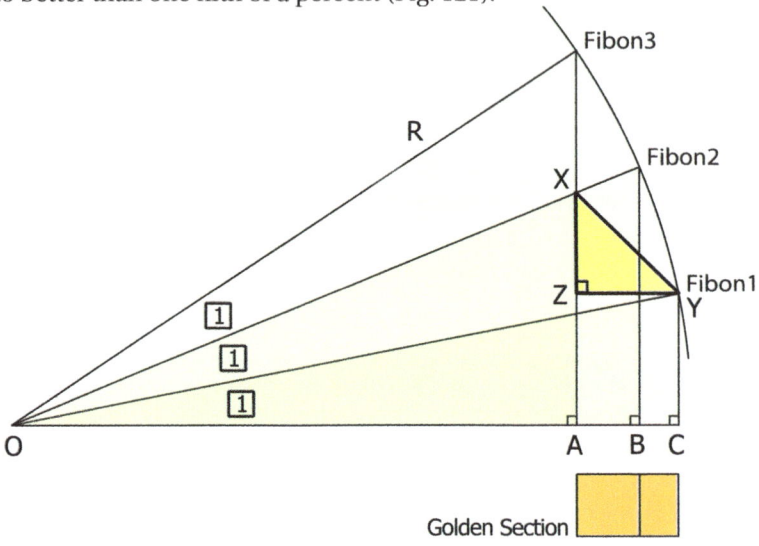

Figure 121: **Ori32—near-Golden section**

We find that B falls between A and C at a location remarkably close to the Golden Section for that interval (shown for comparison). The difference of 0.14% is not visible on this diagram. $\triangle XYZ$ is Fibon4—(related exercises follow at the end of the chapter).

$$AC = R\cos\boxed{1} - R\cos\boxed{3}$$
$$AB = R\cos\boxed{2} - R\cos\boxed{3}.$$

When we take the ratio of $AC : AB$, we lose the radius factor R. Also to evaluate, we recall that $\boxed{1} = 11.25°$, $\boxed{2} = 22.50°$, and so on. Hence

$$\frac{AC}{AB} = \frac{\cos\boxed{1} - \cos\boxed{3}}{\cos\boxed{2} - \cos\boxed{3}} = 1.6158... = \phi - 0.14\%.$$

164

The Nine and ϕ

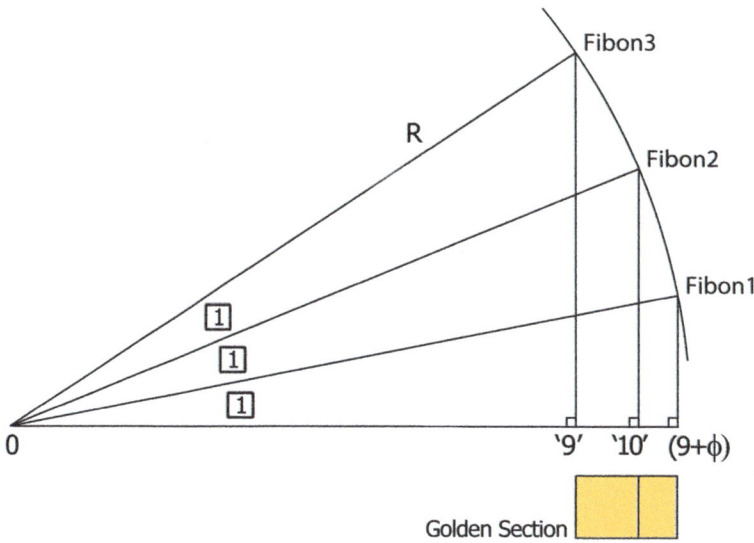

Figure 122: **Ori32—The Nine and ϕ.**

Having looked at the general case (Fig. 121), we shall now look at one specific instance (Fig. 122, and for background as to 'why' see p.455). Here we set the base of the Fibon1 to be $(9 + \phi)$. We then calculate the length of its hypotenuse—which is also the hypotenuse length of both other Fibons. Using this, we calculate the Fibon2 and Fibon3 bases—and we are surprised by how near these are to 9 and 10.

We start with the Fibon1 base

$$(\mathbf{9 + \phi}) = 9 + \frac{(1 + \sqrt{5})}{2} = 10.61803399\ldots$$

From this we deduce the Fibon1 hypotenuse, (which is the arc radius),

$$R = (9 + \phi)/\cos \boxed{1} = 10.82605357\ldots$$

And using R we can simply calculate the Fibon2 and Fibon3 bases,

$$\text{Fibon2, } \textbf{'10'} \text{ base} = R \cos \boxed{2} = \mathbf{10.0020}\ldots$$

$$\text{Fibon3, } \textbf{'9'} \text{ base} = R \cos \boxed{3} = \mathbf{9.0015}\ldots$$

A shoot

Fig. 123 shows an example photo-shoot on the River Goyt, upstream from Roman Lakes near Marple, Cheshire, England.

Figure 123: **The author photographing a weathered rock surface.**
© 2011 Trevor Palin, Palin Images [730].

Exercises

For those now keen to try some Ori32 for themselves, here are a few exercises. Appendix A has a reference summary of Ori32 trigonometry on pp.401–403. Example solutions are offered on pp.406–408.

Exercise 1. In Fig. 121, p.164. find a near-Golden section amongst the vertical lines on *A*, *B*, and *C* (in the same way that we did horizontally in Fig. 121), and show the actual ratio to be 1.613.

Exercise 2. In Fig. 121, p.164, prove that XYZ is exactly Fibon4 (i.e. $ZX = ZY$) by considering angles alone.

Exercise 3. In Fig. 121, p.164, prove that XYZ is exactly Fibon4 by using sines and cosines (Table on page 402).

Chapter 11

The Ori32 Fibonacci circle

An early observation in Ori32 is that $\boxed{5}$ is close to one radian (57.3°) —actually it is 1.8% less. So, in a circle radius R, a sector with angle $\boxed{5}$ will have an arc length of close to R. This basic fact will be very important to us both here and in the next chapter. But to start with, let us think of a circle as comprising four quadrants, and in one of these, we may partition its $\boxed{8}$ right-angle as $\boxed{5}$ (the near-radian sector) and $\boxed{3}$. This split is strongly related to the 3 5 8 triangle we just met—the Fibon3. We only need 'fold in' that triangle's $\boxed{3}$ and $\boxed{5}$ vertices to coincide with the $\boxed{8}$ vertex in order to get the same '3+5 makes 8' arrangement of angles comprising the right angle (Fig. 124). This folding property holds for all right triangles, but here our interest is in the Fibon3.

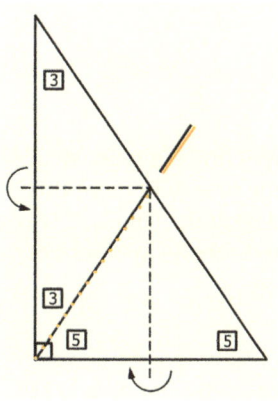

Figure 124: **The (near) Golden Section of a right angle—by folding a Fibon3.**

We recall the Golden Angle from page 52 as being the Golden Section of 360° split into 137.5° and 222.5°. Here we see a $\boxed{3}$ + $\boxed{5}$ split approximating the Golden Section of 90°. Dividing 137.5° by 4, we get 34.377°, while $\boxed{3}$ = 33.75°—a difference of just less than 0.7°. In the diagram the true Golden Section (at a quarter Golden Angle) is shown in pale dots. (Roger Herz-Fischler shares Ozenfant's stated opinion that in art: 'The Golden Number cannot be distinguished in practice from the rational number 5/8.' [363].)

167

As we know well by now, the reason is that $5/8$ is a convergent of $1/\phi$, and therefore such a partitioning will be a close approximation to the Golden Section.[1] Then in Fig. 125, we see that the next quadrant will span $\boxed{8}$, and by also starting with $\boxed{1}$ and $\boxed{2}$, and then by finishing with $\boxed{13}$—we have a perfect Fibonacci sequence that exactly fits. The Fibonacci numbers: $1, 2, 3, 5, 8, 13$, give the angles of the sectors as exact multiples of $(2\pi)/32$ radians. (The circle itself is added to the diagram to show a full turn—there are no curved lines in Ori32 geometry as such.)

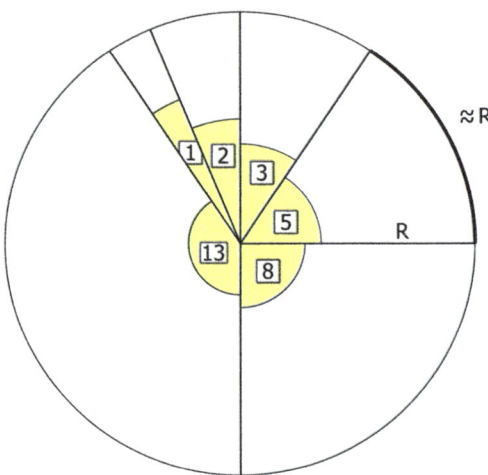

Figure 125: **The Ori32 Fibonacci circle.**

The eternal circle

Looking again at this division of the circle, we might find it has an 'air of rightness' about it. It might even suggest to us the possibility that this growth pattern could carry on, perhaps even indefinitely. That is, if we continue in the same way, adding Fibonacci-sized sectors in blocks of 6, then do we get back to our start angle each time?

---

[1]   (And this really is a footnote…) Looking back at the photo of Nikola Tesla on page 161, we see how the tip of Tesla's shoe is joined by a clear vertical line up to the 3rd radial which marks the 'Golden' $\boxed{3}+\boxed{5}$ split. This is most likely a coincidence, but once noticed, the hint of deliberateness only gets stronger with every viewing.

Let's first construct an expression for this, and then consider it. In our analysis, we shall increasingly refer to Fibonacci numbers by their indices, for example as F_7 rather than simply '13'. We may then summarize Fig. 125 both in terms of MIK angles and in the corresponding Fibonacci numbers. We shall work in groups of 6, and we shall call such sets of 6 Fibonacci numbers 'hexads'. Also we shall use simple 'modular' arithmetic: obtaining remainders after division by 32. Here is the first hexad: [2]

$$\boxed{1} \;+\; \boxed{2} \;+\; \boxed{3} \;+\; \boxed{5} \;+\; \boxed{8} \;+\; \boxed{13} \;=\; \boxed{32}$$

$$F_2 \;+\; F_3 \;+\; F_4 \;+\; F_5 \;+\; F_6 \;+\; F_7 \;=\; 32$$

$$\equiv 0 \quad (\text{mod } 32).$$

To follow this, we conjecture that the sum of the next hexad of Fibonacci numbers will also be divisible by 32, remainder zero:

$$\boxed{21} \;+\; \boxed{34} \;+\; \boxed{55} \;+\; \boxed{89} \;+\; \boxed{144} \;+\; \boxed{233} \;=\; \boxed{576}$$

$$F_8 \;+\; F_9 \;+\; F_{10} \;+\; F_{11} \;+\; F_{12} \;+\; F_{13} \;=\; 576$$

$$\equiv 0 \quad (\text{mod } 32)$$

And finding that it is, we then consider the next hexad, and so on. Also, we may state the relationships more compactly by using summation notation:

$$F_2 + F_3 + F_4 + F_5 + F_6 + F_7 \;=\; \sum_{n=2}^{7} F_n \;\equiv\; 0 \quad (\text{mod } 32) \qquad (11.1)$$

$$F_8 + F_9 + F_{10} + F_{11} + F_{12} + F_{13} \;=\; \sum_{n=8}^{13} F_n \;\equiv\; 0 \quad (\text{mod } 32). \qquad (11.2)$$

If we consider the property expressed in (11.1) and (11.2) as applying to all such hexads, then we may express this conjecture as

$$\sum_{n=2+6k}^{7+6k} F_n \equiv 0 \quad (\text{mod } 32), \qquad k = 0, 1, 2, 3, \ldots \qquad (11.3)$$

where incrementing k by one chooses the next block of six Fibonacci numbers to be summed. Here $k = 0$ will sum the first hexad. We shall see examples of these relationships tabulated in the next section, thinking again of the Fibonacci numbers as amounts of turning— their angles measured in whole numbers of MIK.

---

[2] Mod 32 example: '64 $\equiv$ 0 (mod 32)' is read: '64 is congruent with zero, mod 32.'

Multiple orbits always complete exactly

In Fig. 126 we see a Fibonacci sequence that starts immediately after the seeds $F_0 = 0$ and $F_1 = 1$—i.e. it starts at $F_2 = 1$. We call one complete turn of $\boxed{32}$ an orbit, and we calculate the number of orbits simply as $\text{MIK}/\boxed{32}$—for example $\boxed{10{,}336}/\boxed{32} = 323$. We find that *the result is always a whole number.* Also in passing—it is almost certain that Kepler introduced the term 'orbit' [700].

* * * one hexad per row * * *						MIK for hexad	Orbits for hexad
1	2	3	5	8	13	$\boxed{32}$	1
21	34	55	89	144	233	$\boxed{576}$	18
377	610	987	1597	2584	4181	$\boxed{10{,}336}$	323
6765	...	and so on.					

Figure 126: **Multiple orbits.**

Residue 1

Instead of looking at individual hexads, we may start with the seeds $F_1 = 1$ and $F_0 = 0$—which together will add to 1 (mod 32), and then add in successive hexads—which will each contribute 0 (mod 32) to the result. That is, no matter how big k becomes, then we shall expect that the overall remainder (mod 32) will remain as 1. Such remainders are called 'residues'.

$$\sum_{n=0}^{1+6k} F_n \equiv 1 \quad (\text{mod } 32), \quad k = 0, 1, 2, 3, \ldots \quad (11.4)$$

Now, rather than ponder the validity of this conjectured equivalence at any length, we should expect it to be very easy to check—first by spreadsheet and then using a small computer program. And when we go ahead with this, we find that our guess does seem to be correct (for example in the Appendix B Tables, Figs. 253 and 254, starting page 414). Now confident that probably this *is* a worthwhile line of thought, if we proceed with an inductive proof (full details of which are given on pages 409–413), then we find our conjecture can be

170

confirmed analytically. To reiterate—starting from F_2, we find that by adding successive blocks of 6 Fibonacci numbers, each F_n taken as being a number of units of $\boxed{1}$ of turning; then every time we add the next block, we will always come back to the same orientation—over and over again 'like clockwork', regardless of the rapidly growing number sizes involved.

Residue 2

Now, (11.4) is a cumulative summation, but can it tell us anything about individual Fibonacci numbers? To explore this, we use the standard identity for partial sums of the Fibonacci sequence (11.5) below [490]

$$\sum_{i=0}^{m} F_i \;=\; F_{m+2} - 1 \tag{11.5}$$

which we rearrange as

$$F_{m+2} \;=\; \sum_{i=0}^{m} F_i + 1.$$

We now replace m with $1 + 6k$, so

$$F_{3+6k} \;=\; \sum_{n=0}^{1+6k} F_n + 1, \tag{11.6}$$

and combine this (11.6) with (11.4), to get

$$F_{3+6k} \;\equiv\; 2 \pmod{32}, \quad k = 0, 1, 2, 3, \ldots \tag{11.7}$$

Such residue twos (with period 6) are illustrated in Appendix B—Figs. 255 and 256, page 417.

Even-indexed triads

After a little more exploration and analysis, we may venture to pair off sectors in the 'standard' Fibonacci circle Fig. 125. And as ever— being in the context of the Fibonacci recurrence—if we combine adjacent numbers we get the next number in sequence. So when we combine $\boxed{1}$ and $\boxed{2}$ we get $\boxed{3}$—(i.e. F_4), and $\boxed{3}$ and $\boxed{5}$ give $\boxed{8}$—(i.e. F_6), and $\boxed{8}$ and $\boxed{13}$ give $\boxed{21}$—(i.e. F_8) as in Fig. 127.

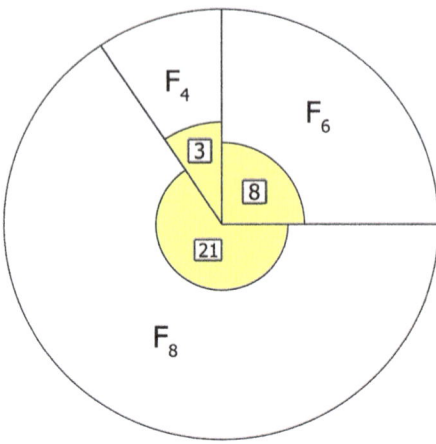

Figure 127: **Even-index version of the Ori32 Fibonacci circle.**

So, we now have an initial triad of even-index Fibonacci numbers. Further, we notice that $F_4 + F_6 = 11 = L_5$, so

$$L_5 + F_8 = 32 \qquad \text{(with Fibonacci indices)} \qquad (11.8)$$

which is one full turn in this context. Alternatively we could group $F_6 + F_8 = 29 = L_7$, so

$$L_7 + F_4 = 32 \qquad \text{(with Lucas indices).} \qquad (11.9)$$

Overall, we now have a simpler scheme than the hexad one, with nothing lost. To complete a whole number of orbits, we only need add the next 3 even-index Fibonacci numbers as MIK rotations, (F_{10}, F_{12}, F_{14}), and we get back to the same point. We may then add the next 3 and the next... and still get back to the starting angle. We therefore write (and prove on page 419)

$$\boxed{F_{6k-2} + F_{6k} + F_{6k+2} \equiv 0 \pmod{32}, \qquad k = 1,2,3,\dots} \qquad (11.10)$$

Finally we may quickly generalize the consecutive pairing that we did to get (11.8). To do this, in (11.10) we let $n = 6k - 2$,

$$F_n + F_{n+2} + F_{n+4} \equiv 0 \pmod{32}$$

172

and combine the first 2 terms with identity $F_n + F_{n+2} = L_{n+1}$ [488]—

$$L_{n+1} + F_{n+4} \equiv 0 \pmod{32}.$$

So,

$$\boxed{L_{6k-1} + F_{6k+2} \equiv 0 \pmod{32}, \quad k = 1, 2, 3, \ldots} \tag{11.11}$$

(noting $6k$ here has -1 and $+2$). And similarly, as we did with (11.9), we can make the alternative consecutive pairing. Therefore using the same identity, but by moving indices two onwards $F_{n+2} + F_{n+4} = L_{n+3}$ [488], we get

$$L_{n+3} + F_n \equiv 0 \pmod{32}.$$

So,

$$\boxed{L_{6k+1} + F_{6k-2} \equiv 0 \pmod{32}, \quad k = 1, 2, 3, \ldots} \tag{11.12}$$

(and $6k$ here has $+1$ and -2). So, to summarize (11.10), (11.11), and (11.12)—we have 3 equivalent expressions of the same relationship:

$$\left. \begin{array}{l} F_{6k-2} + F_{6k} + F_{6k+2} \equiv 0 \pmod{32} \\ L_{6k-1} + F_{6k+2} \equiv 0 \pmod{32} \\ L_{6k+1} + F_{6k-2} \equiv 0 \pmod{32} \end{array} \right\} \quad k = 1, 2, 3, \ldots \tag{11.13}$$

Finally, we may add the last two forms in (11.13) using the identities

$$L_{n+1} + L_{n-1} = 5F_n \quad [486] \text{ and}$$
$$F_{n+2} + F_{n-2} = 3F_n \quad [485]$$

to give us the (simple Fibonacci) sum $5 + 3 = 8$:

$$8F_{6k} \equiv 0 \pmod{32} \quad k = 1, 2, 3, \ldots \tag{11.14}$$

$$F_{6k} \equiv 0 \pmod{4} \quad k = 1, 2, 3, \ldots \tag{11.15}$$

Relaxing the restriction on k

Although we have arrived at these conclusions with $k = 1, 2, 3, \ldots$ (by starting with the Ori32 Fibonacci circle), there does not appear to be any reason why certain of these results—e.g. (11.3) and (11.10)—should not be valid for any integer k (i.e. including zero and negative) —and in Appendix B, we prove them valid for all k in $\mathbb{Z}$ (p.409).

Decads mod eleven

Another example of the divisibility of blocks of Fibonacci numbers is given by Livio [566]. He notes that *any* block of 10 consecutive Fibonacci numbers sum to an exact multiple of 11. That is

$$\sum_{i=n}^{n+9} F_i \equiv 0 \quad (\mathrm{mod}\ 11), \quad n = 0, 1, 2, 3, \ldots$$

He then goes on to note just what the multiplying factor is. For a given starting point in the sequence (index n)—it is F_{n+6}.

F_n divisibility: $\boxed{F_p \text{ may be prime}}$

Looking at the Fibonacci numbers (say in the Table on page 477) we observe that every fourth member ($F_4 = 3$, $F_8 = 21$, ...), is divisible by 3. We write this as

$$3 \mid F_{4n} \qquad n = 1, 2, 3, \ldots \qquad (11.16)$$

and read it as: '3 divides F_{4n}.' It is straightforward to prove (11.16) by induction (p.201), and similarly too: $2 \mid F_{3n}$, $5 \mid F_{5n}$, and $7 \mid F_{8n}$ [804]. And we already have $4 \mid F_{6n}$—this being another way of writing (11.15), p.173. A very well known general divisibility property is that $F_n \mid F_{nk}$. There are a number of ways to prove this.[3] This relationship has an important consequence for the location of primes in the Fibonacci sequence. If we start by regarding the indices of primes as special cases, then we can think of the remaining composites as having indices of the form nk. From this and $F_n \mid F_{nk}$ it follows that if a Fibonacci number has a composite index nk, then it is divisible by F_n and so, clearly cannot be prime (unless $F_n=1$). We can then say of the non-composite cases that the index is only divisible by itself. So, if a Fibonacci number is prime, then its index should be prime (except for prime $F_4=3$, which is divisible by $F_2=1$). Looking back the other way, F_p *may be prime*, but not necessarily. For example[4]

$$F_{19} = 4181 \quad \text{is composite} \quad (37 \times 113).$$

As for the Lucas numbers, in 1985 mathematician Jeffrey Lagarias proved that exactly one third of all the prime numbers will not divide into any of the Lucas numbers (whereas there is no prime that will not divide into some Fibonacci number) [521].

---

[3]   Bicknell & Hoggatt demonstrate four such [57]

[4]   On a scientific calculator a quick way to find F_{19} is to note that in the Fibonacci Binet form (p.25), its β^n term will be very small, so $F_{19} = round\left(\phi^{19}/\sqrt{5}\right)$.

Chapter 12

Five Golden Powers

Here, we look back again at that tantalizing ⑤ sector, with its arc very close to radius length (Fig. 125, page 168). In the previous chapter, we might have remembered that Fibonacci numbers grow from one to the next by a factor of (approximately) the Golden Ratio. Therefore we may say that the sectors adjacent to the ⑤ sector will be smaller and larger by approximately factor ϕ.

From Kepler to constant growth by ϕ

Back in the first chapter, we followed Kepler's thoughts on how such ratios of successive pairs from: 1, 1, 2, 3, 5, 8, ... (the ratios: 1.0, 2.0, 1.5, 1.6666, 1.6, ...) dance either side of 1.6180339887... ever closer, 'honing in' on ϕ, while never reaching its exact value. Starting with the following diagram, Fig. 128, we next consider the differences we get if we specify a constant growth of exactly ϕ per step, rather than the Fibonacci number sequence growth with its repeating 'more than ϕ, less than ϕ'. For example, Fibonacci growth from 5 to 8 represents a factor of 1.6, but if we grow 5 by ϕ (i.e. $5 \times 1.618...$) then we get 8.090...

Let's see where this line of thought takes us.

Right-side semicircle—$\boxed{3}$ $\boxed{5}$ $\boxed{8}$

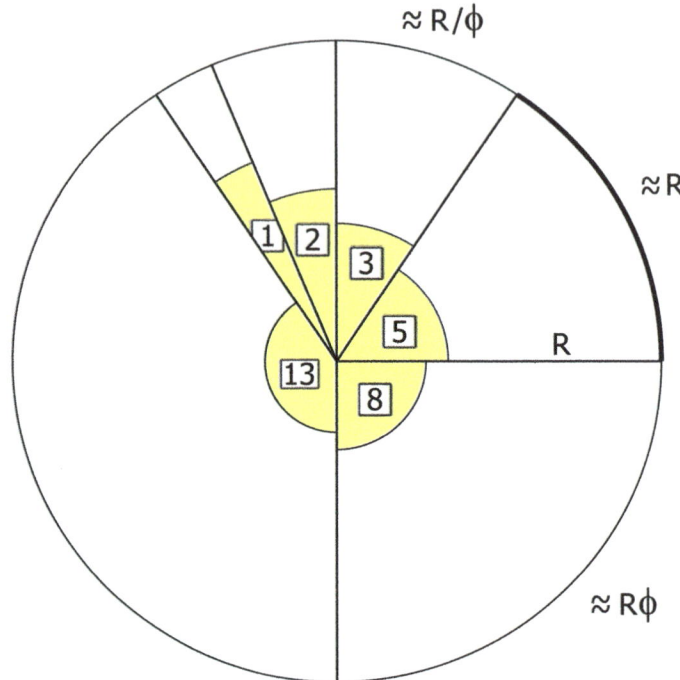

Figure 128: **Sector growth sequence.**

In Fig. 128 we focus on the right-hand semicircle. The circumference arcs here grow (clockwise) by a factor close to the Golden Ratio ϕ. Looking at the ratios of adjacent sector angles (in MIK), we note

$$\frac{\boxed{8}}{\boxed{5}} = 1.6 \quad \approx \left(\phi - \frac{9}{500}\right), \qquad \text{and} \qquad \frac{\boxed{3}}{\boxed{5}} = 0.6 \quad \approx \left(\frac{1}{\phi} - \frac{9}{500}\right).$$

(There is the seed of an idea here, which we shall pick up on in the next chapter.) In the meantime, we see how the whole semicircle arc should have length $(2\pi)R/2$, but by taking the ϕ-based approximations shown, our result

$$\frac{R}{\phi} + R + R\phi = \left(\frac{1}{\phi} + 1 + \phi\right)R \quad \text{exceeds} \quad (2\pi)R/2 \quad \text{by close to} \quad \left(\frac{1}{33}\right)R.$$

Left-side semicircle—$\boxed{1}$ sector squeezed

Next, continuing 'outwards and round' away from the $\boxed{5}$ as before, we move into the left side from both top and bottom to consider the $\boxed{2}$ and the $\boxed{13}$ sectors as well. Their angles will again scale from their neighbours by approximately the Golden Ratio, and based on our experience with the right-side semicircle we might guess that by adding in exactly ϕ-scaled versions of these in too, the total might overshoot by $\approx 1/33$rd. So just what is the cumulative effect of adding all these approximations? Following the pattern, we shall now have

$$\frac{R}{\phi^2} + \frac{R}{\phi} + R + R\phi + R\phi^2 \;=\; \left(\frac{1}{\phi^2} + \frac{1}{\phi} + 1 + \phi + \phi^2 \right) R.$$

Surely all these 'too big' sectors will squeeze the last remaining sector—that is, the $\boxed{1}$—and if so, then by just how much?

Summing powers of ϕ—the 1st 'pentad'

To calculate the result of this squeeze simply, we set the radius to unity, giving a circumference of (2π) (which is also the number of radians in a full circle), and then quietly move back to talking about sector angles instead of arc lengths. Let's start with the $\boxed{5}$ sector and make the small change to bring it to 1 radian. Either side of it, we shall now have the 'overshooting' values of $1/\phi$ and ϕ radians. We then apply the same logic to the former $\boxed{2}$ and $\boxed{13}$ sectors and get the angles $1/\phi^2$ and ϕ^2 radians. What does this geometric sequence of 5 angles—that we shall call a *pentad*—sum to? Well, $\phi^{-2} + \phi^{-1} = 1$, and $\phi^2 = \phi + 1$, so

$$\phi^{-2} + \phi^{-1} + 1 + \phi + \phi^2 \;=\; 2\phi + 3 \;=\; 4 + \sqrt{5}. \qquad (12.1)$$

Surprisingly, this 5-sector scheme just about fills the circle (Fig. 129).

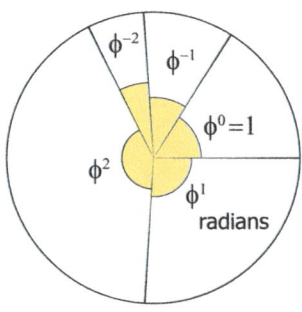

Figure 129: **5 Golden Sectors.**

177

Now, if we re-write the '1' as ϕ^0, then we can say

$$\sum_{n=-2}^{2} \phi^n \;=\; 2\phi + 3 \;=\; 4 + \sqrt{5} \;\approx\; (2\pi) \qquad (12.2)$$

where the Golden Ratio $\phi = (1 + \sqrt{5})/2$. So, the sixth sector $\boxed{1}$ has been squeezed down to $(2\pi) - (4 + \sqrt{5}) = 0.0471$ radians $(\equiv 2.7°)$. However, in the diagram Fig. 129, for artistic reasons, we distribute this residue evenly as 5 tiny gaps between the ϕ-power sectors: each one $2.7/5 = 0.54°$. And because we show these gap sectors 'as is' in black, they taper and vanish towards the centre of the circle.

Analogue and digital

All the time, we must keep reminding ourselves that with these 'powers- of-ϕ' sectors, we are not looking at angles 'in proportion': here their numeric values are already *the actual angles in radians* and we next compare their layout with the Ori32 Fibonacci circle from page 168.

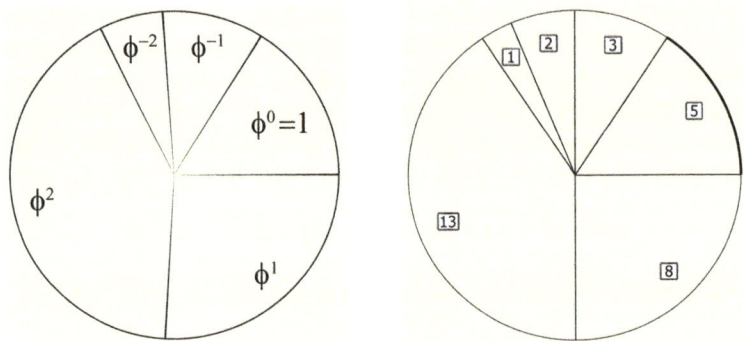

Figure 130: **Analogue and digital**

Also, as the angles on the left are based on ϕ, we should not be surprised that the composition has a 'fragile organic quality' compared with the 'rigid mechanical perfection' of the integer-based Fibonacci circle to the right. Whichever way we show it, there really does seem to be something interesting going on; specifically, the way these five major sectors fit so neatly into the circle—each a power of ϕ—with two pairs of powers (± 2, and ± 1) symmetric about the zero.

Let us therefore name this special class of sectors:

> 'A **Golden Sector** is one that has an angle in radians
> which is an integral power of ϕ, the Golden Ratio,
>
> $$(1 + \sqrt{5})/2.\,'$$

As we have seen, unity (one) is the zeroth power of ϕ, so by our definition, a one radian sector is also seen as Golden, even though it does not explicitly reference ϕ—we shall utilize the '1' Golden Sector in a moment.

5 Golden Powers as (tinted) lengths

Next we ask: 'Does our set of 5 Golden Powers (-2 through $+2$) appear anywhere else—or is it just some arbitrary sequence?' (We recall that in the *In nature* chapter—p.53, we saw 4 out of these 5 powers in the 137.5° Golden Angle calculations—viz. ϕ^{-2}, ϕ^{-1}, ϕ^{+1}, ϕ^{+2}.)

Now, in Fig. 131 we consider groups of lengths. In the top left, we highlight in colour the arcs made by the Golden Sectors. We know that in a circle radius r, for a sector angle θ, the length of its arc a (on the circumference) is simply

$$a = r\theta.$$

And to simplify things even further, we take the radius as unity ($r = 1$), so the arc length of a sector becomes the same as its angle in radians. Again we evenly share the 2.7° shortfall, and this results in 5 equal-length gaps around the circumference. We now notice how these unit circle Golden Sector arc lengths correspond to the five spokes of a single turn of a Golden Spiral.[1] As we saw earlier[2], this special logarithmic spiral grows by a factor of the Golden Ratio ϕ for every quarter turn. The Golden Spiral we see here passes through a point at unity radial distance from its centre—and we take this as our reference spoke—with quadrant spokes ϕ^{-1} and ϕ^{-2} before it in growth, and ϕ and ϕ^2 after. We also recall the parabola diagram from (top of) page 5, and further realize that the key lengths featured in that are again our 5 Golden Powers.

---

[1] Five spokes, as we count both the start and the finish spokes.
[2] Page 42

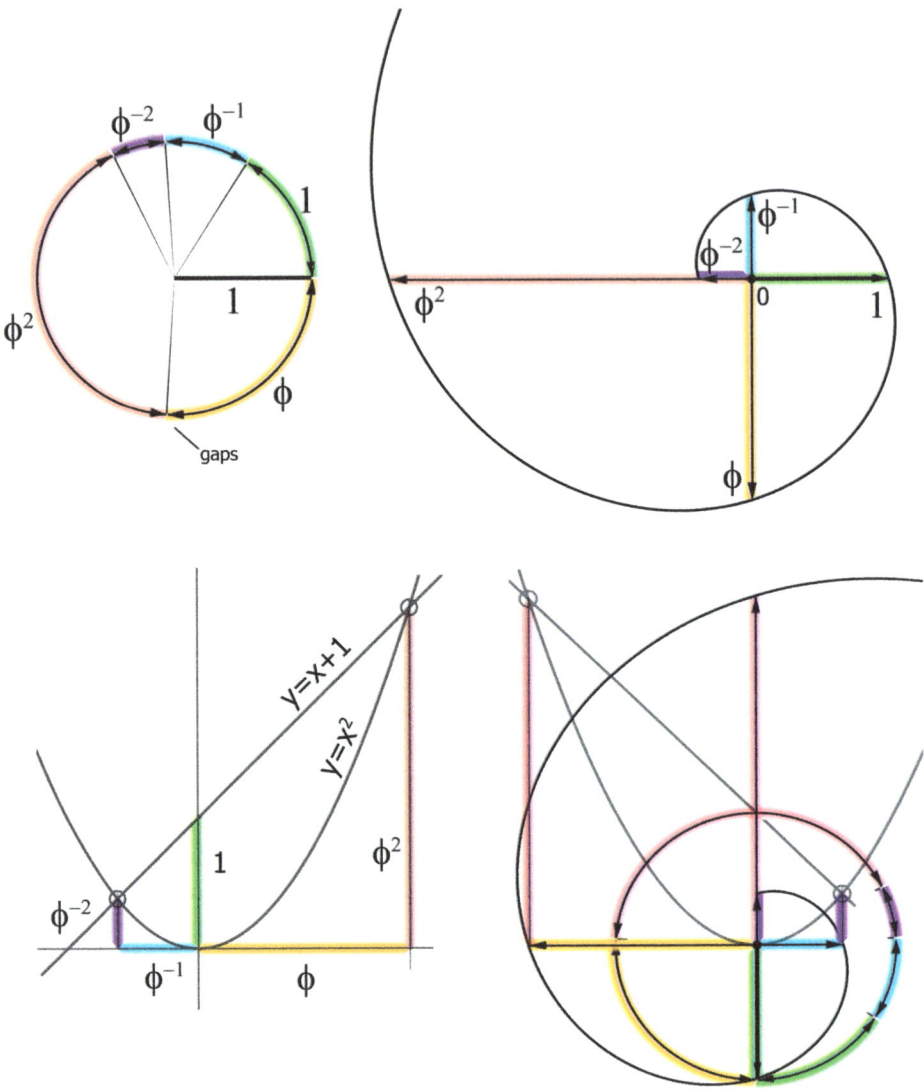

Figure 131: **5 Golden Powers**

Last, when we combine the circle, the spiral, and the parabola (using two rotations and a flip respectively), it is then interesting to see how the colours (i.e. Golden Powers) group together. Also in the combination diagram, the spiral and the parabola look as if they might be crossing (on the right) at 90°. However this is 'not quite' true—there is a difference of 2.43°. (For the calculation, see Appendix D, p.431, footnote 2).

Higher powers of ϕ—the 2nd pentad

Just as we did with the Fibonacci Circle, where we continued, adding groups of 6, and checking the effect; let's carry on here too and see what we get. The indication here though, is that we should group in fives. Also we note that we can simplify higher-power terms by successively applying $\phi^2 = \phi + 1$. So for example, $\phi^3 = \phi^2 + \phi = 2\phi + 1$ (Fig. 132).

Power of ϕ	Simplified	$\sqrt{5}$ equivalent
3	$2\phi + 1$	$(4 + 2\sqrt{5})/2$
4	$3\phi + 2$	$(7 + 3\sqrt{5})/2$
5	$5\phi + 3$	$(11 + 5\sqrt{5})/2$
6	$8\phi + 5$	$(18 + 8\sqrt{5})/2$
7	$13\phi + 8$	$(29 + 13\sqrt{5})/2$
2nd pentad totals	$31\phi + 19$	$(69 + 31\sqrt{5})/2$
add 1st pentad*	$3\phi + 2$	$(8 + 2\sqrt{5})/2$
Both pentads	$33\phi + 22$	$(77 + 33\sqrt{5})/2$

Figure 132: **Golden Powers—the 2nd pentad.** *From (12.2), p.178.

In the centre column we spot the two Fibonacci sequences running vertically in parallel, and from these, it is easy to see the general formula $\phi^n = F_n\phi + F_{n-1}$ [491]. And similarly in the third column we spot a vertical sequence of Lucas numbers next to a vertical Fibonacci sequence—the general form here being $\phi^n = (L_n + F_n\sqrt{5})/2$ [794].

The first 5 primes

We note the bottom line totals in Fig. 132 include multiples of 11. Indeed, $33\phi + 22 = 11 \cdot (3\phi + 2) = 11\phi^4$. And in the $\sqrt{5}$ form, if we factor out the 11, we get

$$\sum_{n=-2}^{7} \phi^n = \frac{11}{2}\left(7 + 3\sqrt{5}\right). \tag{12.3}$$

Now we see that the right-hand side neatly combines the first 5 primes:

$$2, \ 3, \ 5, \ 7, \ \text{and} \ 11.$$

So (whimsically rearranging),

$$\sum_{n=-2}^{7} \phi^n = 11 \cdot \left(7 + \sqrt{5} \cdot 3\right)\big/2 = 11.9995 \cdot (2\pi) \approx 12 \text{ turns}$$

(12.4)

Then recalling that this sum is also equal to $11\phi^4$ (p.181),

$$\phi^4 = \left(7 + \sqrt{5} \cdot 3\right)\big/2$$

(12.5)

Residue nearly vanishes

Surprisingly, our cumulative residue (angle to the nearest full turn) has now reduced to just a tiny 0.0005 of a turn—that is, one two-thousandth turn, or 0.0031 radians, or 0.18°—and only that, after 12 turns—a total of 4,320°. In Fig. 133 we have the longest radial (11 o'clock) as our reference, and we measure clockwise: ϕ^{-2}, ϕ^{-1}, ϕ^0, ϕ^1, ϕ^2 without gaps. We then see and *ignore* the 2.7° thin sector. We continue adding sectors, starting from the black spot, now showing a spiral track. We add a ϕ^3, and then ϕ^4, and so on, to complete the 2nd pentad. And as we already know, the final ϕ^7 brings us almost exactly back to our starting point—the long reference radial. For a zoom view of the 0.18° difference, see Fig. 134.

Pythagorean comma

So, although we do not have an exact fit, there does appear to be some kind of loose or coincidental relationship between the Golden Ratio and the circle constant (2π), at least in the region we are exploring.[3] This 12-turn, ϕ^7 result may remind some of the 'Pythagorean comma' in the maths of music, where the cycle of 12 fifths does not quite match to 7 octaves (the difference is about a quarter of a semitone)—that is,[4]

$$\left(\frac{3}{2}\right)^{12} \neq 2^7.$$

[872]

---

[3]   This is analysed further in Appendix D, p.438.
[4]   The cycle of fifths is: C–G–D–A–E–B–G♭ (F♯)–D♭–A♭–E♭–B♭–F–   [C–G–D–...].

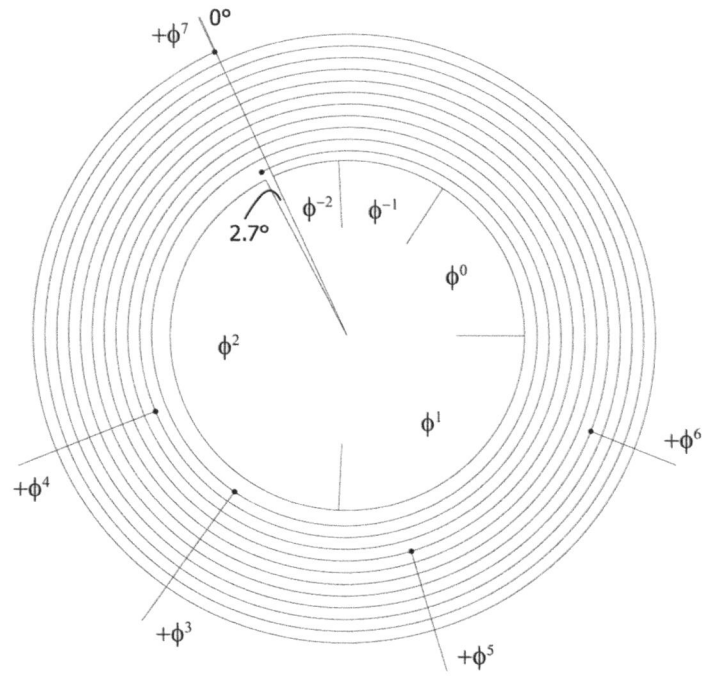

Figure 133: **Golden Sector circle with next 5 higher powers spiral.**

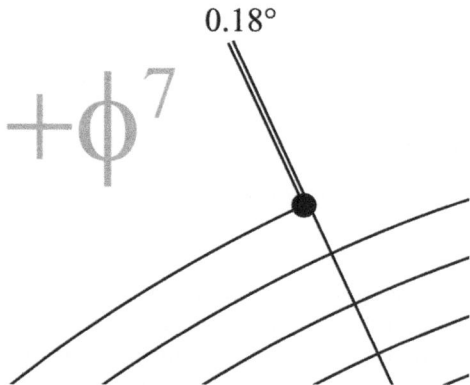

Figure 134: **Ten Golden Powers zoom.** Detail view of Fig. 133.

An observation—6/5ths of ϕ^2

Figure 135: **Athanasius Kircher (1602–1680) Woodcut. Frontispiece of** *Arithmologia* **(detail) 1665 [474].** Bavarian State Library. On the left, in the man's book, we see both a pentagram and a hexagram.

Through his interest in the relationship of five to six (Fig. 135), architect and esoteric pattern researcher Scott Onstott [715] has observed that 6/5ths of ϕ^2 is remarkably close to π—that is,

$$\boxed{\phi^2 \cdot \frac{6}{5} \ = \pi \times 1.000015\ldots \approx \pi}$$ (12.6)

(This is also noted by Richard Dunlap [188].) And if we look at this in a certain way, we find it 'explains' the 2.7° gap we saw earlier. If we rearrange, and move from π to (2π)—a full turn—we shall get

$$2\phi^2 \approx \frac{5}{6} \cdot (2\pi).$$

We recall that $\phi^2 = \phi + 1$, and $\phi^0 = \phi^{-1} + \phi^{-2}$—that is, Golden Powers have the same property as Fibonacci numbers in that each is the sum of the two predecessors. So we may now partition $2\phi^2$ (radians) as

$2\phi^2 = \phi^2 + (\phi + 1)$ and by splitting the 1, again in Golden Powers,

$= \phi^2 + \phi^1 + \phi^{-1} + \phi^{-2} \ \equiv \ 150.00° + 92.71° + 35.41° + 21.89°.$

So, in this way, we account for 4 out of the 5 Golden Powers (here, as Golden Sectors), and these then constitute very close to five-sixths of a full turn (actually it is 300.0046...°), leaving one-sixth—a near-60° sector. In Fig. 130, p.178 (on the left), this corresponds to the location of the ϕ^0 (= 1 radian). And there it is—our 2.7° is revealed (almost entirely) as *the difference between 1 radian (57.3°) and 60°*.

A further observation—11/24ths of ϕ^4

Looking again at that 'six fifths' approximate (but close) relationship

$$\phi^2 \cdot \frac{6}{5} = \pi \times 1.000015\ldots \approx \pi \tag{12.7}$$

we remember something similar occurring in the sum of the first two pentads (Eqns. (12.3) and (12.5), p.181), and we can show that result in the same form—viz.

$$11 \cdot \left(7 + \sqrt{5} \cdot 3\right) \Big/ 2 = 11\phi^4$$

$$11\phi^4 = (2\pi) \times 11.999506\ldots = 12 \times (2\pi) \times 0.999959\ldots$$

$$\boxed{\phi^4 \cdot \frac{11}{24} = \pi \times 0.999959\ldots \approx \pi} \tag{12.8}$$

Now, in order to understand how these two examples are related (as surely they are), let's divide (12.8) by (12.7) and rearrange the result.

$$\frac{\phi^4}{\phi^2} \cdot \frac{11}{24} \cdot \frac{5}{6} = \frac{0.999959\ldots}{1.000015\ldots}$$

$$\phi^2 = \frac{144}{55} \times 0.999944\ldots$$

This, we are pleased to find, all makes sense. But the interesting thing here is the way Fibonacci numbers 144 and 55 show up. And now that they have, we recall that the growth factor from 55 to 89 (the next Fibonacci number in the sequence) is near to the Golden Ratio ϕ, and the same occurs again with 89 growing to 144. This explains why the rational approximation $144/55$ is so close to ϕ^2.

Higher powers of ϕ—the 3rd pentad

Let us consider just one pentad more and add in the powers eight to twelve (Fig. 136):

Power of ϕ	Simplified	$\sqrt{5}$ equivalent
8	$21\phi + 13$	$(47 + 21\sqrt{5})/2$
9	$34\phi + 21$	$(76 + 34\sqrt{5})/2$
10	$55\phi + 34$	$(123 + 55\sqrt{5})/2$
11	$89\phi + 55$	$(199 + 89\sqrt{5})/2$
12	$144\phi + 89$	$(322 + 144\sqrt{5})/2$
3rd pentad totals	$343\phi + 212$	$(767 + 343\sqrt{5})/2$
add 1st and 2nd pentads	$33\phi + 22$	$(77 + 33\sqrt{5})/2$
Three pentads	$376\phi + 234$	$(844 + 376\sqrt{5})/2$

Figure 136: **Golden Powers—the 3rd pentad.**

As before, we consider the cumulative total, expressed in terms of $\sqrt{5}$, and when we halve 844 and 376, we get

$$\sum_{n=-2}^{12} \phi^n = 422 + 188\sqrt{5} = 134.069 \cdot (2\pi) \qquad (12.9)$$

And again, the additions bring us close to a whole number of turns. Also, this '422, 188' expression has a certain appeal—just so long as we remember that the showing of its 1, 2, 4, and 8 digits depends entirely on using the base-10 representation, which was not the case with the 5 primes result just earlier—Equation (12.3). We look in further detail at the Golden Powers in Appendix D on page 431, and Tables are provided in the *Collected Formulæ* section, on page 477.

Angle bracket notation $\qquad \langle F_n \rangle$

We first saw the use of angle brackets to denote sequences in the *Art—Mondrian* chapter (p.107). They are used to show a grouping of elements where order is important. For example we might represent the Fibonacci sequence as $\langle F_n \rangle$. From now on, we shall be making increasing use of these brackets.

Phi powers × 1000—the '10, 12 sequence' $\langle K_n \rangle$

Robert Lawlor, is a painter and sculptor who has studied yoga with Sri Aurobindo, and studied the Egyptology of R A Schwaller de Lubicz. In his book *Sacred Geometry, Philosophy and practice*, he touches upon the fact that the 'converging-to-ϕ' property of the Fibonacci recurrence (Eqn. (1.1) p.1) does not depend on particular starting seeds; any two numbers will do [525]. He examines a few sequences: the Fibonacci starting with 1, 1; the Lucas starting 1, 3; and then one beginning $\langle 1, 5, 6, 11, 17, \ldots \rangle$ (A022095 [839]). This latter he points out as being interesting because when members of this '1, 5' sequence are doubled, they appear to show powers of ϕ (scaled a thousand times). For example, along the way he finds 618, 1618 and 2618 (reminding us of ϕ^{-1}, ϕ, and ϕ^2). Let's see why this should be. For clarity we shall start by doubling the seeds 1 and 5. This factor 2 will then propagate through, doubling all the members to give $\langle 2, 10, 12, 22, 34, \ldots \rangle$ And for ease of memorizing we shall 'start one in' and drop the 2—now taking 10 and 12 as our seeds. The interesting part of the sequence will stay the same—we have merely moved the positional index along one. So now we have

$$K_{n+2} = K_{n+1} + K_n \quad \text{with} \quad K_0 = 10, \quad K_1 = 12, \quad \text{and} \quad n = 0, 1, 2, 3, \ldots$$

In Fig. 137 we compare this sequence (left-hand columns) with the scaled Golden Powers (to the right). The key thing here is that at index $n = 10$, $K_{10} = 1000$. Further, we know that we do not need to look at hundreds of members of a Fibonacci sequence before the ratios of successive numbers converge reasonably close to ϕ. So, here we see $K_9 = 618$, $K_{11} = 1618$, and $K_{12} = 2618$, each a good approximation for the relevant Golden Power—times a thousand. (We 'solve' this $\langle K_n \rangle$ recurrence—that is, we find a Binet-like formula for its nth term—on page 446—Eqn.(F.17).)

Exercises

Exercise 4. In a spreadsheet, using the same format as Fig. 137, show that applying the Fibonacci recurrence to the seeds 51 and 2621 produces a sequence that (in less than 20 iterations) includes the number one million. Compare the accuracy of the significant digits produced with those of powers of ϕ multiplied by 1,000,000.

Exercise 5. What integer power of 10 is included in the sequence with seeds 1880 and 385835? Also, do 311 and 5454 make a good seed pair?

$$K_{n+2} = K_{n+1} + K_n \quad \text{with} \quad K_0 = 10, \ K_1 = 12, \ \text{and} \ n = 0, 1, 2, 3, \ldots$$

p $= n - 10$	K_n	n	ϕ^p ×1000 to one dec. pl.	ϕ^p ×1 to five dec. pl.
−10	10	0	8.1	0.00813
−9	12	1	13.2	0.01316
−8	22	2	21.3	0.02129
−7	34	3	34.4	0.03444
−6	56	4	55.7	0.05573
−5	90	5	90.2	0.09017
−4	146	6	145.9	0.14590
−3	236	7	236.1	0.23607
−2	382	8	382.0	0.38197
−1	618	9	618.0	0.61803
0	1000	10	1000.0	1.00000
1	1618	11	1618.0	1.61803
2	2618	12	2618.0	2.61803
3	4236	13	4236.1	4.23607
4	6854	14	6854.1	6.85410
5	11090	15	11090.2	11.09017
6	17944	16	17944.3	17.94427
7	29034	17	29034.4	29.03444
8	46978	18	46978.7	46.97871
9	76012	19	76013.2	76.01316
10	122990	20	122991.9	122.99187

Figure 137: **Scaled ϕ powers—the 'ten twelve' sequence $\langle K_n \rangle$.**

Exercise 6. Create $\langle K_n \rangle$ for a friend. Starting with just the seeds 10 and 12, repeat adding the previous two numbers to match the K_n column in Fig. 137—at least as far at 2618—using blackboard, whiteboard, envelope, or beach sand (and a suitable stick). Alternatively, you start, but invite your friend to continue.

Exercise 7. Starting with '34', take each result from the previous exercise, and scale it down by 1000, adding decimal points and zeroes as required—e.g. 0.034, 0.056, 0.090, … Then write '0' to the left of 1.000, '1' to the left of 1.618, and so on, completing the 'p' column (of powers) to the left of the values just as in Fig. 137. We omit the less accurate: 10, 12, and 22.

(*No solutions are given for the above four exercises.*)

Part III

THE FIBONACCI RESONANCE

Chapter 13

The penny drops

Now we come to the essence of the 'Fibonacci Resonance' discovery. Having sensitized ourselves by studying the growth of the Fibonacci sequence and the growth of powers of the Golden Ratio—looking at what appears to be an arbitrarily close but nevertheless approximate relationship; we are now in a position to get our first glimpse of a tiny part of what may turn out to be a completely general pattern.

We saw earlier in the '5 Golden Powers' chapter (p.176) that

$$\frac{8}{5} = 1.6 \quad \approx \left(\phi - \frac{9}{500}\right) \qquad \text{and} \qquad \frac{3}{5} = 0.6 \quad \approx \left(\frac{1}{\phi} - \frac{9}{500}\right).$$

Next we look at the neighbouring sectors either side of these (in Fig. 128, p.176), stepping one further round in each direction, while still dividing by 5—our reference for this exercise, and we get

$$\frac{13}{5} = 2.6 \quad \approx \left(\phi^2 - \frac{9}{500}\right) \qquad \text{and} \qquad \boxed{\frac{2}{5} = 0.4 \quad \approx \left(\frac{1}{\phi^2} + \frac{9}{500}\right)}$$

The first three of these results have been well known for hundreds of years. But that fourth one (bottom right)—it does seem interesting and unexpected.

In which case, let's explore this mysterious 'approximately 9/500'—calling it p. What is p precisely? Perhaps the most obvious place to extract a value is from the 8/5 case. We know that

$$\phi = 1.6\,\underline{18033\,9887\,50}\ldots$$

so $\quad \dfrac{8}{5} = 1.6 \ = \ \phi - p.$

Rearranging, $\quad p = \phi - \dfrac{8}{5}$ (13.1)

$$= 0.0\,\underline{18033\,9887\,50}\ldots$$

Therefore p, as defined in (13.1), is exactly the part of the decimal expansion of ϕ that follows the 1.6, (and for that matter, that follows the 2.6 in ϕ^2, and follows the 0.6 in $1/\phi$).[1] Therefore, let's now move on from approximation and restate our first four cases (in size order), while also bringing in the '5' reference itself:

$$2\big/5 \ = \ 0.4 \ = \ \boxed{\phi^{-2} + p}$$

$$3\big/5 \ = \ 0.6 \ = \ \left(\phi^{-1} - p\right)$$

$$5\big/5 \ = \ 1.0 \ = \ \phi^0$$

$$8\big/5 \ = \ 1.6 \ = \ \left(\phi \ - \ p\right)$$

$$13\big/5 \ = \ 2.6 \ = \ \left(\phi^2 \ - \ p\right).$$

And let's look one more time at that Fibonacci circle diagram, both to remind ourselves of the context and to make the above associations— those between Fibonacci numbers and powers of ϕ (Fig. 138).

---

[1] We recall seeing the 0.6 case earlier—in Seurat's *Parade*. There, we saw the 0.018 difference between the 'G' and 'R' sections—the irrational ϕ^{-1}=0.618 and the rational 3/5=0.600 (p.85).

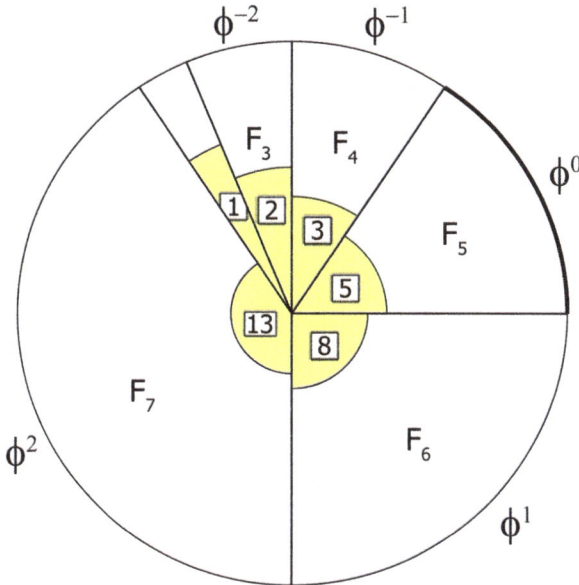

Figure 138: **Sector growth—
associating Fibonacci numbers with Golden Powers.**

In this diagram *all the angles are whole numbers of MIK.* The ϕ powers noted are positioned solely in order to associate them with Fibonacci numbers. As we have seen before, the sector angles in this Ori32 Fibonacci Circle grow by factors increasingly close to the Golden Ratio ϕ. The *key relationship* to notice in the diagram though, is the offset between the indices of the Fibonacci numbers, and the numbers giving the powers of ϕ. To analyse, we start at the reference ⑤ sector and work 'out and round'. This 5 MIK (F_5) corresponds to ϕ^0. Next greater is the ⑧ (F_6)—which we associate with ϕ^1. And on the smaller side we associate the ③ (F_4) with ϕ^{-1}. In each case, we see the offset between index and power is 5. So from Fig. 138, we conclude that in our 'Fibonacci number ÷ 5' formulæ (listed on page 192), if we adopt n as index, then we shall correspond $F_n/5$ with ϕ^{n-5}. This in turn suggests the next logical steps: to make a Table to compare $F_n/5$ with ϕ^{n-5} across a range of n values; and then to divide their differences by p. What we want to know is: 'Does p deserve to be called a "quantum" (i.e. a common measure)—that is, do we always get integer multiples of p in the differences?' (Fig. 139).

193

$$\text{Candidate quantum } p = \left(\phi - \frac{8}{5} \right) = 0.0\underline{18033989}\ldots \approx \frac{9}{500}.$$

n	F_n	$\dfrac{F_n}{5}$	ϕ^{n-5}	difference $\Delta = \phi^{n-5} - \frac{F_n}{5}$	$\dfrac{\Delta}{p} = $ no. of quanta
0	**0**	0.0	0.090169944	0.090169944	+5
1	**1**	0.2	0.145898034	−0.054101966	−3
2	**1**	0.2	0.236067977	0.036067978	+2
3	**2**	0.4	0.381966011	−0.0\underline{18033989}	−1
4	**3**	0.6	0.6\underline{18033989}	0.0\underline{18033989}	+1
5	**5**	1	1	0.0	**0**
6	8	1.6	1.6\underline{18033989}	0.0\underline{18033989}	1
7	13	2.6	2.6\underline{18033989}	0.0\underline{18033989}	1
8	21	4.2	4.236067977	0.036067978	2
9	34	6.8	6.854101966	0.054101966	3
10	55	11.0	11.09016994	0.090169944	5
11	89	17.8	17.94427191	0.144271910	8
12	144	28.8	29.03444185	0.234441854	13
13	233	46.6	46.97871376	0.378713764	21
14	377	75.4	76.01315562	0.613155617	34

...

Figure 139: **Quantized difference calculations.**

After studying the Table in Fig. 139, we may feel fairly confident that p indeed *does function as a 'quantum'*. We quickly find that the difference between the *Fibonacci number divided by 5*, and the Golden Power (ϕ^{n-5}), *is an exact multiple of our quantum p*. Further, we recognize the resulting set of counts (far right column) as itself a Fibonacci sequence(!)—one shifted along by 5 places with respect to our index n. So it now appears that for difference Δ we have

$$\Delta/p = F_{n-5} \quad \text{where} \quad \Delta = \phi^{n-5} - F_n/5.$$

Therefore we conjecture that for all n,

$$\frac{F_n}{5} = \phi^{n-5} - F_{n-5}\, p.$$

But can we prove it? To make a start, we multiply each side by 5

$$F_n = 5\phi^{n-5} - 5F_{n-5}\, p. \tag{13.2}$$

Also, to simplify our expressions we define a new quantum 'q' that is 5 times p—using (13.1),

$$\boxed{\quad q = 5p = 5\left(\phi - \frac{8}{5}\right) = 5\phi - 8 = 0.0901699437495\ldots \quad} \tag{13.3}$$

Those very familiar with ϕ will instantly recognize $5\phi - 8$ as Golden Power ϕ^{-5}. This linear form is just another way of writing it. As we saw in the *5 Golden Powers* chapter (p.181, by repeatedly applying $\phi^2 = \phi + 1$ we can reduce any integer power of ϕ to an expression that is linear in ϕ (i.e. where ϕ^1 is the highest power). The shortcut formula for this process is $\phi^n = F_n\phi + F_{n-1}$ [491]. Here $\phi^{-5} = F_{-5}\phi + F_{-6}$ and we know that negative-indexed Fibonacci numbers have the same absolute magnitude as the positive, but their sign alternates according to $F_{-n} = (-1)^{n+1}F_n$ [482], so $F_{-5} = +5$ and $F_{-6} = -8$. Thus

$$q = \phi^{-5} = 5\phi - 8. \tag{13.4}$$

So using q from (13.4), we restate (13.2) as equation (13.5) below, which we shall next prove.

$$F_n = 5\phi^{n-5} - F_{n-5}\, q \quad \text{where} \quad q = \phi^{-5}. \tag{13.5}$$

Theorem:

$$F_n = 5\phi^{n-5} - F_{n-5}\,q \qquad \text{where} \quad q = \phi^{-5} \tag{13.6}$$

Proof:

In (13.6), first we decompose the left-hand side F_n according to the Fibonacci recurrence formula, then likewise the terms produced.[2]

$$F_n = F_{n-2} + (F_{n-1}).$$

Splitting $F_{n-1} = F_{n-2} + F_{n-3}$ (then splitting F_{n-2})

$$= F_{n-2} + (F_{n-2} + F_{n-3})$$
$$= 2(F_{n-2}) + F_{n-3}$$
$$= 2(F_{n-3} + F_{n-4}) + F_{n-3}$$
$$= 3F_{n-3} + 2F_{n-4}$$

and so on, as far as:

$$F_n = 5F_{n-6} + 8F_{n-5}. \tag{13.7}$$

We now prove the right-hand side of (13.6) is equal to this (13.7). We start by applying $\phi^n = F_n\phi + F_{n-1}$ to the ϕ^{n-5} term, [491]

$$\phi^{n-5} = F_{n-5}\phi + F_{n-6} \tag{13.8}$$

and in passing, when $n = 0$, we again get the linear form for q,

$$q = \phi^{-5} = F_{-5}\phi + F_{-6} = 5\phi - 8. \tag{13.9}$$

Substituting (13.8) and (13.9) back into the right-hand side of (13.6),

$$F_n = 5\left(F_{n-5}\phi + F_{n-6}\right) - F_{n-5}\left(5\phi - 8\right)$$
$$= 5F_{n-5}\phi + 5F_{n-6} - 5F_{n-5}\phi + 8F_{n-5}$$
$$= 5F_{n-6} + 8F_{n-5}, \qquad \text{we show equality with (13.7).} \quad \blacksquare$$

To summarize, we have found a set of quantized adjustments. Each one will get us from an integer Fibonacci number to exactly 5 times a 'Golden Power'. All the adjustments are multiples of our quantum q— and successive quanta counts are found to form a lagged sequence of Fibonacci numbers.

---

[2] Effectively applying Hansen [323], $F_{N+M} = F_M F_{N+1} + F_{M-1} F_N$, with $N = 5$, $M = n - 5$. That is, $F_{5+(n-5)} = F_{n-5}\cdot 8 + F_{n-6}\cdot 5$.

Peer review and search

I showed this 5-based result to a colleague,[3] and after a quick check with a spreadsheet, he suggested that '1' would work in the formula (13.6), in place of 5. I readily agreed with this, but noted that the '1' result was already known in the form $\phi^n = F_n\phi + F_{n-1}$ [491]. That is, if we divide through by ϕ and rearrange we get the 1-based case,

$$F_n = 1 \cdot \phi^{n-1} - F_{n-1}\phi^{-1}.$$

But, his point was not to leave it at these 1 and 5 cases (where $F_n = n$) and instead try to generalize the form—a point well made. However, starting out with paper and pen, I quickly concluded that some computing was required... Let's work through the approach taken and the results obtained.

First we make a trial parameterization of equation (13.6)

$$g_n(s, r, a) = r \cdot \phi^{n-s} + a \cdot F_{n-s}\phi^{-s} \qquad (13.10)$$

with variables:

s : the index offset parameter
r, a : the coefficients of the first and second terms
n : the index of the expected result, Fibonacci number F_n.

We now compute (13.10) for a wide range of possible permutations of integer variables s, r, a in a 'brute-force search'. For each permutation, we perform a block test of a range of n values to check how well the particular formula under test produces the expected results (that is, F_n each time). Oracle Java® [412] BigDecimal and BigNumber classes proved very useful here, and a program was written to make about 50 million permutations and print any set of values s, r, and a that 'worked'. We next consider the results of this exercise, and along the way, we shall describe a formula tested as *valid* when it consistently produces expected Fibonacci numbers when checked for a wide range of n values.

---

[3] Who wishes to remain anonymous.

Analysis of the search results

test case			(reported valid) formulæ found				
s	r	a	\multicolumn	$F_n = r\cdot\phi^{n-s} + a\cdot F_{n-s}\phi^{-s}$			
-1	1	-1	F_n	$=$	ϕ^{n+1}	$-$	$F_{n+1}\phi^{+1}$
1	1	-1	F_n	$=$	ϕ^{n-1}	$-$	$F_{n-1}\phi^{-1}$
2	1	1	F_n	$=$	ϕ^{n-2}	$+$	$F_{n-2}\phi^{-2}$
-3	2	-1	F_n	$=$	$2\phi^{n+3}$	$-$	$F_{n+3}\phi^{+3}$
3	2	-1	F_n	$=$	$2\phi^{n-3}$	$-$	$F_{n-3}\phi^{-3}$
4	3	1	F_n	$=$	$3\phi^{n-4}$	$+$	$F_{n-4}\phi^{-4}$
-5	5	-1	F_n	$=$	$5\phi^{n+5}$	$-$	$F_{n+5}\phi^{+5}$
5	5	-1	F_n	$=$	$5\phi^{n-5}$	$-$	$F_{n-5}\phi^{-5}$
6	8	1	F_n	$=$	$8\phi^{n-6}$	$+$	$F_{n-6}\phi^{-6}$
-7	13	-1	F_n	$=$	$13\phi^{n+7}$	$-$	$F_{n+7}\phi^{+7}$
7	13	-1	F_n	$=$	$13\phi^{n-7}$	$-$	$F_{n-7}\phi^{-7}$
8	21	1	F_n	$=$	$21\phi^{n-8}$	$+$	$F_{n-8}\phi^{-8}$
-9	34	-1	F_n	$=$	$34\phi^{n+9}$	$-$	$F_{n+9}\phi^{+9}$
9	34	-1	F_n	$=$	$34\phi^{n-9}$	$-$	$F_{n-9}\phi^{-9}$

Figure 140: **Search results.**

In Fig. 140 we show only those permutations of s, r, and a which produce valid formulæ, and we note that:

- It is very convenient for our analysis that even though -50 to $+50$ was offered as the range for 'a' values, only $+1$ and -1 are reported valid.

- It is particularly interesting that the r values are the Fibonacci numbers for index s ($s > 0$)—that is, F_s.

- Also, if s is odd, then $a = -1$, and both $\pm s$ give valid formulæ.

- And, if s is even, then $a = +1$, and only $+s$ 'works'.

CHAPTER 13. THE PENNY DROPS

So to summarize, we now have three cases,
(with ± signs linked, here to be read as all plus or all minus):

$$F_n = \begin{cases} F_s\phi^{n\pm s} - F_{n\pm s}\phi^{\pm s} & \text{if } s \text{ is odd,} \\ F_s\phi^{n-s} + F_{n-s}\phi^{-s} & \text{if } s \text{ is even.} \end{cases} \qquad (13.11)$$

In order to combine these two expressions, we note that $(-1)^s$ will provide the correct sign for both odd and even cases. Also, for the sake of harmony, we shall take the minus case in the $\pm$ signs. But as soon as we apply these to get

$$F_n = F_s\phi^{n-s} + (-1)^s F_{n-s}\phi^{-s}$$

we realize that $(-1)^s\phi^{-s} = (-\phi)^{-s}$. So now we have our result:

$$\boxed{F_n = F_s\phi^{n-s} + F_{n-s}(-\phi)^{-s}} \qquad (13.12)$$

And while (13.12) shows a certain symmetry, this becomes a lot more obvious if we express it using the roots α and β from page 3,

$$F_n = F_s\alpha^{n-s} + F_{n-s}\beta^s. \qquad (13.13)$$

We now use (13.13) twice in the well known identity (13.14) below [956]

$$L_n = F_{n-1} + F_{n+1}. \qquad (13.14)$$

This relates Lucas numbers to Fibonacci numbers, and it should allow us to derive an equivalent relation for the Lucas numbers.

$$L_n = F_s\alpha^{n-1-s} + F_{n-1-s}\beta^s +$$
$$F_s\alpha^{n+1-s} + F_{n+1-s}\beta^s$$

$$= F_s\alpha^{n-s}\left(\alpha^{-1} + \alpha^{+1}\right) + \left(F_{n-1-s} + F_{n+1-s}\right)\beta^s.$$

Hence, as $\alpha^{-1} + \alpha = \sqrt{5}$, and $F_{n-1-s} + F_{n+1-s} = L_{n-s}$, we have

$$L_n = F_s\,\alpha^{n-s}\left(\sqrt{5}\right) + (L_{n-s})\,\beta^s. \qquad (13.15)$$

So, showing (13.15) with the earlier F_n result (13.13) to emphasize their complementarity, and with $\mathbb{Z} = \{\dots -2, -1, 0, 1, 2 \dots\}$ we have Equation pair

$$\left.\begin{array}{rcl} F_n &=& F_s\,\alpha^{n-s} + F_{n-s}\,\beta^s \\[2mm] L_n &=& \sqrt{5}\,F_s\,\alpha^{n-s} + L_{n-s}\,\beta^s \end{array}\right\} \text{ for all } n, s \text{ in } \mathbb{Z} \qquad (13.16)$$

Or again, writing pair (13.16) back in terms of ϕ,

$$\left.\begin{array}{rcl} F_n &=& F_s\,\phi^{n-s} + F_{n-s}\,(-\phi)^{-s} \\[2mm] L_n &=& \sqrt{5}\,F_s\,\phi^{n-s} + L_{n-s}\,(-\phi)^{-s} \end{array}\right\} \text{ for all } n, s \text{ in } \mathbb{Z} \qquad (13.17)$$

20/20 Hindsight

'What a wonderful thing.' From our new standpoint, we may look back to page 192, and see that we could have avoided making the detailed Table Fig. 139, p.194. That is, to obtain the further results quickly, we could have applied the Fibonacci recurrence to the 'divide by five' rows, and generalized from there. We only needed to 'add two rows to get the next' (as $F_6/5 + F_7/5 = F_8/5$, and $\phi + \phi^2 = \phi^3$, and so on). Starting from the $F_6 = 8$ row, (and keeping $p = 0.0180\dots$),

$$
\begin{array}{rcccl}
8/5 &=& 1.6 &=& (\phi - p) \\
+13/5 &=& 2.6 &=& (\phi^2 - p) \\
\hline
=21/5 &=& 4.2 &=& (\phi^3 - 2p) \\
\end{array}
$$

$$
\begin{array}{rcccl}
\text{hence}\quad 34/5 &=& 6.8 &=& (\phi^4 - 3p) \\
55/5 &=& 11.0 &=& (\phi^5 - 5p) \\
&\cdots&&& \\
\text{therefore } F_n/5 &=& &=& \phi^{n-5} - F_{n-5}\,p.
\end{array}
$$

Chapter 14

Proving the theorem

In this chapter we shall check the conclusions from the last chapter—taking the F_n and L_n equations from (13.16), stating each as a theorem, and then proving it using standard techniques. As these are key theorems, we shall for the duration switch to a slightly more formal approach. Also, with the goal of making this chapter more self-contained, we shall include brief recaps on the Fibonacci and Lucas numbers and their Binet forms.

Mathematical induction

In order to ensure the proofs are rigorous, we shall apply the technique of mathematical induction. This powerful and mechanical process works in two stages by applying concepts closely related to recurrence sequences. In recurrence we typically have an initial instance—the seed or seeds—and then some rule for getting from one instance to the next. Similarly, in induction we first choose and prove a specific case (usually the simplest)—and we call this the 'base case'. Then, independently of any specific case, we prove in general that if (we assume) one case is true, then it follows that the next case will also be true. This is called the 'inductive step'. The power of the method derives from these two proofs being held as true together. It then follows that all cases (in the step-by-step, one-to-the-next sequence) are true. To extend a proof to those cases that *precede* the base case, the stepping process is reversed (as we shall see shortly).

It might be argued that the reverse stepping is unnecessary in the following two cases, however the idea is introduced here, both to reassure us of the full scope of their validity, and to prepare for later generalization—where we find that certain restrictions apply.

The Fibonacci numbers and their Binet form

The Fibonacci number sequence begins:

$$\langle\, 0, 1, 1, 2, 3, 5, 8, 13, 21, 34, 55, \dots \,\rangle.$$

We recall that starting with seeds 0 and 1, the sequence is formed by adding the preceding two numbers to get the next, that is,

$$F_n = F_{n-1} + F_{n-2}, \quad \text{with} \quad F_0 = 0, \quad \text{and} \quad F_1 = 1 \qquad (14.1)$$

and we have seen that this sequence has the associated characteristic equation $x^2 - x - 1 = 0$ with roots α and β, with values

$$\alpha = \frac{1 + \sqrt{5}}{2}, \qquad \beta = \frac{1 - \sqrt{5}}{2}.$$

From these we see that $\alpha + \beta = 1$, $\alpha - \beta = \sqrt{5}$, $\alpha\beta = -1$. Now, as the Golden Ratio is defined as $\phi = (1 + \sqrt{5})/2$, then $\alpha = \phi$ and $\beta = (-\phi)^{-1}$. Remarkably, it is possible to express the integer Fibonacci numbers in terms of powers of α and β (which are typically irrational). And although we now refer to this result as 'Binet's formula' (1843); it was known to Euler, Daniel Bernoulli, and de Moivre more than a century earlier (p.25),

$$F_n = \frac{\alpha^n - \beta^n}{\sqrt{5}} = \frac{\phi^n - (-\phi)^{-n}}{\sqrt{5}}. \qquad (14.2)$$

In the proofs that follow, we shall use α and β rather than ϕ and $(-\phi)^{-1}$. This is because:

- β is more readable than $(-\phi)^{-1}$
- as a conjugate pair, α and β better expose symmetries
- the use of α and β conforms to a style that is used in more general works in this field

The Fibonacci Resonance—main theorem:

*Let F_n be the nth Fibonacci number, and
let α be the Golden Ratio—that is, $\alpha = (1 + \sqrt{5})/2$, and
let β be the conjugate of α—that is, $\beta = (1 - \sqrt{5})/2$.
Then with integer index offset s,*

$$\boxed{F_n = F_s \alpha^{n-s} + F_{n-s} \beta^s, \quad \text{for all } n, s \text{ in } \mathbb{Z}} \qquad (14.3)$$

Proof:

$\langle F_n \rangle$ **Base case:** We show that (14.3) is true for $n = 0$, that is

$$F_0 = F_s \alpha^{-s} + F_{-s} \beta^s.$$

Using $F_{-s} = (-1)^{s+1} F_s$ (p.29), and $\beta = (-1)\alpha^{-1}$ (from $\alpha\beta = -1$, p.4)

$$F_0 = F_s \alpha^{-s} + F_s(-1)^{s+1} \cdot (-1)^s \alpha^{-s}.$$

And as $(-1)^{2s} = +1$,   then $(-1)^{2s+1} = -1$,

$$F_0 = F_s \alpha^{-s} - F_s \alpha^{-s}$$
$$= 0.$$

This verifies the base case according to the $F_0 = 0$ definition in (14.1).

$\langle F_n \rangle$ **Inductive step for $n \geq 0$:**

(Writing (14.3) in j), if we assume that for any integer j it is true that

$$F_j = F_s \alpha^{j-s} + F_{j-s} \beta^s, \quad \text{for all } s \text{ in } \mathbb{Z} \tag{14.4}$$

then we must show that

$$F_{j+1} = F_s \alpha^{j+1-s} + F_{j+1-s} \beta^s. \tag{14.5}$$

To do this, we let A equal the right-hand side of (14.5), and then we substitute F_s with its Binet form (14.2) $F_s = (\alpha^s - \beta^s)/\sqrt{5}$;   and also substitute F_{j+1-s} similarly

$$A = \left(\frac{\alpha^s - \beta^s}{\sqrt{5}} \right) \alpha^{j+1-s} + \left(\frac{\alpha^{j+1-s} - \beta^{j+1-s}}{\sqrt{5}} \right) \beta^s$$

$$\sqrt{5} A = (\alpha^s - \beta^s) \alpha^{j+1-s} + \left(\alpha^{j+1-s} - \beta^{j+1-s} \right) \beta^s$$

$$= \alpha^{j+1} - \alpha^{j+1-s} \beta^s + \alpha^{j+1-s} \beta^s - \beta^{j+1}$$

$$A = \left(\alpha^{j+1} - \beta^{j+1} \right) / \sqrt{5}.$$

We now identify A as being the Binet form for F_{j+1}, thus demonstrating equality in (14.5). This completes the proof of the theorem for $n \geq 0$. Now to address n negative, we shall make the inductive step *backwards*—this technique is called 'reverse induction'.

$\langle F_n \rangle$ Reverse inductive step for $n < 0$:

If we again consider the main theorem (14.3) written in j, and assume that for any integer j it is true that

$$F_j = F_s \alpha^{j-s} + F_{j-s}\beta^s, \quad \text{for all } s \text{ in } \mathbb{Z} \qquad (14.4), \text{p.203}$$

then (stepping back one) we must show that

$$F_{j-1} = F_s \alpha^{j-1-s} + F_{j-1-s}\beta^s. \qquad (14.6)$$

To do this, we let B equal the right-hand side of (14.6) and substitute F_s with its Binet form (14.2) $F_s = (\alpha^s - \beta^s)/\sqrt{5}$; and similarly F_{j-1-s}

$$B = \left(\frac{\alpha^s - \beta^s}{\sqrt{5}}\right)\alpha^{j-1-s} + \left(\frac{\alpha^{j-1-s} - \beta^{j-1-s}}{\sqrt{5}}\right)\beta^s$$

$$\sqrt{5}A = \left(\alpha^s - \beta^s\right)\alpha^{j-1-s} + \left(\alpha^{j-1-s} - \beta^{j-1-s}\right)\beta^s$$

$$= \alpha^{j-1} - \alpha^{j-1-s}\beta^s + \alpha^{j-1-s}\beta^s - \beta^{j-1}$$

$$B = \left(\alpha^{j-1} - \beta^{j-1}\right)/\sqrt{5}.$$

We now identify B as being the Binet form for F_{j-1}, thus demonstrating equality in (14.6), which proves the (Fibonacci) theorem (14.3) for negative n, and thus completes its proof for all integer n. ∎

The Lucas numbers and their Binet form

The Lucas number sequence begins:

$$\langle 2, 1, 3, 4, 7, 11, 18, 29, 47, 76, 123, \ldots \rangle.$$

As companions to the Fibonacci numbers, the Lucas numbers share exactly the same recurrence relation and second seed, but have their first seed as 2 instead of 0, that is

$$L_n = L_{n-1} + L_{n-2}, \quad \text{with } L_0 = 2, \quad \text{and } L_1 = 1. \qquad (14.7)$$

Lucas numbers too have a Binet form,

$$L_n = \alpha^n + \beta^n = \phi^n + (-\phi)^{-n}. \qquad (14.8)$$

To complement the Fibonacci number formula (14.3), we have one very similar for the Lucas numbers, which follows.

The Fibonacci Resonance—Lucas Corollary:

*Let L_n be the nth Lucas number, and again
let α be the Golden Ratio—that is, $\alpha = (1 + \sqrt{5})/2$, and
let β be the conjugate of α—that is, $\beta = (1 - \sqrt{5})/2$.
Then with integer index offset s,*

$$L_n = \sqrt{5}F_s\alpha^{n-s} + L_{n-s}\beta^s, \quad \text{for all } n, s \text{ in } \mathbb{Z} \tag{14.9}$$

Proof:

$\langle L_n \rangle$ **Base case:**   We show that (14.9) is true for $n = 0$, that is

$$L_0 = \sqrt{5}F_s\alpha^{-s} + L_{-s}\beta^s.$$

Using $L_{-s} = (-1)^s L_s$ (p.29),   and $\beta = (-1)\alpha^{-1}$ (from $\alpha\beta = -1$, p.4)

$$L_0 = \sqrt{5}F_s\alpha^{-s} + (-1)^s L_s(-1)^s\alpha^{-s}$$

$$= \alpha^{-s}\left(\sqrt{5}F_s + L_s\right).$$

We now substitute for F_s and L_s with their Binet forms
(14.2) and (14.8):

$F_s = (\alpha^s - \beta^s)/\sqrt{5}$,   and

$L_s = \alpha^s + \beta^s$, so

$$L_0 = \alpha^{-s}\left[\sqrt{5}\left(\frac{\alpha^s - \beta^s}{\sqrt{5}}\right) + (\alpha^s + \beta^s)\right]$$

$$= \alpha^{-s}\left(\alpha^s - \beta^s + \alpha^s + \beta^s\right)$$

$$= \alpha^{-s}\left(2\alpha^s\right)$$

$$= 2.$$

This verifies the base case according to the $L_0 = 2$ definition in (14.7).

$\langle L_n \rangle$ Inductive step for $n \geq 0$:

If write (14.9) in j, and we assume for any integer j it is true that

$$L_j = \sqrt{5}F_s\alpha^{j-s} + L_{j-s}\beta^s, \quad \text{for all } s \text{ in } \mathbb{Z} \tag{14.10}$$

then we must show that

$$L_{j+1} = \sqrt{5}F_s\alpha^{j+1-s} + L_{j+1-s}\beta^s. \tag{14.11}$$

To do this, we let C equal the right-hand side of (14.11), and we substitute with Binet forms—again with both F_n and L_n forms:

$$C = \sqrt{5}\left(\frac{\alpha^s - \beta^s}{\sqrt{5}}\right)\alpha^{j+1-s} + \left(\alpha^{j+1-s} + \beta^{j+1-s}\right)\beta^s$$
$$= \alpha^{j+1} - \alpha^{j+1-s}\beta^s + \alpha^{j+1-s}\beta^s + \beta^{j+1}$$
$$= \alpha^{j+1} + \beta^{j+1}.$$

And as we identify C as the Binet form for L_{j+1}, we therefore demonstrate equality in (14.11). This completes the proof of the Corollary for $n \geq 0$. Now to address n negative, we use reverse induction (exactly as we did for the Fibonacci case).

$\langle L_n \rangle$ Reverse inductive step for $n < 0$:

If we again consider the Corollary (14.9) written in j, and assume that for any integer j it is true that

$$L_j = \sqrt{5}F_s\alpha^{j-s} + L_{j-s}\beta^s, \quad \text{for all } s \text{ in } \mathbb{Z} \tag{14.10} \text{ copy}$$

then (stepping back) we must show that

$$L_{j-1} = \sqrt{5}F_s\alpha^{j-1-s} + L_{j-1-s}\beta^s. \tag{14.12}$$

To do this, we let D equal the right-hand side of (14.12), and we substitute with Binet forms as before:

$$D = \sqrt{5}\left(\frac{\alpha^s - \beta^s}{\sqrt{5}}\right)\alpha^{j-1-s} + \left(\alpha^{j-1-s} + \beta^{j-1-s}\right)\beta^s$$
$$= \alpha^{j-1} - \alpha^{j-1-s}\beta^s + \alpha^{j-1-s}\beta^s + \beta^{j-1}$$
$$= \alpha^{j-1} + \beta^{j-1}.$$

And as we identify D as the Binet form for L_{j-1}, we therefore demonstrate equality in (14.12), which proves the (Lucas) Corollary (14.9) for negative n, and thus completes its proof for all integer n. ∎

Chapter 15

Visualizing the effect

Analysis of the F_n and L_n formulæ

We recall the equation pair (13.17) from page 200,

$$\left.\begin{array}{rl} F_n = & F_s\,\phi^{n-s} \;+\; F_{n-s}\,(-\phi)^{-s} \\[2mm] L_n = & \sqrt{5}\,F_s\,\phi^{n-s} \;+\; L_{n-s}\,(-\phi)^{-s} \end{array}\right\} \quad \text{for all } n, s \text{ in } \mathbb{Z}. \quad (15.1)$$

But looking at these formulæ, it is not obvious why they may be described as 'resonant'. We need to get inside them somehow—to experience what they are saying. In this chapter we shall relate them to Golden Spirals, and in the next chapter we shall extend the visualization ideas in order to consider a sound analogy.

First terms: Backbone growth

To begin the analysis, it will help us to consider the simplest cases where n, s, and $n-s$ are all greater than zero. As a result, the first terms in (15.1) will dominate ('big'), while the second terms provide adjustments ('small'). We note that both these first terms $F_s\,\phi^{n-s}$ and $\sqrt{5}\,F_s\,\phi^{n-s}$ are based on $F_s\,\phi^{n-s}$, and as such they might be regarded as 'backbone-growth terms'. Granted, they vanish when $s=0$ (as $F_0=0$), but otherwise (s given, $s \neq 0$), they share the same regular growth rate of ϕ per step increment in n. Also, L_n's first term *leads in growth* by a constant factor of $\sqrt{5}$. This must come solely from the kick-start given by the first seed $L_0=2$, (as compared with $F_0=0$), because all else remains the same—that is, the other seed and the recurrence relation.

Second terms: Adjustments using Golden Quanta

In (15.1), if we now focus on the second terms and also consider s as *given*, then we might regard $(-\phi)^{-s}$ as a fixed 'Golden Quantum'. In which case, let us name it q_s, and to visualize it simply, we will prefer it to be without sign—that is, as $|(-\phi)^{-s}|$. This means we must account for its sign separately with $(-1)^s$, writing

$$
\left.
\begin{aligned}
q_s &= \quad \phi^{-s} \\[4pt]
F_n &= \quad F_s\,\phi^{n-s} + (-1)^s\,F_{n-s}\,q_s \\[4pt]
L_n &= \sqrt{5}\,F_s\,\phi^{n-s} + (-1)^s\,L_{n-s}\,q_s
\end{aligned}
\right\} \quad \text{for all } n, s \text{ in } \mathbb{Z}. \qquad (15.2)
$$

In this new light we notice the *pattern of adjustment* for F_n (the q_s term) is just *the sequence itself* written in numbers of quanta and lagged by s places, hence $\langle F_{n-s} \rangle$. And similarly, the adjustments for L_n refer to the Lucas sequence *itself*—repeating *its* pattern, again with a lag of s places.

The discrete and the continuous— again visualized using spirals

As we saw at the end of Chapter 2 (*Spirals*, p.47), by positioning number points near to a spiral having the same growth rate as the underlying growth rate of the number sequence, we can quickly (and compactly) compare continuous growth with discrete integer instances. Let's recap to be completely clear about the technique as we shall be using it extensively from now on. Here we have the discrete sequences $\langle F_n \rangle$ and $\langle L_n \rangle$, and in each, with increasing n, the ratio of one number to its predecessor tends towards a common growth factor, ϕ. Again we shall think of this growth as an underlying continuous function—a logarithmic spiral, specifically the Golden Spiral. We shall then regard integer sequence members as being 'variations from' or 'adjustments to' this reference. Now, we know that the Fibonacci sequence grows at approximately the Golden Ratio per step (Kepler), and we know that the Golden Spiral grows at exactly the Golden Ratio per quadrant—by definition. (This rate is chosen to correspond to the shape that arises when we grow a Golden Rectangle by adding whirling squares—p.36). Accordingly, the growth per full turn of the Golden Spiral will be four such successive ϕ growths compounded—that is, ϕ^4 per full turn.

In the spiral, if we measure the anticlockwise turning of the radial vector with θ, then the fraction of a turn will be $\theta/(2\pi)$. Using (r, θ) polar coordinates, and with k representing a scaling constant (yet to be determined), we choose the Golden Spiral draft form

$$r = k\,(\phi^4)^{\frac{\theta}{(2\pi)}} \;\; = k\,\phi^{\frac{4\theta}{(2\pi)}}. \tag{15.3}$$

Again using polar coordinates, we shall position the $\langle F_n \rangle$ numbers at precise locations near to this spiral, in such a way as to help us study their relationship to it. We place the F_n sequence round and round the origin ('polar Fibonacci', spiral fashion—p.45), at a distance from the origin of $r = F_n$ and on successive quadrant radial axes.

$$\langle F_n \rangle \text{ numbers are placed at} \quad (r,\theta) = \left(F_n, \;\; n \cdot \frac{(2\pi)}{4} \right). \tag{15.4}$$

This 'one quadrant per step of n' corresponds to a 'one power of ϕ' growth in the spiral, and positive growth is anticlockwise.

Anchoring the spiral

Now, to match the backbone-growth terms in (15.2) we want a spiral which is appropriately 'anchored'—that is, a spiral that passes through the placed number point F_s (and there has spoke length $r = F_s$). Looking back at our draft Golden Spiral formula (15.3) above, we see that such an anchoring will happen when $k = F_s$ and the power of ϕ is zero—that is, when $r = F_s\phi^0$ (as $\phi^0=1$). Further, in (15.4) we see that this F_s number point will have been placed at $\theta = s \cdot (2\pi)/4$ (i.e. an angle of s quadrants). Now this, or the same but rearranged as $s = 4\theta/(2\pi)$, in turn fixes the value of θ where the said power of ϕ is required to go to zero—so we can simply say that:

'When $\dfrac{4\theta}{(2\pi)} = s,$ then $\dfrac{4\theta}{(2\pi)} - s$ goes to zero.'

Therefore for Golden Spiral 'number s' we let

$$r = F_s\,\phi^{\frac{4\theta}{(2\pi)} - s}. \tag{15.5}$$

Hence, from now on, we will think of index s as our *spiral identifier*, and every placed Fibonacci number will have its own Golden Spiral anchored onto it. Also, in diagrams we shall mark the anchor point with a green square.

And another way of looking at the ϕ^0 situation, is to look in (15.2) at the first term of the F_n formula (that is, $F_s\,\phi^{n-s}$), and see that ϕ^0 occurs when $n-s=0$ (i.e. when $n=s$). This means that the second term F_{n-s} will go to zero because $F_0=0$, which (reassuringly) confirms that the anchor point is the unique position on the spiral where there are no adjustment quanta. This is why sometimes we shall call the anchor point the 'zero-q' (zero quanta) point.

Motivation

Our overall purpose here is to demonstrate that using this configuration—with a spiral given by (15.5) and quadrant axes according to (15.4)—then the distance along an axis between the spiral's crossing point and a number point F_n is a (hereby known) whole number of quanta. Let's consider the 1-based and 5-based examples in terms of spirals.

The s=1 Golden Spiral—with quantum q_1

As we saw earlier (p.197), the identity $\phi^n = F_n\phi + F_{n-1}$ may be restated in quantum terms:

$$F_n = \phi^{n-1} - F_{n-1}\,q_1 \qquad where \ q_1 = \phi^{-1} \tag{15.6}$$

We explore this relationship in Fig. 141. If we compare this (15.6) with the equation that we have based on q_5 (Eqn. (13.6), p.196), we find the quantum here, q_1, is much bigger—in fact, ϕ^4 times bigger, and this change of scale is reflected for example in the reciprocal/quadratic relations (which we cover in more detail in Appendix C, page 421).

$$\frac{1}{q_1} = q_1 + 1, \qquad \frac{1}{q_5} = q_5 + 11 \tag{15.7}$$

Fibonacci number	equals	s=1 spiral spoke	less	quantized difference
$F_0 = 0$	=	ϕ^{-1}	–	$(+1)\,q_1$
$F_1 = 1$	=	1_{Anchor}	–	$(0)\,q_1$
$F_2 = 1$	=	ϕ	–	$(1)\,q_1$
$F_3 = 2$	=	ϕ^2	–	$(1)\,q_1$
$F_4 = 3$	=	ϕ^3	–	$(2)\,q_1$
$F_5 = 5$	=	ϕ^4	–	$(3)\,q_1$
$F_6 = 8$	=	ϕ^5	–	$(5)\,q_1$
$F_7 = 13$	=	ϕ^6	–	$(8)\,q_1$
$F_8 = 21$	=	ϕ^7	–	$(13)\,q_1$
$F_9 = 34$	=	ϕ^8	–	$(21)\,q_1$
$F_{10} = 55$	=	ϕ^9	–	$(34)\,q_1$
...				

Figure 141: **Relating the F_n sequence to the 1-anchored Golden Spiral**

Here we see the quantized differences for a 1-anchored Golden Spiral, with $s=1$ and $F_s=1$. With each new row, we turn a quadrant of the Golden Spiral, and the spoke length (radial vector) grows by ϕ. The quantization effect is shown, with Golden Quantum $q_1=\phi^{-1}$. The Fibonacci-count sequence (column on right) lags by 1 place with respect to the first column.

Dilation by 5   ≡   rotation by 58. 99 00 33 11°

Next, we again consider the equivalence of dilation and rotation for a log spiral (p.38). In Fig. 142, we start with a Golden Spiral that passes through polar coordinates $(1,0)$—this anchor point being marked with a green square. We then compare it with a (different) Golden Spiral, green-square anchored at polar coordinate $(5,0)$. One way to get from the 1-anchored spiral to the 5-anchored case, would be simply to scale it 5× bigger—that is, dilate it. But another way to achieve this same result, would be to rotate the 1-anchored spiral by 58.99003311°—so that its point A (already 5 from centre) coincides with the 5 anchor. The reason for this is that rotating the spiral 58.99003311° in one direction is equivalent to rotating it 301.009669° in the opposite direction. This is important because if we consider a radial vector, and we move along the curve in the direction of growth, until it has grown 5 times, then our vector will have turned 301.009669°. To calculate this angle we use a ratio of logarithms—comparing growth by 5 with the ϕ^4 growth of a full turn.

$$\frac{log(5)}{4\,log(\phi)} \cdot (2\pi) \;=\; 5.2536098\ldots \quad \text{radians} \qquad\qquad (15.8)$$

$$\equiv\; 301.0096689\ldots° \;=\; 360\,-\,\textbf{58.99003311}\ldots°$$

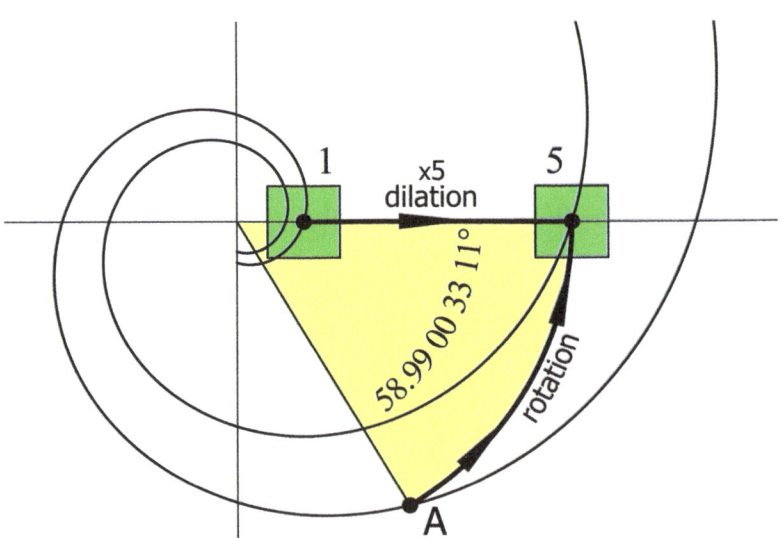

Figure 142: **Scaling a unit-anchored Golden Spiral by 5.**

Lucas axes offset   150.505°

We recall the quantization equation set (15.2) from page 208—that is,

$$\left.\begin{array}{l} q_s = \quad \phi^{-s} \\[2mm] F_n = \quad F_s\,\phi^{n-s} + (-1)^s\, F_{n-s}\, q_s \\[2mm] L_n = \sqrt{5}\, F_s\,\phi^{n-s} + (-1)^s\, L_{n-s}\, q_s \end{array}\right\} \quad \text{for all } n, s \text{ in } \mathbb{Z}.$$

In it we see that L_n's first term has a $\sqrt{5}$ scaling factor. We also recall devising a continuous expression to represent the backbone growth of the Fibonacci numbers (the $F_s\,\phi^{n-s}$ term)—in the form of an anchored Golden Spiral—Equation (15.5)

$$r = F_s\,\phi^{\frac{4\theta}{(2\pi)}-s}.$$

Now, in this context of log spiral growth (and starting to think about the Lucas numbers), we ask the question: *'If one spoke is longer than another by a factor of $\sqrt{5}$, then what is the angle between them?'* Again we note that a Golden Spiral grows by ϕ per quadrant of $(2\pi)/4$ radians, therefore, using the same logic as before (in dilation/rotation on page 212), we find the angle in (15.9), and we show it in Fig. 143.

$$\text{Angle} \quad \omega = \frac{log(\sqrt{5})}{4\,log(\phi)} \cdot (2\pi) \;=\; 2.62680\ldots \;\equiv\; \mathbf{150.505\ldots^{\circ}} \qquad (15.9)$$

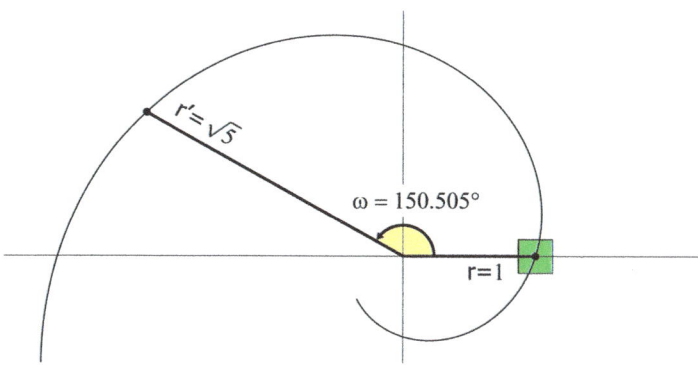

Figure 143:
In a Golden Spiral, $\sqrt{5}$ growth is equivalent to rotation by 150.505°.

Same spiral, same quantum size

We note that ω is constant, independent of n and s.[1] So, if we keep the existing 'F_s anchored spiral' fixed in place and rotate a set of quadrant axes underneath it by angle ω in the direction of increasing growth (see Fig.145, p.217), then the $\sqrt{5}$ scaling will be taken care of for all combinations of n and s. Hence, in this way, we may use the same spiral 'backbone' for both $\langle F_n \rangle$ and $\langle L_n \rangle$ and use the same quantum size q_s for adjustments. Accordingly we offset the 'Lucas axes' by $+\omega$ from the 'Fibonacci axes'.

$$
\left.
\begin{aligned}
\langle F_n \rangle \text{ numbers are placed at} \quad (r, \theta) &= \left(F_n, \ n \cdot \frac{(2\pi)}{4} \right) \\[2em]
\langle L_n \rangle \text{ numbers are placed at} \quad (r, \theta) &= \left(L_n, \ n \cdot \frac{(2\pi)}{4} + \omega \right)
\end{aligned}
\right\}
$$

$$(15.10)$$

Equation pair (15.10) provides both the quadrant axes and the companion (offset) axes. The Fibonacci numbers' quantized adjustments take place along the primary quadrant axes—as we specified in Equation (15.4) on page 209. And now the Lucas numbers demonstrate a complementary pattern, with their quantized adjustments occurring along the offset quadrant axes.

This combined scheme shows

Fibonacci number = Golden Spiral quadrant spoke
+ quantized adjustment

Lucas number = Golden Spiral offset quadrant spoke
+ quantized adjustment

---

[1] That is, the n and s of F_n and q_s as in (15.2) on page 208.

Relating 5× dilation to the Lucas offset angle

We saw earlier that scaling a Golden Spiral by a factor of 5 was equivalent to rotating it by 301.01° in the direction of growth, and that was equivalent to a rotation of 58.99° in the opposite direction—by 'equivalent' here, we mean resulting in the same outcome. And we have just seen that the required offset of the Lucas axes is 150.505...° Could there be some link here? Well yes certainly, as the growth needed for the Lucas offset was $\sqrt{5}$ needing 150.505° rotation, and for 5× growth we need $\sqrt{5} \times \sqrt{5}$—in other words, we must rotate 150.505° twice, which comes to 301.01°.

A numeric coincidence   2.222°

The previous result means that the angle between the offset axes and their nearest main quadrant axes is $180 - 150.505 = 29.495°$. As an aside, what if we compare this angle with the diagonal angle of a Golden Rectangle, viz. $arctan(1/\phi) = 31.717°$? We find the difference

$$arctan\left(\frac{1}{\phi}\right) \quad - \quad (180° - 150.505...°) \quad = \quad 2.222...° \qquad (15.11)$$

(although with more decimal places, the twos do not recur).

Example: s=5 Golden Spiral

Let's choose an example Golden Spiral configuration with say, $s = 5$. In this case s is odd, so $(-1)^s = -1$. Therefore we may rewrite (15.2) as

$$\left. \begin{aligned} q_5 &= \phi^{-5} \\ F_n &= F_5 \phi^{n-5} - F_{n-5} q_5 \\ L_n &= \sqrt{5} F_5 \phi^{n-5} - L_{n-5} q_5 \end{aligned} \right\} \quad \text{for all } n \text{ in } \mathbb{Z}. \qquad (15.12)$$

We explore this sample case in detail in the diagram Fig.145 (p.217) and in the Tables in Figs. 144 and 146.

215

Fibonacci number	=	spiral Fspoke	−	quantized difference
$F_0 = \underline{\mathbf{0}}$	=	$5\phi^{-5}$	−	$(+5)\,q_5$
$F_1 = 1$	=	$5\phi^{-4}$	−	$(-3)\,q_5$
$F_2 = 1$	=	$5\phi^{-3}$	−	$(+2)\,q_5$
$F_3 = 2$	=	$5\phi^{-2}$	−	$(-1)\,q_5$
$F_4 = 3$	=	$5\phi^{-1}$	−	$(+1)\,q_5$
$F_5 = 5$	=	5_{Anchor}	−	$\underline{\mathbf{0}}\ q_5$
$F_6 = 8$	=	5ϕ	−	$(\,1\,)\,q_5$
$F_7 = 13$	=	$5\phi^2$	−	$(\,1\,)\,q_5$
$F_8 = 21$	=	$5\phi^3$	−	$(\,2\,)\,q_5$
$F_9 = 34$	=	$5\phi^4$	−	$(\,3\,)\,q_5$
$F_{10} = 55$	=	$5\phi^5$	−	$(\,5\,)\,q_5$
...				

Figure 144: **Fibonacci: One spiral (s=5), many number points F_n.**

The Table Fig. 144 demonstrates the example Equation (15.12), p.215, with its constant quantum size of $q_5 = \phi^{-5} = 5\phi - 8$. We see how the Fibonacci sequence $\langle F_n \rangle$ 'refers to itself' in its quantized-difference terms—that is, the *Fibonacci numbers* have associated quantized differences in a lagged *Fibonacci-number* sequence. This lagging reveals negative-indexed members of the Fibonacci-number sequence—ones whose values alternate in sign (p.29). Now, while these are sometimes thought of as purely theoretical 'possible extensions' to the sequence in $\mathbb{Z}_{\geq 0}$; here they extend quite naturally and seamlessly into $\mathbb{Z}_{<0}$. In this example with s=5, the lag is 5 places and the 'backbone' spiral is anchored on $F_5=5$. For comparison, an s=4 spiral would pass through the placed $F_4=3$, and the adjustments in that case would then have quantum size ϕ^{-4}. There exist an infinity of such spiral relationships—one for each Fibonacci number.

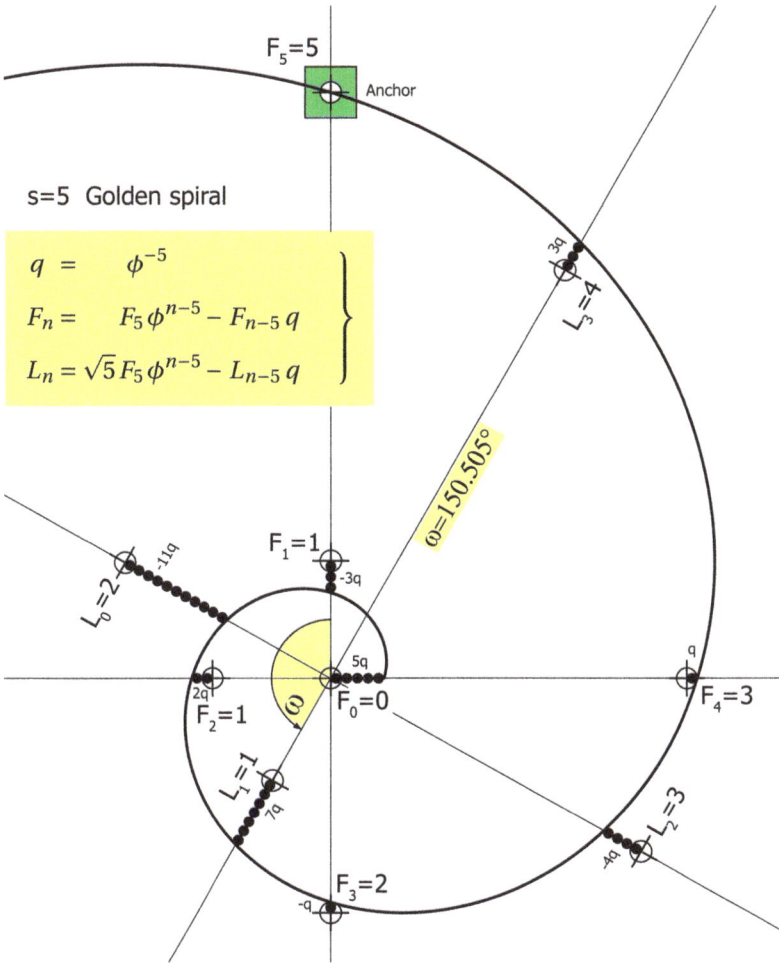

$F_5 = 5$

Anchor

s=5 Golden spiral

$$q = \phi^{-5}$$

$$F_n = F_5\,\phi^{n-5} - F_{n-5}\,q$$

$$L_n = \sqrt{5}\,F_5\,\phi^{n-5} - L_{n-5}\,q$$

$\omega = 150.505°$

$F_1 = 1$

$F_2 = 1$

$F_0 = 0$

$F_4 = 3$

$L_0 = 2$

$L_1 = 1$

$L_2 = 3$

$L_3 = 4$

$F_3 = 2$

Figure 145: **Example of the Fibonacci Resonance (s=5).**

In Fig. 145 we see Fibonacci and Lucas numbers and their quantized relationships to a Golden Spiral according to Equation (15.12), p.215. This particular Golden Spiral is anchored at $F_5 = 5$ (which is therefore its zero-q point). $\phi = (1 + \sqrt{5})/2 = 1.618...$ the Golden Ratio. Offsetting the 'L' axes by $\omega = 150.505°$ corresponds to a spiral growth of $\sqrt{5}$ which matches the underlying growth lead of the Lucas numbers. The unfilled circles locate number points F_n and L_n. The filled circles—'abacus beads'—measure out the adjustment by counting quanta.

Lucas number	=	offset Lspoke	−	quantized adjustment
$L_0 =$ **2**	=	$5\sqrt{5}\,\phi^{-5}$	−	$(-11)\,q_5$
$L_1 =$ 1	=	$5\sqrt{5}\,\phi^{-4}$	−	$(+7)\,q_5$
$L_2 =$ 3	=	$5\sqrt{5}\,\phi^{-3}$	−	$(-4)\,q_5$
$L_3 =$ 4	=	$5\sqrt{5}\,\phi^{-2}$	−	$(+3)\,q_5$
$L_4 =$ 7	=	$5\sqrt{5}\,\phi^{-1}$	−	$(-1)\,q_5$
$L_5 =$ 11	=	$5\sqrt{5}$	−	**2** q_5
$L_6 =$ 18	=	$5\sqrt{5}\,\phi$	−	$(1)\,q_5$
$L_7 =$ 29	=	$5\sqrt{5}\,\phi^{2}$	−	$(3)\,q_5$
$L_8 =$ 47	=	$5\sqrt{5}\,\phi^{3}$	−	$(4)\,q_5$
$L_9 =$ 76	=	$5\sqrt{5}\,\phi^{4}$	−	$(7)\,q_5$
$L_{10} =$ 123	=	$5\sqrt{5}\,\phi^{5}$	−	$(11)\,q_5$
$\cdots$				

Figure 146: **Lucas: One spiral (s = 5), many number points L_n.**

This Table (Fig. 146) is the L_n companion for the F_n Table (Fig. 144, p.216). Again it demonstrates the (s=5) example Equation (15.12) p.215, with fixed quantum size of $q_5 = \phi^{-5}$. In this case however, we see the Lucas sequence referring to *itself* in the quantized-differences column. The *Lucas numbers* have their adjustment quanta in a lagged *Lucas number* sequence. And again, the lagging reveals negative-indexed sequence members which (as we know) also have values that alternate in sign.

No anchor: Lucas number points are never used to anchor spirals in this visualization scheme.

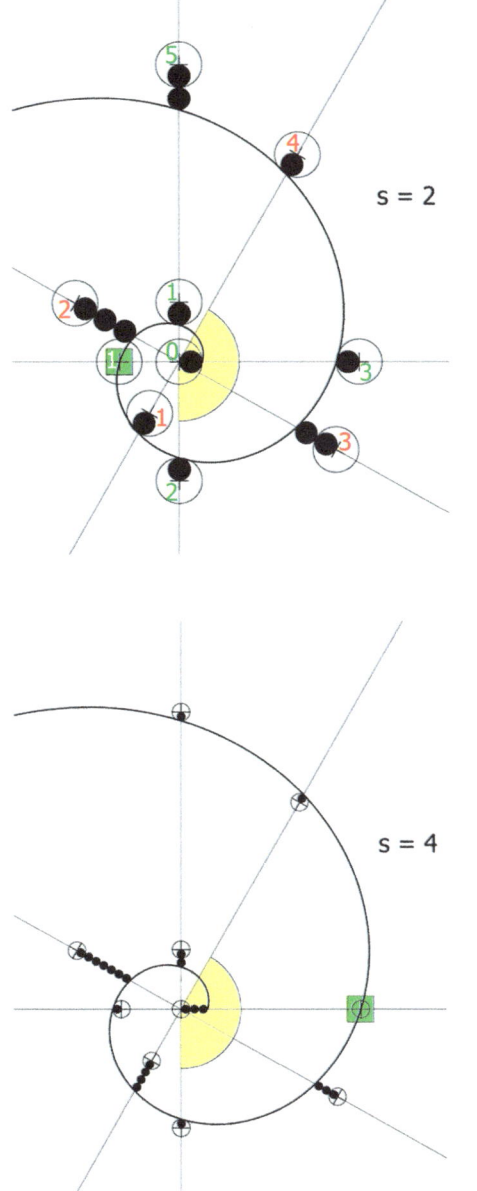

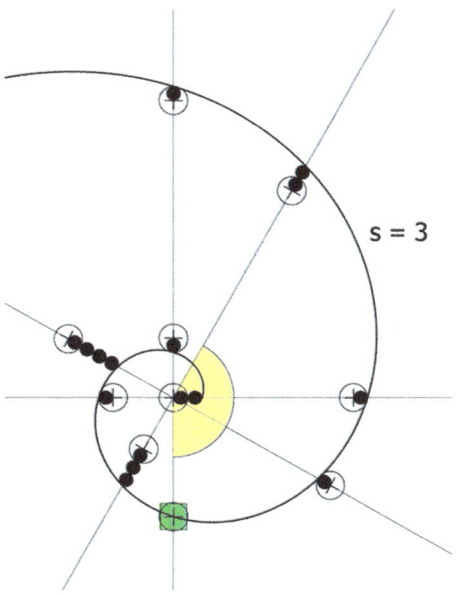

Figure 147:
Each Golden Spiral has its own associated quantum size.

Here we see three Golden Spirals, differing only by their anchor points:
$s = 2$, anchored at $F_2 = 1$,
$s = 3$, anchored at $F_3 = 2$, and
$s = 4$, anchored at $F_4 = 3$.

Using (15.4) from page 209 these green-square anchors are placed at

$$(r, \theta) = \left(F_s, \quad s \cdot \frac{(2\pi)}{4} \right).$$

So, for example (top left) the '$s = 2$, $F_2 = 1$' anchor is placed at radial distance 1 from centre and rotation of two quadrants—i.e.

$$(r, \theta) = \left(1, \quad 2 \cdot \frac{(2\pi)}{4} \right).$$

Self-similar, fractal-like replication

When we choose different values for s as in Fig.147 (p.219), we see (relative to the anchor) self-similar patterns of quantized adjustments when s is even, and the same (but sign-reversed) when s is odd. Each spiral has a different associated quantum size—which scales one to the next by the Golden Ratio ϕ. Let us consider the anchor point on the spiral where the spoke length is equal to a particular Fibonacci number F_s. From this zero-q reference location, we look along the spiral and see a pattern of quantized adjustments with quantum size q_s. Now, if we move say two quadrants anticlockwise and anchor a spiral at F_{s+2}, and then we look along that spiral, we shall see the same quantized pattern of adjustments relative to *that* point, but this time with quanta ϕ^2 times smaller. We see this in Fig.147 if we compare the $s=2$ and $s=4$ spirals—looking at the adjustment sequence out either side of the anchor. Thus we realize this self-similarity of adjustment patterns repeats at smaller and smaller quantum sizes without end. We shall return to this fractal-like behaviour shortly, and consider it in more detail.

Unity sized quanta

But what about the $s=0$ case of $F_s=0$? The spiral $s=0$ is *degenerate* because all its spoke lengths have gone to zero; it is just a point—the origin. Also, as the $s=0$ quanta (q_0) are size one, the distance in quanta of a Fibonacci or Lucas number from the 'point spiral' is just the number itself—that is, the Fibonacci and Lucas numbers are at radial distances (measured in quanta) equal to the numbers themselves.

New context

We may now think of this previously very familiar relationship (the simple number-distance from zero) in the context of all the adjacent *non-degenerate* spirals and their relationships. In fact, this $F_s=0$ case is the *only one* where the 'backbone spiral plus quantized adjustments' visualization of growth is hidden from us.

Chapter 16

Hearing it

A sound analogy

As a way to hold a picture in our minds of the adjustment pattern along one spiral, and then many, it will surely help to relate the effect to something familiar. Indeed the Fibonacci Resonance takes its name after a sound analogy that we shall now explore. In our 'two sets of quadrant axes' configuration, if we consider any particular anchored Golden Spiral, we know it will have a quantized relationship with all the Fibonacci and Lucas number points. A sound engineer might look at (say) Fig. 148, and imagine the groups of same-sized abacus beads as 'standing waves' on a string. Thought of this way, the act of anchoring a Golden Spiral onto a Fibonacci number point will 'bring the spiral into resonance' with all the other Fibonacci number points and all the Lucas number points at the same time—that is, into resonance with an infinity of such points.

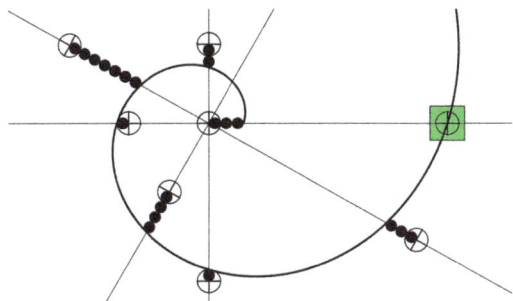

Figure 148: **Considering quanta as resonant standing waves.**

More on resonance—a taut string

In the *Art—Mondrian* chapter (p.110) we considered water resonating in a bathtub. Here, we imagine a string stretched between two points—a violin string for example—then in Fig. 149 we see a few of the possible ways in which it might vibrate. These different modes of vibration are called 'harmonics'.

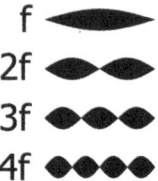

Figure 149: **Harmonics.** The higher harmonics have multiples of the fundamental frequency 'f'—the first harmonic.

The first harmonic is shown as 'f' (also called the fundamental), and the second as '2f', and so on. The frequency of vibration (number of vibrations per second) depends on the tension and density of the string, but if we keep these constant, then the '2f' and '3f' (and so on) vibrations will be at double and triple the frequency. Such shapes of vibration are called 'standing-wave' patterns. We see the same effect close to sea cliffs. As waves come in, they are reflected from the cliff, and the combination of incoming and reflected waves produce a steady pattern of 'node' lines (where the water is still) and in between these, 'anti-nodes', where the water rises and falls. No travelling waves are seen near the cliff. And just the same thing happens in the string—waves are travelling along the string, but because they reflect at each end, bounce back and add and subtract, we see a single elongated almond 'envelope' shape that contains the rapidly moving string. If the vibration of the string is being driven by say sound from a loudspeaker, or the drawing across of a bow, the energy in the vibration builds up and up as the wave is strengthened and reflected back on itself again and again. In the '2f' case, we see two connected envelopes, and if the string in one is 'up', then in the other it will be 'down', and vice versa. Together these make one full wave cycle ('zero, up, back to zero, through and down, back to zero'), and so the end-to-end distance is called a 'wavelength'.

Now, if instead of considering the possibilities of multiple frequencies, we consider the same frequency but with multiples of the string length, we shall arrive at Fig. 150.

We imagine the string used continues to be of the same material, and it is tightened to the same tension, but its length is multiplied up exactly. The shape of the fundamental on the short string is repeated to constitute a higher harmonic on the longer string—but— the frequency of vibration remains the same. If all these four strings were still and the shortest was plucked at its centre, then if the other strings were close, they would start to vibrate in response just as in Fig. 150 but with reduced strength. This effect is called 'sympathetic resonance'.

Figure 150: **'Standing-wave' resonance**—same type of string, same tension, same frequency, but different string lengths. These lengths are an exact multiple of half a wavelength. We conventionally denote a whole wavelength as λ.

This situation where different-length strings all vibrate at the same frequency is the one that provides us with an analogy for the Fibonacci Resonance (Fig. 151). Each anchored spiral has its own associated half-wavelength (which we have called a quantum or measure) and in our polar-spiral scheme, the number points are spaced away from a spiral at an exact multiple of these half-wavelengths. We might consider just one layer and imagine we hear the spiral 'resonating' with its infinity of Fibonacci and Lucas number points: 'hmmmmmmmmmmm... '

Figure 151: **Quantized distances visualized as standing-wave patterns—then 'heard as sound'.**

Golden Noise

Now, one spiral has one associated 'wavelength'. What if we consider all the possible resonances at once—in a thought experiment—the 'sound' of the whole structure. We know that *white noise* is the sound of all audible frequencies combined. In this case we will have a comb spectrum[1] with each tooth representing a frequency that is ϕ times the previous—that is, $8\frac{1}{3}$ semitones apart.[2] Perhaps this combination should be called 'Golden Noise'. All that is required is for us to choose a starting reference frequency. A=432 is an interesting choice in that it produces near integers when multiplied by powers of ϕ (in the audio range) and we know, a similar-sized Fibonacci or Lucas number will have this property too, but the same cannot be said for A=440 [405]. The closeness to whole numbers of the A=432 frequencies depends entirely on our definition of the duration of a second. This means that if we hold the pitch of a note constant, but define a 'new second' to be 1.5 old seconds, then the note's new frequency will be 1.5 times as big—and typically not near a whole number any more. But regardless of our time measurements, if we start with a reasonably sized Fibonacci number, and we scale it successively in steps of ϕ up and down, then we shall expect to produce near integers.

To sum up, we may say: 'Every placed Fibonacci number has its own Golden Spiral, which *in a sense*, "resonates" with all the other placed Fibonacci numbers, plus all the placed Lucas numbers—as if it had its own associated pitch.'

Aside: Music from Fibonacci-related sequences

In a footnote on p.94 we mentioned the Padovan sequence, and in Appendix C we consider Gold, Silver, Bronze, and Copper Ratios, and their related sequences. From all of these, Dale Gerdemann, a computational linguistics researcher in Tübingen, Germany, has generated some surprisingly interesting computer music.[3]

---

[1] On a chart, individual frequency spikes show up like the teeth on a comb.

[2] To find this interval we consider it as a fraction of an octave—a doubling of frequency. That is, we compare ϕ with 2.0 using logs, then multiply this fraction by 12 to get the number of semitones: $[log(\phi)/log(2)] \times 12 = 8.33090$ semitones $\equiv 833$ cents. But given the irrationality of ϕ (p.422), we had better not expect sweet harmony when ϕ-based pitches are combined. Though this possibility did not deter communications engineer Heinz Bohlen, who went on to propose an 833-cents scale [71].

[3] e.g. (2011) *Tuba Fantasy*—www.youtube.com/watch?v=H7BkwoYLVSM.

Chapter 17

3D model

Fig. 152 shows a 3D model of a small region of the Fibonacci Resonance pattern near the origin. The complete structure is, as we have discussed, unbounded. We are looking downwards onto the model, which is built in plane layers. Each layer has a spiral in its plane—corresponding to an integer value of s, its spiral number. This s increases vertically, causing the ϕ^{-s} quantum sizes to shrink layer on layer by ϕ. The Fibonacci numbers are represented as thin vertical rods intersecting each layer at the same plane-coordinate positions. For example, the polar origin ($r=0$) on each level is located by the $F_0=0$ vertical rod; and the polar reference direction ($\theta=0$) is given by the direction that the F_0 beads line up—as the F_0 adjustments take place along this axis in each layer. We omit the $s=0$ level as this would simply consist of unit-size beads radiating out along the axes from the centre rod to each Fibonacci number. As we know, this is because spiral $s=0$ is degenerate—just a point at the origin. The anchor for each Golden Spiral is marked with a small green cube—fixing (one to one) a unique spiral to its own Fibonacci number vertical rod. Arranged vertically up the centre rod ($F_0=0$) the beads get smaller and smaller. The count for each size follows the Fibonacci sequence: $\langle 1, 1, 2, 3, 5, 8 \rangle$. These beads demonstrate the quantized adjustment (along a quadrant axis) required to get from a spiral to the F_0 rod. We see numbered spiral anchors progress in quarter turns according to spiral layer: $s = 1, 2, 3, 4$ (as in Fig. 2, p.45). Above the '1' anchor cube we see the same vertical pattern of beads as with the centre rod, but this time shifted vertically, starting on the next spiral up (the red, $s=2$). Similarly this repeats again above the '2' anchor.

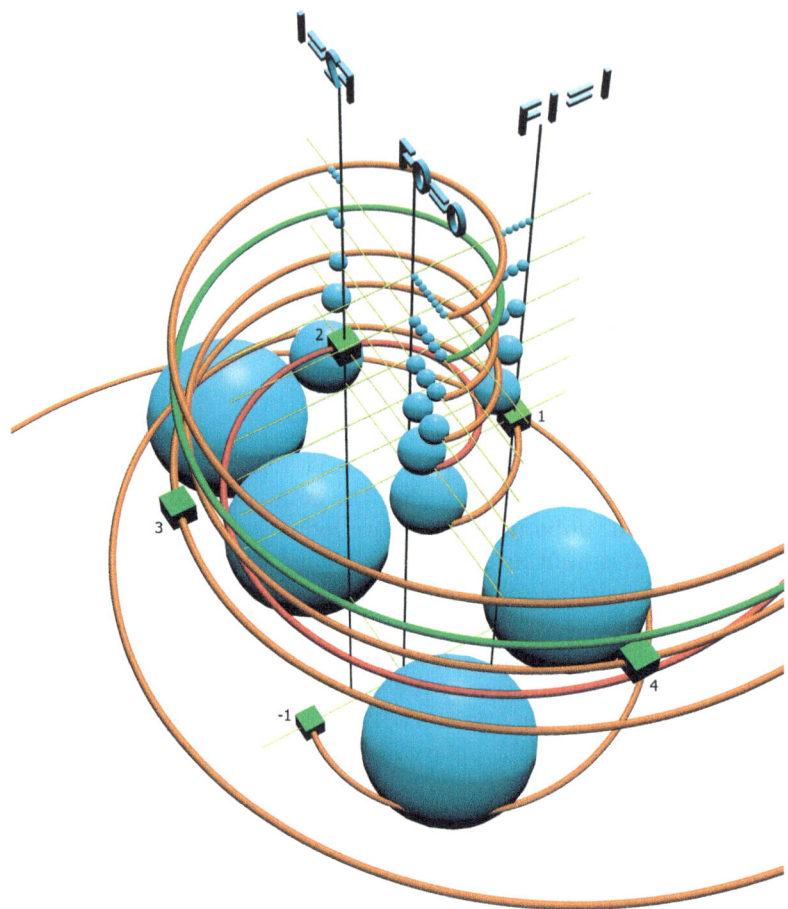

Figure 152: **Fibonacci Resonance 3D model, s = −1, 1, 2, 3, 4, 5, 6. The s=2 spiral is red; the s=5 is green; (s increases vertically).**

To the right of the green '−1' cube at the $s=-1$ spiral level, we see a (largest-size) blue bead, then 3 more same-size beads: $\langle 1,1,2 \rangle$. In order to accommodate varying quantum sizes, the vertical spacing layer to layer is changed. As the diameters of the abacus beads (i.e. quantum sizes q_s) grow by ϕ layer by layer, we see how each bead diameter is also the sum of diameters of the beads one and two sizes smaller—that is, $q_s = q_{s-1} + q_{s-2}$. Fig. 153 shows the complementary Lucas adjustment patterns—with their quadrant axes offset by $150.505°$.

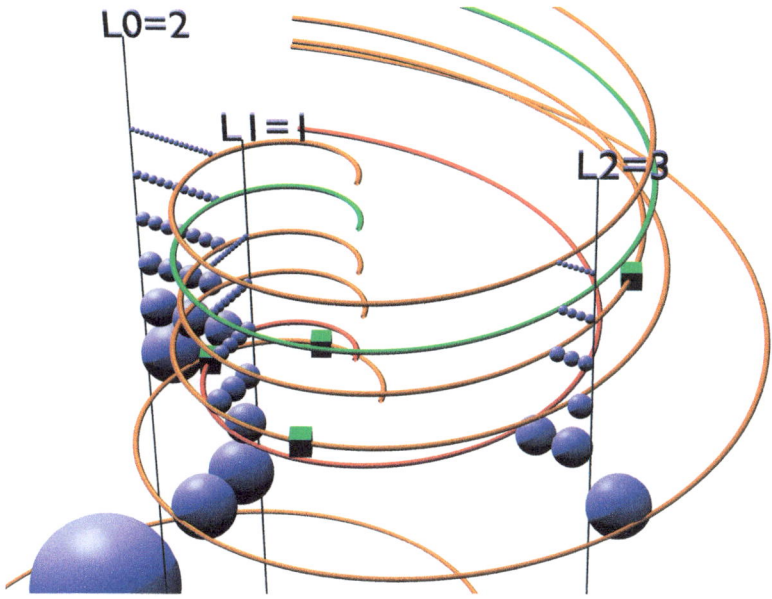

Figure 153: **Fibonacci Resonance 3D model, s = $-1, 1, 2, 3, 4, 5, 6$,** showing the Lucas (i.e. companion sequence) relationship.

Only a few example Lucas number rods and their adjustment beads are shown, as the full view (especially when shown together with the Fibonacci rods) appears mystifyingly complex. However, by looking at just these examples, we can understand the model as a whole. The green anchor cubes only fix onto Fibonacci rods; they never connect with Lucas rods. In this case, upwards along the $L_1=1$ rod (centre-left), we may read off the Lucas sequence: $\langle\, 2, 1, 3, 4, 7, 11 \,\rangle$.

Over the page, in Fig. 154 we see odd- and even-s spiral layers—separately, then combined. The top spiral has $s=1$, and is anchored on $F_1=1$ with adjustments from the spiral pointing in towards the centre. Next down is the $s=2$ spiral, anchored on $F_2=1$, with its adjustment quanta pointing out. Away from the degenerate case $s=0$ (p.220), these two patterns alternate indefinitely. Bead sizes for even layers scale one to the next by ϕ^2, with the same rule for odd layers too.[1]

---

[1] The sharp eyed will have noticed this effect is not very evident in Figs. 152 and 153. This is because they mostly show cases where $s > n$, so $n-s$ is negative, and the alternating signs of F_{n-s} are cancelling the effect of the $(-1)^s$ 'in-out' parameter.

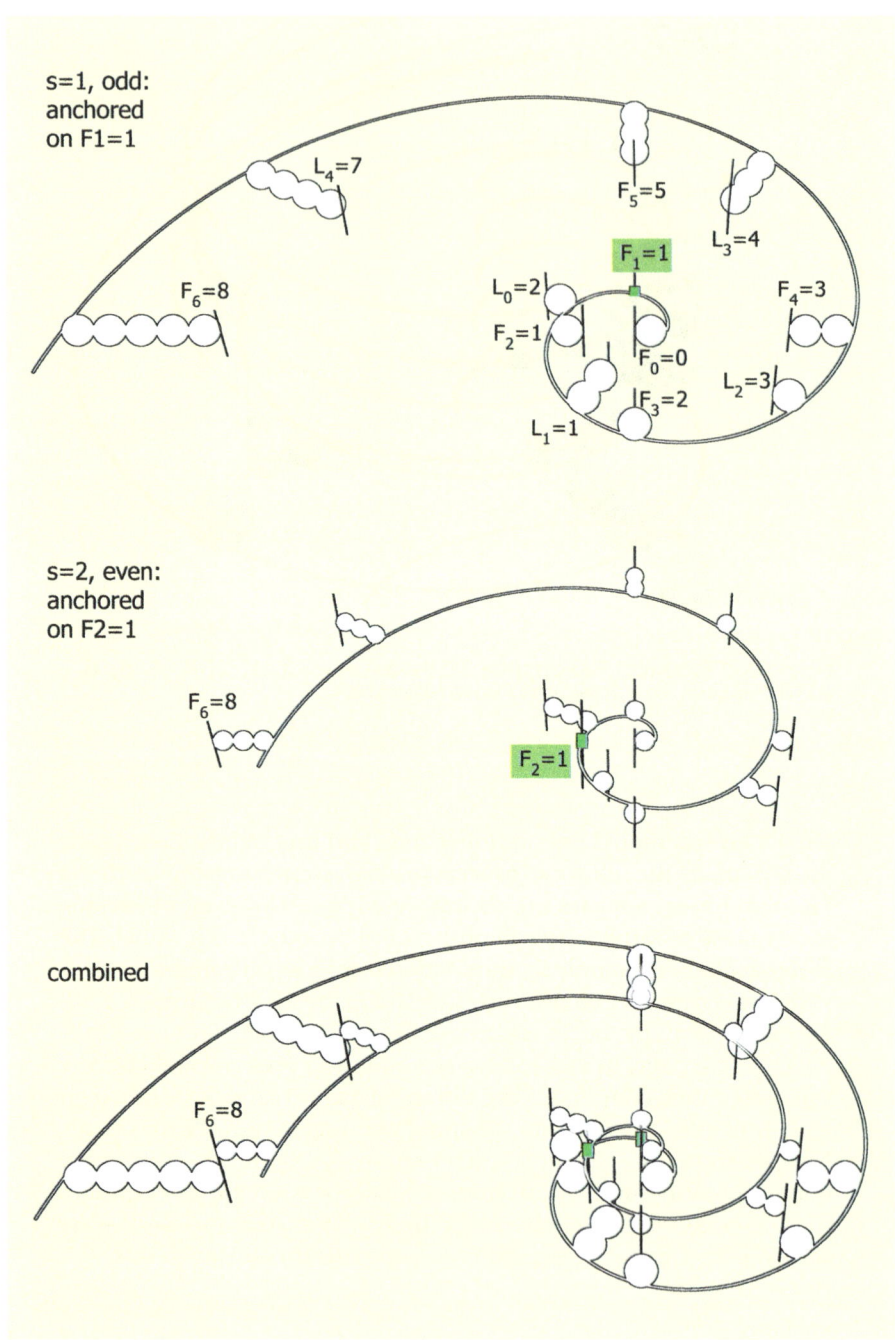

Figure 154: **Odd-in, even-out: The effect of the $(-1)^s$ sign parameter.**

Chapter 18

Is it fractal?

Figure 155: **Benoît Mandelbrot (1924–2010).** Photo: Rama [773, 141].

In this chapter our main purpose is to understand enough basic fractal geometry that we are equipped to consider the question: 'Is the Fibonacci Resonance (in its spiral visualization) a fractal?' And to do that, we shall start with another question.

'How long is a coastline?'

This (at first sight trivial) problem was first considered by English mathematician, physicist, meteorologist, and pioneer of modern maths-based weather forecasting, Lewis Fry Richardson. His work was later quoted by mathematician (and founder of fractal geometry) Benoît Mandelbrot (Fig. 155) in a more specific paper, 'How long is the coastline of Britain?' [780, 622]. In Fig. 156 we begin to see why this problem is not at all trivial. We find that 'the shorter the measuring rod, the longer the coastline.' In the extreme, we might imagine measuring all the tiny details of rocks and pebbles—and at these finer and finer levels there are more and more 'indents', and all these further increase the cumulative length. As we shall see, situations like this appear to have 'fractional dimension', and Benoît Mandelbrot (in the 1980s) coined the term 'fractal', to refer to objects or systems having non-integer dimension [623]. We know that a line has dimension one, and a rectangle has dimension two. But, strange as it first may seem, a fractal could have dimension 1.58.

Figure 156: **The British coastline—different length measuring rods.**

Qualities of fractals

Here are some typical (and at times overlapping) qualities of fractals:

- They are infinitely intricate, infinitely 'complex'—they can be magnified indefinitely yet still reveal further interesting results —perhaps previously unseen patterns—yet retain a 'family style' (Figs. 157 and 158 are celebrated examples).
- They exhibit self-similarity—the subsets resemble or match the whole—in never-ending patterns of recognizable shapes that repeat at different scales—like the Romanesco broccoli which we saw in the *In nature* chapter (p.55).
- They include detailed structure at arbitrarily fine scales.
- They may arise from a simple iterative or recursive definition.
- They have fractional dimension (more on this shortly).
- They have irregularity which typically is not easily described in Euclidean geometry.
- They provide a way for mathematics to describe natural forms as varied as coastlines, trees and ferns, and blood vessels. This fractal-maths approach in turn assists the creation of efficient computational algorithms for generating such forms in graphics applications such as architectural visualization and gaming.

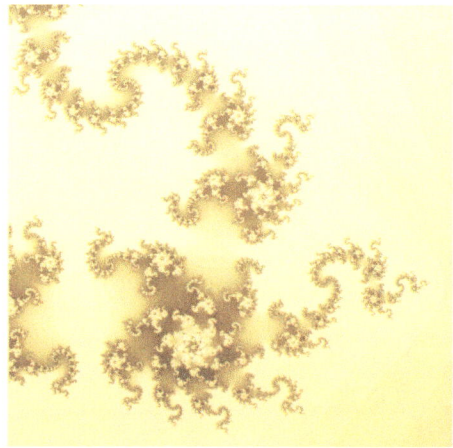

Figure 157: **A Julia set.**

Figure 158: **The Mandelbrot set.**

What are fractional dimensions?

The normal concept of whole-number-sized dimensions is deeply ingrained in most people. If we halve the length of a line, we expect that adding two such new lines will give the original length. If we halve the sides of a square, we expect to have a new square that is one quarter the area of the original. And similarly, halving the edges of a cube should give us one eighth of the volume. However, a key concept in fractals is that of fractional dimension. In the coastline example, we may have assumed that measuring the coast with a half-sized measuring stick would give us twice the number of length measurements. But, as we have seen, a smaller stick finds more detail. The measured path gets more kinks, and the total perimeter measurement is *longer than expected*. It is precisely this effect that gives rise to fractional dimension.

Let's create a formula for integer dimension and then see how this can be applied in the fractional case. We let r be the rod-size reduction factor (e.g. $r=2$ will halve the rod length), and let N count the number of times we use the new (shorter) rod in making the measurement. In the example of the straight line, halving the measuring rod ($r=2$) gave a new count of twice as many pieces ($N=2$). This is the simplest one-dimensional case, and we say dimension $D=1$. For the square, again we halve the measure ($r=2$), but this time 4 small squares will fit into the original (so $N=4$)—this is the 2-dimensional case, with $D=2$. For the cube, $r=2$ gives $N=8$, with $D=3$. Hence, to generalize these three cases, we say that when we reduce a linear dimension by a factor of r to become our new measure, the resulting count N will become r^D. We have

$$N = r^D.$$

And to make D the subject, we take logs

$$\log(N) = D \log(r)$$

and rearrange, which gives us a formula for dimension in terms of r and consequent N,

$$D = \frac{\log(N)}{\log(r)}. \tag{18.1}$$

But before we move on to fractional dimensions, let's be completely clear about the integer cases first.

Simple non-fractional dimensions

For a straight line, if we cut our measuring rod into two pieces ($r=2$), and we measure again, we shall match its length twice (so $N=2$). And from our formula (18.1) we get dimension one—that is,

$$D = \frac{\log(2)}{\log(2)} = 1. \qquad \text{(line)}$$

For a square, if we cut our measuring rod into two pieces ($r=2$), and we measure again, we shall fit 4 'half side' squares into the original (so $N=4$). Again using (18.1) we get a dimension of two,

$$D = \frac{\log(4)}{\log(2)} = 2. \qquad \text{(square)}$$

For a cube, if we cut our measuring rod into two pieces ($r=2$), and we measure again, we shall fit 8 'half side' cubes into the original (so $N=8$). From which we get a dimension of three,

$$D = \frac{\log(8)}{\log(2)} = 3. \qquad \text{(cube)}$$

To move on to fractals, we shall first study the 'Koch curve' in some depth, and then briefly the Sierpiński triangle. Both these are examples of Iterated Function System ('IFS') fractals. We shall then briefly review the 'reducing-branch-tree' fractal that depends on ϕ, and the 'AMD' which is Fibonacci generated but involves both Golden and Silver Ratios.

Koch curve

In Fig. 159 we see one of the earliest examples of a fractal [624]. This was originally constructed by Swedish mathematician Helge von Koch (Fig. 160). Our analysis will closely follow the logic we used in the coastline example. The rule to create this curve is essentially 'search and replace—forever'. At the bottom of Fig. 159, we have a straight line that is called the 'initiator'. Next up is a shape called the 'generator'. Then to create each next iteration working upwards, we search the previous shape for initiator sections, and replace each found with a suitably scaled and rotated version of the generator. This produces an intricate wiggly outline somewhat like the coastline we saw just earlier.

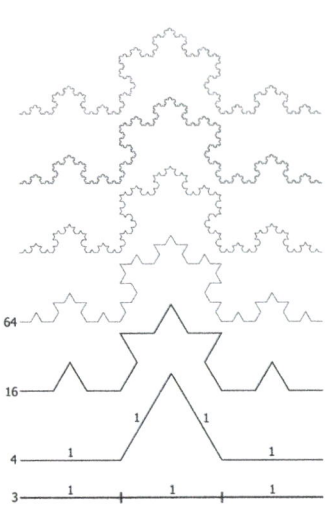

64

16

1 1

4 1 1

3 1 1 1

Figure 159: **The Koch fractal**

Figure 160: **Helge von Koch (1870–1924).** Olof Edlund.

If we start by using a measure that is the same length as the initiator straight line—then that will be 'all it can see'. But, if we shrink this measure by a factor of 3 (r=3) and measure again, we find the shorter rod can now 'get inside the curve' and resolve 4 sections—so N=4 in this case (Fig. 161).

Figure 161: **Koch curve—different sized measuring rods.**

To find the dimension in this case, we again apply Equation (18.1) (p.232), but we now expect the result to be non-integer. This generalized concept of dimension is called the 'Hausdorff dimension',

$$D_H = \frac{\log(4)}{\log(3)} \qquad = 1.2618\ldots \qquad \text{(Koch)}$$

In Fig. 162, we see a closed version of this curve (Fig. 162). This is called the 'Koch snowflake' or the 'Triadic Koch Island' [624]. It starts with an equilateral triangle composed of 3 initiators which then generate in the same way as the single-line initiator.

Figure 162: **The Koch snowflake.**

Accordingly, the fractal dimension is the same again (1.2618...) Although closed, the resulting curve is *infinitely long.* This is because every time the subdivision rule is applied, the length increases by 4/3rds—so after n iterations it will have grown by $(4/3)^n$. We then let n go to infinity. However, the *area* enclosed by this infinite perimeter is finite, and it may be calculated by summing the areas of the equilateral triangles that are added at each iteration. The final total is 8/5ths of the original triangle's area. (We recognize 8 and 5 as Fibonacci numbers, and 8/5 as a convergent of ϕ).

In a 2003 paper, John Turner checked how this dimension would change if the 1/3rd and 2/3rd vertices were instead located at $1/\alpha^2$ and $1/\alpha$—that is, at $(1 - 0.618)$ and 0.618 (the 'Goldpoints'), and with each straight section being $1/\alpha^2$ [907]. For this case, he found dimension

$$D_H = \frac{\log(4)}{2\log(\alpha)} \qquad = 1.4404\dots \qquad \text{(Goldpoint-Koch)}$$

The Sierpiński triangle, gasket, or sieve

Figure 163: **Sierpiński fractal.**

Figure 164: **Wacław Sierpiński (1882–1969).** La Société des Sciences et des Lettres de Varsovie.

In Fig. 163 we see Polish mathematician Wacław Sierpiński's well-known (and now many-named) fractal [625]. To understand how this fractal scales, we imagine holding the bottom-left corner fixed. Then we shrink the shape back to half its width and height. If we think of the original 'tissue' as being constructed in three large pyramids, then by halving back, we will shrink all 3 pyramids into just one—the lower-left pyramid. This thought experiment does make the non-integer dimensionality strikingly obvious to us. We scale the figure by one-half, but end up with one-third of the tissue. And looking at this figure in terms of r and N, then the bottom-left triangle is the result of an $r = 2$ reduction (halving), and using this 'measure', we count 3 such triangles in the original figure—top, bottom left, bottom right—so here $N = 3$, and the (fractional) dimension is

$$D_H = \frac{\log(3)}{\log(2)} = 1.5850\ldots \qquad \text{(Sierpiński)}$$

A reducing-branch tree fractal

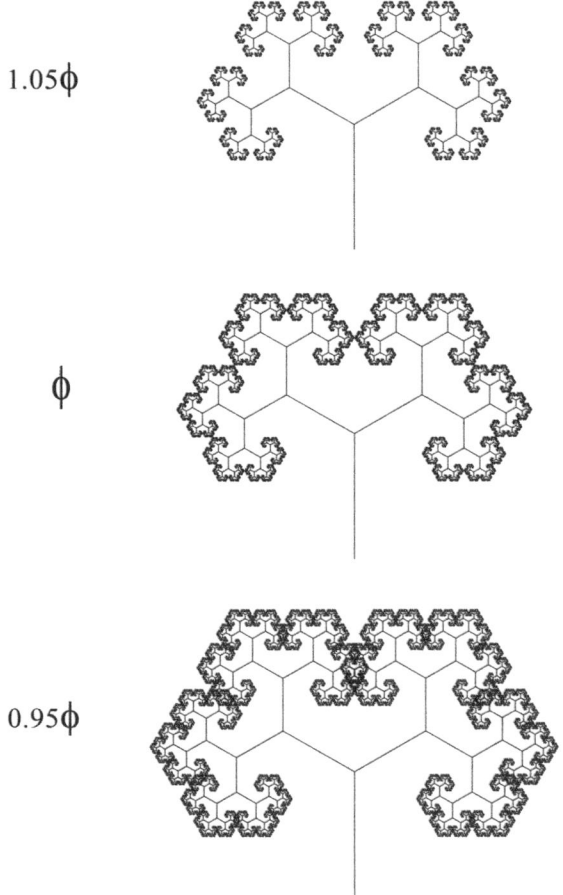

1.05ϕ

ϕ

0.95ϕ

Figure 165: **Reducing-branch tree fractal.**

We see in Fig. 165 a very simple tree fractal. In this fractal, line segments reduce in size at each bifurcation by a constant factor, and they fan out with $(2\pi)/3$ radians separation. The reduction factor of ϕ (centre) gives 'just-touching' branches. A reduction factor of 1.05ϕ (top) causes gaps, whereas a factor of 0.95ϕ (bottom) produces overlaps. Posamentier & Lehmann have studied this fractal, and they give an analytic proof of ϕ being the critical reduction factor [763].

The Fibonacci word

The Fibonacci word, also known as the Golden String or chain, is a Fibonacci recurrence with two seeds, where the previous two elements are added to get the next member of the sequence. The difference from its numerical counterpart is that the elements are strings of ones and zeros (or equivalently, strings of As and Bs, or ones and zeros swapped,[1] etc.) that are added by shunting together. There are subtleties about the way different authors get the process started which we cover in the final chapter ('-*Onics*'), but in the meantime, we shall briefly consider the simplicity of a '1, 0' construction, generation ('G') by generation, using the recurrence

$$G_{n+2} = G_{n+1} \frown G_n, \qquad \text{where the '}\frown\text{' means concatenate.}$$

Starting with seeds 1 and 0, the next member will be 01 (as looking back from the result position: predecessor = 0, then one before that = 1). Hence, the sequence progresses:

1

0

01

010

01001

01001010

0100101001001 ...

The resulting infinite string of zeros and ones that is produced is called the Fibonacci word (A003849) [967].

The AMD Fibonacci word fractal

The AMD fractal was discovered in 2008 by independent French researcher Alexis Monnerot-Dumaine. It provides an exotic visualization of the non-periodic orderliness of the Fibonacci word (Fig. 166). Its construction is controlled directly by the ones and zeros in the word: initialize an index for digit position in the word n=0 and draw a segment. If the digit at n=0 is 0 then 'turn left if n is even (right if odd)'. Then increment n, draw next segment, test the next digit, turn if required, and so on [661].

---

[1] Swapping ones and zeros in the Fibonacci word produces the complement which sometimes called 'the rabbit sequence' A005614.

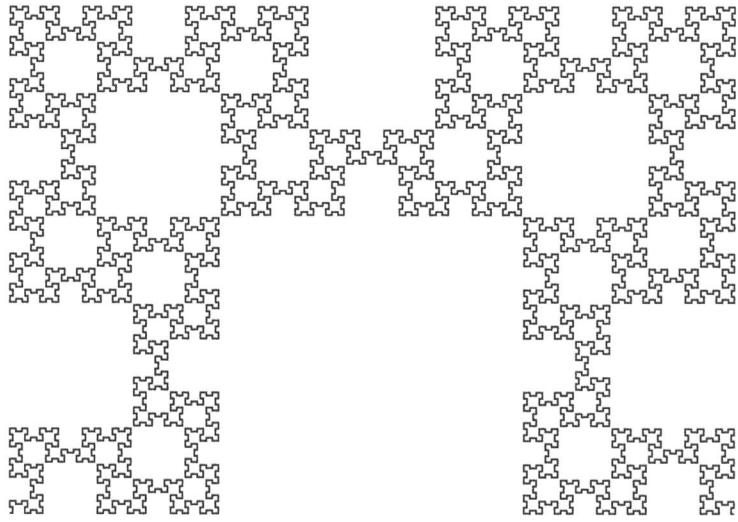

Figure 166: **AMD Fibonacci word fractal—after F_{23} (28,657) steps.**

In his preprint paper Monnerot-Dumaine derives the Hausdorff dimension for this fractal—with $\phi = (1 + \sqrt{5})/2$, and $\delta = 1 + \sqrt{2}$—as

$$D_H = 3\frac{\log(\phi)}{\log(\delta)} = 1.6379\ldots \qquad \text{(Monnerot-Dumaine [662])}$$

This expression depends on *both* the Golden Ratio ϕ and Silver Ratio δ. Monnerot-Dumaine goes on to explore different methods of construction and other turning angles—for example $60°$ gives a Koch-like fractal with a very elegant expression for D_H (when expressed in terms of continued fractions) which complements the one above involving ϕ and δ [663].

Silver scaling

A Silver Rectangle may be partitioned into two end squares leaving a Silver Ratio × smaller Silver Rectangle between. This follows directly from the definition of the Silver Ratio, viz. $\delta = 1 + \sqrt{2} = 2 + (1/\delta) = 1 + (1/\delta) + 1$. The central ($\delta×$ smaller) Silver Rectangle may again be partitioned: square, Silver Rectangle, square—and again and again in an infinitely repeated pattern of self-similar subdivision. In Fig. 167 we see local Silver Ratio scaling directly—by observation.

(The Silver Rectangles—blue and red—were initially discovered by eye and then drawn as overlays.) One diagonal of the large Silver Rectangle, shown in blue, is vertical. Hence its long side is offset from vertical by 22.5°—i.e. ② in Ori32 geometry—the 'naming vertex' of the 'Silver' Fibon2, p.163. The first subdivision blue Silver Rectangle is repeated above in red. This red Silver Rectangle has one diagonal horizontal.

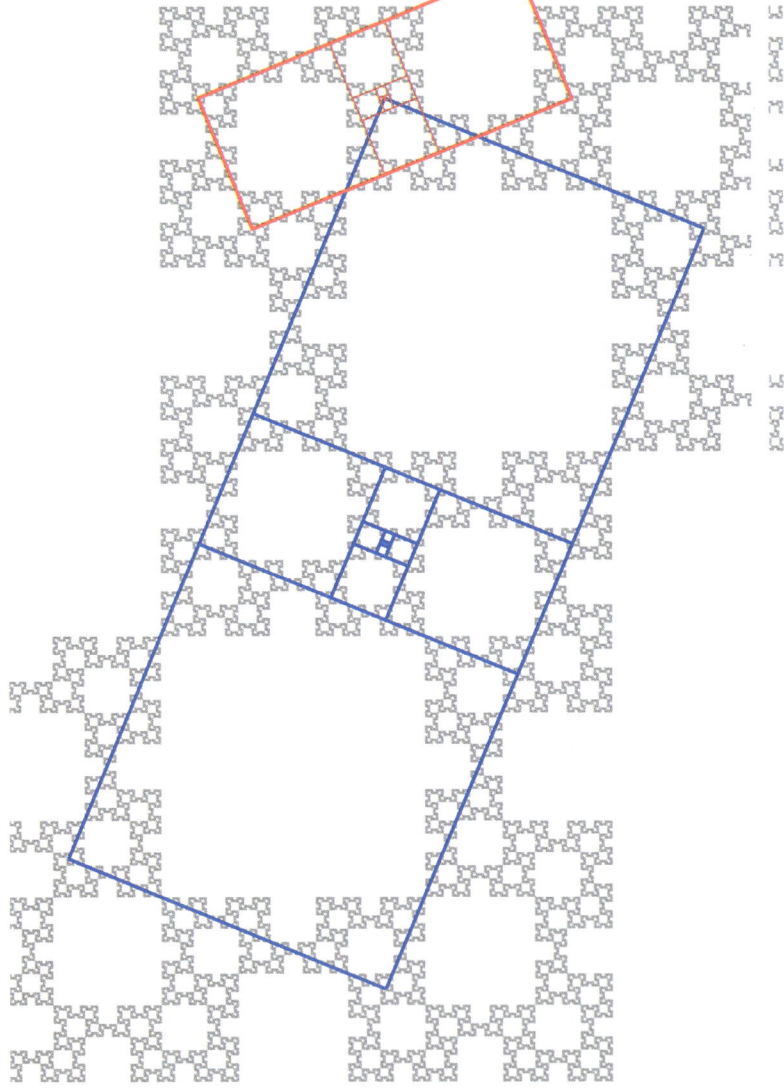

Figure 167: **AMD fractal—showing local Silver Ratio scaling.**

So, is the Fibonacci Resonance a fractal?

Let's check through the properties we listed earlier and see which match and which do not:

- We do seem to have self-similarity—patterns made by the quanta are the Fibonacci and Lucas sequences themselves, lagged and repeated in every quadrant on successive Golden Spiral 'backbones' (if the sign flip of the adjustments is ignored). Also, these patterns repeat at different scales—scaled by ϕ from one to the next (or by ϕ^2 if we do take into account the sign-flip).

- We also have detailed structure at arbitrarily fine scales— quanta become arbitrarily small, but finer scales do not reveal new complexity, so we do not have a truly 'infinitely intricate' structure (such as may be explored in the Mandelbrot set).

- The number point positioning appears irregular, but it is nevertheless completely 'systematic' as we saw on p.45.

- We do not have any kind of 'search and replace' generation rule. Subsequent scaled versions exist in their own right, offset by a quarter turn one to the next (or again, half turn—if we take into account the 'in-out' adjustment sign flips).

- And, do we have an iterative or recursive definition? At a stretch the process of considering a succession of anchored spirals might be considered as an iterative process... but not in the same sense as typical iterated-function-system fractals. Then again, we could say the growth rule is: 'turn one quadrant, grow the spiral radial vector (spoke) by ϕ, and repeat the F_n and L_n patterns all along the (infinite) spiral using quanta that are ϕ times *smaller*, adjusting using the correct "in-out" sign'.

- Finally, in order to consider fractional dimension, let's rehearse these last actions in detail forwards and back...

Growth step—growing the spiral

First we ask: 'What does it mean to "grow" a log spiral?' We have already discussed the equivalence of dilation and rotation (p.212). So here, in order to be more specific, let's define a unique 'size' for our spiral (in terms of a given s value). Let's simply define it as the distance from the origin (spiral centre), to its anchor point—which will be F_s.

This way if we make one spiral growth step, we shall see the following:

- The spiral number s increases by one.
- The 'size' (spoke length) increases from $r=F_s$ to $r=F_{s+1}$ (approximately ϕ growth).
- The spoke angle increases one quadrant.
- The 'in-out' adjustment sign flips.
- The quantum size *decreases* by a factor of exactly ϕ.

Growth step—growing the quanta

Alternatively what if we grow the quantum size instead? Well, given the constraints of the formulæ and our spiral configuration, the only possible result will be a reversal of the above case. Here, in order to grow by one step, we need to:

- increase quantum size by exactly ϕ
- decrease spoke angle by one quadrant
- decrease the spiral number s by one
- flip the 'in-out' adjustment sign
- decrease spoke length from $r=F_s$ to $r=F_{s-1}$ (reduction of approximately ϕ)

Thinking aloud (just about the quanta)... this is the strange thing—if we 'grow' an anchored spiral, the quantum size shrinks by close the growth factor, and if we grow the quantum, the spiral shrinks—as if (whichever way we look at it) the dimension were close to *minus one*. And, as s increases, this fractional dimension will tend closer and closer to minus one. Getting back to our original question—it has to be said that the results of our enquiry have been less than conclusive. But we have noted a high degree of self-similarity (excluding the degenerate case), and also there is the (negative) fractional dimension when considering the pattern scaling (with respect to spiral number). So at best, we might call the Fibonacci Resonance 'fractal-like'.

Negative dimension?

The concept of fractional dimension was first introduced by German mathematician Felix Hausdorff in 1918 [324], and it was developed by A S Besicovitch. One aspect of this development was that it had no place for negative values. Nevertheless, this was not a constraint on Mandelbrot—who yet further extended the concept by introducing 'fractal dimensions' that *could* be negative [626].

Chapter 19

Fibonacci Resonance summary

As a brief summary (and certainly *not* a proper definition)—the Fibonacci Resonance provides a way to visualize in detail how Fibonacci and Lucas numbers grow. It is based on a set of Golden Spirals and employs 2 sets of quadrant axes upon which Fibonacci and Lucas numbers are successively placed, spiral fashion. It gets its name from a sound analogy, where quantized distances along these axes (spiral to number point)[1] are filled exactly with the standing waves of some imagined vibration. A spiral is said to be 'anchored' when scaled and/or rotated such that it passes through a chosen reference point. This way, each anchored Golden Spiral has its own associated unique 'wavelength' whereby it may be said to 'resonate' with, or be 'tuned to' all the placed Fibonacci and Lucas numbers. The scheme has an infinity of such spirals, and the measures (wavelengths) scale by ϕ one spiral to the next .

> *The combined visual configuration and equations here described unify the Fibonacci numbers, Golden Spirals, and Golden Quanta, and also the Lucas numbers—locking all together into one infinite fractal-like structure.*

Anatomy—growth and adjustment terms

According to this spiral growth visualization, we may think of the key formulæ pair as follows, with explanations A–F (Fig. 168):

---

[1] That is, distances that share a common measure.

243

A   The Lucas numbers get a kick-start by having seeds of $\langle 2, 1 \rangle$ compared with the Fibonacci seeds of $\langle 0, 1 \rangle$, and this gives them a constant underlying growth lead of $\sqrt{5}$. In any Golden Spiral this growth is equivalent to $150.505°$ of turning.   (p.213)

B   We anchor each Golden Spiral to its own Fibonacci number point at a distance of F_s from the origin, where s is the spiral number. For example spiral number 4 will be anchored at $F_4 = 3$ units from the origin. This distance, the radial vector, we call a spoke. At the anchor point, the power of ϕ is zero, so $n = s$ (at angle $n = s$ quadrants).
                                                                              (p.209)

C   With s given, the ϕ^{n-s} terms supply a growth of ϕ per step of n, and hence the corresponding spiral radial vector r grows at ϕ per quadrant (making the spiral Golden). In polar coordinates,   (p.209)

$$\text{Golden Spiral 'number } s\text{',} \qquad r = F_s \phi^{\frac{4\theta}{(2\pi)} - s}.$$

D   Each spiral has a set of quantized adjustments that relate it to the number points. With increasing s, these adjustments alternate inside then outside the spiral, as we saw earlier. This sign term reflects the parity of the spiral number, and we see (for Fibonacci and Lucas numbers) 'odd in'—when the sign is negative, the adjustment is a reduction (in towards origin). And conversely, when the sign is positive—the adjustment is an increase: 'even out'.   (p.228)

E   We find that each pattern of adjustment (one for Fibonacci and one for Lucas) repeats its own parent sequence—and it does so with a lag of s steps. When $n = s$ the Fibonacci adjustment vanishes: we get zero-q (no quanta), as $F_{n-s} = F_0 = 0$. This is the anchor point.   (p.208)

F   For a given spiral (identified as spiral number s), all the adjustments, for all Fibonacci and all Lucas numbers are distances that are a whole number of a single measure that we call a quantum (or abacus bead diameter). For a given s, the quantum size is ϕ^{-s}.   (p.208)

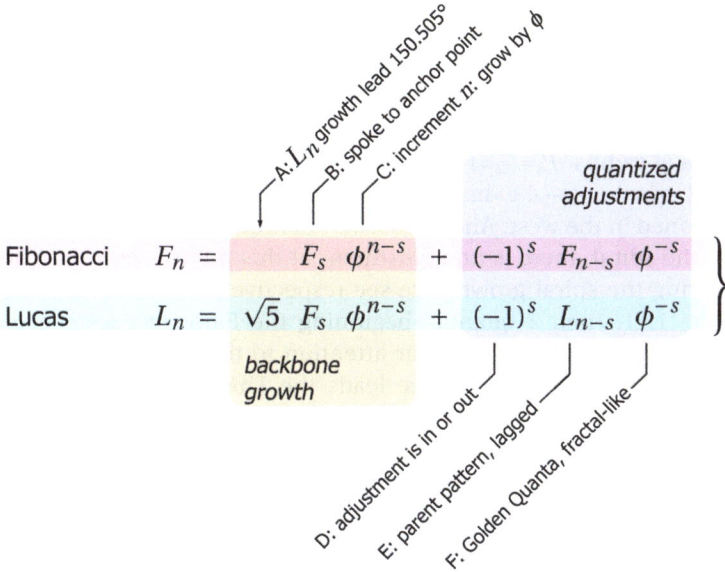

Figure 168: **Fibonacci resonance formula pair. In the visualization, radial distance to number point = spiral spoke + abacus beads.**

But does it really work on a grand scale?

Our intuition alone might suggest it is completely reasonable that as quanta get smaller and smaller (and number points get very very close to the spirals anyway), then 'everything fits' (even though this is not how it works). However such 'fitting' may be more surprising when we think about increasingly negative spiral numbers, where the quanta get *arbitrarily* big. If we have any doubts on this, we may use both the Table on page 414 and a calculator to run a quick example check. Let's say $s=-23$, $n=5$. We know that odd-indexed Fibonacci numbers are positive, so from the Table, $F_{-23}=28657$. We shall also need $F_{n-s}=F_{28}=317811$. Plugging these into the formula and doing the calculation, we do indeed confirm the expected result that

$$F_5 = 28657\phi^{28} - 317811\phi^{23} = 5.$$

The second-term 'adjustment' here (in our scheme, the distance from spiral to number point), is made up of around a third of a million quanta (317,811), with each quantum sized close to 64,000 units ($\phi^{23} = 64,079.00002 \approx L_{23}$).

Seeing a single layer

To discuss Fig. 169, it will help us to borrow north-south-east-west from map navigation (using east as our zero-angle polar reference.) We start with the green-square anchor. For the $s=2$ spiral, this anchor will be at radius $F_s=F_2=1$ from the polar origin, and at a rotation of $n=s=2$ quadrants—i.e. half a turn anticlockwise from east: and so be positioned in the west. And because by definition this point coincides with the spiral, there is no adjustment at this point ('zero-q'). Then, following the spiral growth, we see respectively at south, east, then north: 1, 1, then 2 quanta—beginning the Fibonacci sequence of adjustments. Next we turn our attention to the offset axes, and we recall that the Lucas sequence leads the Fibonacci by 150.505° of Golden-Spiral turning. So, from the green anchor, we look anti-clockwise round to the south-east to the start of the Lucas number sequence adjustments, with its 2 quanta. In the north-east we see the next Lucas adjustment of 1 quantum. And looking back the other way from the Lucas start, we see adjustments in negative Lucas number sequence: south-west with -1 quantum (i.e. inwards) and north-west with +3 quanta (outwards).

Seeing the whole thing

Is it possible to capture the whole effect in our imagination? If we look at say, a '3, 4, 5' triangle, we instantly feel that we have apprehended all its key points. With the Fibonacci Resonance however, we had best make our approach in stages. Having got a basic idea from the 3D model (pages 226 and 227), we start by thinking of just one spiral, its anchor point and the Fibonacci pattern of adjustments running out either side of the anchor, quadrant by quadrant. We then consider the offset Lucas adjustments on the offset axes in between. Once this is very clear in our minds, we consider the next spiral with its ϕ-times smaller quanta, with all its in-out patterns reversed, and that whole layer rotated by one quadrant. And when we can hold the image of one pair of layers, we may start to consider more pairs rotated and stacked above and below. The complete conception will also include the special $s=0$ layer with its 'point spiral' (which is no spiral at all) along with its unity-sized quanta. With patience and luck, we may find it possible to experience fleeting impressions of the full (infinite) relationship all at once, but then afterwards, not be sure if we really did glimpse it or not.

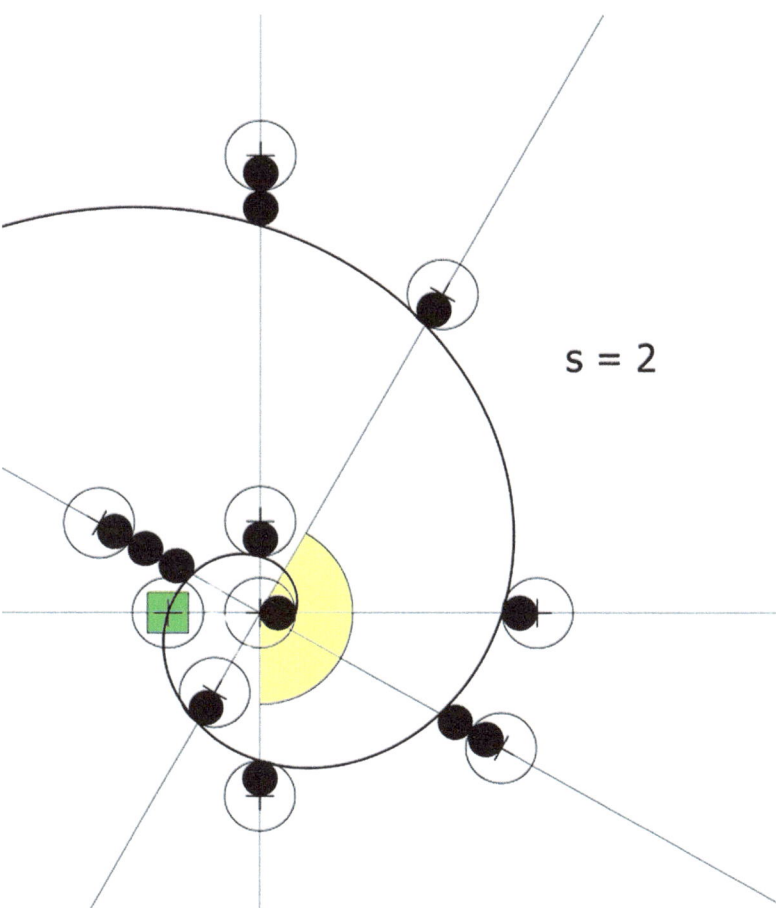

Figure 169:
Fibonacci resonance (s=2). Open circles locate number points (F_n and L_n). Filled circles (abacus beads) show adjustment quanta.

After all, to imagine this, we must think in terms of the microcosm and the macrocosm. If we start with quantum ϕ^0 as (e.g.) our own height, then the smaller quanta will be ϕ-times smaller and ϕ-times smaller again, and so on, until they are smaller than a cell (very approximately 25 steps), an atom (50 steps), and then still carry on getting smaller indefinitely. And the larger ones will grow ϕ-times bigger, and ϕ-times bigger, and so on, until wider than the Earth (30 steps), wider than the Solar System (60), our Galaxy (100), and the observable Universe (125), and still keep getting bigger indefinitely...

Sumer & Babylon? $\langle F_n \ (\text{mod } 10) \rangle$

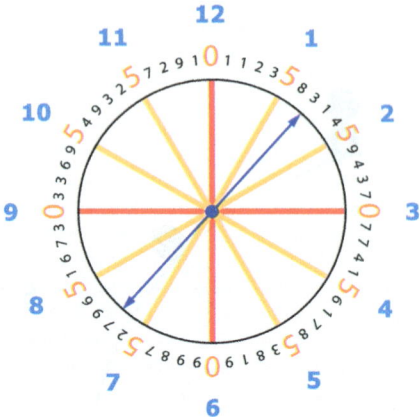

Figure 170: **Fibonacci sequence: 60 final digits (*clockwise* from top).**
(For next—add 2 prior, but discard tens.)   © 2013 Lucien Khan, shown with kind permission.

Finally, before we get into generalization, what if we consider just the end digits of the Fibonacci numbers? We find the units repeat their pattern in a cycle of 60.[2] Such sequences are called 'Pisano periods' (in honour of Fibonacci) and their lengths vary according to the divisor used.[3] Also, by counting residue instances, we find there are 4 each of the 5 even residues: (0, 2, 4, ..., 8), and 8 each of the 5 odd ones. We also know the ancient Sumerians and Babylonians used a combined 10- and 60-based number system, and while 60 is useful for its divisibility by: 2, 3, 4, 5, 6, 10, and 12; this in itself is not a convincing reason for its adoption. However, independent researcher Lucien Khan has recently made a remarkable discovery by laying out the cycle of 60 residues $\langle F_n \ (\text{mod } 10) \rangle$ in a circle. He was delighted to find the modern 'clock' pattern with zeros at the quarter hours, and fives at the remaining 5-minute markers—giving an overall division into 12 sectors (Fig. 170). Khan also showed that sums of opposites (blue arrow points) all end in zeros—that is, $F_n + F_{n+30} \equiv 0 \ (\text{mod } 10)$. Then by overlaying copies of his number circle, Khan analysed the *vesica piscis* (2 circles, p.88), the 'Seed of Life' (a 7-circle honeycomb structure), and more [469]. Perhaps the Sumerians knew of this aspect of the Fibonacci sequence and revered it as sacred geometry. Perhaps it then inspired: their sexagesimal number system, their measurement of time in 12-hour blocks, and their 12-house zodiac.

---

[2] Lagrange found this period 60 in 1774 [886, 522]—c.f. $L_n (\text{mod } 10)$ with period 12.

[3] For example, when the modulus is 32 (p.170), the Pisano period is 48 (also p.409).

248

Part IV

TO PELL AND BEYOND

Chapter 20

Generalization to Lucas Sequences

Before we start, we must first note the difference between *Lucas Sequences* and the Lucas-number sequence. We have already met the Lucas-number sequence and have seen how closely it is related to the Fibonacci-number sequence (p.27). However, *Lucas Sequences* 'sit above' both of these. They generalize to a next higher level. In addition to Fibonacci and Lucas numbers, *Lucas Sequences* include: Pell numbers, Jacobsthal numbers, and both their companion sequences. They also include Mersenne $2^n - 1$ numbers and Fermat $2^n + 1$ numbers (which themselves are general cases respectively of $2^p - 1$, p prime, and $2^{2^k} + 1$ numbers). We shall discuss these latter names at the end of the next chapter in a specific example. First we show how the Fibonacci Resonance formulæ, (which, so far, have been Fibonacci/Lucas specific), may be generalized to cover all Lucas Sequences. (Lucas Sequences now have important applications in cryptography, for example in public-key cryptosystems.)

Emphasis

Our main concerns in this chapter are to:

- introduce generalized Fibonacci Resonance formulæ
- check their mathematical consistency (rigorously)
- generalize the 'quanta and spirals' framework for their visualization

Only in the next chapter, when we look at a range of specific cases along with relevant diagrams, shall we expect to grasp the general forms fully.

251

Lucas Sequences—PQ generalization

Recurrence formulæ for the Fibonacci numbers and the Pell numbers have coefficients fixed by definition:

$$Fib_n = 1 \cdot Fib_{n-1} + 1 \cdot Fib_{n-2} \quad \text{with seeds} \quad Fib_0 = 0, \quad Fib_1 = 1$$

$$Pell_n = 2 \cdot Pell_{n-1} + 1 \cdot Pell_{n-2} \quad\quad\quad Pell_0 = 0, \quad Pell_1 = 1.$$

These coefficients (1, 1) and (2, 1) may be generalized as integers P and *minus* Q [1] [592, 595, 175, 377]. For the primary sequence we retain the seeds 0 and 1:

$$U_n = P \cdot U_{n-1} - Q \cdot U_{n-2} \quad\quad U_0 = 0, \ U_1 = 1. \quad\quad (20.1)$$

This $\langle U_n \rangle$, is known as the *Lucas Sequence* [2], and in exactly the same way that the Fibonacci numbers have a companion sequence (the Lucas numbers), $\langle U_n \rangle$ has a companion sequence $\langle V_n \rangle$, with seeds 2 and P:

$$V_n = P \cdot V_{n-1} - Q \cdot V_{n-2} \quad\quad V_0 = 2, \ V_1 = P. \quad\quad (20.2)$$

For given P and Q, we do not know if $\langle U_n \rangle$ will settle towards a steady growth rate; but if it does, we shall call the rate x. Then in Equation (20.1) we shall have $U_n = x^n$, and $U_{n-1} = x^{n-1}$, and $U_{n-2} = x^{n-2}$, that is,

$$x^n = Px^{n-1} - Qx^{n-2}$$

and the same result would be obtained by letting (companion member) $V_n = x^n$ in (20.2)—because only the seeds change for V_n, not the recurrence itself. Dividing through by x^{n-2} we get

$$x^2 = Px - Q. \quad\quad (20.3)$$

This equation (20.3) is the characteristic quadratic:

$$x^2 - Px + Q = 0. \quad\quad (20.4)$$

That is, it is the PQ-generalized version of Fibonacci-specific characteristic equation which we derived on page 31.

---

[1] It has been conventional (since Lucas 1876) that in equations such as (20.1), Q has a minus sign (though a number of modern writers write $+p$ and $+q$), and it is also conventional that P and Q be relatively prime integers [177]. However, for the special case of $(6, 8)$, we shall ignore the 'relatively prime' convention.

[2] Sequences produced by relations of the type $G_n = P \cdot G_{n-1} - Q \cdot G_{n-2}$ are also known as Generalized Secondary Fibonacci Sequences—GSFS's [854].

Round-trip—from sequence to quadratic

To understand this generalization fully, we shall also consider the 'round-trip' approach taken by Swiss geometer Hans Walser [934].

We have just seen (20.1) where the Fibonacci sequence 'PQ generalizes' to become

$$U_{n+2} = PU_{n+1} - QU_n. \tag{20.5}$$

Let $\langle r_n \rangle$ denote the sequence of growth rates: the ratios of successive terms. So

$$r_n = \frac{U_{n+1}}{U_n}. \tag{20.6}$$

And for the next index

$$r_{n+1} = \frac{U_{n+2}}{U_{n+1}}. \tag{20.7}$$

We substitute the numerator of (20.7) with (20.5)

$$r_{n+1} = \frac{PU_{n+1} - QU_n}{U_{n+1}}$$

$$= P - \frac{QU_n}{U_{n+1}}. \tag{20.8}$$

And by taking reciprocals in (20.6) and applying the result to the Q term in (20.8)

$$r_{n+1} = P - \frac{Q}{r_n}. \tag{20.9}$$

Now if (as we saw with the Fibonacci sequence), the growth rate settles and converges to a non-zero limit (there the Golden Ratio), then we can say that in the limit r_n and r_{n+1} will both become arbitrarily close to a particular value, let us call it x. In this case

$$x = P - \frac{Q}{x}$$

which rearranged becomes the characteristic equation we just saw—

$$x^2 - Px + Q = 0. \tag{20.4}, \text{p.252}$$

And back—from quadratic to sequence

On the return journey, we start with the characteristic quadratic equation

$$x^2 - Px + Q = 0 \qquad\qquad (20.4), \text{p.252}$$

and derive the associated recurrence sequence. The roots of (20.4) are conventionally called α and β, which here refer to such roots in general: the $\alpha=\phi$ and $\beta=-\phi^{-1}$ we have used so far for Fibonacci/Lucas will from now on be regarded as special instances, as we shall soon see. Now, Equation (20.4) has discriminant

$$\Delta = P^2 - 4Q. \qquad\qquad (20.10)$$

And in the following, we shall consider only cases where $\Delta > 0$, which ensures roots α and β are real and distinct. Thus from the quadratic formula:

$$\alpha = \frac{P + \sqrt{\Delta}}{2} \qquad\qquad (20.11)$$

$$\beta = \frac{P - \sqrt{\Delta}}{2} \qquad\qquad (20.12)$$

$$\alpha + \beta = P \qquad\qquad (20.13)$$

$$\alpha\beta = Q \qquad\qquad (20.14)$$

$$\alpha - \beta = \sqrt{\Delta}. \qquad\qquad (20.15)$$

Back on page 181 in the Table for the 2nd pentad, we saw the concrete workings of the identity $\phi^n = F_n\phi + F_{n-1}$ [491]. And later we noted how this could be used to reduce any power of ϕ to a linear expression. Here, with a slight rearrangement of the characteristic equation (20.4), we have the basis for an equivalent linearizing formula:

$$x^2 = Px - Q \qquad\qquad (20.16)$$

here thinking of x as being one of the two solutions $x=\alpha$, or $x=\beta$. Let us represent the general case as

$$x^n = a_n x + b_n. \qquad\qquad (20.17)$$

Hence taking the next higher power

$$x^{n+1} = a_{n+1} x + b_{n+1}. \qquad\qquad (20.18)$$

But also, plainly,

$$x^{n+1} = x \cdot x^n$$

so using x^n from (20.17)

$$x^{n+1} = x(a_n x + b_n)$$
$$= a_n x^2 + b_n x. \tag{20.19}$$

And applying x^2 from (20.16) in (20.19)

$$x^{n+1} = a_n(Px - Q) + b_n x$$
$$= (a_n P + b_n)x - a_n Q. \tag{20.20}$$

Still following Walser's approach, we identify the coefficients in (20.20) and equate them with those in (20.18)

$$a_{n+1} = (a_n P + b_n) \tag{20.21}$$
$$b_{n+1} = -a_n Q. \tag{20.22}$$

Taking the next index in (20.21)

$$a_{n+2} = (a_{n+1} P + b_{n+1}) \tag{20.23}$$

and substituting for b_{n+1} from (20.22) we get

$$a_{n+2} = Pa_{n+1} - Qa_n$$

which is the PQ-generalized Fibonacci recurrence. The key thing here (as Walser notes [935]) is to see exactly how the characteristic quadratic is associated with the linear recurrence sequence and vice versa.

Binet forms and negative indices

From Lucas, 1876 [592], we have the Binet forms:

$$U_n = \frac{\alpha^n - \beta^n}{\alpha - \beta}, \tag{20.24}$$

$$V_n = \alpha^n + \beta^n. \tag{20.25}$$

From (20.14) it follows that $\alpha^{-n} = Q^{-n}\beta^n$ and $\beta^{-n} = Q^{-n}\alpha^n$, hence using (20.24)

$$\begin{aligned}
U_{-n} &= (\alpha^{-n} - \beta^{-n})/(\alpha - \beta) \\
&= (Q^{-n}\beta^n - Q^{-n}\alpha^n)/(\alpha - \beta) \\
&= -Q^{-n}U_n. \tag{20.26}
\end{aligned}$$

Similarly,

$$\begin{aligned}
V_{-n} &= \alpha^{-n} + \beta^{-n} \\
&= Q^{-n}\beta^n + Q^{-n}\alpha^n \\
&= Q^{-n}V_n. \tag{20.27}
\end{aligned}$$

We shall take care if and when applying these formulæ, as terms such as Q^{-n} will typically be non-integer. In other words, we may need to restrict the scope of formulæ according to the value of Q, if we wish to continue discussing integer sequences.

Generalizing the Fibonacci Resonance

We conjecture that the Fibonacci Resonance formula—(13.17), p.200

$$F_n = F_s\alpha^{n-s} + F_{n-s}\beta^s$$

very simply generalizes 'as is' (with F_n becoming U_n), and we now prove by induction that indeed it does.

Theorem:

Let P and Q be non-zero integer coefficients satisfying $P^2 - 4Q > 0$, and let U_n be the nth term of a Lucas Sequence defined by the recurrence $U_n = P \cdot U_{n-1} - Q \cdot U_{n-2}$ where $U_0 = 0$, $U_1 = 1$, and let α and β be the roots of the characteristic quadratic equation associated with $\langle U_n \rangle$, which is $x^2 - Px + Q = 0$ [592]. Then with integer index offset s,

$$U_n = U_s\alpha^{n-s} + U_{n-s}\beta^s, \quad \begin{cases} \text{for all } n, s \text{ in } \mathbb{Z}, \text{ if } |Q| = 1; \\ \text{for all } n, s \text{ in } \mathbb{Z}, \ 0 \le s, \ s \le n, \\ \text{if } |Q| \ne 1. \end{cases}$$

(20.28)

The restriction of $n, s, (n-s) \ge 0$ when $|Q| \ne 1$ arises because (as mentioned) reversing the recurrence relation (to obtain negative-indexed members) requires division by Q which typically gives non-integer results.

Proof:

$\langle U_n \rangle$ **Base case:** We show that (20.28) is true when $n=0$, that is

$$U_0 = U_s \alpha^{-s} + U_{-s} \beta^s.$$

Using $\quad \alpha^{-s} = Q^{-s} \beta^s \quad$ from $\alpha\beta = Q$, (20.14), p.254

and $\quad U_{-s} = -Q^{-s} U_s \quad$ i.e. (20.26), p.256

$$U_0 = U_s \cdot Q^{-s} \beta^s - Q^{-s} U_s \cdot \beta^s$$
$$= U_s \cdot Q^{-s} (\beta^s - \beta^s)$$
$$= 0.$$

This verifies the base case according to the $U_0=0$ definition in (20.1), p.252.

$\langle U_n \rangle$ **Inductive step:** If we write the theorem (20.28) in j and assume for any integer j it is true that

$$U_j = U_s \alpha^{j-s} + U_{j-s} \beta^s$$

then we must show that

$$U_{j+1} = U_s \alpha^{j+1-s} + U_{j+1-s} \beta^s. \qquad (20.29)$$

To do this, we substitute for U_s in the right-hand side of (20.29) using its Binet form $\quad U_s = (\alpha^s - \beta^s)/(\alpha - \beta) \quad$ i.e. Eqn. (20.24). We then substitute U_{j+1-s} similarly. So, letting G be equal to the right-hand side of (20.29)—after substitution we have

$$G = \left(\frac{\alpha^s - \beta^s}{\alpha - \beta} \right) \alpha^{j+1-s} + \left(\frac{\alpha^{j+1-s} - \beta^{j+1-s}}{\alpha - \beta} \right) \beta^s$$

$$(\alpha - \beta)\, G = \left(\alpha^s - \beta^s \right) \alpha^{j+1-s} + \left(\alpha^{j+1-s} - \beta^{j+1-s} \right) \beta^s$$

$$= \alpha^{j+1} - \alpha^{j+1-s} \beta^s + \alpha^{j+1-s} \beta^s - \beta^{j+1}$$

$$G = (\alpha^{j+1} - \beta^{j+1})/(\alpha - \beta).$$

We now identify G as being the Binet form for U_{j+1}, thus proving equality in (20.29).

$\langle U_n \rangle$ Reverse inductive step:

If we again consider the theorem (20.28) written in j, and we assume for any integer j it is true that

$$U_j = U_s \alpha^{j-s} + U_{j-s} \beta^s$$

then (stepping back) we must show that

$$U_{j-1} = U_s \alpha^{j-1-s} + U_{j-1-s} \beta^s. \tag{20.30}$$

To do this, we let H equal the right-hand side of (20.30), then we substitute for U_s with its Binet form (20.24): $U_s = (\alpha^s - \beta^s)/(\alpha - \beta)$; and then substitute U_{j-1-s} similarly:

$$H = \left(\frac{\alpha^s - \beta^s}{\alpha - \beta} \right) \alpha^{j-1-s} + \left(\frac{\alpha^{j-1-s} - \beta^{j-1-s}}{\alpha - \beta} \right) \beta^s$$

$$(\alpha - \beta) H = \left(\alpha^s - \beta^s \right) \alpha^{j-1-s} + \left(\alpha^{j-1-s} - \beta^{j-1-s} \right) \beta^s$$

$$= \alpha^{j-1} - \alpha^{j-1-s} \beta^s + \alpha^{j-1-s} \beta^s - \beta^{j-1}$$

$$H = (\alpha^{j-1} - \beta^{j-1})/(\alpha - \beta).$$

We now identify H as being the Binet form for U_{j-1}, thus demonstrating equality in (20.30), which proves the $\langle U_n \rangle$ theorem (20.28), p.256, for negative n—and thus completes its proof for all integer n.

■

Companion form

In the Fibonacci Resonance we found the Lucas numbers' underlying growth lead (ahead of the Fibonacci numbers) to be $\sqrt{5}$, as in

$$L_n = \sqrt{5} F_s \alpha^{n-s} + L_{n-s} \beta^s. \tag{13.17), p.200}$$

Tracing back, we find this $\sqrt{5}$ first appears just prior in (13.15), where it results from $(\alpha^{-1} + \alpha)$. Now, if we think of this in terms of ϕ, but order reversed and with its reciprocal double negated—that is, as $(\phi - (-\phi)^{-1})$ then this appears to suggest the generalization to $(\alpha - \beta)$. We shall now prove that $(\alpha - \beta)$ is indeed the required growth-lead term.

Corollary:

Again, let P and Q be non-zero integer coefficients satisfying $P^2 - 4Q > 0$, *and let* V_n *be the nth term of a companion Lucas Sequence defined by the recurrence* $V_n = P \cdot V_{n-1} - Q \cdot V_{n-2}$ *where* $V_0 = 2$, $V_1 = P$, *and let* α *and* β *be the roots of the characteristic quadratic equation associated with* $\langle V_n \rangle$, *which is* $x^2 - Px + Q = 0$ [592]. *Then with integer index offset* s,

$$V_n = (\alpha - \beta)\, U_s\, \alpha^{n-s} + V_{n-s}\beta^s, \quad \begin{cases} \text{for all } n, s \text{ in } \mathbb{Z}, \text{ if } |Q| = 1; \\ \text{for all } n, s \text{ in } \mathbb{Z}, \ 0 \le s, \ s \le n, \\ \quad \text{if } |Q| \ne 1. \end{cases}$$

$$(20.31)$$

Proof:

$\langle V_n \rangle$ **Base case:** We show that (20.31) is true when $n{=}0$—that is

$$V_0 = (\alpha - \beta) U_s \alpha^{-s} + V_{-s}\beta^s.$$

Using $\quad V_{-s} = Q^{-s}V_s \qquad$ i.e. (20.27), p.256

and $\quad \beta^s = Q^s \alpha^{-s} \qquad$ from $\alpha\beta = Q$, (20.14), p.254

$$V_0 = (\alpha - \beta)U_s\alpha^{-s} + Q^{-s}V_s \cdot Q^s \alpha^{-s}$$

$$= \alpha^{-s}\left[(\alpha - \beta)U_s + V_s\right].$$

We now substitute U_s and V_s with Binet forms (20.24) and (20.25)
$U_s = (\alpha^s - \beta^s)/(\alpha - \beta) \quad$ and $\quad V_s = \alpha^s + \beta^s$:

$$V_0 = \alpha^{-s}\left[(\alpha - \beta)\left(\frac{\alpha^s - \beta^s}{\alpha - \beta}\right) + (\alpha^s + \beta^s)\right]$$

$$= \alpha^{-s}(\alpha^s - \beta^s + \alpha^s + \beta^s)$$

$$= \alpha^{-s}(2\alpha^s)$$

$$= 2.$$

This verifies the base case according to the $V_0{=}2$ definition in (20.2), p.252.

$\langle V_n \rangle$ **Inductive step:** If we assume for any integer j it is true that

$$V_j = (\alpha - \beta)U_s\alpha^{j-s} + V_{j-s}\beta^s$$

then we must show that

$$V_{j+1} = (\alpha - \beta)U_s\alpha^{j+1-s} + V_{j+1-s}\beta^s. \qquad (20.32)$$

To do this, we substitute for U_s in the right-hand side of (20.32) using its Binet form (20.24) and similarly substitute V_{j+1-s} using (20.25). So, letting J be equal to the right-hand side of (20.32)—after substitution we have

$$J = (\alpha - \beta)\left(\frac{\alpha^s - \beta^s}{\alpha - \beta}\right)\alpha^{j+1-s} + \left(\alpha^{j+1-s} + \beta^{j+1-s}\right)\beta^s$$

$$= \left(\alpha^s - \beta^s\right)\alpha^{j+1-s} + \left(\alpha^{j+1-s} + \beta^{j+1-s}\right)\beta^s$$

$$= \alpha^{j+1} - \alpha^{j+1-s}\beta^s + \alpha^{j+1-s}\beta^s + \beta^{j+1}$$

$$= \alpha^{j+1} + \beta^{j+1}.$$

Here, we identify J with the Binet form for V_{j+1}, thus proving equality in (20.32).

$\langle V_n \rangle$ **Reverse inductive step:** If we again assume for any integer j it is true that

$$V_j = (\alpha - \beta)U_s\alpha^{j-s} + V_{j-s}\beta^s$$

then (stepping back), we must show that

$$V_{j-1} = (\alpha - \beta)U_s\alpha^{j-1-s} + V_{j-1-s}\beta^s. \tag{20.33}$$

To do this, we let K equal the right-hand side of (20.33), then we substitute for U_s with its Binet form (20.24), and then similarly substitute V_{j-1-s} using (20.25):

$$K = (\alpha - \beta)\left(\frac{\alpha^s - \beta^s}{\alpha - \beta}\right)\alpha^{j-1-s} + \left(\alpha^{j-1-s} + \beta^{j-1-s}\right)\beta^s$$

$$= \left(\alpha^s - \beta^s\right)\alpha^{j-1-s} + \left(\alpha^{j-1-s} + \beta^{j-1-s}\right)\beta^s$$

$$= \alpha^{j-1} - \alpha^{j-1-s}\beta^s + \alpha^{j-1-s}\beta^s + \beta^{j-1}$$

$$= \alpha^{j-1} + \beta^{j-1}.$$

Here, we identify K with the Binet form for V_{j-1}, thus demonstrating equality in (20.33), which proves the $\langle V_n \rangle$ corollary (20.31), p.259, for negative n—and thus completes its proof for all integer n.

■

Visualization

Again we shall use quanta ('abacus beads') and logarithmic spirals for visualization:

Sequence member = Spiral quadrant spoke

+ quantized adjustment.

We shall follow through the same step-by-step logic that we used previously for the Fibonacci and Lucas numbers; and in order to avoid the repeated need to refer back, we shall duplicate the format. As a result, if some sections sound familiar, then we shall regard them as 'convenience copies'.

First terms: Backbone growth

Summarizing (20.28) and (20.31) as (20.34) below, we see both first terms are based on $U_s \alpha^{n-s}$,

$$
\left.\begin{aligned}
U_n &= && U_s\alpha^{n-s} + U_{n-s}\beta^s \\
V_n &= (\alpha - \beta)\,U_s\alpha^{n-s} + V_{n-s}\beta^s
\end{aligned}\right\}
\quad
\begin{cases}
\text{for all } n, s \text{ in } \mathbb{Z}, \text{ if } |Q| = 1; \\
\text{for all } n, s \text{ in } \mathbb{Z},\ 0 \leq s,\ s \leq n, \\
\quad \text{if } |Q| \neq 1.
\end{cases}
$$

$$(20.34)$$

These $U_s \alpha^{n-s}$ and $(\alpha - \beta)\,U_s\alpha^{n-s}$ might be regarded as 'backbone-growth terms'. We note these vanish when $s=0$ (as $U_0=0$), but otherwise, they share the same regular growth rate of α per step increment in n. Also, V_n's first term *leads in growth* (compared with U_n's) by a constant factor of $(\alpha - \beta)$.

Second terms: Quantized adjustment

Again looking at (20.34), we focus on the second terms (in β^s) and consider s as *given*. We may think of the second term as an adjustment to the first. Hence we may regard β^s as a fixed quantum of adjustment—which we now name 'q_s'. And in order to visualize this quantum we shall prefer it to be without sign—i.e. as the modulus (absolute value) of β^s:

$$q_s = |\beta^s|. \qquad (20.35)$$

This means that we shall need to preserve the sign of β^s separately. Also we will need to understand this sign when we look at specific examples—as it determines when adjustments are added and when subtracted. So to make this sign explicit, we introduce t with possible values -1 or $+1$.

$$t = \begin{cases} -1, & \text{if } \beta < 0 \text{ and } s \text{ is odd,} \\ +1, & \text{otherwise.} \end{cases} \tag{20.36}$$

We incorporate t into (20.34) along with q_s (20.35) to give us

$$\left. \begin{aligned} t &= \begin{cases} -1, & \text{if } \beta < 0 \text{ and } s \text{ is odd,} \\ +1, & \text{otherwise.} \end{cases} \\ \\ q_s &= |\beta^s| \\ U_n &= \quad U_s \alpha^{n-s} + t\, U_{n-s}\, q_s \\ V_n &= (\alpha - \beta)\, U_s \alpha^{n-s} + t\, V_{n-s}\, q_s \end{aligned} \right\} \begin{cases} \text{for all } n, s \text{ in } \mathbb{Z}, \text{ if } |Q| = 1; \\ \text{for all } n, s \text{ in } \mathbb{Z}, \ 0 \le s, \ s \le n, \\ \qquad \text{if } |Q| \ne 1. \end{cases}$$

$$\tag{20.37}$$

Here we notice the pattern of adjustment for $\langle U_n \rangle$ is just *the sequence itself* written in numbers of quanta and lagged by s places, hence 'U_{n-s}'. And similarly, the V_n expression refers to *itself* too—repeating its own sequence, again with a lag of s places.

Logarithmic spirals

Here we have the discrete sequences $\langle U_n \rangle$ and $\langle V_n \rangle$, and in each, with increasing n, the ratio of one number to its predecessor tends towards a common growth factor, α. And as before, we shall move on to think of this growth as an underlying continuous function—a logarithmic spiral. We may then regard sequence members as being 'variations from' or 'adjustments to' this reference. But what should be the formula for such a spiral? We have already seen that the Fibonacci sequence grows at approximately the ϕ per step, and that the Golden Spiral grows at exactly ϕ per quadrant. This ϕ-per-quadrant definition had its origin in the spiral growth seen when growing a Golden Rectangle by adding whirling squares (p.36). So, to maintain compatibility in the general case (for all our Lucas Sequences), we shall adopt the 'backbone growth' of α per quadrant for our spiral growth. This in turn means that the growth per turn will be four such successive growths compounded—that is, α^4 per full turn. In the spiral, if we measure the angle of turning of the radial vector anticlockwise with θ, then the fraction of a full turn will be $\theta/(2\pi)$. Using (r, θ) polar coordinates, and with k, a constant yet to be determined, we choose the logarithmic spiral draft form

$$r = k(\alpha^4)^{\frac{\theta}{(2\pi)}} = k\alpha^{\frac{4\theta}{(2\pi)}}. \tag{20.38}$$

Again using polar coordinates, we shall position the $\langle U_n \rangle$ numbers at precise locations near to this spiral, in such a way as to help us study their relationship to it. We place the $\langle U_n \rangle$ sequence of numbers round and round the origin (spiral fashion), at a distance from the origin of $r=U_n$ and on the quadrant axes—i.e. where θ is a multiple of $(2\pi)/4$. This 'one quadrant per step' corresponds with the 'one-power-of-α' growth in the spiral. Hence

$$\langle U_n \rangle \text{ numbers are placed at } \quad (r,\theta) = \left(U_n, \ n \cdot \frac{(2\pi)}{4} \right). \tag{20.39}$$

Now, to match the backbone-growth terms in (20.37) we want a spiral that is appropriately 'anchored'—that is, a spiral that passes through the number point U_s. Looking back at our draft spiral formula (20.38), we see that such an 'anchoring' will happen when $r=U_s\alpha^0$. In (20.39) we see that this U_s number point will be located at $\theta=s \cdot (2\pi)/4$ (that is, at s quadrants), and this locates just where that power of α goes to zero. We may rearrange this as $s=4\theta/(2\pi)$, and as before (p.209) say for log spiral 'number s',

$$r = U_s \, \alpha^{\frac{4\theta}{(2\pi)} - s}. \tag{20.40}$$

Hence, we shall continue to think of index s as our *spiral identifier* (as we did in the 3D models, p.226 and following). Our overall purpose here is to demonstrate that using this configuration, the distance along an axis between a spiral's crossing point and a number point U_n is a (hereby known) whole number of $|\beta^s|$ quanta.

Offset axes

In (20.34) V_n's first term has an $(\alpha - \beta)$ scaling factor giving it a growth lead. This, in the context of log spiral growth, raises the question: '*If one spoke is longer than another by a factor of $(\alpha - \beta)$, then what is the angle between them?*' We know our spiral grows by α^4 per full turn of (2π) radians. Therefore this growth lead is equivalent to an offset angle

$$\omega = \frac{log(\alpha - \beta)}{4 log(\alpha)} \cdot (2\pi) \tag{20.41}$$

which, using Eqns. (20.15), (20.10) and (20.11), may be expressed as

$$\text{offset} \quad \omega = \frac{log(P^2 - 4Q)}{8\left[log\left(P + \sqrt{P^2 - 4Q}\right) - log(2)\right]} \cdot (2\pi) \qquad (20.42)$$

In passing, we note that for $P^2 >> 4Q$,

$$2log(P) / \left[8\left(log(2P) - log(2)\right)\right] = 1/4,$$

so $\omega = 90°$. Alternatively we could consider this as $\Delta \to P^2$, $\alpha \to P$, $\beta \to 0$; so the growth lead $(\alpha - \beta)$ tends to α—a quarter turn of growth.

Now, just as we saw with the Fibonacci/Lucas case (p.214) for given P and Q, ω is constant—independent of n and s. So, if we keep the existing 'U_s anchored spiral' fixed in place and rotate a set of quadrant axes underneath it by angle ω in the direction of increasing growth (as we did in Fig.145, p.217)—then the $(\alpha - \beta)$ scaling will be taken care of for all combinations of n and s.

> *Hence, in this way, we may use the same spiral backbone for both $\langle U_n \rangle$ and $\langle V_n \rangle$ and also use the same quantum size q_s for all adjustments.*

Accordingly we offset the companion quadrant axes by ω:

$$\left.\begin{array}{l} \langle U_n \rangle \text{ are placed at} \quad (r, \theta) = \left(U_n, \; n \cdot \dfrac{(2\pi)}{4}\right) \\[2em] \langle V_n \rangle \text{ are placed at} \quad (r, \theta) = \left(V_n, \; n \cdot \dfrac{(2\pi)}{4} + \omega\right). \end{array}\right\} \qquad (20.43)$$

To recap: The U_n numbers are placed on successive quadrant axes, and we see their related quantized adjustments (as in (20.37)) taking place along these axes (20.39). Similarly, the companion V_n numbers are placed on offset quadrant axes (20.43) and their quantized adjustments occur along those axes.

Chapter 21

Pell, Jacobsthal and Mersenne

We have seen that both $\langle U_n \rangle$ and $\langle V_n \rangle$ Lucas Sequences satisfy the following recurrence with seed pairs $\langle 0, 1 \rangle$ and $\langle 2, P \rangle$ respectively:

$$x_n = Px_{n-1} - Qx_{n-2}. \tag{21.1}$$

In the Table Fig. 171 (p.267) we see values for P and Q that generate the better known Lucas Sequences [378], plus several others, including a special case $(6, 8)$ that we explore towards the end of this chapter. For given P and Q, both the primary sequence and its companion share a common underlying growth rate. For the Fibonacci and Lucas sequences, this is ϕ; for the Pell and its companion it is δ the Silver Ratio. The growth rates of certain other sequences have also been accorded metallic nicknames, and these numbers have some interesting properties (see p.429).

Fractal-like structure (or not?)

As we check through these cases, we shall find that for some, the sizes of their quanta depend on s. Such quanta then scale in a self-similar way according to β^s, resulting in the fractal-like structure that we discussed earlier (p.241). However there are other cases where the quantum size is unity; so in these cases, there is no successive scaling (as 1^s remains equal to 1)—hence their structure is not fractal-like. We shall first summarize the results we have already—now thinking of the Fibonacci and Lucas numbers as being specific Lucas Sequences. This will also give us a further check on our generalized formulæ. Then we shall move on to new cases, where the values of α and β will depend on the particular sequence that we are discussing.

Specific Case: (1,− 1) Fibonacci numbers, Lucas numbers and the Golden Ratio

For the Fibonacci and Lucas numbers, we set $P = 1$ and $Q = -1$ in (21.1), to obtain the recurrences

Fibonacci numbers $\quad F_n = F_{n-1} + F_{n-2}, \qquad F_0 = 0,\ F_1 = 1$

Lucas numbers $\qquad L_n = L_{n-1} + L_{n-2}, \qquad L_0 = 2,\ L_1 = 1$

and using (20.10), p.254, $\quad \Delta = 5, \quad$ therefore

using (20.11), p.254, $\qquad \alpha = (1 + \sqrt{5})/2 \ = \ \phi$ $\qquad$ (21.2)

using (20.12), p.254, $\qquad \beta = (1 - \sqrt{5})/2 \ = \ (-\phi)^{-1}$ $\qquad$ (21.3)

using (20.36), p.262, $\qquad t = (-1)^s.$ $\qquad$ (21.4)

Applying equations (21.2) through (21.4) in (20.37), p.262, gives

$q_s = \phi^{-s} \quad \Longrightarrow$ fractal-like structure

$F_n = \quad F_s \phi^{n-s} + (-1)^s F_{n-s} q_s$ $\qquad$ for all n, s in $\mathbb{Z}$. $\quad$ (21.5)

$L_n = \sqrt{5} F_s \phi^{n-s} + (-1)^s L_{n-s} q_s$

From (20.40), p.263, the spiral is Golden, growing ϕ per quadrant:

$$r = F_s \phi^{\frac{4\theta}{(2\pi)} - s}.$$ $\qquad$ (21.6)

From (20.41), p.263, the Lucas axes offset angle is

$$\omega = \frac{log(\sqrt{5})}{4\,log(\phi)} \cdot (2\pi) \ = \ 2.62680\ldots \ \equiv \ 150.505\ldots^\circ$$ $\qquad$ (21.7)

And from (20.43), p.264, the polar coordinates for placing the number points are

$\langle F_n \rangle$ are placed at $\quad (r, \theta) \ = \ \left(F_n, \ n \cdot \dfrac{(2\pi)}{4} \right)$

$\langle L_n \rangle$ are placed at $\quad (r, \theta) \ = \ \left(L_n, \ n \cdot \dfrac{(2\pi)}{4} + \omega \right)$ $\qquad$ (21.8)

As an example of this Fibonacci Lucas case, Fig. 145 on page 217 shows an s=5 Golden Spiral with its quantized relationships to the number points. Also, on page 477 there is a Table of Fibonacci and Lucas numbers and powers of ϕ. Next, Fig. 171 lists some of the better known Lucas Sequences and 'metallic means'. The offset ω is defined in Eqn. (20.42), p.264. Other means include the Bronze Ratio P=3, Q=−1, (p.470); and the Nickel Ratio P=1, Q=−3, (p.472) [493, 854].

ω for selected (P, Q) Lucas Sequences

P	Q	$\langle U_n \rangle$ *sequence*	*ω offset*	*companion $\langle V_n \rangle$*
1	-1	**Fibonacci**	150.505...°	**Lucas**
		0, 1, 1, 2, 3, 5, 8,		2, 1, 3, 4, 7, 11, 18,
		13, 21, 34, 55, ...	**Golden: ϕ**	29, 47, 76, 123, ...
		A000045		A000032
2	-1	**Pell**	106.169...°	**Pell-Lucas**
		0, 1, 2, 5, 12, 29,		2, 2, 6, 14, 34, 82,
		70, 169, 408, ...	**Silver: δ**	198, 478, 1154, ...
		A000129		A002203
4	-1	**$\sqrt{5}$ convergents' denominators**	98.38...°	**Even Lucas** (L_{3n})
		(0), 1, 4, 17, 72,		2, 4, 18, 76, 322,
		305, 1292, 5473, ...		1364, 5778, ...
		A001076		A014448
1	-2	**Jacobsthal**	142.646...°	**Jacobsthal-Lucas**
		0, 1, 1, 3, 5, 11, 21,		2, 1, 5, 7, 17, 31, 65,
		43, 85, 171, 341, ...	**Copper: 2**	127, 257, 511, ...
		A001045		A014551
3	2	**'Mersenne'** $(2^n - 1)$	0.0°	**'Fermat'** $(2^n + 1)$
		0, 1, 3, 7, 15, 31, 63,		2, 3, 5, 9, 17, 33, 65,
		127, 255, 511, ...		129, 257, 513, ...
		A000225		A000051
6	8	**Perfect host** $2^{n-1} \cdot (2^n - 1)$	45°	**Perfect companion** $2^n \cdot (2^n + 1)$
		0, 1, 6, 28, 120, 496,		2, 6, 20, 72, 272,
		2016, 8128, 32640, ...		1056, 4160, ...
		A006516		A161168

Figure 171: **Lucas Sequences.**

267

Specific Case: (2, −1) Pell numbers, Pell-Lucas numbers and the Silver Ratio

Just as the Fibonacci numbers are closely related to the Golden Ratio ϕ, so the Pell numbers are similarly related to the Silver Ratio $\delta = 1 + \sqrt{2} = 2.41421\ldots$ In Fig. 172 we see the Pell and Pell-Lucas growth rates, for comparison with the Fibonacci and Lucas growth chart we saw earlier, (Fig. 28 on page 30). The Pell[1] numbers 1, 2, 5, 12, ... were well known to the ancients and form the denominators of the convergents of $\sqrt{2}$ (Babylonian method—p.74):

$$\frac{1}{1}, \frac{3}{2}, \frac{7}{5}, \frac{17}{12}, \frac{41}{29}, \frac{99}{70}, \frac{239}{169}, \ldots$$

The numerators of these same convergents are half the values of the companion sequence: the Pell-Lucas numbers. This numerator sequence has Sloane reference A001333 [839].

> *Again to avoid confusion, we had best note that letters P and Q are commonly used to denote both the recurrence coefficients and the Pell and Pell-Lucas sequences.[2] However, as sequence members are subscripted, and recurrence coefficients are not, this should not pose us any problems.*

From the Table on page 267, for Pell numbers and Pell-Lucas numbers we set coefficients $P=2$ and $Q=-1$, giving us the recurrences

Pell numbers $\qquad P_n = 2P_{n-1} + P_{n-2}, \qquad P_0 = 0, \ P_1 = 1$

Pell-Lucas numbers $\quad Q_n = 2Q_{n-1} + Q_{n-2}, \qquad Q_0 = 2, \ Q_1 = 2.$

Also, we denote the Silver Ratio as $\delta = 1 + \sqrt{2} = 2.4142\ldots$ Hence,

using (20.10), p.254, $\qquad \Delta = 8, \quad$ therefore

using (20.11), p.254, $\qquad \alpha = 1 + \sqrt{2} \ = \ \delta \qquad\qquad$ (21.9)

using (20.12), p.254, $\qquad \beta = 1 - \sqrt{2} \ = \ -\delta^{-1} \qquad$ (21.10)

using (20.36), p.262, $\qquad t = (-1)^s. \qquad\qquad\qquad$ (21.11)

---

[1] It is often said that both Pell Equation and Pell numbers were named because of a misunderstanding by Euler (e.g. Tattersall [885] p.255). However [MacTutor] O'Connor & Robertson—in addition to pointing out that the so-called Pell's equation $y^2 = ax^2 + 1$ was studied in India 1000 years prior by mathematician astronomers Brahmagupta and Bhaskara II—note that it appears in a book 'by Rahn', co- or fully written by Pell [705]. The full theory was later worked out by Lagrange.

[2] For example, Weisstein [959].

Applying equations (21.9) through (21.11) in (20.37), p.262, gives

$$
\left.
\begin{aligned}
q_s &= \delta^{-s} \quad \implies \text{ fractal-like structure,} \\
P_n &= \quad P_s\delta^{n-s} + (-1)^s P_{n-s}\,q_s \\
Q_n &= 2\sqrt{2}\,P_s\delta^{n-s} + (-1)^s Q_{n-s}\,q_s
\end{aligned}
\right\} \quad \text{for all } n,s \text{ in } \mathbb{Z}. \quad (21.12)
$$

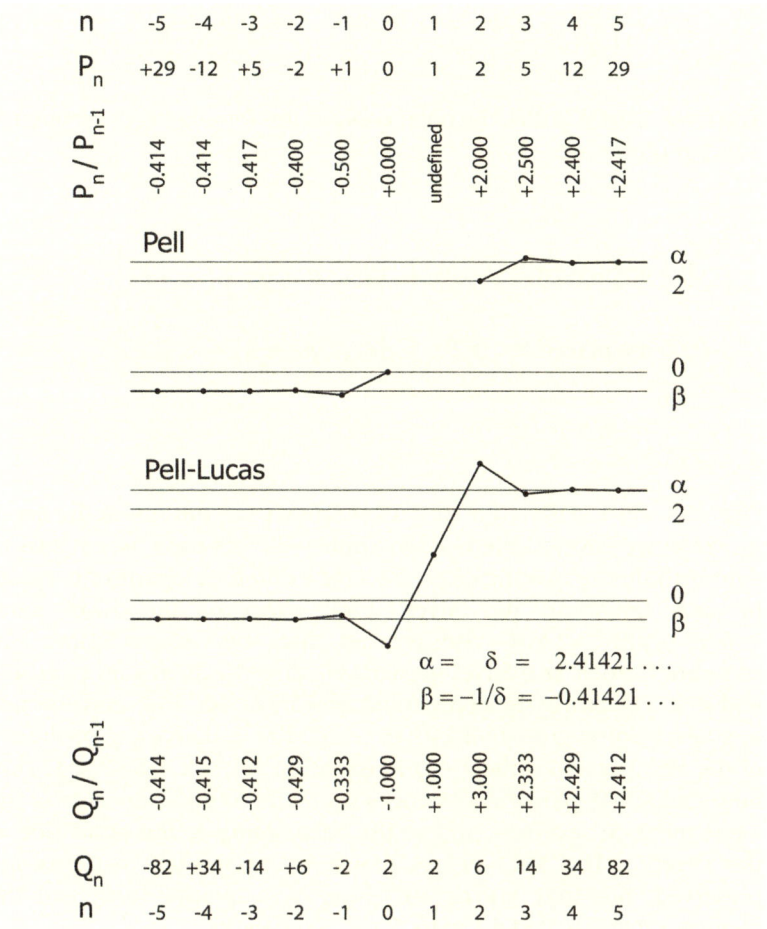

Figure 172: **Pell and Pell-Lucas step-by-step growth rates.**

With $\alpha = \delta$ (the Silver Ratio) in (20.40), p.263, we see the spiral here is Silver, growing δ per quadrant:

$$r = P_s\,\delta^{\frac{4\theta}{(2\pi)} - s}.$$

(21.13)

From (20.41), p.263, the Pell-Lucas axes offset angle is

$$\omega = \frac{log(2\sqrt{2})}{4\,log(\delta)} \cdot (2\pi) \; = \; 1.853005\ldots \; \equiv \; 106.169\ldots^{\circ}$$

(21.14)

And from (20.43), p.264, the polar coordinates for placing the sequence number points are

$$\langle P_n \rangle \text{ are placed at } \;\; (r,\theta) \; = \; \left(P_n, \; n \cdot \frac{(2\pi)}{4} \right)$$

$$\langle Q_n \rangle \text{ are placed at } \;\; (r,\theta) \; = \; \left(Q_n, \; n \cdot \frac{(2\pi)}{4} + \omega \right).$$

(21.15)

Figs. 173 and 174 illustrate this Silver Spiral case, but not in the detail in which we covered the Golden Spiral—the purpose here is just to demonstrate the quantization effect on a different logarithmic spiral. In each quadrant, the Silver Spiral grows by the Silver Ratio $\delta = (1 + \sqrt{2}) \approx 2.414$—visibly faster than the Golden Spiral. To compare Golden and Silver growth, see also the Tables on page 477 and 479. Offsetting the companion axes by $\omega = 106.169°$ corresponds to a Silver-Spiral growth of $2\sqrt{2}$ to match the underlying growth lead of the Pell-Lucas numbers—as in equation (21.12). In Fig. 174 we note that as the parity of s changes from odd to even, the sign of the adjustments is reversed, and so the beads jump to the other side of the curve. The Pell and Pell-Lucas quanta-count patterns are preserved, but they are moved along by a quarter turn, and the quantum size is scaled down by δ, the Silver Ratio.

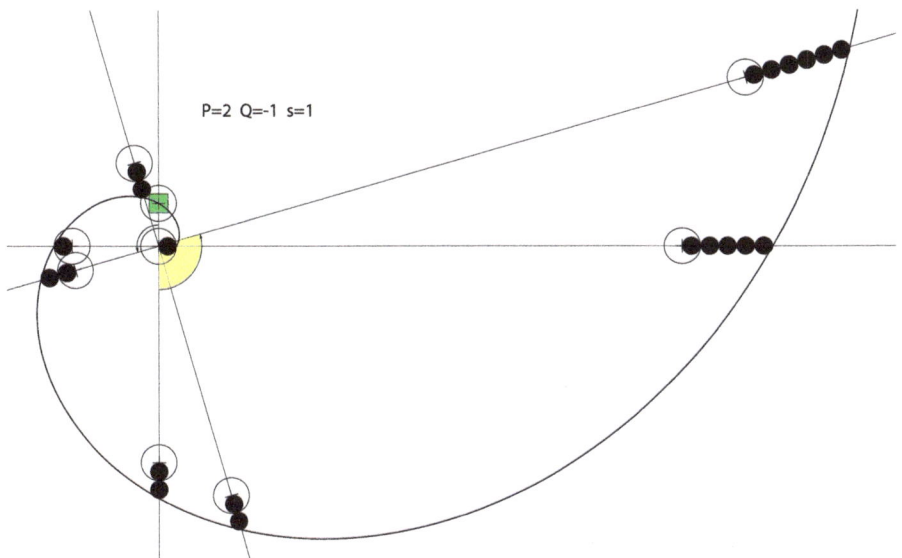

P=2 Q=-1 s=1

Figure 173: **Pell numbers, s=1 Silver Spiral, example showing Pell numbers (cardinal axes) and their 'Pell-Lucas' companions (offset axes).** This s=1 spiral is anchored at point P_1=1, its zero-q point, (green square).

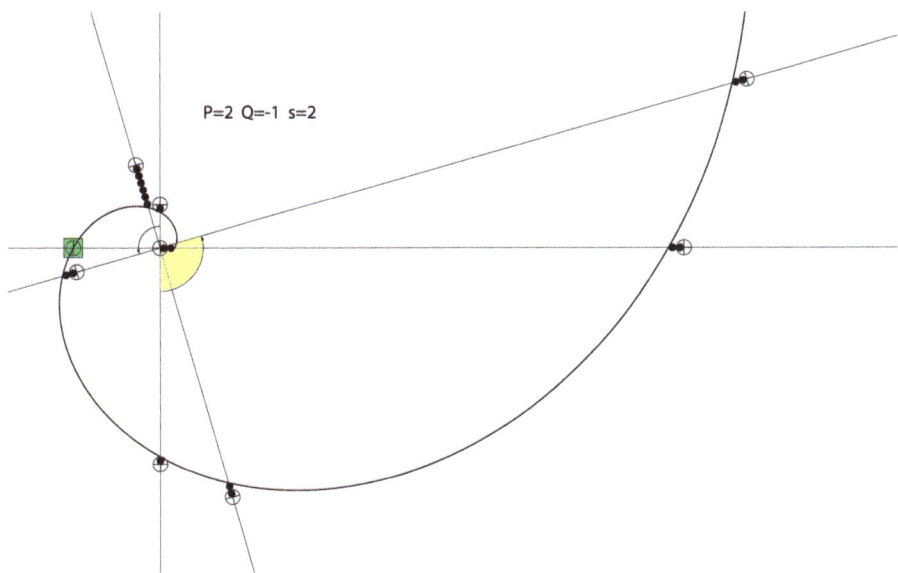

P=2 Q=-1 s=2

Figure 174: **Pell numbers, s=2 Silver Spiral, for comparison with above.** In this case we have an s=2 spiral anchored at point P_2=2, (green square).

271

Aside—Moon Gate: Skip-Pell and its generalization

The Tiwanaku gates[3] are also interesting when considered in terms of Pell and Pell-Lucas numbers (Table on page 267). Looking at the numbers in Fig. 175, we see 5, which is both a Pell and a Fibonacci prime, and 29, which is both a Pell and a Lucas prime. Indexed in the Pell sequence, 5 and 29 are P_3 and P_5; and these add to give Pell-Lucas number $Q_4=34$. Thus we see a demonstration of 'skip-Pell', with $P_4=12$ being skipped. We recall that Pell-Lucas numbers have an underlying $\approx 2\sqrt{2}$ growth lead ahead of the Pell numbers—(21.12) on page 269—so here we have $34 \approx 2\sqrt{2} \times 12$. Hence for Pell and Pell-Lucas numbers $\langle P_n \rangle$ and $\langle Q_n \rangle$, the generalization appears to be

$$Q_n = P_{n+1} + P_{n-1}. \tag{21.16}$$

Now, we could set out to prove this using the Pell equivalents of the Fibonacci roots α and β, viz. the Silver Ratio δ and its conjugate. But instead, let's quickly check for ourselves the general case identity (valid for all Lucas Sequences)—using Binet substitutions as we have previously, but without conducting a full inductive proof. Then we can come back to the Pell/Pell-Lucas case by taking account of its specific P and Q values.

As we saw back on 255, from Lucas, 1876 [592], we have the two Binet forms

$$U_n = \frac{\alpha^n - \beta^n}{\alpha - \beta} \tag{21.17}$$

$$V_n = \alpha^n + \beta^n. \tag{21.18}$$

The general identity is[4]

$$V_n = U_{n+1} - Q U_{n-1}. \tag{21.19}$$

Substituting for each U in (21.19) with (21.17), and multiplying through by $(\alpha - \beta)$ we get

$$(\alpha - \beta) V_n = \alpha^{n+1} - \beta^{n+1} - Q\left(\alpha^{n-1} - \beta^{n-1}\right).$$

---

[3] Pages 9 and 115.
[4] Along with equivalent identity $V_n = 2U_{n+1} - P U_n$.

Now, as $Q = \alpha\beta$

$$(\alpha - \beta)\, V_n = \alpha^{n+1} - \beta^{n+1} - \alpha^n \beta + \alpha \beta^n$$
$$= \alpha^n (\alpha - \beta) + \beta^n (\alpha - \beta).$$

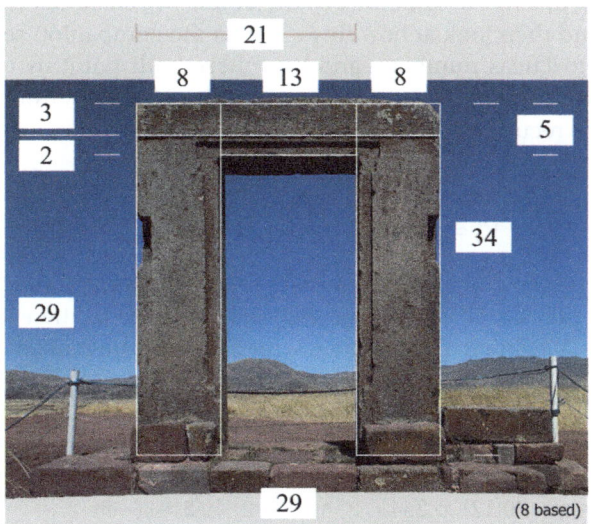

Figure 175: **The Gate of the Moon. Skip Pell: $29 + 5 = 34$.**
Original gate image by Daniel Maciel [605, 140].

Which brings us back to the Binet form for V_n—Eqn. (21.18)

$$V_n = \alpha^n + \beta^n. \qquad (21.20)$$

Hence according to identity (21.19), the simple skip form (for example (21.16)) is valid for all Lucas Sequences where $Q = -1$, which includes:

- (P=1) Gold/Fibonacci,
- (P=2) Silver/Pell,
- (P=3) Bronze,
- (P=4) the denominators of $\sqrt{5}$ convergents.[5]

$$\begin{cases} U_n &= P{\cdot}U_{n-1} + U_{n-2}, & U_0 = 0,\ U_1 = 1, & P \text{ in } \mathbb{Z} \\ V_n &= P{\cdot}V_{n-1} + V_{n-2}, & V_0 = 2,\ V_1 = P. \end{cases}$$

$$V_n = U_{n+1} + U_{n-1}, \qquad (Q = -1). \qquad (21.21)$$

---

[5] Pages 266–267.

Specific Case: $(4, -1)$ CF $\sqrt{5}$ Denom's / even Lucas

The case of $P=4$ and $Q=-1$ (see Table on p.267) is reviewed by Argentinian mathematician Vera de Spinadel [854] who finds the underlying growth rate of this Lucas Sequence to be ϕ^3. Instead of analysing this in the same way as the others, we shall first check why $\alpha = \phi^3$, and then look at how this explains the companion sequence's relation to Lucas numbers, and how these both point us to similar scenarios where the ϕ powers are a multiple of 3. So, setting $P=4$ and $Q=-1$, the recurrences are

$$\left.\begin{array}{ll} U_n = 4U_{n-1} + U_{n-2}, & U_0 = 0, \; U_1 = 1 \\ V_n = 4V_{n-1} + V_{n-2}, & V_0 = 2, \; V_1 = 4. \end{array}\right\}$$

The $\langle U_n \rangle$ sequence here consists of the denominators of the convergents of $\sqrt{5}$.[6] And the $\langle V_n \rangle$ is comprised of the even Lucas numbers. We find α and β as before then express these in terms of ϕ:[7]

using (20.10), p.254, $\qquad \Delta = 20, \quad$ therefore

using (20.11), p.254, $\qquad \alpha = 2 + \sqrt{5} = \phi^3$ $\qquad$ (21.22)

using (20.12), p.254, $\qquad \beta = 2 - \sqrt{5} = (-\phi)^{-3}.$ $\qquad$ (21.23)

Also, from (20.41), p.263, the companion axes offset angle is

$$\omega = \frac{log(20)}{8\,(log(4+\sqrt{20}) - log(2))} \cdot (2\pi) \;=\; 1.629803829\ldots$$

$$\equiv \; 93.3\,808\,808\,58\ldots^{\circ} \qquad (21.24)$$

In the 'Golden' cases $\langle F_n \rangle$ and $\langle L_n \rangle$, we had $\alpha = \phi$ and $\beta = -\phi^{-1}$. But here, $\alpha = \phi^3$ and $\beta = (-\phi)^{-3}$. So, in the companion Binet form,

$$V_n \;=\; \alpha^n + \beta^n \;=\; \phi^{3n} + (-\phi)^{-3n} \;=\; L_{3n}. \qquad (21.25)$$

And in the Table on p.477, we see the even Lucas numbers are located at indices: $0, 3, 6, 9, \ldots$ This then, is why the $3n$ index in (21.25) selects the even members. And further, by consulting the Table on page 478 we propose the next 'ϕ-power-is-multiple-of-3' occurrence as being $\alpha = \phi^6 = (9 + 4\sqrt{5})$ and $\beta = \phi^{-6} = (9 - 4\sqrt{5})$. This would require $P = (\alpha + \beta) = 18$ and $Q = (\alpha\beta) = 1$, which would then generate $\langle L_{6n} \rangle$.

---

[6] These convergents may be obtained from successive stages of refinement of the continued fraction for $\sqrt{5}$. This is discussed in Appendix C, p.428.

[7] A Table of ϕ powers, listed in terms of $\sqrt{5}$, is shown on p.478.

Specific Case: $(1, -2)$ Jacobsthal numbers, Jacobsthal-Lucas, and the Copper Ratio

These numbers are named after German mathematician Ernst Jacobsthal (1882–1965) [954, 699]. Jacobsthal studied at the University of Berlin under Georg Frobenius and Issai Schur. In Fig. 176 we see the quantization effect with respect to a log spiral that doubles its radial vector length every quarter turn. Here, not only are the quanta much bigger (size one), but also they remain constant, in contrast to the Fibonacci and Pell cases where they scaled from one s value to the next. In Fig. 176 the $s=1$ spiral is anchored at point $J_1=1$, its zero-q point (in green square).

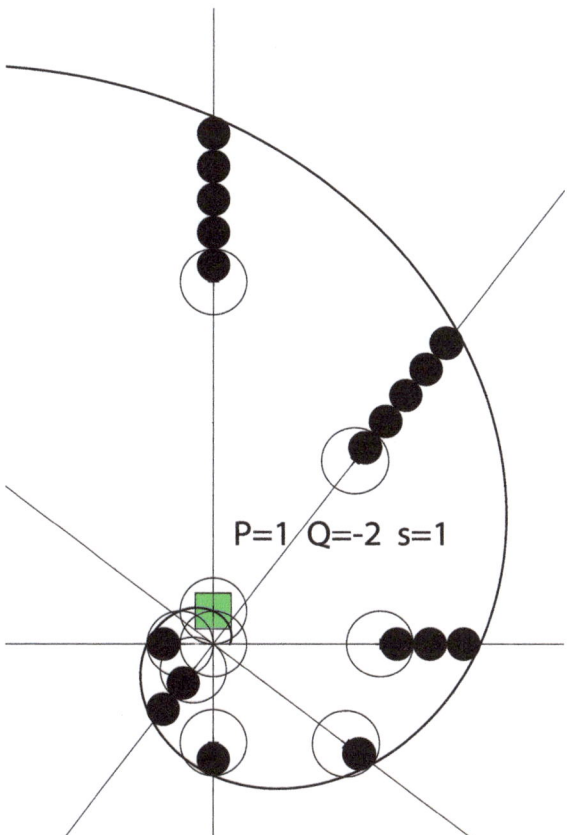

Figure 176: **Example showing Jacobsthal numbers (cardinal axes) and their companions (offset axes). Spiral 'doubles-per-quadrant'.**

Offsetting the companion axes by ω=142.647° corresponds to a spiral growth of 3× to match the underlying growth lead of the companion numbers—as in equation (21.29). Near the origin we see the square (anchor) at radial vector length (spoke) 1. Anticlockwise ('west') we see one quantum, size 1. Then proceeding in quadrants 'south', 'east', and 'north', we see: 1, 3, and 5 quanta—the start of the Jacobsthal sequence. Now, starting back in the south west, and working anti-clockwise by quadrants (SW, SE, and NE) we have quanta counts: 2, 1, 5—the start of the companion sequence. From the Table on page 267, for Jacobsthal and Jacobsthal-Lucas we set P=1 and Q=−2, so

Jacobsthal numbers $\qquad\qquad J_n = J_{n-1} + 2J_{n-2}, \qquad J_0 = 0, \; J_1 = 1$

Jacobsthal-Lucas numbers $\quad j_n = j_{n-1} + 2j_{n-2}, \qquad j_0 = 2, \; j_1 = 1$

and

$$\begin{array}{llll}
\text{using (20.10), p.254,} & \Delta = 9, \quad \text{therefore} & \\
\text{using (20.11), p.254,} & \alpha = 2 & (21.26) \\
\text{using (20.12), p.254,} & \beta = -1 & (21.27) \\
\text{using (20.36), p.262,} & t = (-1)^s. & (21.28)
\end{array}$$

Applying equations (21.26) through (21.28) in (20.37), p.262, gives

$$\left.\begin{array}{l}
q = |\beta^s| = 1, \quad \text{(so, not fractal-like)} \\[4pt]
J_n = \; J_s 2^{n-s} + (-1)^s J_{n-s} \, q \\[4pt]
j_n = 3 J_s 2^{n-s} + (-1)^s j_{n-s} \, q
\end{array}\right\} \quad \text{for all } s \leq n, \text{ and } n, s \text{ in } \tilde{\mathbb{N}}.$$

$$(21.29)$$

With $\alpha = 2$ in (20.40), p.263, we see that in this log spiral, the radial vector doubles per quadrant:

$$r = J_s 2^{\frac{4\theta}{(2\pi)} - s}. \qquad\qquad (21.30)$$

From (20.41), p.263, the companion axes offset angle is

$$\omega = \frac{log(3)}{4\,log(2)} \cdot (2\pi) \; = \; 2.48965327\ldots$$

$$\equiv \; 142.646625\ldots° \qquad\qquad (21.31)$$

And from (20.43), p.264, the polar coordinates for placing the sequence number points are

$$\langle J_n \rangle \text{ are placed at } (r,\theta) = \left(J_n, \ n \cdot \frac{(2\pi)}{4} \right)$$

$$\langle j_n \rangle \text{ are placed at } (r,\theta) = \left(j_n, \ n \cdot \frac{(2\pi)}{4} + \omega \right) \qquad (21.32)$$

We have seen how the Jacobsthal numbers (and their companion sequence) have an underlying growth rate (α) of 2, arising from P=1, Q=−2. It is in this context that this 2 is named the Copper Ratio. It is a member of the Metallic Means family, but not of the Silver means family. For more details on this see Appendix C, in particular p.429).

Specific Case: (3, 2) $2^n - 1$ 'Mersenne' and $2^n + 1$ 'Fermat' numbers

As with the distinction between the Lucas number sequence and Lucas Sequences, here too we need to avoid confusion. In the context of Lucas Sequences, a Mersenne number may be regarded as being of the form $2^n - 1$ (A000225 [839]); and a Fermat number may be taken as being of the form $2^n + 1$ (A000051 [839]).   However outside this context, Mersenne numbers are usually considered to be only the special cases $2^p - 1$ where p is prime (A001348 [839]). Also, Fermat numbers are usually regarded as being just the cases where the power of 2 is itself a power of 2, so having the form $2^{2^n} + 1$ (A000215 [839]) [950]. In Fig. 177 we see the same spiral that we had for the Jacobsthal numbers (in which the spoke length doubles per quadrant), and this time we see the quantized relationships of the spiral to both $2^n - 1$ and $2^n + 1$ numbers. Again the quantum size is unity. From the Table on page 267, for $2^n - 1$ and $2^n + 1$ numbers we set P=3 and Q=2, so

$$\text{'Mersenne' numbers} \quad f_n = 3f_{n-1} - 2f_{n-2}, \qquad f_0 = 0, \ f_1 = 1$$

$$\text{'Fermat' numbers} \quad g_n = 3g_{n-1} - 2g_{n-2}, \qquad g_0 = 2, \ g_1 = 3$$

and

using (20.10), p.254, $\qquad$ $\Delta = 1,$ therefore

using (20.11), p.254, $\qquad$ $\alpha = 2$ $\qquad\qquad$ (21.33)

using (20.12), p.254, $\qquad$ $\beta = 1$ $\qquad\qquad$ (21.34)

using (20.36), p.262, $\qquad$ $t = +1.$ $\qquad\qquad$ (21.35)

Applying equations (21.33) through (21.35) in (20.37), p.262, gives

$$\left.\begin{array}{l} q = |\beta^s| = 1, \ \text{(not fractal-like either)} \\[1em] f_n = f_s 2^{n-s} + f_{n-s}\, q \\[1em] g_n = f_s 2^{n-s} + g_{n-s}\, q \end{array}\right\} \quad \text{for all } s \leq n, \text{ and } n, s \text{ in } \tilde{\mathbb{N}}.$$

$$(21.36)$$

Again with $\alpha = 2$ in (20.40), p.263, we have the same log spiral that we had for Jacobsthal—one which doubles its spoke length every quadrant:

$$r = f_s\, 2^{\frac{4\theta}{(2\pi)}-s}. \qquad\qquad (21.37)$$

From (20.41), p.263, the companion axes offset angle is

$$\omega = \frac{log(1)}{4\, log(2)} \cdot (2\pi)$$

$$= 0 \quad \text{(zero offset).} \qquad\qquad (21.38)$$

So, from (20.43), p.264, and with $\omega{=}0$ this time, the polar coordinates for placing the sequence number points are

$$\left.\begin{array}{l} \langle f_n \rangle \text{ are placed at } (r,\theta) \;=\; \left(f_n, \; n \cdot \dfrac{(2\pi)}{4} \right) \\[2em] \langle g_n \rangle \text{ are placed at } (r,\theta) \;=\; \left(g_n, \; n \cdot \dfrac{(2\pi)}{4} \right). \end{array}\right\} \qquad (21.39)$$

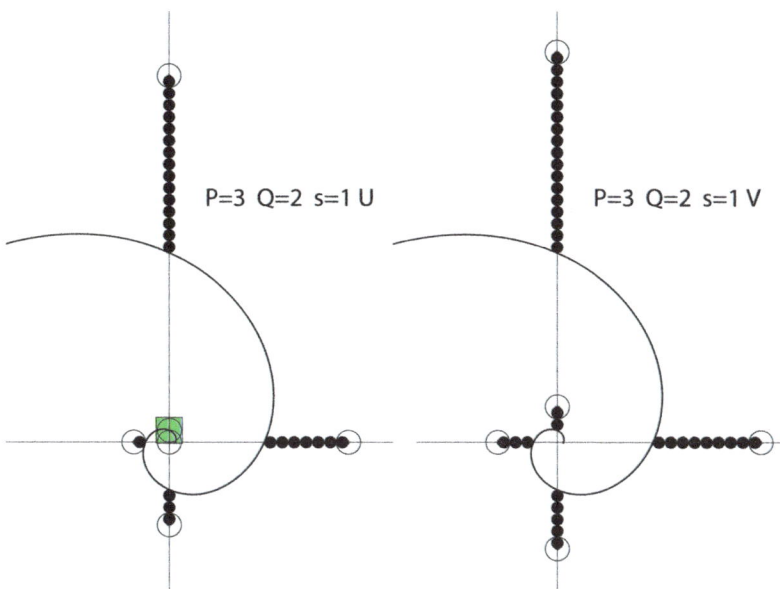

Figure 177: $s=1$, double-per-quadrant spiral—
'Mersenne numbers' and their companions ('Fermat').

In Fig. 177 the $s=1$ spiral is anchored at point $f_1=1$, its zero-q point (green square). In equation (21.36) we see there is no growth lead—both first terms have the same (unity) coefficient. This means that the offset angle is zero and the companion axes coincide with the primary axes. Therefore, in the diagram, we show the 'U' and 'V' patterns separately to avoid overlap.

Other examples of $\omega = 0$

This instance of $\omega=0$ when $P=3$ and $Q=2$ is not unique. We see that it results from having $log(1)$ in the numerator of (20.42), p.264, and this will occur whenever $P^2 = 4Q + 1$, for example when

$$(P, Q) \;=\; (5, 6), \;\; (7, 12), \;\; (9, 20), \;\; (11, 30), \;\; \text{and so on.} \qquad (21.40)$$

Specific Case: (6, 8)—'The perfect host'

In exploring the ω offset formula ((20.42), p.264), it is interesting to note that for $P=6$ and $Q=8$, the companion axes offset angle

$$\omega = \frac{log(2)}{4 \cdot 2 \cdot log(2)} \cdot (2\pi) = \frac{(2\pi)}{8}$$

$$\equiv 45° \text{ (exactly).} \tag{21.41}$$

As such, $(6, 8)$ seems to be rather special. First, we note that the 45° offset places the companion quadrant axes (in terms of relative angles) in a perfectly balanced configuration exactly half way between the primary quadrant axes. And also, as soon as we check the number sequence that coefficients $(6, 8)$ generate, we find that it is host to the *even perfect numbers*, which are located at *prime indices*.

Perfect numbers have intrigued mathematicians since ancient times. A perfect number is one which is the sum of its divisors, other than itself. (Such divisors are also known as 'aliquot divisors' or 'aliquot parts' [712].) The first perfect number is 6, having the aliquot divisors: 1, 2, and 3, which sum to 6. The three perfect numbers that follow 6 are: 28, 496, and 8128, and these four smallest have been known for thousands of years [706].[8] Number theorist Florian Luca has shown that neither $\langle F_n \rangle$ nor $\langle L_n \rangle$ contain any perfect numbers [591]. Now, as just mentioned, we find that by using $P=6$ and $Q=8$ in the Lucas Sequence recurrence— Eqn. (20.1), p.252,

$$U_n = P \cdot U_{n-1} - Q \cdot U_{n-2}, \qquad U_0 = 0, \ U_1 = 1$$

we generate the sequence:[9]

$\langle U_n \rangle \quad = \quad \langle$ 0, 1, <u>6</u>, <u>28</u>, 120, <u>496</u>, 2016, <u>8128</u>, 32640, 130816, 523776, 2096128, 8386560, <u>33550336</u>, 134209536, 536854528, 2147450880, <u>8589869056</u>, ... $\rangle$

which (as indicated) includes the even perfect numbers:[10]

<u>6</u>, <u>28</u>, <u>496</u>, <u>8128</u>, <u>33550336</u>, <u>8589869056</u>, ...

---

[8] A perfect number is a triangular number. It always ends in 6 or 8 (last digit: Sloane A094540). The reciprocals of all its divisors sum to 2 [963]. If it is greater than 6, then it can be written as a sum of cubes [649]. It has a 'digital root' (result of repeatedly adding its decimal digits) of 1 (e.g. 4+9+6=19, 1+9=10, 1+0=1) [885]. Fermat numbers $2^{2^n} + 1$ (p.277) have digital roots of: 3 (n=0), 5 (n odd), else 8—(Koshy/Beedassy, A000215 [839]).

[9] Sloane A006516 [839].

[10] Sloane A000396 [839].

with these (and those that follow) being found at prime indices: [11]

$$n = 2, 3, 5, 7, \quad 13, 17, 19, \quad 31, \ldots$$

Prime-index 11 is omitted because $U_{11}=2096128$ is not perfect, and similarly, 23 and 29 are the next prime-index omissions. The companion sequence $\langle V_n \rangle$ is given by Eqn. (20.2), p.252, with seeds of 2 then P, thus beginning[12]

$$\langle V_n \rangle \; = \; \langle 2, 6, 20, 72, 272, 1056, 4160, 16512, 65792, 262656, \ldots \rangle.$$

Let's analyse this (6, 8) case in the same way we did the others. We have

$$\left. \begin{array}{lll} \text{perfect host} & U_n = 6U_{n-1} - 8U_{n-2}, & U_0 = 0, \; U_1 = 1 \\ \text{perfect companion} & V_n = 6V_{n-1} - 8V_{n-2}, & V_0 = 2, \; V_1 = 6 \end{array} \right\}$$

and

$$\begin{array}{llll} \text{using (20.10), p.254,} & \Delta = 4, & \text{therefore} & \\ \text{using (20.11), p.254,} & \alpha = 4 & & (21.42) \\ \text{using (20.12), p.254,} & \beta = 2 & & (21.43) \\ \text{using (20.36), p.262,} & t = +1. & & (21.44) \end{array}$$

Applying equations (21.42) through (21.44) in (20.37), p.262, gives

$$\left. \begin{array}{l} q \; = |\beta^s| = 2^s, \;\; \Longrightarrow \;\; \text{fractal-like} \\ U_n = \; U_s 4^{n-s} + U_{n-s} q \\ V_n = 2 U_s 4^{n-s} + V_{n-s} q \end{array} \right\} \quad \text{for all } s \leq n, \text{ and } n, s \text{ in } \tilde{\mathbb{N}}.$$

$$(21.45)$$

In this case, with $\alpha = 4$ in (20.40), p.263, and s given, we see that this log spiral's radial vector quadruples every quadrant:

$$r = U_s 4^{\frac{4\theta}{(2\pi)} - s}. \tag{21.46}$$

As we previously noted (p.280), and from (20.41), p.263, the companion-axes offset angle here is a special case: it is exactly 45°. This result now makes immediate sense, as in (21.45) the underlying growth lead for V_n over U_n is factor 2, and the per-quadrant growth is

---

[11] Sloane A000043 [839].
[12] Sloane A161168 [839].

4. So as $\sqrt{4} = 2$, the V_n axis will be mid-way between the U_n quadrant axis and the U_{n+1} quadrant axis—that is, at an offset of one eighth of a turn. Finally, from (20.43), p.264, the polar coordinates for placing the sequence number points are

$$\langle U_n \rangle \text{ are placed at } (r,\theta) = \left(U_n, \ n \cdot \frac{(2\pi)}{4} \right)$$

$$\langle V_n \rangle \text{ are placed at } (r,\theta) = \left(V_n, \ n \cdot \frac{(2\pi)}{4} + \frac{(2\pi)}{8} \right). \tag{21.47}$$

Perfect host—Binet form

Having found α and β for the $(6, 8)$ sequence, let's apply these in the general U_n Binet form (20.24) that we saw on page 255 [592]:

$$U_n = \frac{\alpha^n - \beta^n}{\alpha - \beta}.$$

By using $\alpha = 4$ and $\beta = 2$ from (21.42) and (21.43), we have

$$U_n = \frac{4^n - 2^n}{4 - 2} = \frac{2^n \cdot 2^n - 2^n}{2} = \frac{2^n(2^n - 1)}{2}.$$

$$\boxed{U_n = 2^{n-1}(2^n - 1)} \tag{21.48}$$

Now, the form of (21.48) is important in the study of even perfect numbers and Mersenne primes (primes of the form $2^n - 1$). It explains why the $(6, 8)$ sequence is host to the even perfect numbers. Euclid proved in his *Elements Book IX* that: 'If $2^n - 1$ is prime for some n, then $2^{n-1} \cdot (2^n - 1)$ is an even perfect number.' [329, 631]. Some 2,000 years later, Leonard Euler (1707–1783) proved that *all* even perfect numbers are of this form—that is: 'If N is an even perfect number, then N can be written in the form $N = 2^{n-1}(2^n - 1)$, where $(2^n - 1)$ is prime.' [176, 163]. To reflect this one-to-one correspondence, we now say: 'An even positive integer is a perfect number, (a number that equals the sum of its aliquot divisors), *if and only if* it has the form $2^{n-1}(2^n - 1)$, for some n such that $2^n - 1$ is prime.' Consequently this combined statement is known as the Euclid-Euler theorem. And further, (as we discuss on p.447): 'If for some $n > 0$, $2^n - 1$ is prime, then n too is prime.' which, in the perfect host sequence, positions the perfect numbers at prime indices. For more on perfect numbers, see (for example) O'Connor & Robertson [706] and Weisstein [962].

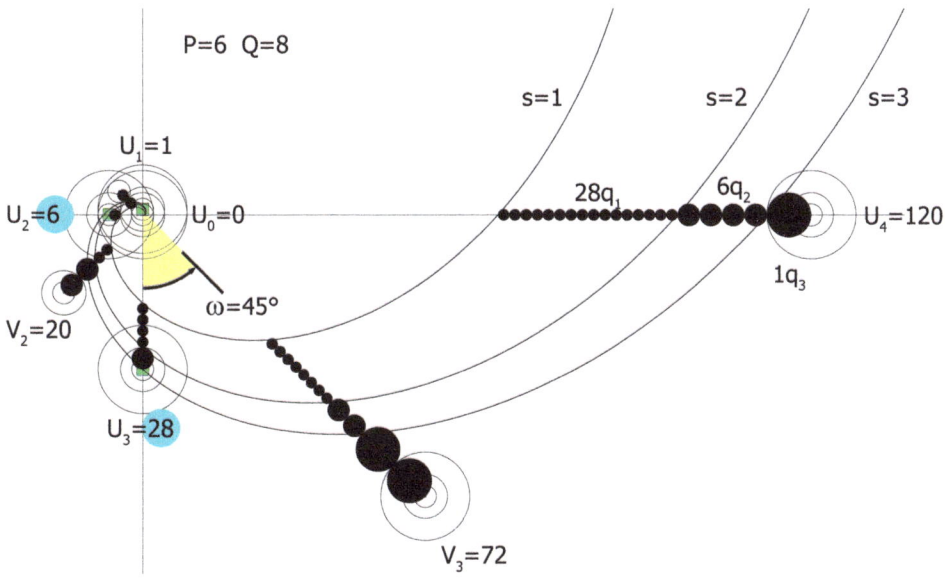

Figure 178: **Lucas Sequence (6, 8)**
including perfect numbers 6 and 28: a composite of s =1, 2, 3.

Positive dimension

In the examples we have seen so far, the absolute magnitude of β has been 1 or less. This has meant that with increasing spiral number s, when 'less than one' was raised to successively higher powers, it caused the quantum size to shrink, hence demonstrating negative fractal (-like) dimension. However, in this case, $\beta=2$ and so quantum size will increase with s, in fact, it will double when s increments by 1. In Fig. 178 we see 3 sizes of quantum superposed; these correspond to s values 1 to 3, with quantum sizes: 2 (for $s=1$), 4 (for $s=2$), and 8 (for $s=3$), with the larger beads covering the smaller ones. Number points are located using concentric circles with U_0 to U_4 placed centre, north, west, south, and east. Points V_2 and V_3 are placed south-west and south-east. As we know, the $\langle V_n \rangle$ quadrant axes are offset by $\omega=45°$. Counting north as polar angle $(2\pi)/4$, then to the east, we observe three sets of adjustments. From the $s=3$ spiral to the

U_4 number point is one quantum q_3 of size 8. This covers 2 quanta size 4, and adding a further 4 of these, we get $6q_2$ being the adjustment for the $s=2$ spiral, and lastly these $6q_2$ are obscuring the $12q_1$ (being half the size), and adding the other $16q_1$ that are visible, we get 28. As before, the adjustments for the other number points and spirals work in exactly the same way.

Specific Case: (4, 1)—and Mersenne primes

In his book about Fibonacci and Lucas numbers, Steven Vajda (a pioneer of linear programming) relates $P=4$, $Q=1$ to Mersenne numbers [913]. We explore this case in Appendix G, p.447.[13]

Reader exercise—OPNs

Find one or more odd perfect numbers,
else show why this is not possible.

So far, we have discussed even perfect numbers. But the question of whether or not odd perfect numbers exist is still unsolved [709]. This problem has been considered at least since Descartes, who corresponded with his friend Marin Mersenne on the subject from 1638 onwards [270]. Since then, a great deal of work has been done restricting possible properties of odd perfect numbers. For example in 1888 Catalan proved that if not divisible by 3, 5, or 7, an odd perfect number must have at least 26 distinct prime factors, and therefore have at least 45 digits [798]. And in 1991, Brent, Cohen, and Riele proved that if an odd perfect number were to exist, then it would have to be greater than 10^{300} [82].

Full circle

With this mention of Mersenne and perfect numbers, we might also note that immediately prior to his rabbits example in *Liber Abaci* [221], Fibonacci has a section entitled *On the finding of perfect numbers.*[14]

---

[13] This (4, 1) case is important because it provides the basis for the Lucas-Lehmer test of primality.

[14] In the original 1202 rabbits problem (in Fibonacci's *Liber Abaci*): Adult pair A have 'baby' pair B, giving total $A + B$, and for next generation, the baby B pair mature to become a new adult A pair, while the original A pair produces again [221]. This is, of course, an extremely contrived model of rabbit behaviour. However, the pattern (when applied stage by stage into the past) does describe bee ancestry *perfectly* [42, 812].

Part V
ϕ SCIENCE

Chapter 22

Introduction to ϕ science

We noted in the Preface that even today, after thousands of years, the Golden Ratio and the Fibonacci numbers are still very much 'hot topics'. And, as we shall see in the following chapters, they are contributing at the core of a major revolution in science and technology built on new theories and the exotic properties of ingenious designer materials. New ways of thinking are breaking through several 'glass ceilings'—long-held beliefs that until now have imposed significant constraints. For example:

- the 'crystallographic restriction'—a supposed limit on the types of crystal structure possible

- the 'diffraction limit'—the 'law' that light cannot be focused to a dot smaller than half a wavelength

- the wild dream of 'invisibility cloaking'—surely the stuff of science fiction!?

- the impossibility of controlling different types of wave using structures much smaller than a wavelength

- the apparently unbridgeable chasm that communications and computing engineers have faced: that between fibre optics (carrying light waves with micron-scale wavelengths) and silicon electronics (2 orders of magnitude smaller—at tens of nanometres)

So, to bring our coverage of Fibonacci and ϕ up to date, here in Part V we shall briefly review relevant discoveries in different areas of science and mathematics—noting how these have in turn sparked intense research activity worldwide. In the following eight chapters, we shall touch upon:

- phyllotaxis—both botanic and 'non-'

- quantum mechanics and ϕ-based magnetic resonance— a simple overview

- molecular-level ϕ—icosahedral order in supercooled metals and clusters of water molecules; ϕ in atomic bonding

- Penrose tilings—a mathematical 'jigsaw puzzle' that took hundreds of years to complete

- quasicrystals—a dramatic discovery which came as a great shock to traditional crystallographers

- medieval Islamic architecture and ornament—the remarkable new insights afforded by quasicrystal studies

- superlattices and the metamaterials revolution—breakthroughs that are turning science fiction into science fact

- the '-onics' technologies—how quasicrystals (especially those based on Fibonacci and ϕ) significantly extend capabilities in photonics, plasmonics, and phononics

Now, as certain of these developments have (in a very short period) inspired *thousands* of scientific papers, our choice of particular examples is to a great extent arbitrary. Therefore, while aiming at the ideal of a truly representative selection; it is hoped (at very least) that the choices made will be indicative. For further study, a fair number of the many papers here referenced have their own extensive lists of references—especially those written as review articles.

Chapter 23

Phyllotaxis

There have been many theories attempting to explain the plant phyllotaxis that we saw earlier in the *In nature* chapter (e.g. in the cactus on page 53), and next we look at recent research on the plant hormone auxin. After that, we shall expand the 'leaf arrangement' concept of phyllotaxis to include non-botanical examples.

Plant hormone—auxin

The latest theories of plant phyllotaxis focus on the role of the hormone auxin in governing growth in the meristem (where new plant tissue is formed). The auxin story itself starts with evolutionist Charles Darwin and his son Francis and their experiments on plant phototropism—the tendency of plants to grow towards the light. And although Darwin had deduced that some signalling agent was responsible, it took until 1926 (Dutch botanist F W Went) and 1934 (Went & English-born plant physiologist and microbiologist K V Thimann) for it to be described and isolated [543]. The newly discovered hormone was named auxin. In the case of phyllotaxis, it is now known that new growth in a flower head (for example of new florets) is determined by access to this hormone. Growth takes place outwards from the centre in some initial direction. This 'uses up' the local hormone in that direction, and the next new growth starts in a different region: currently the most auxin-rich. Once the process is started, there is reinforcing positive feedback, and growth usually settles into stable divergence at the Golden Angle of 137.5°, or sometimes at the Lucas Angle of 99.5°(p.53). Theoretical explanations for this mechanism are checked with computer simulations, which in turn then suggest further experiments.[1]

---

[1] There is a lot of activity in this field, for example see Bainbridge et al. [25] who give many further references.

Polypeptide phyllotaxis

Here is our first non-botanic example—we move from plants to proteins. Proteins are polymers of amino acids with a backbone structure of polypeptide chains. Such chains are formed by 'condensation' of amino acids—a process where many amino acid molecules string together. Their linkages are called peptide bonds, and one water molecule is shed for each new bond made. Once locked into a chain, the amino acid units are referred to as 'residues', and chains of 10 or 11 of these are very common [51]. The resulting backbone chain is a waltz of nitrogen and carbon atoms:

$$\ldots N\,C\,C\,N\,C\,C\,N\,C\,C\,N\,C\,C\,\ldots$$

In Fig. 179 we see an important example (found in keratin, hair, horn, fingernails, haemoglobin, and muscle): the alpha-helix. The diagram shows how its telephone-cord structure is stabilized; how the helix maintains a particular shape. Key to this are the dotted-line bonds that link from one turn of the helix to the next.[2] American chemist and biochemist Linus Pauling and biochemist Robert Corey, (and later Bamford et al.) studied example chain structures using X-ray crystallography. The structure of the alpha-helix was discovered by Pauling when he sketched the molecule on a sheet of paper and then folded it along parallel lines to make a tube helix and adjusted it to 3.6 residues per turn—this correctly positioned hydrogen bonds locking one turn to the next [732]. This, along with other important work on chemical bonding led to Pauling being awarded the 1954 Nobel Prize in Chemistry.

Bamford's group, in their work, found helices with '18 residues in five turns', '29 residues in 8 turns', and '47 residues in 13 turns', and proposed closest packing as the explanation [28]. Then, soon after the Bamford group's paper, Albert Frey-Wyssling commented in *Nature* that he had spotted a pattern in the numerators, and (without explicit mention of the Fibonacci sequence) he noted that the appearance of numbers: 5, 8, and 13, 'remind one of the divergences of phyllotaxis' [228]. At the time, Frey-Wyssling was one of perhaps a very few researchers who *could* have made such a comment. He was a leading Swiss botanist whose interest in the underlying structure of plants had taken him beyond the (light) microscope—to become one of the founders of molecular biology [633].

---

[2] Hydrogen bonds between N-H and C=O groups.

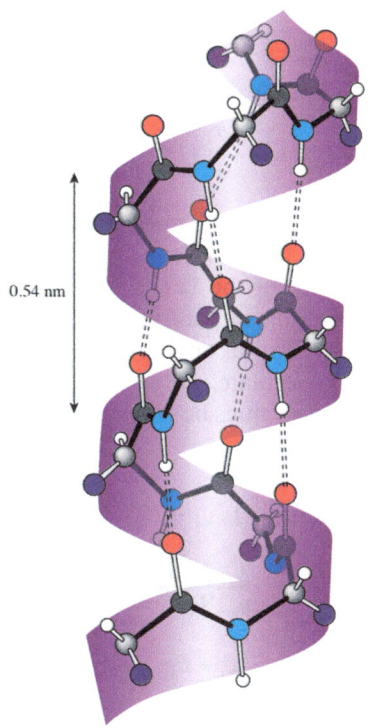

0.54 nm

pitch 3.6 residues, 0.54nm

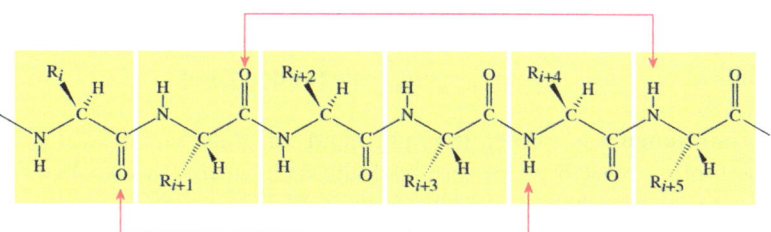

H-bonds between peptide units: N-H of nth
residue and C=0 of $(n$-$4)^{th}$ residue.

Figure 179: **Polypeptide alpha helix.** Grey (light and dark): carbon. Blue: nitrogen.
Red: oxygen. White: hydrogen. Designs by Steve Trevett [904], © Institute of Physics.
Reproduced with permission.

Frey-Wyssling listed the Bamford group's ratios: '18 residues in five turns' etc.—while adding (in brackets) his own back-extrapolation:

$$\text{helix sequence} \quad \left(\frac{1}{3}, \frac{1}{4}, \frac{2}{7}, \frac{3}{11},\right) \ \frac{5}{18}, \frac{8}{29}, \frac{13}{47}, \ \cdots$$

In this we may also note, (as did mathematician, biologist, and symmetry specialist R V Jean [417]), a parallel pattern in the denominators—namely that of the Lucas number sequence 18, 29, 47. And from this observation, we may write the divergence ratios in the general form F_n/L_{n+1}. But for analysis, it will help us to consider reciprocals. We thus calculate, $18/5=3.6$, $29/8=3.625$, and $47/13=3.61538$. Could these values be converging onto $2+\phi$?

Let's check by using the identity $L_{n+1} = F_n + F_{n+2}$ [488]:

$$\frac{L_{n+1}}{F_n} = \frac{F_n + F_{n+2}}{F_n}.$$

By decomposing F_{n+2} into $F_{n+1} + F_n$, we get

$$\frac{L_{n+1}}{F_n} = \frac{F_n}{F_n} + \frac{F_{n+1} + F_n}{F_n}$$

$$= 2 + \frac{F_{n+1}}{F_n}.$$

And as $n \to \infty$, $F_{n+1}/F_n \to \phi$, the Golden Ratio.

So our hunch was correct: the right-hand side does become $2+\phi$ in the limit. Now, if we reverse back the reciprocal we took, then we get $1/(2+\phi)$ which we recall as one of the expressions for the 99.5° Lucas Angle on page 58. Indeed if we check back in the divergences and calculate angles in degrees, $360/3.6=100°$, $360/3.625=99.31°$, and $360/3.61538=99.57°$.

In addition to reassuring us of the consistency of it all, this one example demonstrates all together: non-botanic phyllotactic number sequences, convergence ratios, and divergence angles.

Superconductor flux-lattice phyllotaxis

Superconductors are famous for having zero electrical resistance at low temperatures. They also have a property that, as they are cooled and they start to superconduct, they can expel weak magnetic fields that had previously passed through them. The external magnetic fields cause 'supercurrents' to be generated in the surface of the superconductor which then cancel the field in the body of the material. It is this property (called the Meissner effect) that allows a small magnet to be levitated above a superconductor. However, this only works for weak fields, and at higher strengths the magnetic flux penetrates the material—though not evenly—it gets confined into tubes called vortices. Further, as a result of flaws in the material, these vortices then form into a lattice pattern [602]. In his 1991 experimental study of layered superconductors, L S Levitov of the L D Landau Institute for Theoretical Physics in Moscow observed phyllotactic patterns in such flux lattices. He then went on to explain these in terms of energy minimization. Levitov had been looking at how changing the magnetic field caused a lattice to adjust itself, and he was surprised to find patterns 'very similar to those known in botanics,' with successive pairs of Fibonacci numbers appearing [549]—just as we saw earlier when we looked at sunflowers and pineapples (page 56).

Ferrofluid phyllotaxis

In the early 1990s, researchers Stéphane Douady and Yves Couder (at the Laboratory for Statistical Physics in Paris) performed a series of experiments and numerical simulations during their study of self-organizing processes. In one experiment, they let drops of magnetic liquid (a ferrofluid) fall into the centre of a magnetized dish containing silicone oil. As the drops travelled out to the dish edge, they repelled each other and (remarkably) self-organized into phyllotactic patterns—paired Fibonacci numbers of spiral whorls (parastichies) [182, 184, 227]. In their paper, Douady & Couder mention a general phyllotaxis formula

$$\Omega_p = 360/(\phi + p), \quad p = 1, 2, 3$$

which gives the divergence angles of 137.5°, 99.5°, and 77.955° for the first through third parastichies [417, 2]. As with Levitov's experiment, the drops find their least-energy configuration. Video of this experiment is (at time of writing) available on the internet [183].

Magnetic-cactus phyllotaxis

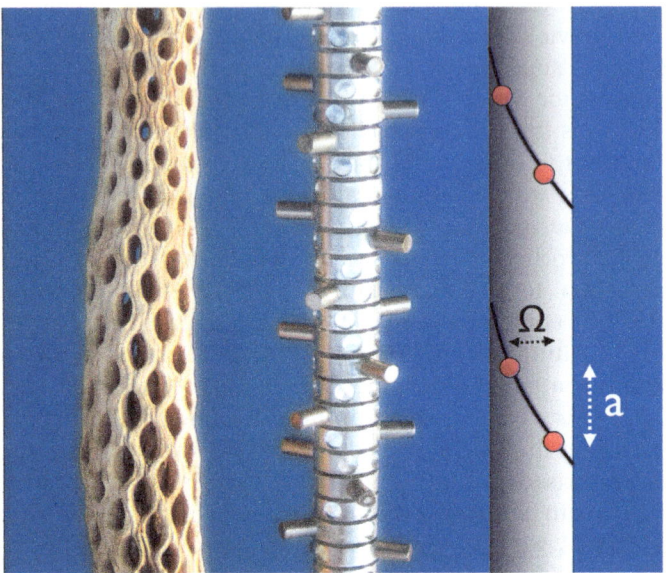

Figure 180: **Magnetic cactus.** Image courtesy of Cristiano Nisoli, Los Alamos National Laboratory.

In a further energy minimization experiment in 2009, Cristiano Nisoli of Los Alamos National Laboratory, and colleagues from Cornell and Pennsylvania State Universities, built a 'magnetic cactus' (Fig. 180) with 50 spines made of iron-neodymium magnets. These were stacked on, and could rotate about, a non-magnetic stem. Before each of their experimental measurements, the team used a novel 'athermal annealing' process that included winding the bottom-most magnet to twist the spines into a tighter-and-tighter spiral until an 'explosive release of energy' disordered the system. Then with some additional less dramatic influences, they caused it to adopt a stable ordered state. Their results showed phyllotactic patterns with examples of both Fibonacci and Lucas numbers [682, 683].

294

Packing geometry

When considering possible 3D-packing structures, it is a truism to say that in a given situation only certain arrangements make sense. As phyllotaxis becomes more widely recognized, different disciplines from biochemistry to physics are seeing how the dense packing of (sufficiently deformable) units—for example soft spheres on an axis or cylinder—leads to essentially the same well-known geometric configurations. Accordingly, Levitov regarded his results as suggesting that phyllotaxis: 'must occur in all soft lattices subjected to strong deformation' [549]. His overall message is that phyllotaxis is not (as such) coded into the genetic material of plants, any more than it is a special property of peptide chemistry—no, it is about spatial geometry and optimized close packing: what is possible and what is not. In his book on phyllotaxis, R V Jean describes the discovery of the 3D structure of protein molecules as a landmark in molecular biology [420]. In addition to reviewing the work of Pauling & Corey and Frey-Wyssling that we mentioned just earlier (p.290), Jean particularly considers the helical-packing work done by Crick (1953) [145]—that is, Nobel Laureate Francis Crick of DNA fame. Jean notes that authors such as Crick were apparently not familiar with phyllotaxis—yet had they been, it would have usefully informed both their experiment design and the precision of their analysis [421]. This is because in phyllotaxis, variations in divergence angle as small as 0.1° can make obvious differences in the resulting patterns (Fig. 181). Then again, some might say that Crick (et al.) did rather well without a knowledge of phyllotaxis—but that would be to miss Jean's point.

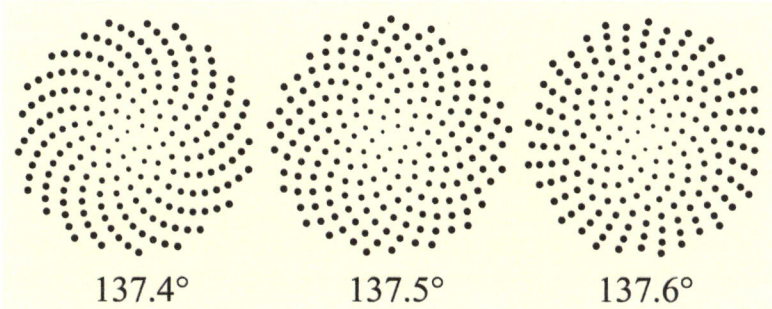

137.4°   137.5°   137.6°

Figure 181: **Divergence variation, $137.5 \mp 0.1°$: moiré-like sensitivity.**
(After R V Jean, 1994).

Harvesting sunlight

Figure 182: **Abengoa Solar PS10 tower plant, Seville (11 megawatt).**
Photo: Courtesy of and © Abengoa Solar S.A.

In Fig. 182 we see how the principle of phyllotaxis has been applied in a solar power plant to collect sunlight efficiently. The PS10 was the first commercial tower in the world; it is located at the Solúcar Complex in Sanlúcar la Mayor, Seville, Spain. The sunflower-like arrangement of reflectors is held to bring with it the natural qualities of dense packing and shadow minimization, thereby maximizing the collection of sunlight. Abengoa has applied for patent protection for this phyllotactic distribution of the 'heliostats'—the combined mirror-plus-steering devices that track the sun and keep the reflected rays directed onto the target [234]. The power output of 11 megawatts matches the electricity usage of approximately 5,500 households, and the company calculates their 'clean/green' solution prevents the release of 6,000 tonnes of CO_2 into the atmosphere every year (when compared with equivalent fossil fuel power generation) [1].

Chapter 24

ϕ in quantum mechanics

Quantum mechanics is one of the greatest achievements of modern science. It concerns the physics of atoms and the subatomic world, and it includes the ideas of wave-particle duality and quantization. The theory can predict the results of many and diverse experiments with extraordinary accuracy [218]. However, a proper understanding of why it works has proved elusive. Richard Feynman (Nobel Prize for Physics 1965) freely admitted that he did not understand it, and said that (in his opinion) *nobody* understood it [219]. Solid-state physicist N David Mermin counted himself among those uncomfortable with this lack of conceptual foundation, which he summed up in the words: 'Shut up and calculate!' [643].[1]

Entanglement: 'Spooky action'

When two particles are created at the same time and place they may share an 'entangled' state—a kind of shared existence. And even when they travel far apart, taking a measurement on one will instantly affect its partner. From the outset, Einstein found this idea problematic. So much so, that at the 6th Solvay conference in 1930, he presented a challenge in the form of a thought experiment which involved a box that was open for an arbitrarily short time, with a photon escaping from it.[2] The purpose of each of these 'experiments' was effectively to refute quantum mechanics—saying this kind of instant remote effect

---

[1] Mermin is also famous for applying Lewis Carroll's *The Hunting of the Snark* term 'boojum' to superfluidity—as something that causes an effect to: 'softly and suddenly vanish away'. We shall return to Lewis Carroll later, for help from his disappearing cat.

[2] This now appears to be his first such challenge [680], although it is normally reported that his collaboration to produce the EPR paradox was the first in 1935.

was impossible according to special relativity. Einstein famously referred to entanglement as 'spooky action at a distance.' This problem was only resolved in 1964 when Northern Irish physicist John Bell (after his sabbatical from CERN) introduced the concept of 'non-locality' [48]. In this interpretation, one particle may influence the other, but not in a way that communicates classical information—thus fixing the apparent problem with special relativity. Having said that, some researchers hope there is a way round this limitation, and they would be very pleased to develop a near-instantaneous communications device. In 1993 theoretical physicist Lucien Hardy devised a powerful thought experiment which tests this aspect of quantum mechanics—and this has since been physically performed in the laboratory [304, 996, 601]. A pair of particles (for example atoms) in an entangled state are ejected from a source, one to the left and one to the right, each towards its own detector. Each detector can be set to a vertical or tilted orientation. The experiment is configured in such a way that classical physics would predict nothing interesting happening at all—but quantum mechanics predicts 'something sometimes'. In practice, the experiment clearly demonstrates non-classical behaviour, and the measured result (the probability of entanglement) accurately matches the theory. So, what is Hardy's probability of quantum entanglement as a number? Surprisingly, it is based on the Golden Ratio and turns out to be Golden Power $\phi^{-5} = 0.09017$ or 9.017%—(by coincidence, this is the value of q_5 that we first met as q on page 195). Theoretical physicist Dan Styer has published an overview of Hardy's test, including the relevant ϕ mathematics [877].

ϕ in quantum-mechanical magnetic resonance

In an experiment devised by Professor Alan Tennant, researchers from Helmholtz-Zentrum Berlin, Oxford and Bristol Universities, and Rutherford Appleton Lab., UK, have observed a nanoscale symmetry in solid-state matter. Using low temperatures and precise magnetic fields they put a cobalt niobate sample into a 'Schrödinger cat' quantum-critical state, and when they: 'tuned it like a nanoscale guitar string', they found resonances with pitch-ratio ϕ. Radu Coldea is convinced that this is evidence of hidden 'E8 symmetry', which *might* be important in the search for a 'Theory of Everything' [120, 333, 561].

Chapter 25

Molecular ϕ

In this chapter we shall look at several examples where ϕ appears in the study of the molecular structure of solids and liquids, especially where icosahedra are involved.[1] First we consider how spheres may pack together, and then we look at the pioneering work of crystallographer Alan Mackay. After discussing ϕ-based (icosahedral) clusters in noble metals and gases, we shall consider the same in the structure of water. We shall then review research relating icosahedral structure to the supercooling of certain metals and alloys, and finally we shall consider ϕ in atomic bonding.

Kepler, Newton, Gregory: Sphere packing[2]

When spheres are arranged in a straight line (i.e. in 1 dimension), one sphere may touch 2 others at most. And just as 6 equal-sized circles will fit exactly around one circle in 2 dimensions,[3] the same is true for spheres resting on a plane. These counts of 2 and 6 are therefore referred to as 'kissing numbers'. But in three dimensions, although we can easily fit 12 hard spheres around a single sphere,[4] the fit is loose and we are left with gaps between spheres. Might there even be room for a 13th? Kepler considered the density of sphere packing in relation to heaps of cannonballs [294], and later, both Isaac Newton and Scottish astronomer-mathematician David Gregory considered

---

[1] The icosahedron is one of the Platonic solids we discussed on page 15.

[2] Édouard Lucas was also interested in this subject: from a number theoretic point of view. In 1875 he conjectured that there is only one way to stack a flat square of cannon balls into a square pyramid—that is when 70^2 balls are stacked as 24 layers. Lucas' conjecture was proved correct by G N Watson in 1918 [958].

[3] To check this, arrange 7 coins (all the same) on a table.

[4] Considering all as having the same diameter.

the 3D kissing number directly. In a famous dialogue of 1694, Gregory asserted to Newton that 13 should be possible, while Newton maintained that 12 was the maximum [101, 749]. Yet although we now know for certain that Newton was right, a correct mathematical proof for 'the thirteen-spheres problem' was not achieved until 1953 [810]. The difficulty is that there are an infinite number of ways to arrange the 12 balls around the central one [123].

ϕ—Alan Mackay and Mackay icosahedra

Notwithstanding this, when atoms do pack together around a central sphere, they may achieve an even distribution in space and thereby may form the regular icosahedral structure we saw on page 15 [226]. We recall the icosahedron as being one of the two ϕ-based Platonic solids. This may be further 'coated' with a new shell of atoms—just one atom deep, and such layering may then repeat. This growth process creates a succession of completed structures named 'Mackay icosahedra' after Alan Lindsay Mackay (Fig. 184), a British crystallographer and a pioneer of 5-fold symmetry studies.[5] The journal *Structural Chemistry* dedicated an issue to him in 2002 [307]. Starting with a central atom, we add 12 atoms to make the first shell, then add 42 to get the next, and 92 the next.[6] From these additions we build a sequence of cumulative totals (partial sums):

$$\langle\, 1, 13, 55, 147, 309, 561, 923, 1415, 2057, 2869, 3871, 5083, \ldots \,\rangle.^7$$

These are called 'the centred icosahedral numbers',[8] and chemists refer to them as 'magic number combinations' [910]. Fig. 183 shows the 1st, 2nd, 3rd, and 10th shells. Small numbers of shells occur most commonly. The top-right example (when filled) totals 55 atoms —this is so common that it has been called *the* Mackay icosahedron: 'MI' [518]. The icosahedral geometry becomes increasingly clear with larger structures, for example the (bottom-right) tenth shell. The count of atoms along a triangle side ignores sharing and simply considers a flat triangle and counts atoms along one edge.

---

[5]   Dr. Alan Lindsay Mackay (b. 1927), Fellow of the Royal Society (London), is Professor Emeritus at the Department of Crystallography, Birkbeck College, University of London [305].

[6]   The number of atoms added per shell is given by $10n^2+2$, Sloane A005901 [213].

[7]   These partial sums are given by $a_n = (2n+1)(5n^2+5n+3)/3$, Sloane A005902.

[8]   Also the 'cuboctahedral numbers'; also the 'crystal ball sequence for the fcc lattice'.

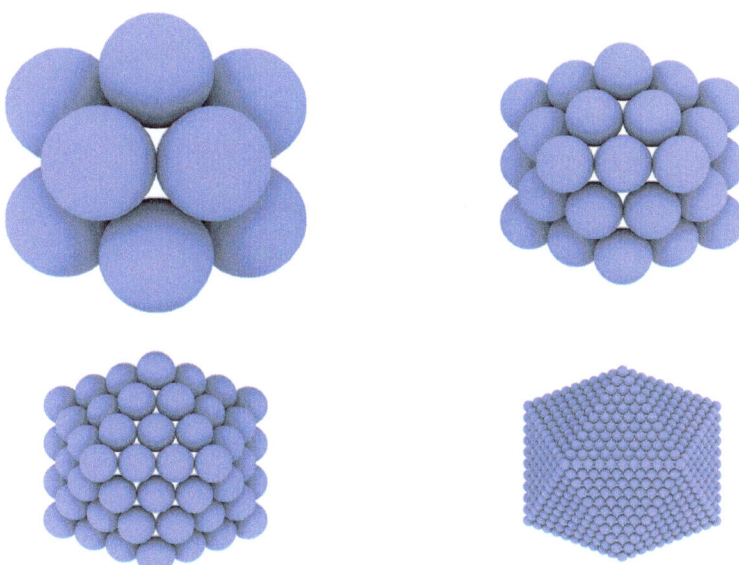

shell n	$\triangle$ side atoms $n+1$	shell atoms $10n^2+2$	(magic number) sum total atoms
1	2	12	13
2	3	42	55
3	4	92	147
4	5	162	309
5	6	252	561
...	...	...	...
10	11	1002	3871
...	...	...	...

Figure 183: **Mackay Icosahedral shells.** Image created in *Blender* [787].

The edge count makes it trivial to determine the shell number (which is then just one less). This building process provides a good example of 'gnomonic expansion' where the old form is contained within the larger new, and the shape remains the same—a reminder of similar triangles. Here the 'gnomon' is the current shell being added. We saw 2D gnomons earlier in this book (Mondrian p.102 and Tiwanaku p.118), and here we have a 3D example.

Figure 184: **Alan Lindsay Mackay.**

Alan Mackay's doctoral advisor was John D Bernal (known to his colleagues as 'Sage'), Professor of Physics and Head of Crystallography at Birkbeck College, another leading crystallographer. Bernal had worked with William Bragg at the Royal Institution, and he was one of the first to become disenchanted with (what he saw as) the overly restrictive way crystallographers around the world decided what was crystal and what was not. As early as 1976, Alan Mackay, took up this cause in earnest—that is, of effecting a paradigm shift that would free crystallography from (what he then called) the 'hog-tie formalism' of the *International Tables of Crystallography* [607]—with their 230 space groups *(only!)*. He envisioned: how crystals might cease to depend on translational symmetry; how they could be built from the repeated application of simple rules; and how such 'generalized crystals' could be highly ordered, but at the same time, not be mentioned in the *Tables*. Mackay cited John Horton Conway's *Game of Life* as an example of the kind of rule set he was talking about [607, 237].[9] Further, (and again after Kepler), he discussed snowflake growth and form in these terms.

---

[9] For an epic video demonstration of the capabilities this deceptively simple cellular automaton see Ascani [19]. We return to John Conway on p.304.

In 1981, as a prime example of structural orderliness not recognized as 'crystal', Mackay wrote a paper about 5-fold symmetry entitled *De Nive Quinquangula* [609, 608].[10] This is a title that Kepler would have much appreciated—as it referred not only to his 1611 essay *Strena Seu De Nive Sexangula* (about the 6-fold symmetry of the snowflake) [467] but also to Kepler's own very particular interest in 5-fold symmetry.

ϕ—Buckminster Fuller and icosahedra

As Alan Mackay reported in a short paper of 1962—R Buckminster Fuller (neo-futurist architect and inventor of the geodesic dome) found a very close relationship between the icosahedron and the cuboctahedron [606]. It was that by building the cuboctahedron with rigid rods jointed at their vertices, it is possible to convert the figure into an icosahedron just by rotating the triangles about their normals. This turning has somewhat the same effect as winding a clock spring and the whole figure shrinks by about 5%—a reversible process.[11] Buckminster Fuller applied both the cuboctahedron, and later the icosahedron to mapping the world [558]. His 'Dymaxion' maps[12] approximate the surface of the globe as the facets of a polyhedron. The icosahedron is unwrapped to flatness as joined triangles and this may be done in such a way that continents almost join to form a single land mass. The normal distortions of mapping ('Greenland too big' and so on) are very much reduced at the expense of a rather awkward overall shape.

Icosahedral nobles

In 1977 Gaspard, Hodges, & Gordon were studying the stability of icosahedral clusters of noble-metal atoms (such as gold), and they found the icosahedral form clearly the most stable (compared with cubic form) for magic numbers: 13, 55, and 147, (the number of atoms per cluster) [255]. Regarding noble gases, Jim Baggott, in his article about krypton clusters, describes work done at the University of Sussex by Paul Lethbridge & Tony Stace who cooled krypton atoms by passing them through a 200-micron hole into a vacuum [24, 547]. As the (now slowly moving) atoms collided, the weak attractive forces

---

[10] '5-pointed snowflake'.

[11] So now we see why the 'centred icosahedral numbers' are also called the 'cuboctahedral numbers'—the shapes interchange, but the number of vertices remains the same.

[12] Dymaxion is a trademark of the Buckminster Fuller Institute.

between them were greater than the thermal energy that would have otherwise bounced them apart. In this way, orderly icosahedral clusters could form by accretion. In fact, all the stable noble gases will form clusters, and Mackay icosahedra dominate the smaller size clusters [627]. In his 1981 paper, Mackay presented well-ordered icosahedral gold clusters with: 13, 55, 147, 309, and 561 atoms— perfect regular solids that 'the man on the Clapham omnibus'[13] would have no hesitation in confirming as crystal. He argued strongly that these should not be regarded as non-crystal or somehow *defective* (as would be the case from the traditional crystallographic point of view) [608, 609]. Finally, in passing, we may note that some not-so-noble viruses take the icosahedral form too.

John Horton Conway

John Conway has made major contributions to mathematics, but he is perhaps mostly widely known as being the creator of *The Game of Life*[14]—which we mentioned just earlier as an example of repeated applications of simple rules. We now see how strongly such ideas overlap with those of fractals—with their typically simple but repetitive creation algorithms, that Benoît Mandelbrot and ('L-Systems') Aristid Lindenmayer were working on around this time [623, 770].[15] Iterated Function System fractals and L-Systems are now studied under the general heading of Production Systems where the repeated application of a small set of rules can generate natural forms, can model the development of systems over time, and can also generate totally unexpected results and extraordinary complexity [985, 102]. We shall shortly be interested in Conway's work in developing the patterns discovered by Roger Penrose, but here we may note that his first published work followed on from John Leech's study of the close packing of spheres [122]. The topic is effectively a modern take on Kepler's cannonball stacking and similarly relates to the kissing number discussions of Newton and Gregory. The basic ideas from 1D, 2D and 3D packing remain the same, but instead Conway finds results in the 24-dimensional 'Leech lattice'. The kissing number for this configuration is a staggering 196,560 [1004]. As Conway remarked: 'There is a lot of room up there.' [237].[16]

---

[13] The 'reasonable person' concept used in English law courts for over 100 years.
[14] A 'cellular automaton' [237].
[15] That we glimpsed earlier—Chapter 17, p.229.
[16] Perhaps recalling Richard Feynman's invitation to nano-technology: *There's plenty of room at the bottom* [217].

Icosahedral symmetry

The icosahedron (p.15) has 20 sides—all equilateral triangles. And if we look down a normal (a perpendicular axis) onto the centre of one of these faces, we realize that if we rotate the figure by one third of a turn (about this normal) then it will appear unchanged. The same would be true if we rotated two thirds or three thirds. This property is called 3-fold rotational symmetry. If we consider a pair of opposite edges, with an axis through their midpoints, then we get 2-fold rotational symmetry, and with an axis through opposite vertices we see 5-fold. In all, the icosahedron's rotational symmetry axes comprise:

- six × 5-fold
- ten × 3-fold
- fifteen × 2-fold

Buckyballs, fullerenes, and nanobuds

It was research into astronomical dust—particularly carbon-rich grains from old stars—that led to the discovery in 1985 of C_{60} 'buckminsterfullerene' also known as the 'buckyball'. Researchers Robert Curl, Harold Kroto, and Richard Smalley jointly shared the Nobel Prize in Chemistry 1996 for this work (and for the discovery of other fullerenes as well).[17] In their letter the journal *Nature*, they said of the superstable structure: 'We suggest a truncated icosahedron... This object is commonly encountered as the football.' [514]. One way to make a truncated icosahedron is to start with a simple icosahedron and draw the largest possible hexagon that will fit, onto each of its 20 faces. Next, grind down all the 12 initial vertices so that they are now flat pentagons whose sides are also the sides of the drawn hexagons. These 20 hexagons along with the 12 pentagons add to make a total of 32 faces.[18] Given that it is based on the icosahedron, we will not be surprised that the truncated icosahedron has full icosahedral symmetry. Another fullerene discovered is the carbon nanotube, a single honeycomb layer formed into a hollow cylinder, typically about 1 nanometre diameter. Such tubes can now be made in a length-to-diameter ratio in excess of a hundred million. Recently buckyballs have been joined onto nanotubes as 'nanobuds'—plant-like growths out of the tube stem [668].

---

[17] Three persons maximum may be cited for a Nobel prize—hence graduate students: James Heath, Sean O'Brien, and Yuan Liu, were not included [acs.org].

[18] The ratio of hexagons to pentagons is $5/3$, a convergent of ϕ. The rectangles formed by the opposite edges are 1:3ϕ—recalling Le Corbusier's Villa Savoye (p.125).

Icosahedral water, $20 \times 14 \times H_2O$ (and Plato)

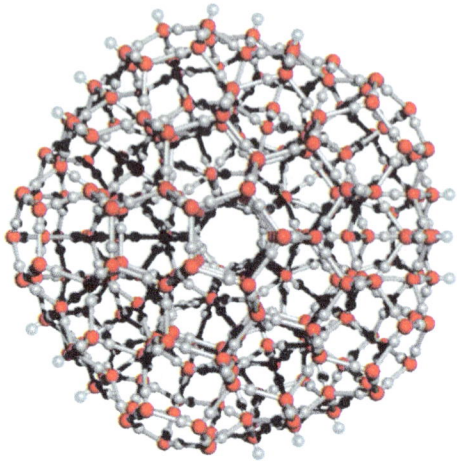

Figure 185: **Icosahedral water.** Image courtesy of Martin Chaplin [105].

Water has many anomalous properties when considered alongside similar compounds [108]. For example, the boiling point of water is much higher than for similar compounds, and this is due to relatively strong hydrogen bonds between the water molecules. These connections are being made and broken all the time, but at any time the overall 'holding together' effect is significant. And this tendency to clump together is not random—it is strongly influenced by the bond-angle geometry of the water molecules and their electric charge distribution. This means that specific 3D patterns tend to arise in the aggregations [106]. In 1999, biochemist Martin Chaplin proposed one such possible clustering: an icosahedral pattern—Fig. 185 [105].[19] A complete example of the proposed structure would comprise 20 units of a subgroup of 14 molecules of water—a total of 280 molecules [109]. At first sight, the construction might remind some of the 'buckyball' C_{60} molecule (p.303). However, this structure has 14 times as many atoms, and it is not hollow.[20] As observed by both Chaplin [110] and Hargittai [313]—Plato associated water with the icosahedron in his dialogue *Timæus* [753].

---

[19] On this, chemist István Hargittai notes J D Bernal's earlier conjecture that icosahedral coordination was the key to understanding the structure of liquid water [312].

[20] The '14' numerical coincidence comes from 20 units × 3 atoms per molecule = 60.

Supercooling levitated metallic liquids

The first thing to say here is that supercooling is not necessarily to do with making things very cold. In fact, in the molten metal examples we shall consider, it will be about a relatively small lowering of a high temperature. Let's therefore define this term more accurately. Super-cooling is the process of cooling a liquid in such a way that its temperature goes below its normal freezing point but *without* solidification taking place. Under the right conditions, water can be cooled as low as $-48°C$ without it beginning to crystallize into ice. The first account of this effect was published by Fahrenheit in 1724 (who got down to $-9°C$).

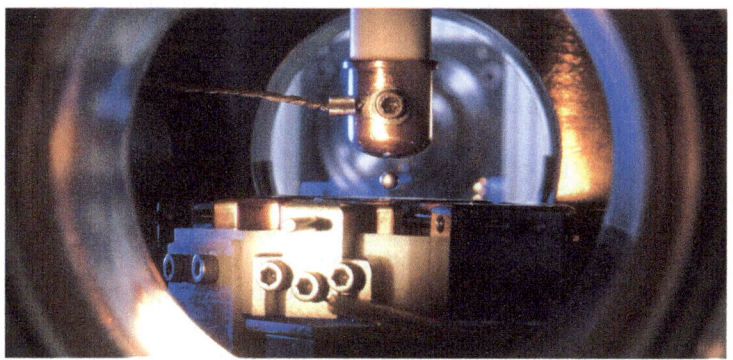

Figure 186: **ESL—Electrostatic levitation of molten metal drop, a target for X-ray diffraction.** Image courtesy of NASA Marshall Centre.

In 1949, materials scientist David Turnbull was first to supercool a liquid metal—with surprising results [906]. Previously, it was assumed that a metal had a similar structure both before and after it solidified. But Turnbull's results said otherwise. A few years later, in 1952 Charles Frank was attempting to explain supercooling in liquid metals, and he conjectured that within a supercooled liquid, order develops in-herently and that this ordering affects crystallization [225, 462]. Frank suggested that while liquid, the metal adopted predominantly icosahedral clustering, but in order to crystallize, it would need to break this structure. Such an added necessity would explain the nucleation barrier—an energy hurdle that must be overcome before a metal can solidify [461].

Physicist and materials scientist Kenneth Kelton and his group set out to observe these phenomena, and published their results in 2003 [463]. As a way to avoid container contact—which would trigger solidification—they used the Electrostatic Levitator at NASA's Marshall Center (Fig. 186) to suspend droplets of molten metal. The results they obtained were in good agreement with Frank's predictions. Looking back in a recent review, Kelton details work over the last decade that has provided structural evidence of icosahedral short range order (ISRO)—that is, ϕ-based geometry [462]. This has come from X-ray and neutron-diffraction studies of supercooled liquids, using electromagnetic and electrostatic levitation techniques. Kelton goes on to say that based on these studies, we are now confident that ISRO is *dominant* in transition metal element and transition alloy liquids—even in certain non-transition metal liquids. Sadoc and Mosseri list further research in this area [796].

ϕ in atomic bonding

In 2005 Czech physicist Raji Heyrovská published his paper *The Golden ratio, ionic and atomic radii and bond lengths* in the Journal of Molecular Physics [364]. In it, he shows the Golden Ratio has a significant role in atomic and molecular physics—in particular, determining the way that electron distributions change shape and set the bond lengths. Crystal engineers are concerned with how molecules interact and how these interactions govern packing. They may then use this knowledge to create new 'designer' solids. In 2006 Chinese crystallographer and biochemist Daqiu Yu and colleagues published their paper *Golden ratio and bond-valence parameters of hydrogen bonds of hydrated borates.* [998]. (In it, they cite Heyrovská's work). Yu and his colleagues have been particularly interested in hydrogen bonding and the ratios of distances in strong and weak bonds. On his website, Martin Chaplin discusses further work by Heyrovská, and he concludes that in water-based chemistry, ϕ shows up as the ratio between atomic and ionic diameters, and simple functions of ϕ relate ion-water distances to covalent radii [107, 365, 366]. Chaplin also refers to Kelton's supercooling work, and further, refers to the icosahedral clusters (e.g. of 13 spherical) atoms of the kind found in liquid argon, liquid krypton and liquid xenon—the configurations we saw earlier (p.301) which were studied by Alan Mackay.

Chapter 26

Penrose tiling

This chapter and the next two form a unit—one with diverse, yet highly related content, which is summarized in an infographic on page 357.

Aperiodic tilings

We are all familiar with domestic and ornamental walls and floors that feature square or rectangular tiles, hexagonal tiles or even equilateral-triangle tiles. Using these shapes it is easy to create a repeating pattern with 100% coverage (i.e. leaving no gaps). Such regularly repeating patterns are called 'periodic'. We note that these tilings have translational and rotational symmetries. For example, a copy of a square grid can be shifted such that it appears to coincide with the original, or it could be rotated about a grid point: 0°, 90°, 180°, or 270° to get the same effect (i.e. 4-fold symmetry). But what about 5-sided tiles—that is, pentagons? It does not help us to remember that the surface of a sphere can be perfectly tiled with 12 pentagons thereby making a dodecahedron (p.15). This is because, as soon as we start to unwrap and flatten this, unavoidable gaps open up. It was exactly this 'problem' that spurred the early experiments of Dürer then Kepler which then led to Kepler's 'Aa' tiling (p.22). But, in terms of the mathematical studies of the Western world, little further progress was made on such tiling problems until the 1960s, when Chinese American logician, philosopher, and mathematician Hao Wang, and his student, applied-mathematician Robert Berger, considered the possibility of non-periodic tilings. Such tilings would lack translational symmetry—that is, their patterns could not be created by simply repeating some basic unit cell. Wang and Berger were considering tiling schemes as including not just the shapes themselves, but also restrictive rules for combining them (for example, colouring the edges and requiring colour matches). Such schemes can give either periodic or non-periodic results. Hao Wang had in 1961 conjectured that any set of tiles that can tile the plane can do so periodically [936]. However, in 1966, Wang's student, Robert Berger, made the remarkable discovery that it is possible to have a special scheme that 'admits' *only* non-periodic tiling. That is, by strictly adhering to the rules, all possible resulting tilings will be non-periodic. For this special case, we say the non-periodicity is 'forced' and to distinguish, we call the tiling 'aperiodic'. Berger had produced a set of 20,426 different 'Wang dominoes' (square jigsaw pieces with edge matching rules) that would *only* admit non-periodic tiling [52]. This was a very significant breakthrough and alone sufficient to show Wang's conjecture was false. However, Berger's result was actually received rather more as a challenge—that of: 'Who can produce increasingly elegant aperiodic solutions (but not limited to square shape dominoes) comprising fewer and fewer tiles and rules?' The '20,426 tiles needed' became 104, then 92 and 40, then 6.

Roger Penrose's *two-tile* solutions

Eventually, it was theoretical physicist Roger Penrose who, in the mid-1970s, (and by stages) achieved aperiodicity using just two shapes. Roger Penrose is well known not just for his major contributions to general relativity and cosmology, but also for his interest in recreational mathematics. (In the 1950s he independently rediscovered the 'impossible triangle' that appears to be solid [960]. Any pair of joints 'makes sense' but when all 3 are taken together, they are not consistent.) Having found the first two tile shapes, he effectively repeated his discovery using another two related shapes. We note that the latter pair—his 'rhomb' version—was also discovered separately by independent researcher Robert Ammann in 1976 [250]. As a topic of recreational mathematics, Penrose tiling was introduced to a wide audience by popular maths and science writer Martin Gardner in his comprehensive 1977 article in Scientific American [238]. This was reprinted just over a decade later as Chapter 1 of his book *Penrose Tiles to Trapdoor Ciphers*. In Chapter 2, Gardner went on to review progress in the field over the intervening years, and he acknowledged the significant contributions made by John Horton Conway [239].

Figure 187: **Roger Penrose and his aperiodic rhomb tiling.** (Foyer of the Mitchell Institute for Fundamental Physics and Astronomy, at Texas A&M University.) Image by Solarflare100 [847, 143].

In Fig. 187 we see Roger Penrose standing on an example of his rhomb tiling. There are now many examples of such architectural use including a decorated rhomb pavement at Wadham College, Oxford, and others all over the world [800, 742]. Because Penrose could see potential commercial applications in puzzles and games, he took out a United States Patent to cover his invention [740].

311

Penrose P1 pentagonal aperiodic tiling

Penrose took inspiration from Kepler and significantly progressed Kepler's ideas. In 1973 he came up with his first aperiodic tiling [241], the 'P1'[1] using just 6 tiles [241, 737]. We saw pentagram stars surrounded by pentagons earlier (p.22)—in Kepler's 'Aa' tiling, and we compare 'Aa' and 'P1' in Fig. 188. Penrose also added diamonds (rhombs) and 'boats' to complete his P1 tiling. The particular P1 instance shown is a special case—having 5-fold rotational symmetry. The difference between the number of shapes (4) and the number of tiles (6) arises from the need to have 3 kinds of pentagon, each with its own set of matching rules (only allowing certain adjacent placings). The 3 types of pentagon are shown in different shades. The job of the matching rules is to prevent periodic arrangements. However, a consequence of this is that placing the next tile may be a matter of choice, or otherwise there may be only one legal next step. In that case, we say the placing is 'forced'.

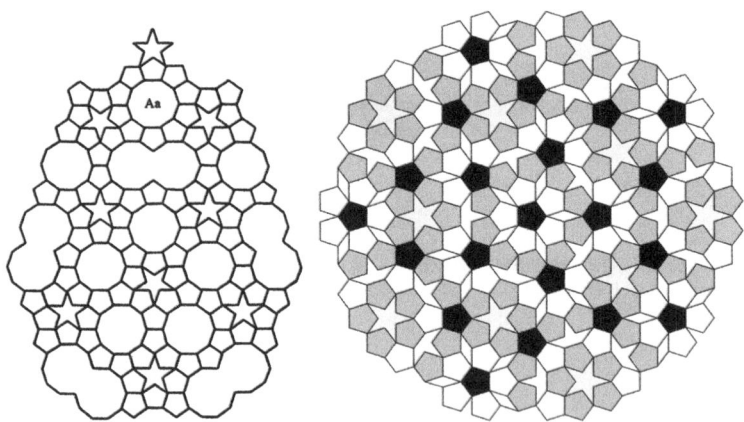

Figure 188: **Kepler Aa compared with Penrose P1 tiling (1973).**
Original P1 image by Inductiveload [402].

---

[1] The 'P1, P2, P3' classification is due to Grünbaum & Shephard [279].

Penrose P2 kite and dart aperiodic tiling

In 1974, a year after his P1 discovery, Penrose reduced the 6 tiles to 4, and then soon after to just two—the 'kite' and the 'dart'—names coined by Conway [739]. This pair of shapes is used to produce the P2 tiling (Fig. 189).

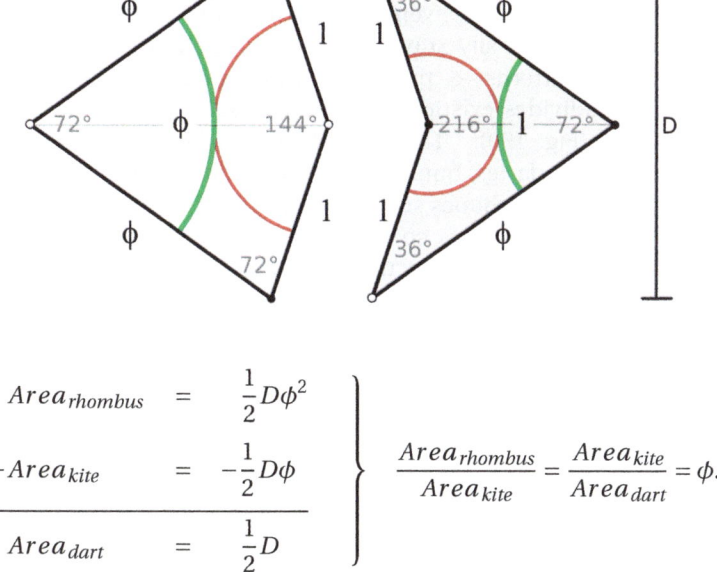

$$Area_{rhombus} = \frac{1}{2}D\phi^2$$

$$-Area_{kite} = -\frac{1}{2}D\phi$$

$$Area_{dart} = \frac{1}{2}D$$

$$\left. \begin{array}{c} \\ \\ \\ \end{array} \right\} \quad \frac{Area_{rhombus}}{Area_{kite}} = \frac{Area_{kite}}{Area_{dart}} = \phi.$$

Figure 189: **Penrose P2 tiles (1974)—kite and dart.** Image courtesy of Geometry Guy [257, 144], annotated using lengths from Edwards [196].

Two ways of implementing the matching rule are shown (though one is sufficient). Either the arc decorations must join at the edges, or, the vertices must be matched (here marked with hollow or filled circles, sometimes referred to as H and T—heads and tails). Matching rules can also be enforced by bumps and dents on the edges, just as in any jigsaw puzzle. It is easy to see that the areas of the kite and the dart are in the ratio ϕ (the area of a kite is half the product of its diagonals). In Fig. 189 we imagine the kite and dart shunted together to form a rhombus (which is strictly forbidden in P2, by the way), and we call its short diagonal (the vertical extent) D. Hence, as the long diagonal (here horizontal) is $1 + \phi = \phi^2$, the areas are seen to be in Golden Proportion—that is, (rhombus : kite) = (kite : dart) = ϕ.

313

The size of patterns that can be built using these tiles is unlimited, and as the number of tiles becomes large, that ratio of kites to darts tends to φ, the Golden Ratio [242]. This is perhaps counter-intuitive as we might guess a need for more of the (smaller) darts. Respecting the matching rules, patterns are built out from a point by joining vertices and abutting edges to leave no gaps, then adding more tiles, thus making a bigger-and-bigger patch.

One of the most important discoveries Penrose made about these tilings was that from a given initial scale, any kite and dart (P2) pattern could be locally subdivided (indefinitely) to produce new, finer-grained patterns—a process Conway called 'deflation' [243]. Deflation subdivides existing tiles and regroups the parts into new smaller tiles (Fig. 190). This transformation does not respect the original tile boundaries, but it does respect half-tile boundaries. The sides of the two tile shapes scale down by φ. The reverse process of 'inflation' subdivides tiles so that they may be aggregated into a φ-times-coarser-grained structure. So here we have a potentially infinitely intricate structure, with self-similar scaling—where the scaling factor from generation to generation is φ [244].

What configurations are possible then, if we rigorously conform to the matching rules? We find for instance that there are only seven ways to add kites and darts round a vertex. They are called: star, ace (also known as fool's kite), sun, king, jack, queen, and deuce [280]. Where a vertex configuration forces adjacent tile choices, the resulting patch is called an 'empire'—for example we might refer to the 'king's empire' [248]. By making the correct tiling choices, each of the sun and star patterns can be tiled to infinity such that it maintains 5-fold rotational symmetry—thereby becoming the 'infinite sun' and the 'infinite star'. These two patterns exactly complement each other—inflating or deflating one gives the other (perhaps reminiscent of the Platonic duals on page 15—but here with scaling too). In Fig. 191 we see (1) the sun pattern. This is deflated to give (2) the star pattern, with tiling elements φ times smaller. The next stage of deflation (3) is seen to be the sun pattern again, but note that it is rotated by one tenth of a turn. Deflating this we get (4) which is another star pattern, and again we notice the $(2\pi)/10$ rotation. So whereas it is normally reported that double deflation brings back the same pattern but ϕ^2 times smaller; we see that taking the rotation into account as well, we shall require 4 generations to obtain a rotationally equivalent match which is then ϕ^4 scaled. (We recall the Golden Spiral scaling by ϕ^4 after 4 quadrants.)

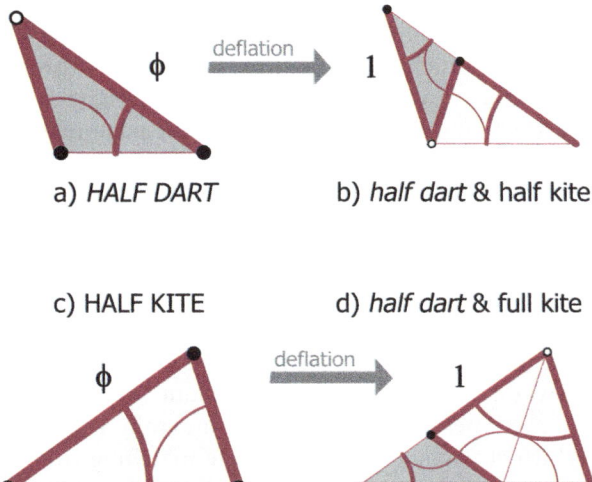

a) *HALF DART*   b) *half dart* & half kite

c) HALF KITE   d) *half dart* & full kite

Figure 190: **Penrose P2 deflation.** Layout based on Griffiths [273, 144], original tiling components by Tovstra [902, 144].

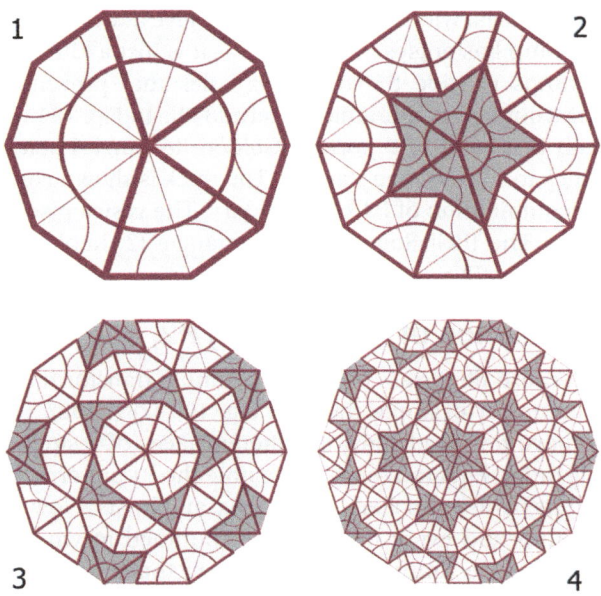

Figure 191: **Penrose P2 successive deflation: sun, star, sun, star, and so on.** Tiling patches by Tovstra [902].

The sun and star are the only two P2 tilings that have whole-plane 5-fold rotational symmetry. Those familiar with Penrose tilings will have also seen the chicken-shaped jigsaw tiles that Penrose created for building patterns in the style of artist M C Escher. Based on P2 rhombs, the shape outlines cleverly enforce the edge-matching rules [210]. And this link with Escher was not a coincidence. Penrose's father was a psychologist who studied visual paradoxes, and Penrose and his father were in contact with Escher and contributed ideas to him for designs [816].

Penrose P3 rhomb aperiodic tiling

A Penrose P2 tiling pattern of kites and darts may be 're-subdivided' in different ways, and Penrose found a unique local transformation yielding just two types of rhombus—'thick and thin' also known as 'fat and skinny' which (again with appropriate matching rules) produce the P3 tiling (Figs. 192 and 193). This latter Figure is a special case—as was Fig. 188, p.312—both have the property of 5-fold rotational symmetry. Also, for large P3 tilings, the ratio of the numbers of thick to thin rhombs used tends to the Golden Ratio ϕ. The thick rhomb has internal angles 72° and 108° and is made by joining the kite and the dart whose axial widths were ϕ and 1, to give a width of $\phi + 1 = \phi^2$. The thin rhomb has angles 36° and 144°—it is just a kite rearranged. Arcs or bumps provide the matching rules that prevent combinations that would give a periodic pattern [840]. In Fig. 194 we see the thin rhomb is related to 2 portrait Golden Rectangles stacked 'GG', and the thick to 2 landscape stacked. The thick is shown containing a forbidden combination of P2 kite and dart. The vesica piscis (p.88) is seen centre. The [RG] Root Golden Rectangle (2×Kepler triangles), and approximant [RG−] show up here too (p.74).

The equivalence of Penrose P1, P2, and P3

All three Penrose tilings are equivalent: simple transformations exist between them, for example those in Fig. 195 [281, 282]. The three tilings we have reviewed show us different faces of a single abstract underlying order. First (on the left), we note the P1 pattern outlined in black, with its pentagram stars, pentagons, and boats. Now we realize that every thin and every thick rhomb (outlined in white) has the same black line decoration. We could say that it is these rhomb decorations that join to give the P1 pattern. And on the right, it easy to see that the thin rhomb comprises two half kites; and the thick comprises two half kites and a dart.

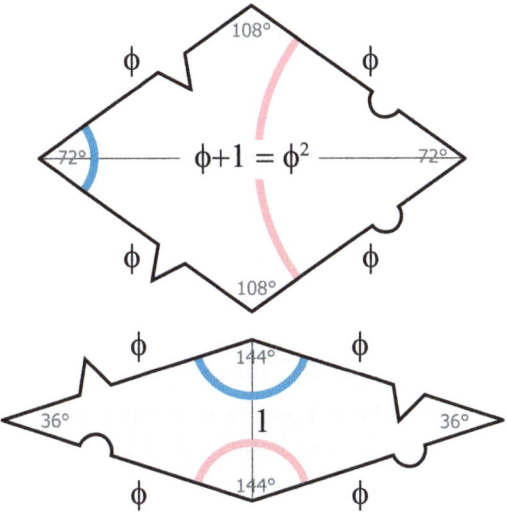

Figure 192: **Penrose P3 tiling (1974)—thick and thin rhombs.**
Image courtesy of Geometry Guy [257, 144], annotated using lengths from Edwards [196].

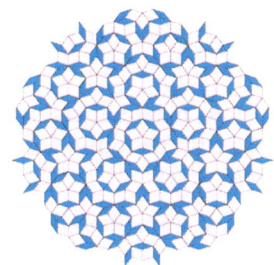

Figure 193: **Penrose P3 tiling example—thick and thin rhombi.**
Image courtesy of Inductiveload [403].

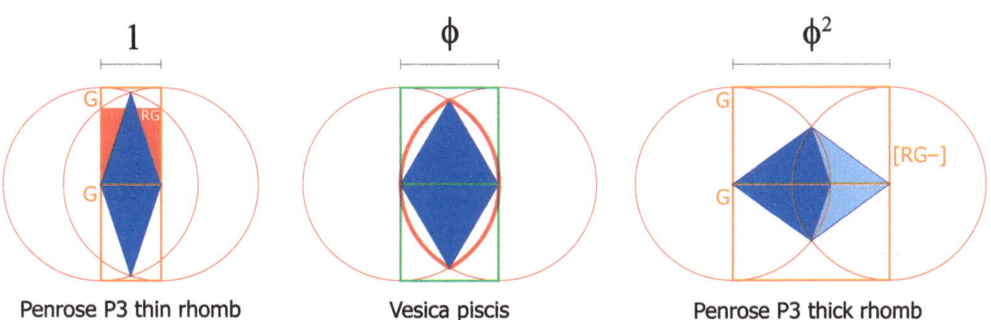

Penrose P3 thin rhomb   Vesica piscis   Penrose P3 thick rhomb
filled with P2 kite & dart

Figure 194: **Vesicæ—the circles are all radius ϕ.**

317

Properties of Penrose tilings

- Penrose tilings may be aggregated into coarser-grained Penrose patterns with larger tiles (inflated), or subdivided into finer-grained Penrose patterns with smaller tiles (deflated).

- Inflation and deflation are each unique local processes producing new tiles scaled up and down (respectively) by ϕ (Fig. 190, p.315) [283].

- Double inflation (or double deflation) takes us back to the same pattern—but rotated $(2\pi)/10$, and scaled by ϕ^2 (Fig. 191, p.315).

- (As we noted earlier), 4 generations are required to get back to the same pattern unrotated, producing a scaling of ϕ^4 compared with the original.

- A Penrose tiling, taken together with its infinity of inflation and deflation generations, may be viewed as an infinitely intricate fractal structure [244].

- Penrose tilings conform to orientational geometry similar to Ori32, (p.159). Each tile edge orients parallel to one of a set of star axes [95].

- Penrose tilings exhibit 'maximal cross-similarity'—every finite region in one pattern can be found an infinite number of times in every other pattern [284].

- Penrose tilings show exceptionally dense self-similarity—starting from the edge of a region diameter d, the distance to the edge of an exact copy will never be more than

$$d \times \phi^3/2 = 2.11d \qquad [245].$$

- Penrose tilings have a direct relation to quasicrystals. They can be 'lit' in such a way as to produce 'diffraction patterns' (more on this shortly).

- They may have reflection symmetry.

- They include two special patterns (the infinite sun and the infinite star) which have 5-fold rotational symmetry (Fig. 191, page 315).

- The ratio of Penrose P2 kites to darts tends to ϕ as the number of tiles increases—as does the ratio of P3 thick rhombs to thin.

 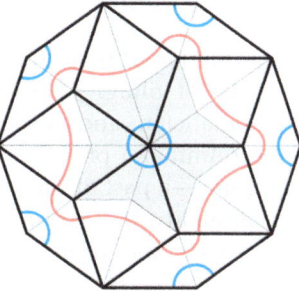

Figure 195: **Penrose P1-P3 and P2-P3 equivalence.** Image (left) courtesy of Inductiveload [404], (right) Geometry Guy [258, 144].

A particularly strange property has been called: 'indeterminacy of the construction process' [95]. We may believe (for example) that we have been provided with an infinite sun pattern—but by checking the local pattern, and then checking outwards as far as we can, we shall never be able to decide if we *are* actually on the infinite sun, or on some other, indeed *any* other pattern. Certainty may only be achieved 'after' an infinite check… [246]. In 1978 Penrose provided a detailed review of his tilings in an article titled *Pentaplexity* [738].

No single master pattern

Once we accept that any finite patch in one pattern is found infinitely many times in every other pattern, we might begin to wonder if there could exist just one master pattern—and that we have been considering only local views of it. However, an easy way to be certain that this is not the case, is to consider the infinite sun and infinite star patterns (Fig. 191). Each provides a centre of 5-fold symmetry and there is a simple proof that no tiling can have two centres of 5-fold symmetry at the same time.[2] In fact, there is an uncountable infinity of Penrose tiling patterns, and each is different from every other in an infinity of different ways [247].

---

[2] Gardner quotes Conway who credits Barlow (d. 1862) for this proof. We choose two centres that are the nearest possible, and start by positioning them so that we may draw a horizontal line between them. We pivot the line at its left end and swing a copy of it up to be at 72° from horizontal (making an italic *L* shape). We do the same pivoting at the right end (to give a mirrored *L*). We note that the free ends are now closer together than the original line length, and that they are non-coincident, so there is a contradiction ('closer than closest') [254]—for comparison 60° would have worked fine here.

Ammann Bars

Earlier we noted that Robert Ammann independently discovered the Penrose P3 tiling. We understand from an article by American mathematician and historian of science Marjorie Senechal that Robert Ammann probably had many of the symptoms of Asperger's syndrome—a part of the autism spectrum [820].[3] Though once seen as a disability, this syndrome is instead increasingly considered as a 'difference', and more attention is being given to its associated capabilities. It is claimed for instance, that about 10% of autistics have: 'striking islands of ability (i.e. savant abilities)' [466]. And typically autistics have above average skills in pattern recognition and pattern understanding [665]. Evidently, Ammann had an exceptional skill in this area—both in seeing and in processing patterns. He had become interested in non-periodic tilings after thinking about the printed dots used to show colour photos in newspapers. He recalled drawing criss-cross lines at angles that he considered appropriate, and remarked that: 'tiles popped out' [819].

Printers have used the 'halftone' dot technique since about 1850. Instead of attempting to adjust the ink intensity, full intensity is used, but as larger or smaller dots in a regular grid pattern. Seen from a distance the dots blur together to give the impression of continuous tone. Further, when process colours (such as the CMYK set: Cyan, Magenta, Yellow, and blacK) are combined in this way, an acceptably wide range of hues and saturations may be seen. However, if this system is applied without forethought, and the patterns of dots for different colours are casually combined, a strong moiré interference pattern will typically result. This happens when dots periodically synchronize with each other to create objectionable bands of dark and light. In Fig. 196 the right-side circular inset shows an example of this quilt-like moiré pattern.[4] The solution is to rotate the dot patterns ('screens') against each other at well chosen angles. A standard technique is to choose screen-to-screen angles that have irrational tangent values. This way, the dots from one screen can never 'synchronize' with the dots of the other. This is the same as saying there can be no common measure between dot periods when projected along any particular line through the pattern. Just as we have seen so many times with ϕ, irrationality is key to the result here [331]. In Fig. 196 we see the non-periodic rosette effect generated by an industry-standard set of screen angles [330]. These

---

[3] Which we mentioned in connection with Mondrian, page 108.

[4] Here just two screens with an arbitrarily chosen mutual offset of 6.8°.

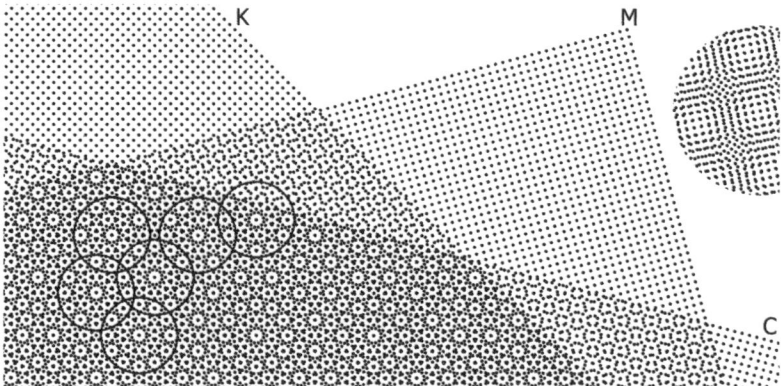

Figure 196: **C, M, and K colour printing screens.** The 'rosettes' are much easier to see when the colour screen dots are all rendered in black. (Top right: Moiré pattern.)

are: Cyan (105.0°), Magenta (165.0°), and blacK (45.0°), while the Yellow (0.0°) screen is omitted for clarity. The function of the yellow is to change hue, for example changing magenta to red—it has little tonal effect (that is, on light/dark). Several rosettes are circled, and a multitude of similar areas may be seen overlapping each other. (Try staring at another centre dot, and a rosette should appear around it, then perhaps its overlaps.)

One of Robert Ammann's discoveries was that for certain tilings—for example the P3 rhombs—the tiles may be decorated in such a way that the line segments on the tiles will consistently join up to produce continuous straight lines that extend fully across the pattern—as we see in Fig. 197 [318]. For the Penrose pattern there are 5 line families angled from each other by a multiple of $(2\pi)/5 \equiv 72°$. At a given tiling scale, the perpendicular distance between neighbouring parallel bars is one of only two possibilities which we call S (short) and L (long), and we find that $L = \phi \cdot S$. The sequence of these distances forms a special pattern itself—a kind of one-dimensional analogue of the Penrose tiling. Mathematicians Grünbaum and Shephard call these 1D patterns 'musical sequences', and note how they may be inflated or deflated in a very similar way to the 2D patterns [285]—we discuss such sequences (and their relation to the Fibonacci recurrence) in a later chapter (p.383) [817]. The apparent equivalence of the (global) Ammann bars approach and that of (local) tile placing rules led some authors to ask whether tiles or bars were more fundamental to the pattern [286]. Either way, Ammann bars clearly demonstrate the great underlying order.

S   L   L   S   L   L   S   L   S

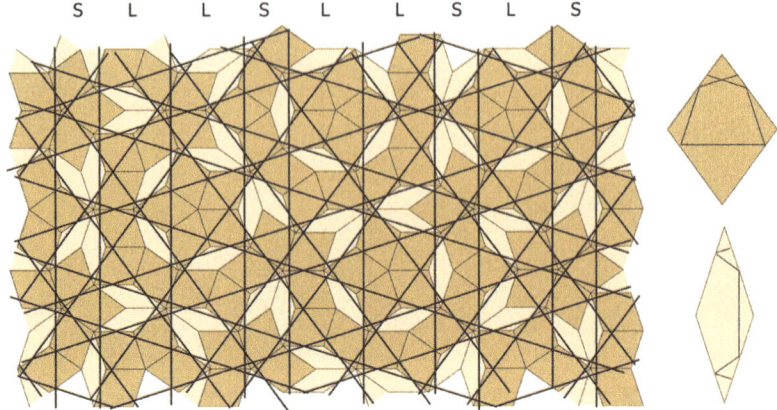

Figure 197: **Ammann bars.**
Diagram by Dirk Frettlöh, source: Tilings Encyclopedia [318]. 'SLLS. . . ' added.

Non-periodic tiling in 3D

Another very important discovery by Ammann (in 1976) was a way to 'tile' (fill) space non-periodically with just two shapes. These he gave as particular obtuse and acute rhombohedra [251]. Commenting in 1987, geometer H S M Coxeter called these 'Golden Rhombohedra' [792]. Gardner noted that both these shapes had been studied by Kepler [252]. Ammann showed how by adding face-matching rules these two rhombohedra could tile 3D space non-periodically [160]. As mentioned at the beginning of this section, we can only begin to scratch the surface of this subject. There are many other interesting configurations of Penrose tiles, going by more or less descriptive names, such as: decapod, cartwheel, Batman, buzzsaw, starfish, and Asterix, plus bowties (long and short), and Conway worms. The Penrose P3 rhomb tiling has a variation discovered by American number theorist R M Robinson—Robinson triangles [319].[5] Also N G de Bruijn introduced the idea of 'pentagrids' (at first sight similar to Ammann bars) upon which he based his algebraic theory of non-periodic tilings [89].

Finally, the 'Pen-' in Penrose might remind us of *penta-*, the Greek prefix for 5, (as in pentagon, pentangle, pentad).

---

[5]   In 1954 Robinson held the record for the 5 then largest Mersenne primes [784]. In 1941 Robinson married Julia Bowman, best known for her work on Hilbert's 10th (p.450).

Chapter 27

Quasicrystals

As hinted at earlier in the *Molecular phi* chapter, traditional crystallographers (pre-1980) were absolutely fixed in their views about what constituted a crystal and what did not. This white and black certainty was only possible through their rigid application of the mathematical framework of 230 space groups. These are the natural geometric consequence of periodic crystals having translational symmetry—of their being built up from simply repeated unit cells. In this scheme the allowed symmetries are: 2, 3, 4, and 6, but no others... (Fig. 198).[1] This is what all the textbooks said, and this is how crystallographers were trained to think.

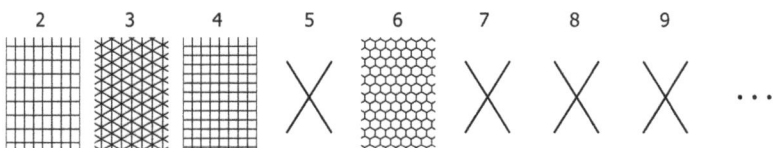

Figure 198: **The crystallographic restriction**

Hence, approached from this point of view (predicated entirely on translational symmetry)—the idea that a crystal could have a rotational symmetry such as 5, 8, 10, or 12-fold, was utterly unthinkable. Nevertheless, this conviction was not shared universally —there were at least 2 people on the planet who thought differently.

---

[1] Haüy's (1784) *Law of Rationality* [325, 1003]—for rotational axes of n-fold symmetry, with n integer, and $\omega = (360°/n)$—this requires that $\cos(\omega)$ be rational. Budalev (1867) showed that the only possible rational values for $\cos(\omega)$ are 0, $\pm\frac{1}{2}$, ±1, and that these occur when $n = 1, 2, 3, 4, 6$ [180].

323

Penrose predicts quasicrystals

In 1974 Alan Mackay first heard of Penrose tilings via Judith Daniels at the University College Computing Centre. He arranged to meet Penrose, and they discussed kites and darts. Penrose's main concern was aperiodicity, whereas that of Mackay was hierarchical structuring [310]. In his *Penrose Tiles to Trapdoor Ciphers*, Martin Gardiner quotes from a letter sent to him in 1976 by Roger Penrose. In it Penrose considered the dodecahedral and icosahedral growth of viruses and conjectured that one way this could happen would be based on Ammann's non-periodic solids. Following the logic through, he visualized the result of such growth as quasi-periodic 'crystals', but noted that these would seem impossible *crystallographically* [253].[2]

Mackay predicts quasicrystals

In his 1981 *Nive Quinquangula* paper ('5-pointed snowflake', p.303), Mackay too predicted the discovery of what we now call 'quasi-crystals', (and with the title of his paper, he provided a name for the photograph Fig. 200, long before it was taken). But, he warned, these special crystals might go unrecognized if they were not expected [611]. He referred to such structures as being based on a 'quasi-lattice'—a term he coined [610, 614]. Again we see Mackay trying to enlarge the traditional crystallographic mindset, so as to prepare crystallographers for the possible physical discovery of quasicrystals. Unfortunately, Mackay's wisdom was not heard. He first published in Russian, in a Russian journal [608], and although this was translated in the following year [609], the paper did not achieve its intended impact. As a result, when the big event came—in 1982 materials scientist Dan Shechtman found a metal alloy that *was* quasicrystal-line—and in 1984 he reported it in a paper [829]—the news was met with widespread shock, disbelief, and vociferous denial. We shall shortly discuss this dramatic development in detail. Mackay had initially been intrigued by the possible connection between Penrose tiling and his quasi-lattice, and he designed an experiment to produce the optical Fourier transform of a Penrose pattern (Fig. 199) —effectively predicting the characteristic 'diffraction pattern' of a quasicrystal (again, more on this shortly) [612, 308]. His result clearly shows 10-fold symmetry. Mackay presented this finding at the crystallography congress in Ottawa in 1981 [315].

---

[2]   As previously discussed, because dodecahedra and icosahedra do not appear in any of the 230 space groups, then they were not considered to fall into the scope of formal crystallography at that time.

Figure 199: **Optical Fourier transform of Penrose pattern 1981** (made for Alan Mackay by George Harburn). Image ©1982 Elsevier, reproduced with permission [612].

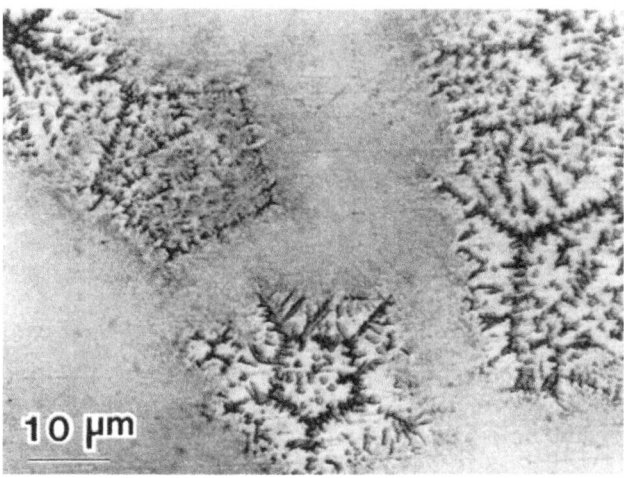

Figure 200: **'Nive quinquangula' 1986—a 5-pointed metal 'snowflake' and pentagonal crystal in Al-Mn alloy.** Image courtesy of Dr. Leonid A Bendersky, National Institute of Standards and Technology (NIST), Gaithersburg MD [802].

Diffraction patterns—Bragg spots

In order to understand the discovery of quasicrystals, we shall need a basic appreciation of 'diffraction patterns'. These are formed when suitable rays (e.g. X-rays or electron 'rays') are deflected by an orderly arrangement of atoms, and these rays then mutually add and subtract ('interfere'). It was German physicist Max von Laue who discovered the diffraction of X-rays by crystals, and for this he was awarded the Nobel Prize for Physics in 1914. And it was (Australian-born) British physicist William Lawrence Bragg who discovered the law of X-ray diffraction (now bearing his name) from which crystal structure may be deduced. Both he and his father William Henry Bragg, were awarded the 1915 Nobel Prize in Physics for their work on the X-ray analysis of crystals.[3] Now let's imagine shining X-rays (of a single wavelength λ) onto a mineral crystal lattice as in Fig. 201.[4] Our interest is in how these waves are scattered off the layers of atoms (spaced by distance d). As the incoming rays (top left) impinge on the crystal, they seem destined to scatter chaotically. However, because of their wave nature, the rays may add or subtract. In the upper left, rays at angle θ_1 cancel—resulting in the black background of the diffraction image (right). In the lower left, the rays at angle θ_2 add to make a fixed-direction beam that shines a bright spot onto photographic film or CCD sensor—a 'Bragg diffraction peak'. Single crystals typically produce highly ordered patterns of distinct, evenly spaced, light spots that may be detected directly by a sensor or photographic film without need of lenses—giving a result in the style of Fig. 202. But, in order for a patterning effect to be achieved, it is essential that the wavelength λ (of the X-rays or the electrons) be comparable with the distance d—the gap between the layers of atoms in the crystal. This explains why visible light is unsuitable for atomic-level work—its wavelengths are of the order of 1,000 times too long. To move on to see electrons as being useful for diffraction experiments would be quite correct. However, this would be somewhat like looking down the wrong end of a telescope. In fact, it was by using diffraction effects that the wave nature of electrons was first demonstrated. Two crystallographers shared the 1937 Nobel Prize for Physics for this work—they were Englishman George Thomson and American Clinton Davisson.[5]

---

[3]   Melvyn Bragg (p.34) is a distant cousin of William and Lawrence.

[4]   We do this to understand the general principle—we shall discuss the use of electrons instead of X-rays later.

[5]   French physicist Louis DeBroglie had proposed this wave/particle duality in 1924.

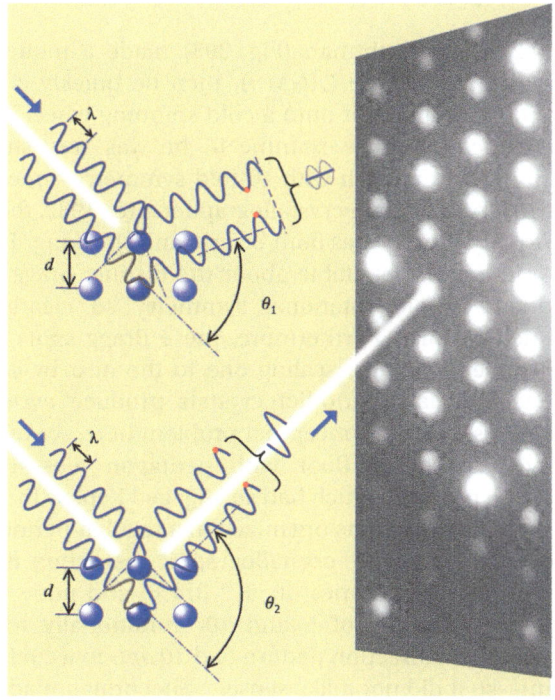

Figure 201: **Bragg diffraction.**
Top: Wave phases cancel.   Bottom: Phases add, ray propagates.
Diagram incorporates: *Bragg diffraction diagram* by Cdang & Gregors [104, 144].

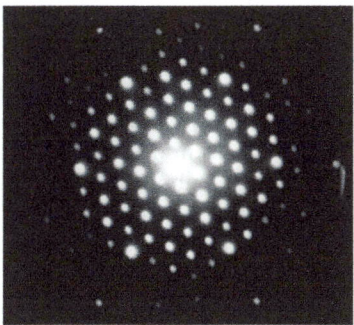

Figure 202: **Electron diffraction pattern for a periodic crystal.   Bragg spots are evenly spaced.**   Image courtesy of MATTER, University of Liverpool [266].

The quasicrystal discovery—Dan Shechtman

In April 1982, Dan Shechtman (Fig. 203) made a molten alloy of aluminium and manganese (Al_6Mn), then he quickly chilled it by 'melt spinning'—dropping it onto a cold spinning wheel [15]. Using his electron microscope to examine it, he was very surprised to observe a diffraction pattern with 10-fold symmetry—one similar to that in Fig. 205 (p.330). For a crystallographer circa 1982, the very idea of 5- or 10-fold symmetry was both bizarre and shocking. But there is nothing borderline or debatable about the distinct Bragg spots and the 'forbidden' 10-fold rotational symmetry so clearly seen in examples such as this. Furthermore, these Bragg spots line up in successive Golden Sections, scaling one to the next by φ (Fig. 206, p.330)—while 'normal' (periodic) crystals produce *evenly* spaced spots. To understand the conceptual problem here, we only have to imagine trying to tile a floor with pentagon tiles—leaving no gaps—again the problem which had so engaged Kepler. In 1982, to be a crystal meant having atoms organized in a regular 'periodic' lattice. As we discussed earlier, the crystallographic restriction then allows only the rotational symmetries of: $1, 2, 3, 4, 6$, and none other. We again note the omission of 5 and 10—traditionally regarded as forbidden. But this diffraction pattern had *10 dots* in a circle around a central point—so it did not make sense. Shechtman made a note in his lab notebook (Fig. 204) '10 fold' and added three question marks.[6] Shechtman recalls thinking to himself: 'There's no such animal.' (a scattering pattern with *10-fold* symmetry). He later wrote up his discovery [829]—announcing that he had discovered a new kind of physical structure. This structure is neither regularly periodic, nor random/amorphous and 'glassy'. The material was first called a 'quasi-periodic crystal'; this name was then shortened to just 'quasicrystal'. Now, we might imagine that once Shechtman had made this amazing new discovery, his immediate scientific community would be very impressed. Perhaps they would joyously celebrate his success as being the start of a brand new chapter in their field, and he would promptly be awarded a Nobel Prize. If only that had been the case. His first paper reporting his discovery (co-authored with fellow Technion materials scientist Ilan Blech) was rejected outright by the *Journal of Applied Physics* [316]. JAP replied: 'Not of interest to physicists... ' and suggested a metallurgical journal instead. And although Shechtman published in *Metallurgical Transactions* in 1985, much was to happen in between [828].

---

[6] 'SAD' stands for 'selected area of diffraction' [840].

Figure 203: **Dan Shechtman** 2011 Nobel Laureate in Chemistry.
Photo by David Blumenfeld [67] courtesy of, and © 2011 Nobel Media.

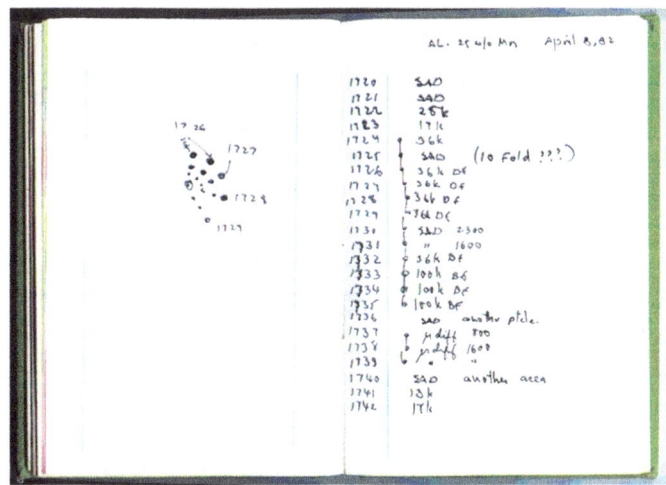

Figure 204: **Dan Shechtman's laboratory notebook.**
Image courtesy of, and © Dan Shechtman.

329

He was disbelieved, laughed at, insulted and asked to leave his research group. Friend and Nobel Laureate Linus Pauling was the foremost critic of his work, with its *unthinkable* non-periodic interpretation—although this was doubly strange, as Pauling had both visited Bernal and worked on icosahedral structures himself [314]. Pauling would declare to packed conferences: 'Danny Shechtman is talking nonsense. There is no such thing as quasicrystals, only quasi-scientists.' [830]. It took Shechtman five years to marshal enough evidence to gain widespread acceptance, and eventually in 2011 (nearly 30 years after his 'no such animal' moment), he *was* awarded the Nobel Prize in Chemistry for: 'The discovery of quasicrystals.' [829, 793]. Although (we now know that) these had been found by other researchers earlier, those researchers had not recognized and pursued their findings (just as Mackay had warned [611]). Shechtman commented later that he did not then know of Mackay's work, and that it was others who made the link between physical quasicrystals and Penrose patterns. The key thing is though, that Shechtman had both the results and the conviction of their great importance.

Figure 205: **A quasicrystal electron diffraction pattern.** As produced by icosahedral Zn-Mg-Ho (zinc magnesium holmium). Image: Materialscientist [632, 144].

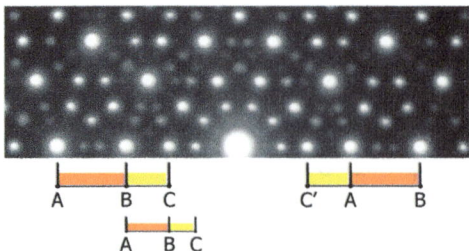

Figure 206: **Repeated Golden Sections:** ABC **and** $C'AB$ (Fig. 205 detail.)

Why 1982? Why Shechtman?

Giving a lecture at Liverpool University in 2014, Professor Shechtman addressed these questions, along with some related ones [562]. 'So why did it take until 1982 for quasicrystals to be discovered?', he asked, 'Was it because they:

- are rare? —No, there are hundreds of types.
- are not stable? —Some are, some are not.
- are hard to make? —They are easy to make.
- need rare elements? —No, their ingredients are readily available and inexpensive: aluminium, manganese, nickel, cobalt, ... '

(So, 'none of the above'.) Instead, he concluded, there were two main reasons. First, at that time, samples were around 1 micron across— far too small for use as X-ray diffraction targets when studying rotational symmetries. Rather, the discovery could only have been made using TEM (transmitting electron microscope) diffraction techniques. Such equipment and procedures were not at all common at the time. Most TEM work was done by students, but then only occasionally. Second, Dan Shechtman was from Technion—a world leader in electron microscopy since 1967 [305]. He combined specialist expertise, 15 years of electron microscopy experience, and a personal tenacity—a tenacity that would see him through the major battle he then had to fight for acceptance of his results.

Pauling's rational argument

The present author was very pleased to meet Professor Shechtman at the above Liverpool University event—having been invited by Professor McGrath, Head of the School of Physical Sciences (an expert in quasicrystal surfaces). Over lunch our conversation briefly touched on the Golden Ratio in perception and aesthetics. Professor Shechtman turned to me, and with a slight earnestness he asked: 'Of course, you know that there is no such thing as a Golden Rectangle?' He was referring to the mathematical definition of ϕ as an irrational ratio, with its decimal representation that goes on forever without falling into any pattern. One consequence of this definition is that it would be impossible to draw or engineer a true Golden Rectangle —this shape having such a sublime specification. But, as soon as we give up the dream of exactness, then the practical question arises: 'In a given context (for example an experiment in aesthetics or a theory of crystals)—what would be an acceptable range of values either side of target?' And as we do this, we admit rational values in our range.

We have already seen (as did Kepler) how the convergents of ϕ—the rationals F_n/F_{n-1} and F_{n+1}/F_n—sit arbitrarily close either side of ϕ (p.22). It was just this distinction between rational and irrational that drove (Nobel laureate) Linus Pauling on to the end of his life trying to prove that quasicrystals did not exist. In his unshakeable opinion, they were just 'twinned' crystals.[7] Pauling saw them as fundamentally periodic, and reasoned that with a sufficiently large unit cell, he could explain near-icosahedral symmetry in rational terms. The problem for Pauling was that as successive experiments produced more and more perfect icosahedral results—his corresponding unit cells needed to grow to incredible sizes [29].

This debate really goes to the core of our understanding of just what the irrationality of ϕ means in practice. For a start, in a 'Golden Rectangles on a sheet of paper' aesthetics experiment we would not expect the results to change significantly if a rectangle had one pixel (say $1/600$th of an inch) more or less width. In Le Corbusier's definition of his Modulor system, we notice measurements here and there 1mm away from apparent ideals (e.g. Fig. 104, p.129). In each of these cases there is a practical limit to the precision required. However, in the case of crystals, with such enormous numbers of atoms, a very great precision must apply for them to exhibit long-range order. If we think back for a moment to the Pythagoreans—they had use of ϕ in the shape of the pentalpha (the 5 point star). The ϕ ratio in their case was not 'fed in' as calculated Golden Section measurements with a certain precision. Rather, it arose—without calculation—directly from the geometry. And if a thousand pentalphas were (fairly accurately) drawn and suitably measured, we should expect to be able to derive a good approximation for ϕ as an output. In another example, graphene has a flat repeated-hexagon structure that will (in theory) extend indefinitely, systematically preserving its atomic-scale honeycomb pattern.[8] If we measured any particular hexagon's dimensions, then with thermal vibrations, it would not be perfect. Yet the overall geometric structure is not distorted by these local variations, and it demonstrates an extremely high degree of long-range order. It is exactly this kind of long-range 'geometric harness' that characterizes quasicrystal structures—their irrational basis is 'built in' (just as with the pentalpha)—it is not some periodic approximation or an accident of crystal imperfection.

---

[7] 'Twinning' describes an imperfection where periodic crystals have grown 'joined together', thus sharing some lattice points.

[8] We saw this structure in the '6' of Fig. 198 on p.323.

Figure 207: **Ho-Mg-Zn dodecahedral quasicrystal** grown by using the self-flux method (excess Mg) with slow cooling from 700°C to 480°C. The edges are 2.2 mm long. Image courtesy of AMES Laboratory, US Department of Energy, Fisher et al. [222].

Theoretical work

The requirements for three-dimensional quasicrystals closely parallel those for Penrose's tiles in two dimensions. They should fill (as well as possible) a space in a highly ordered way, but have no translational symmetry. They should also give rise to a point diffraction pattern. Inspired by Penrose's work, mathematicians Paul Steinhardt & Dov Levine worked on formalizing such structures [548]. Dov Levine has described how important Ammann's work was to them in this task [317]. Penrose's locally restrictive definition (forcing non-periodicity by disallowing certain joins) was not suited to generalization into 3D. However, Ammann's bars readily generalized to Ammann planes in 3D. By 1984 Levine & Steinhardt had got as far as proposing an example structure and had worked out its theoretical diffraction pattern. In his Perimeter Institute talk, Paul Steinhardt recalls meeting physicist David Nelson in 1984 at IBM Research. Nelson showed him Shechtman's 1984 paper along with its diffraction pattern—a pattern remarkably similar to the one he had produced with Levine [863].[9] It was Steinhardt who coined the term 'quasi-crystal' in 1984 [548, 684]. In 2010, Alan Mackay, Dov Levine and Paul Steinhardt were awarded the Buckley Condensed Matter Physics Prize: 'For pioneering contributions to the theory of quasicrystals, including the prediction of their diffraction pattern.' [92].

---

[9] Steinhardt had gone to IBM to find out if they could physically make materials with the 'quasi' structure.

More spookiness?

If we carefully build Penrose patterns using the edge-matching rules exactly as specified, then at some distance from our starting point, we may find there is no legal next step, and backtracking is necessary. A similar situation occurs when modelling how quasicrystals might form in three dimensions. The fact that observed real-world crystals *do* achieve long-range order might therefore suggest some non-local mechanism being involved [864]. The simple view that crystals grow by aggregation according to local rules appeared insufficient to explain the observed long-range structures. This unsettling problem led Penrose to consider whether long-range quantum interactions might be at play [741]. However, by focussing on what happens at the vertices (instead of at the edges), Onoda, Steinhardt, DiVincenzo & Socolar (1988) found rules that *could* guarantee 'no-mistake' tilings of the plane. Their solution was to add tiles only at 'forced sites' (places where there was no choice to be made between possible tiles), and then, where there was no obvious next step, to add a thick tile to either side of a 108° corner. This approach turned out to make perfect quasicrystals [714, 865].

Counting crystal types

The relatively simple periodic crystals are classified in 7 systems, 32 classes, and a total of 230 space groups. But how many types of quasicrystals could there be? This question has been considered by English-born theoretical physicist and mathematician, Freeman Dyson. For example, at the time he wrote (in 2009), it seemed fairly well established that only icosahedral structures were possible in three dimensions. Then again, in two dimensions we may count one type for each regular polygon—the Penrose tilings being associated with the pentagon. When it comes to one-dimensional quasicrystals though,[10] Freeman asserts that the crystals, freed from the requirements of rotational symmetries, will have vastly increased and rich possibilities that we have only just begun to explore [194].[11] Only a year later, it was found that at least one other 3D quasicrystal structure was possible [527]—we shall come back to that discovery later (page 388).

---

[10]   Here we mean a physical 3-dimensional device, typically built in layers—that is, composition only varies in one direction (along the normal to the layers).

[11]   Dyson now conjectures that the study of 1D quasicrystals may help in proving the Riemann hypothesis [892].

Covering with decagons

Figure 208:
Dr. Petra Gummelt. Photo © 2011 University of Greifswald / Marcus Vollmer.

As German physicist Knut Urban pointed out, it was condensed-matter physicist Sergei Burkov who in 1991 first demonstrated that the plane could be covered with overlapping decagons [909, 94]. Then in 1996 German mathematician Petra Gummelt (Fig. 208) showed how a Penrose tiling could be generated with a single decagon and a new overlap rule [289]. Her rule requires decoration into shaded and non-shaded areas—then overlap is only allowed if shaded portions overlap. In (Fig. 209) a decagon is inscribed in a circle radius ϕ, giving side length one. Comparing Gummelt's with Kepler's pattern (Fig. 21, p.22), we see the overlapping decagons here have vertex tops, whereas Kepler's 'monsters' have edge tops. By allowing the overlaps, and now by talking about covering the plane rather than tiling it (Fig. 210), the two-component model (kite/dart or thick/thin rhombs) could be reduced to single decagons that overlap in two possible ways. While there exist direct mappings from the Gummelt scheme back to say kite/dart—the advance here is that we can relate the overlaps to physical structures where they then represent *the sharing of atoms by decagonal 'quasi unit cells'*. This new way of thinking about aperiodic structures provided the crucial opportunity to apply the geometry to the physics and structural chemistry.

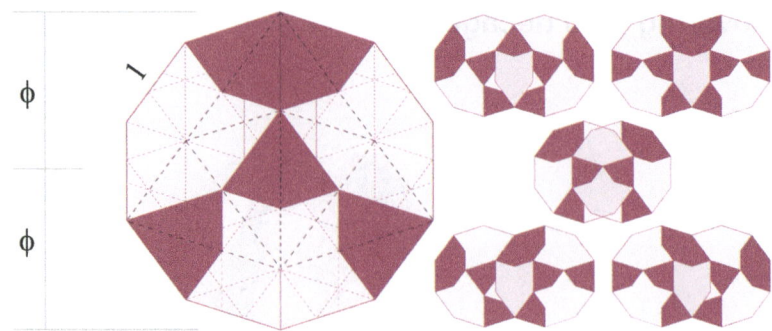

Figure 209: **Petra Gummelt's decorated decagon and its allowed overlaps.** Diagram derived by SharkD & Geometry Guy from original by Ael2 [8]. Relative dimensions added.

Figure 210: **Example coverage pattern from overlapping Gummelt decagons.** Diagram based on output from Greg Egan's website applet [197].

Also in 1996, Steinhardt and Jeong published a paper in which they moved on from matching rules to think in terms of maximizing the density of chosen clusters. Then by considering energy minimization arguments, they offered: 'a simple, physically motivated explanation of why quasicrystals form' [866]. Two years later, Steinhardt, Jeong et al. went on to provide strong experimental support for their model [867]. Then in 2000 a study by Yan and Pennycook mapped Gummelt's overlapping decagons to aluminium nickel cobalt quasicrystals [995]. And in 2001, following on from the work of Steinhardt et al., Lord & Ranganatham used the Gummelt approach as: 'a simple way of labelling important sites in a certain class of quasicrystals' [576]. As we have seen, Petra Gummelt's decorated decagon provides covering by overlapping with others the same. However, one question remains, namely: 'Might there be just one tile that will tile the plane without gaps or overlaps—a single shape with matching rules[12] which would force a non-periodic pattern?' Theorists have already named the sought-after tile as the *'einstein'*—a German pun on 'one stone', but although various candidates have been proposed, none has yet been fully accepted [845].

Quasicrystals based on the Silver Ratio

In 1977 Robert Ammann discovered several important tilings, and one of these, his 'A5' is based on the Silver Ratio $\delta = 1 + \sqrt{2}$ (Fig. 211).[13] A defining property of a quasicrystal is that although it lacks translational symmetry (it is non-periodic), it can produce sharp Bragg scattering peaks when built as (or considered as) a physical structure. Penrose tilings have this property, and so too does this A5 tiling. In the A5's case the diffraction pattern shows the combination of long-range order with an underlying 8-fold symmetry, and in 1987 Wang, Chen, & Kuo found just such a physical quasicrystal formation in V-Ni-Si and Cr-Ni-Si alloys [937].

Figure 211: **A5 Silver-Ratio-based tiling—Ammann-Beenker.**
Image courtesy of Claudio Rocchini [786, 143].

---

[12] Or a definition based on Ammann bar equivalents.
[13] Pisot numbers (p.72) such as ϕ and δ appear as self-similarity ratios in quasi-crystalline structures [856].

New, third type of solid

Regardless of Alan Mackay's sustained campaign for a generalized approach to crystallography, until 1982 most scientists knew only two types of solid: simple crystalline materials with their ordered lattice structures (such as quartz) and randomly structured amorphous materials (such as glass). The discovery of quasicrystals was momentous because (in context) such new materials constituted a *third type of solid matter*—with non-periodic structures that nevertheless demonstrated perfect long-range order by giving sharp Bragg spots. Most importantly, quasicrystals show rotational symmetries (e.g. 5, 8, 10, and 12-fold) which are not consistent with periodic structures. This led many traditional crystallographers and journalists, rather sensationally, to describe such structures as 'forbidden', or even 'impossible'.

Crystals redefined

As the evidence (worldwide) for quasicrystals became overwhelming and sufficient consensus was achieved, in 1992 the International Union of Crystallography took the unprecedented step of *redefining the crystal*. The old 'input' definition requiring the crystal structure to be periodic (and so belong to one of 230 space groups) had to be abandoned; and in its place an 'output' definition was agreed, namely that for a material to be called a crystal, it need only produce well-defined Bragg spots in its diffraction patterns.

Quasicrystal properties and applications

Many different kinds of quasicrystal are now known—typically they are alloys of two or three metals, one of which is often aluminium. In terms of their structure, groups include (φ-based) 'icosahedrals' (3D) and 'polygonals' having rotational symmetry. These latter are simply periodic along an axis of rotation and so are sometimes called axial quasicrystals (2D+1D). Quasicrystals are non-periodic, non-glassy, 'third type' substances not characterized by some bland blend of the properties of periodic crystals and amorphous substances. Instead, they bring exciting, almost magical new properties, and there has been a worldwide rush to explore and apply these. As ever, we shall make the briefest of surveys—but sufficient to assure ourselves of the great importance of Dan Shechtman's discovery and to remind us of his triumph in getting quasicrystals accepted.

Wear-resistance

Quasicrystal particles have been successfully dispersed in HDPE (high-density polyethylene thermoplastic) to obtain a (compression moulded) wear-resistant composite [508].

High mechanical strength

In the making of hardened steels, certain formulations and heat treatments can produce carbon steel frameworks with embedded quasicrystal precipitates, resulting in substantially increased strength and flexibility [408].[14] Alloys of this type are especially suited to usage where the metal will be contact with human body (see bio-compatibility below).

Exceptional hardness

In normal metals, periodic atomic layers slide over each other under sufficient strain. However icosahedral quasicrystals lack these layer planes, resulting in exceptional hardness [508].[15]

Brittleness

Quasicrystals are typically brittle at room temperature, but at very high temperature they become ductile.[16] In some cases, a ceramic-like brittleness can be an advantage—for example making it easy to crush quasicrystals to a fine powder at low cost for use in coatings [508]. Alternatively, brittleness can be addressed by dispersing nanograin quasicrystals in an alloy to produce a combined material or coating [666].

Applicability in protective coatings

Coatings for engine components, injection moulds, mining, and agricultural equipment can be sprayed on using industry-standard plasma-arc techniques [265]. Aluminium nickel cobalt (Al-Ni-Co) alloys are particularly favoured—forming stable decagonal quasi-crystals [508].[17]

---

[14] E.g. Nanoflex® (stainless steel) is a registered trademark of Sandvik, Sweden.

[15] Hardness $HV_{0.02}{\sim}800$.

[16] For example icosahedral AL-Pd-Mn, 690°C for strain rate $10^{-5}s^{-1}$ [903].

[17] $Al_{72}Ni_xCo_{(28-x)}$ with x from 8 to 20.

Low coefficient of friction

The low coefficient of friction should be of considerable importance in industry.[18] Productivity and material losses due to friction (wear and tear) cost developed countries several percent of their GDP—in the USA for example, hundreds of billions of dollars [666].

'Non-stick'

The qualities of low adhesion/low surface energy are well known and applied notably in cookware, but explanations as to why quasicrystals have this non-stick property remain controversial [772].[19]

Oxidation and corrosion resistance

Experiments suggest that quasicrystals have an inherent resistance to oxidation and corrosion [621].

Proven bio-compatibility

Low-adhesion and corrosion resistance make quasicrystal materials suitable for bone repair implants, prostheses, shavers, and surgical equipment such as acupuncture needles and dental reamers. In particular, the quasicrystal *TiZrNi* develops oxides of titanium and zirconium on its surface which lock the nickel underneath, thereby ensuring bio-compatibility [528].

High electrical and thermal resistance

These suit quasicrystals to thermoelectric applications where heat is converted directly into electricity [628]. They are also being applied as thermal barrier coatings (e.g. on diesel engine components [45]).

Structure suitable for hydrogen storage

Both nickel and hafnium form quasicrystalline alloys with titanium and zirconium which can reversibly store large quantities of hydrogen at low temperatures and pressures—astonishingly even *exceeding the density of liquid hydrogen* [471, 414].[20] Hydrogen (once produced) is the ultimate clean 'green' fuel. It burns to produce just heat and water, or it can be passed through a fuel cell to produce just electricity and water.

---

[18] Coefficient of friction: 0.05–0.2 [508].
[19] Surface energy: ~25 mJ/m^2 [508].
[20] *TiZrNi*—e.g. $Ti_{45}Zr_{38}Ni_{17}$ and *TiZrHf*. However, aluminium alloys are unsuitable for hydrogen storage.

Superplasticity at high temperatures

With good corrosion resistance, low thermal conductivity and the ability to accommodate the thermal expansion of the component it is protecting, quasicrystal coatings can be of benefit in rocket motors and aero-engine turbines [413].

Potential as catalyst

Quasicrystals have the potential to replace precious metals in some catalysis applications. Good results have been obtained in converting methanol to the easy-to-liquefy fuel DME (dimethyl ether) [9, 884].

Extraterrestrial quasicrystals

As of 2012, even though a hundred types of quasicrystal had been made in the laboratory, the questions remained: 'Could these structures have long-term stability? Could they be found to occur naturally?' Paul Steinhardt spent ten years searching, and finally he discovered a fragment of 'icosahedrite'—$Al_{63}Cu_{24}Fe_{13}$. Its icosahedral quasicrystals were found in a millimetre sized rock grain at the Museum of Natural History in Florence, Italy. The fragment was mainly composed of khatyrkite—crystalline $(Cu, Zn)Al_2$, and it was labelled as coming from the Koryak Mountains of Far Eastern Russia. Subsequent analysis suggested that the quasicrystals that Steinhardt found were from a 4.5 billion years old meteorite which crashed into the Earth some 15,000 years ago. So, natural quasicrystals are very rare, and a highly stable example does exist [868].

Soft-matter quasicrystals

Many compounds (synthetic and natural) can self-organize into ordered structures in one, two, and three dimensions—a simple example being soap and water. Up until 2004 though, all such structures were found to conform to one of the 17 plane symmetry[21] or the 230 space groups—the 'traditional' crystal structures. It was then that Zeng et al. published a paper showing that soft matter could also self-organize into quasicrystal structures [1001]. Israeli physicist Ron Lifshitz believes these new materials will enable novel applications—ones based on self-assembling nanomaterials [560].

---

[21] We discuss these 'wallpaper' groups in the next chapter.

Interdisciplinary

Just as we saw with phyllotaxis, the quasicrystal discovery has attracted researchers from widely different fields: materials scientists, crystallographers, chemists, mathematicians, physicists, mechanical engineers, tribologists—and specialists in catalysis, cookware, and surgical instruments.

The long view

In the 17th century, important scientists and mathematicians were not only fascinated by the external forms of crystals—their greater interest was in how these outer forms might relate to internal structure. Figures included: Johannes Kepler, René Descartes, Robert Hooke, Christian Huyghens, then in the 18th century, René Just Haüy [879]. We mentioned Kepler's 1611 *Snowflake* essay earlier (p.22). In it he discussed the hexagonal close packing of spheres and he speculated that this was the origin of the 6-fold symmetry of the snowflake. We also recall from p.322 that Kepler had even studied Golden Rhombohedra—these 'only' need Ammann's face matching rules in order to fill space aperiodically [251, 252, 792]. Hooke then extended certain of Kepler's packing ideas to other crystal shapes [302]. Ron Lifshitz has studied quasicrystals for over 20 years, and while introducing his article *Nanotechnology and Quasicrystals: From self assembly to photonic applications,* he notes how Kepler was quick to realize that the triangle, the square, and the hexagon were the only regular polygons that could tile the plane (without gaps). But that nevertheless, Kepler still experimented with tilings based on 5- and 10-fold symmetric shapes—as we saw in his 'Aa' pattern which included pentagons, 5-point stars and (overlapping) decagons—his 'monsters' (p.22) [458]. As Lifshitz says, Kepler had discovered the basics of aperiodic order without realizing it—though with Haüy's mathematical formulation of crystallography (based on periodicity) —Kepler's pentagons and decagons were forgotten [559]. Lifshitz goes on to refer to Mackay's comment about the quasicrystal discovery (made during an interview by István Hargittai [306]), that:

> 'It's a kind of legalistic discovery. It's a discovery of a material which breaks the laws that were artificially constructed. They were not laws of nature; they were laws of the human classificatory system.'

> Alan Mackay, London, October 1994.

Chapter 28

Islamic tiling patterns

Quasicrystals and medieval Islamic art

We briefly considered complex Islamic ornamental designs at the beginning of the book (p.20), mainly in order to place them in their historic context. Here we shall explore them in a little more detail. Initial research interest in such designs was very subjective, relying on pictures, descriptions and comparisons of graphic elements. An important work by Jules Bourgoin in 1879 showed nearly 200 patterns [78]. Bourgoin's study is prefaced with a discussion of how the Mohammedan religion forbids representation of the human figure, and how this restriction had set Islamic art on its own path, separate from that of European art. The mathematics required to begin an objective study were provided in 1891—shortly after Bourgoin's work—by Russian mineralogist and crystallographer, Evgraf Fedorov who proved there exist only 17 'wallpaper' (plane symmetry) groups [212]. These are so-called because the patterns repeat in two or more directions, just as they do in wallpaper (both across adjacent strips and lengthwise down the roll). Fedorov's result was independently proved by George Pólya in 1924 [757]. We also noted George Pólya earlier—as being a key influence on artist M C Escher (p.71). Coincidentally, Escher had been interested in Islamic pattern designs since his visit in 1922 to the Alhambra Palace (Granada, Andalusia) [693]. And in 1937, when a professor introduced Pólya's 1924 paper to Escher, it was a turning point for him. Prior to that, his design work had been experimental. Now he had the definitive raw materials for his very particular style of geometric art.

It had been said that all 17 wallpaper groups could be found applied in the Alhambra. And although this was for a long time thought to be a myth (as only 13 out of the 17 had been recognized)—the missing four were eventually found in 1986 by R Pérez-Gómez [744]. The total 17 are the only possible *periodic* wallpaper patterns, and by definition, all exhibit translational symmetry. However, not all the Islamic designs are based on these 17 groups, and in particular, some are found to have non-repeating, decagonally based patterns. Danish geologist and crystallographer Emil Makovicky at the University of Copenhagen studied such patterns over a period of two decades. He was, in 1992, the first to compare these advanced patterns with Penrose tilings [617]. The example he initially focused on was the Blue Tomb (Fig. 213)— Gunbad-i-Quābūd at Marāgha in Western Iran. (Gunbad-i-Quābūd may be spelt in a number of other ways.)

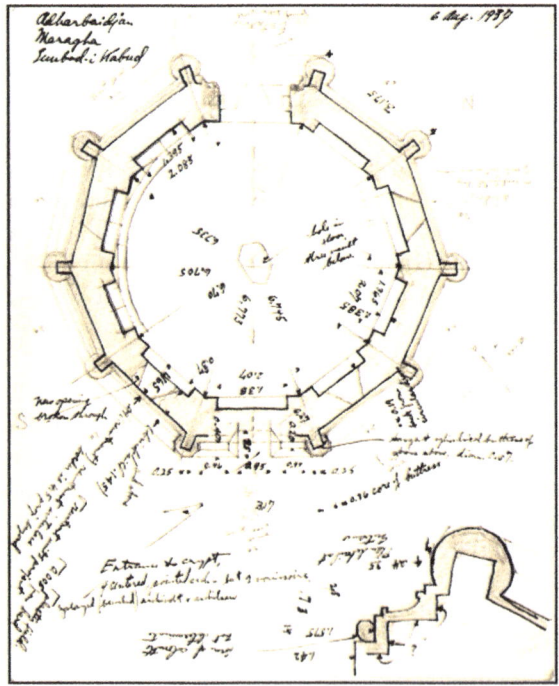

Figure 212: **Myron Bement Smith (1897–1970) Plan of Gunbad-i-Quābūd, Marāgha, Iran 1937.** Myron Bement Smith Collection, Freer Gallery of Art and Arthur M Sackler Gallery Archives, Smithsonian Institution. Gift of Katherine Dennis Smith [844].

Image by Ayub Farabi Assul, ايوب فارابى اصل

Figure 213: **Gunbad-i-Quābūd 1197 AD, 593 AH.** Marāgha, Iran [20, 143].

This prism tower has been described widely as being octagonal. However, as Carol Bier has demonstrated (using the detailed 1937 plan shown in Fig. 212), it is, in fact, decagonal—both inside and out [58]. This decagonal architectural basis is then consistently echoed in the geometry of its ornamentation.

Makovicky concluded that the Marāgha pattern could be simply derived from the Penrose P1 tiling, and that the overall design was based on (what Conway dubbed) the cartwheel pattern [249]. Makovicky also considered how the strapwork patterns were probably built up using a set of equilateral polygonal tiles (now called 'girih tiles'), based on decagonal angles—with each being decorated with line fragments that would then link across to matching lines on adjacent tiles during composition [618].

Peter Lu & Paul Steinhardt significantly extended Makovicky's ideas and in 2007 identified a girih tile set [587, 579, 590] which included tiles seen in the Topkapı scroll [673].[1] This scroll is a 'pattern book' containing 114 geometric tiling fragments for use in architectural decoration. It is about 33 centimetres high and nearly 30 metres long and although it has no text, it does however show construction lines [146]. For example, Panel 50 (Fig. 214, right) clearly shows use of bowtie and hexagon girih tiles, and it suggests decagon edges top left and bottom right. We see how the strapwork pattern lines advance across the tile boundaries. Such graphic corroboration adds great weight to the theory that the intricate and sometimes extensive patterns were realized in practice using girih tiles, rather than (as some had previously thought) by straightedge and compass.[2]

Figure 214: **Left: Girih tiles—pentagon, rhomb, hexagon, bowtie, and decagon. Right: Example strapwork pattern**—Panel 50 from the Topkapı scroll. Left: Image by Cronholm144 [156, 144]. Right: Topkapı scroll [590].

---

[1] Persia, late 15th or early 16th century, Timurid dynasty. Ink and colours on paper. Collection of Topkapı Palace, Istanbul, Turkey.

[2] Lu & Steinhardt argue that girih tiles were used from (at latest) 1200 AD, noting that Persian Abū'l-Wafā' Al-Būzjānī 's (940–998 AD) treatise, *On the geometric constructions necessary for the artisan*, contributed required mathematical tools [587].

Figure 215: **Tile from the Tomb of Buyanquli Khan.** Uzbekistan, Bukhara, c.1358. V&A London. Many pentagons are formed in this symmetric strapwork pattern.

With these building blocks, artisans did not need to understand the relevant mathematics. Instead, they worked with a simple set of tiles that would fit together to produce very detailed and interesting designs—such as the tile in Fig. 215. Here is yet another example of unexpected complexity being achieved by the combination and repetition of simple components and rules—just as we have seen in generalized crystallography, fractals, and cellular automata (such as the Game of Life). As with Penrose tilings, there are recursive subdivision systems for girih tiles [580, 152]. However, in practice these are invariably found in the form of one larger pattern which is then subdivided only once.

Lu & Steinhardt found such a subdivision in the decagonal girih pattern at the Darb-i-Imam shrine, built in 1453 in Isfahan, Iran. They were particularly interested in one panel, the top-right spandrel, (which we covered briefly in the first chapter, p.20). In it, they showed how the large-scale strapwork lines were a cropped view of a wallpaper-periodic pattern of decagons and bowties, and how each of these elements could be subdivided into much smaller decagons, bowties and hexagons (Fig. 216) [589]. This subdivision then perfectly matched the underlying smaller-scale pattern in the original.[3] In his 2007 Harvard colloquium talk, Lu presents the work he did with Paul Steinhardt on this [582]. That is, how they took the analysis to the next step—that of proposing a subdivision for a hypothetical 'large hexagon' (hypothetical, as the Darb-i-Imam large pattern has only decagons and bowties). In doing this, they now had a complete set of tiles which could be used for self-similar subdivision. And this transformation could then be repeated indefinitely (just as with Penrose subdivision).

---

[3] After fixing the (very few) local 'mistakes'—as we noted on p.20.

Using upper-case letters for LARGE tiles, and lower case for small—their full subdivision rule set comprised:

   1 BOWTIE   making 14 bowties, 14 hexagons and 6 decagons[4]
   1 HEXAGON making 22 bowties, 22 hexagons and 10 decagons
   1 DECAGON making 80 bowties, 80 hexagons and 36 decagons

[583]

Lu and Steinhardt then applied linear algebra to show that such a subdivision, successively applied, will produce a pattern having quasicrystal properties. Let's follow their logic through,[5] while noticing for ourselves: ϕ, Lucas, and Fibonacci numbers popping up along the way.

Using simple matrix multiplication,[6] we can find the total numbers for each of the small elements (b, h, d) from some given numbers of large elements (B, H, D):

$$\begin{pmatrix} b \\ h \\ d \end{pmatrix} = \begin{pmatrix} 14 & 22 & 80 \\ 14 & 22 & 80 \\ 6 & 10 & 36 \end{pmatrix} \begin{pmatrix} B \\ H \\ D \end{pmatrix}. \qquad [583]$$

Eigenvalues and eigenvectors

Let's call this transformation matrix **T** and then think about the effect it should have if applied repeatedly. For the successive subdivisions to be self-similar, each application should produce the same change from generation to generation, and at each stage, the relative numbers of tiles should stay the same. In terms of linear algebra this is equivalent to saying a column vector shall only be scaled (and not change direction): $\mathbf{T}v = \lambda v$. This is a special situation (as vectors are typically changed in direction), and a vector v that satisfies this requirement is called an 'eigenvector' for **T**, and the associated scaling λ is called an 'eigenvalue' [873]. Solutions for λ and v may be found using determinants.[7]

---

  4   E.g. see the 14 (red) small bowties in Fig. 216: BOWTIE 'D', bottom left
  5   Transformation discussion quoted with kind permission [582].
  6   See page 485.
  7   Using determinants, we shall work a simple example shortly.

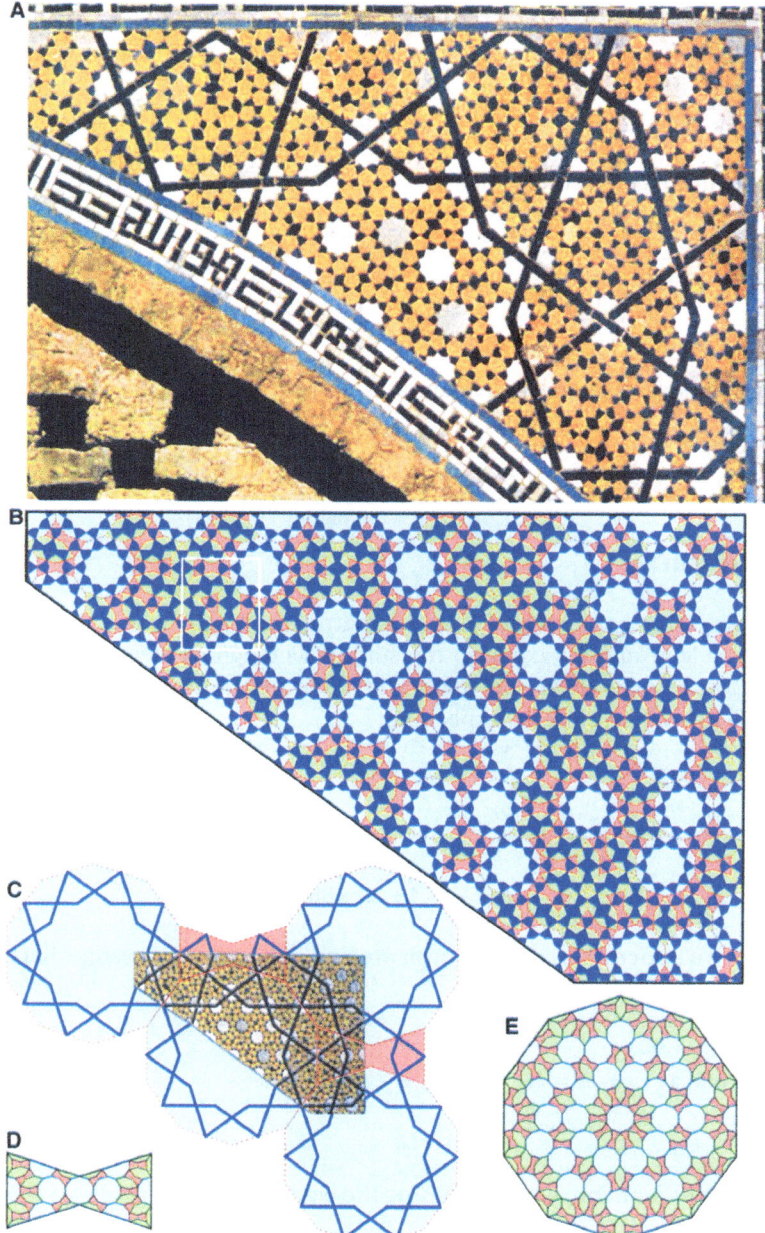

Figure 216: **Darb-i-Imam subdivision.** Diagram © 2007 Peter Lu reproduced with kind permission [589].

Lu & Steinhardt note that the transformation matrix

$$\mathbf{T} = \begin{pmatrix} 14 & 22 & 80 \\ 14 & 22 & 80 \\ 6 & 10 & 36 \end{pmatrix} \qquad [583]$$

has three eigenvectors which are:

$$v_1 = -\sqrt{\tfrac{5}{11}} \begin{pmatrix} 1 \\ 1 \\ \frac{1}{\sqrt{5}} \end{pmatrix} \qquad v_2 = -\sqrt{\tfrac{5}{11}} \begin{pmatrix} 1 \\ 1 \\ \frac{-1}{\sqrt{5}} \end{pmatrix} \qquad v_3 = -\sqrt{\tfrac{1}{11}} \begin{pmatrix} 1 \\ 3 \\ -1 \end{pmatrix}$$

and their associated eigenvalues are:

$$\lambda_1 = 4\phi^6 \qquad\qquad \lambda_2 = 4\phi^{-6} \qquad\qquad \lambda_3 = 0. \qquad [584]$$

The trace

An interesting property of a square matrix (such as **T**) is that the sum of its eigenvalues is equal to the sum of its diagonal elements. This diagonal sum is called the 'trace', and it may be abbreviated as $Tr(\mathbf{T})$ or $tr(\mathbf{T})$ [874]. Therefore we expect that

$$tr(\mathbf{T}) = \lambda_1 + \lambda_2 + \lambda_3.$$

Let's check. Here we have

$$tr(\mathbf{T}) = 14 + 22 + 36$$

and, we notice, this sum can be expressed in terms of Lucas numbers:

$$tr(\mathbf{T}) = 2L_4 + 2L_5 + 2L_6.$$

Further, as $L_4 + L_5 = L_6$, then

$$tr(\mathbf{T}) = 4L_6.$$

On the other hand, when summing the eigenvectors, we get

$$\lambda_1 + \lambda_2 + \lambda_3 = 4\phi^6 + 4\phi^{-6} + 0$$
$$= 4\left(\phi^6 + \phi^{-6}\right).$$

Here we notice the Binet form for L_6 (in the bracket)—strictly this is best written $\phi^6 + (-\phi)^{-6}$, which is equivalent in this case—given the even power, so

$$\lambda_1 + \lambda_2 + \lambda_3 = 4L_6$$

thus confirming equality with the trace of **T**.

Looking back a page at the third eigenvector v_3, we see it has only integer elements with one negative. Lu & Steinhardt show that this vector is telling us that we may replace one decagon (-1 as 'take away'), with 1 bowtie and 3 hexagons. This means that the transformation may be simplified—from the 3×3 transformation matrix **T** to a 2×2—as now we need count only bowties and hexagons:

$$\begin{pmatrix} b \\ h \end{pmatrix} = \begin{pmatrix} 20 & 32 \\ 32 & 52 \end{pmatrix} \begin{pmatrix} B \\ H \end{pmatrix}$$

$$= 4 \begin{pmatrix} 5 & 8 \\ 8 & 13 \end{pmatrix} \begin{pmatrix} B \\ H \end{pmatrix}. \qquad [585]$$

And here (as they point out), the matrix elements are all Fibonacci numbers:

$$\begin{pmatrix} b \\ h \end{pmatrix} = 4 \begin{pmatrix} F_5 & F_6 \\ F_6 & F_7 \end{pmatrix} \begin{pmatrix} B \\ H \end{pmatrix}. \qquad [585]$$

And here again, we see L_6 in the trace—in this case, it is given by the well known identity $F_{n+2} + F_n = L_{n+1}$ [488]. So, we have just seen L_6 formed in 3 different ways:

- by Lucas recurrence sum $L_4 + L_5 = L_6$ and
- by Lucas Binet form, $\phi^6 + (-\phi)^{-6} = L_6$ and here
- by Fibonacci sum (with mid-index F_6 'skipped'), $F_5 + F_7 = L_6$

Relative numbers of tiles

But how are eigenvalues and eigenvectors actually found, and what more can they tell us? For simplicity, let's consider the above 2×2 'Fibonacci' matrix, and call it **M**:

$$\mathbf{M} = \begin{pmatrix} F_5 & F_6 \\ F_6 & F_7 \end{pmatrix} = \begin{pmatrix} 5 & 8 \\ 8 & 13 \end{pmatrix}.$$

(As we proceed, it will also become clear how to treat larger square matrices—such as the 3×3 matrix **T** we saw earlier.)

In the present context of tilings, we are interested in self-similar subdivision, and this requires that the relative numbers of tile types be maintained from one subdivision to the next. We have seen this described by a matrix equation $\mathbf{M}v = \lambda v$, and now to solve this for λ and v, we introduce the identity matrix whose diagonal entries are ones, with all other elements zero. For example the 3×3 identity matrix is

$$\begin{pmatrix} 1 & 0 & 0 \\ 0 & 1 & 0 \\ 0 & 0 & 1 \end{pmatrix}.$$

However, for our simple example, we only need the 2×2 version:

$$\mathbf{I} = \begin{pmatrix} 1 & 0 \\ 0 & 1 \end{pmatrix}.$$

Now, using this $\mathbf{I}$, we can rearrange $\mathbf{M} v = \lambda v$ in the form

$$(\mathbf{M} - \lambda \mathbf{I}) \, v = 0$$

$$\left(\begin{pmatrix} 5 & 8 \\ 8 & 13 \end{pmatrix} - \lambda \begin{pmatrix} 1 & 0 \\ 0 & 1 \end{pmatrix} \right) v = 0$$

$$\left(\begin{pmatrix} 5 & 8 \\ 8 & 13 \end{pmatrix} - \begin{pmatrix} \lambda & 0 \\ 0 & \lambda \end{pmatrix} \right) v = 0$$

$$\begin{pmatrix} (5 - \lambda) & 8 \\ 8 & (13 - \lambda) \end{pmatrix} v = 0.$$

And for non-zero v, this means that the resulting matrix $(\mathbf{M} - \lambda \, \mathbf{I})$ is singular [876], and therefore its determinant[8] is zero, which allows us then to derive values for λ.

$$\begin{vmatrix} 5 - \lambda & 8 \\ 8 & 13 - \lambda \end{vmatrix} = 0$$

$$(5 - \lambda)(13 - \lambda) - 8 \cdot 8 = 0$$

$$\lambda^2 - 18\lambda + 1 = 0.$$

---

[8] Formula on page 485.

Hence, using the quadratic formula we find the eigenvalues, and then using the Table of powers of ϕ (p.478), we can express these $\sqrt{5}$-based values in terms of ϕ:

$$\lambda_1 = 9 + 4\sqrt{5} \quad = \phi^6$$

$$\lambda_2 = 9 - 4\sqrt{5} \quad = \phi^{-6}.$$

If we now substitute λ_1 back into the matrix, the null result of the transform will expose the relationship between the bowties and hexagons

$$\begin{pmatrix} (5 - \phi^6) & 8 \\ 8 & (13 - \phi^6) \end{pmatrix} \begin{pmatrix} B \\ H \end{pmatrix} = 0.$$

That is, considering the dot product using the top row of the matrix:

$$(5 - \phi^6)B + 8H = 0$$

$$\frac{B}{H} = \frac{-8}{(5 - \phi^6)}.$$

Then using $\phi^n = F_n\phi + F_{n-1}$ [491], $\phi^6 = F_6\phi + F_5 = 8\phi + 5$, so

$$\frac{B}{H} = \frac{-8}{(5 - (8\phi + 5))} \qquad = \frac{1}{\phi}.$$

This same ratio is obtained using the bottom row of the matrix instead—then giving

$$8B + (13 - \phi^6)H = 0.$$

In the above we have worked through the mechanics of how Lu & Steinhardt got their result—a result which demonstrates that with repeated application of the subdivision process, the ratio of hexagons to bowties will tend to ϕ, the Golden Ratio. The same ratio is obtained on repeated subdivision of Penrose P2 and P3 tilings (using their respective deflation rules) [586].

In considering tile-to-tile ratios, Peter Cromwell points out that it is the eigenvector with the greatest eigenvalue which contains the relative tile frequencies for a 'full tiling' [151]. This limit is the result of an indefinitely repeated inflation process of subdivision and scaling back up to original side lengths, done in order to produce an infinite tiling of the plane [149]. Therefore, our key conclusion from the above analysis is that it demonstrates irrationality—the signature of quasicrystallinity.

Subdivision scaling factor

In his detailed 2008 review and further analysis, Cromwell also considers: the Gunbad-i-Quābūd, several other sites, and an expanded set of girih tiles [148, 147]. He notes that the largest eigenvalue of a subdivision transformation matrix is the square of the scale factor (between the original elements and those in the subdivision) [150]. If we apply this to the above Darb-i-Imam matrix **T**, the largest eigenvalue is $\lambda = 4\phi^6$, and so the scale factor $\sqrt{\lambda} = 2\phi^3$. Cromwell revisits this spandrel, and although his choice of hexagon (aka bobbin) subdivision is slightly different, in this case it does not affect the overall scheme, as he also finds the scale factor to be $2\phi^3$ [153]. Using the same technique, Cromwell also analyses the Topkapı scroll where he discovers a different subdivision. This is consistent across 4 panels and has a scale factor of $2\phi^2$ [154].

The old and the new

In Fig. 217 we make an informal visual comparison between the 10-fold rotational symmetry found in an example quasicrystal diffraction pattern and a medieval Islamic decagonally based tiling pattern from Samarkand, Uzbekistan. Top left we see again the electron diffraction pattern of an icosahedral Zn-Mg-Ho quasicrystal and top right, a decoration at the Tuman Aqa complex at Samarkand, Shah-i-Zinda, Uzbekistan. Centre top is an example overlay, and the larger image below is another overlay (rotated). Given that we are concerned here with 5- and 10-fold symmetry and these examples are chosen to reflect that, then it is hardly surprising that this aspect of the design 'lines up'. However, it is somewhat more surprising how well (subjectively speaking) the key points in the Islamic design apparently correspond to the radial sequences of *ABC* Golden proportions that we highlighted in Fig. 206 on page 330.

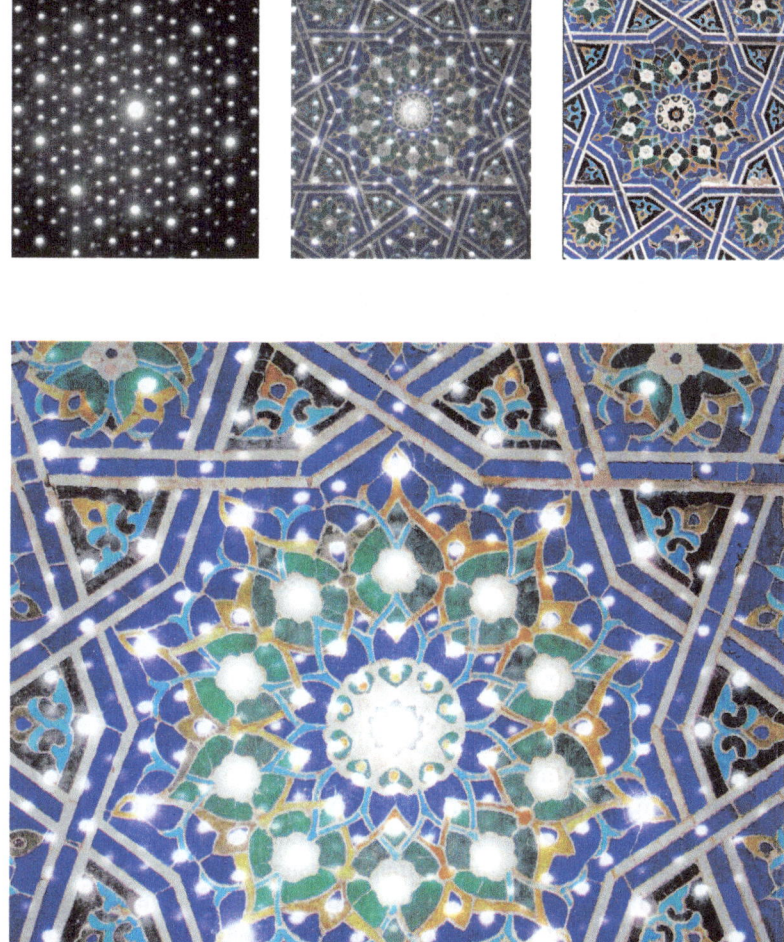

Figure 217: **Quasicrystal diffraction & medieval Islamic tiling.** Diffraction image by Materialscientist [632, 144]. Samarkand image by Patrickringgenberg [731, 144].

8-Fold symmetry too

In the conclusion of his review, Peter Cromwell mentions Islamic designs in Spain and Morocco that are built with local 8-fold symmetry [155].   He also notes a strong resemblance to the Ammann-Beenker quasiperiodic tiling which we saw back on page 337. Also, researchers Fernando & Molina associate 8-fold designs (such as that in Fig. 218) with Moroccan tradition in the Maghreb (Arab regions of North-western Africa) and 9- and 10-fold with Ottoman and Persian influences in the Arabian Gulf area [214].

Figure 218: **Example 8-fold design.** Image © 2011 Nomad Inception [214]

Advanced capability

In summary, it will by now have become clear to us that the Darb-i-Imam designers knew exactly what they were doing in order to get advanced designs such as their 10-fold locally symmetric tiling with its subdivision scheme. And although some have suggested that the Darb-i-Imam pattern was arrived at by accident, the component elements of the design are found in earlier works, and these are brought together to give a potentially quasicrystal structure. Having said that though, we would not expect the designers to have described their great achievement using the machinery of modern mathematics. And as a consequence of that, we would surely not expect them to have developed the kind of advanced analysis and theorems (such as those of Penrose and Conway)—that we summarized under *Properties of Penrose tilings* on page 318.

Penrose, Mackay, quasicrystals, and Islam

Fig. 219 provides a pictorial summary of the last three chapters.

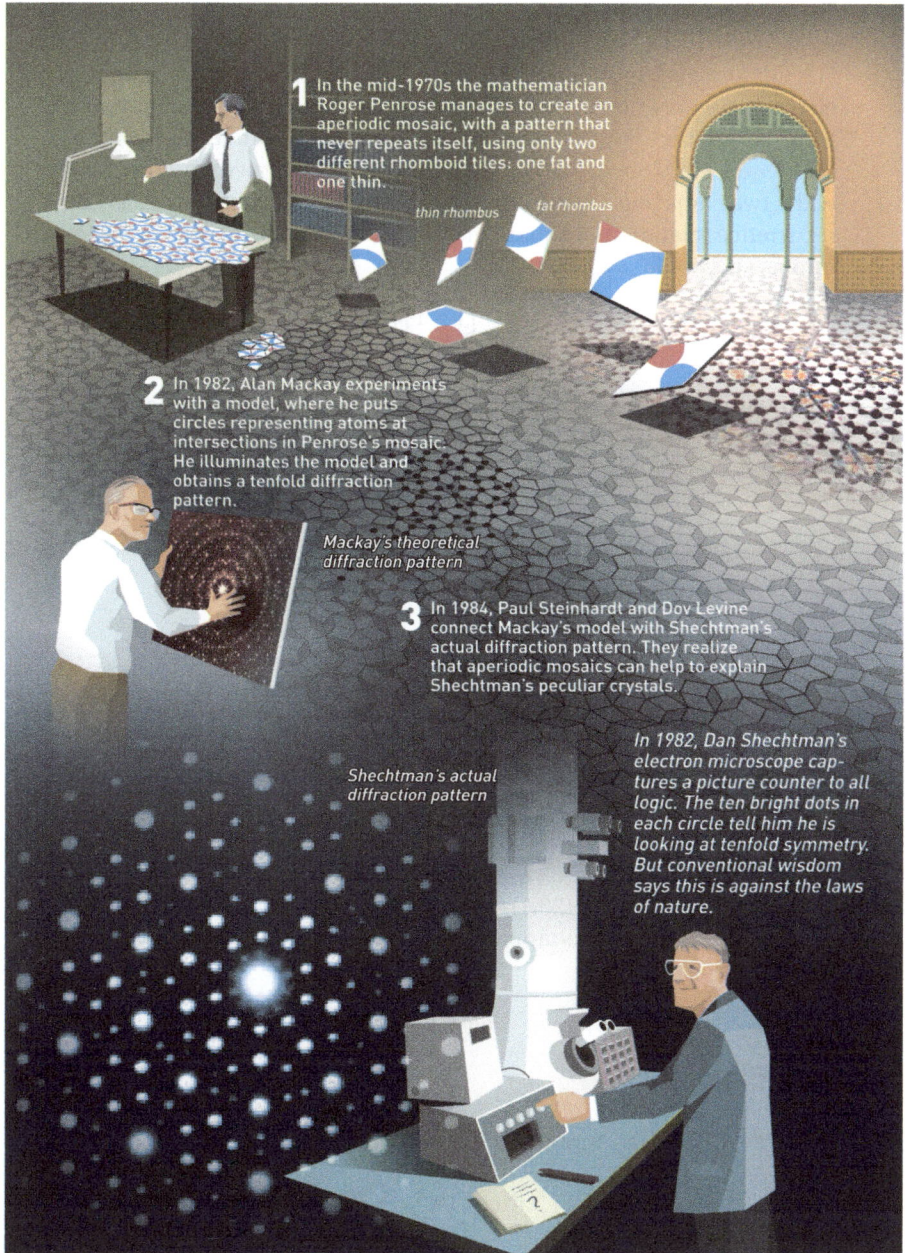

1 In the mid-1970s the mathematician Roger Penrose manages to create an aperiodic mosaic, with a pattern that never repeats itself, using only two different rhomboid tiles: one fat and one thin.

thin rhombus   *fat rhombus*

2 In 1982, Alan Mackay experiments with a model, where he puts circles representing atoms at intersections in Penrose's mosaic. He illuminates the model and obtains a tenfold diffraction pattern.

Mackay's theoretical diffraction pattern

3 In 1984, Paul Steinhardt and Dov Levine connect Mackay's model with Shechtman's actual diffraction pattern. They realize that aperiodic mosaics can help to explain Shechtman's peculiar crystals.

Shechtman's actual diffraction pattern

In 1982, Dan Shechtman's electron microscope captures a picture counter to all logic. The ten bright dots in each circle tell him he is looking at tenfold symmetry. But conventional wisdom says this is against the laws of nature.

Figure 219: **Dan Shechtman, Nobel Prize in Chemistry 2011.** Illustration © Johan Jarnestad, courtesy of The Royal Swedish Academy of Sciences [215].

357

The Fibonacci matrix

Having touched upon matrices, eigenvalues, and eigenvectors in this chapter, we may now consider the Fibonacci recurrence in its matrix form. As we know, this recurrence depends on two predecessor values for its next step. So, to encapsulate these, we make a column vector v with two adjacent values F_{k+1} and F_k, and then express the recurrence relation as a square matrix $\mathbf{A}$ that transforms such a vector into its successor, $v_{j+1} = \mathbf{A}v_j$:

$$\begin{pmatrix} F_{k+2} \\ F_{k+1} \end{pmatrix} = \begin{pmatrix} 1 & 1 \\ 1 & 0 \end{pmatrix} \begin{pmatrix} F_{k+1} \\ F_k \end{pmatrix}.$$

To find the eigenvalues of this matrix $\mathbf{A}$, we use the identity matrix $\mathbf{I}$ as before, with a generalized vector w (that need not have integer elements), so that $(\mathbf{A} - \lambda \mathbf{I})w = 0$,

$$\left(\begin{pmatrix} 1 & 1 \\ 1 & 0 \end{pmatrix} - \lambda \begin{pmatrix} 1 & 0 \\ 0 & 1 \end{pmatrix} \right) w = 0 \qquad (28.1)$$

$$\begin{vmatrix} 1 - \lambda & 1 \\ 1 & -\lambda \end{vmatrix} = 0$$

$$(1 - \lambda)(-\lambda) - 1 \cdot 1 = 0$$

$$\lambda^2 - \lambda - 1 = 0.$$

Again, this is that (so familiar) quadratic whose roots are the Golden Ratio and its conjugate. Therefore the eigenvalues of the Fibonacci matrix $\mathbf{A}$ are $\lambda_1 = \phi$ and $\lambda_2 = -\phi^{-1}$. Then to find the eigenvectors, we simply plug λ_1 into (28.1)—with say w= column vector (x, y), and solve the resulting simultaneous equations (then repeat similarly for λ_2) to get

$$w_1 = \begin{pmatrix} \phi \\ 1 \end{pmatrix} \quad \text{and} \quad w_2 = \begin{pmatrix} -\phi^{-1} \\ 1 \end{pmatrix}.$$

Exercise

Exercise 8. Prove by induction that if

$$\mathbf{A} = \begin{pmatrix} 1 & 1 \\ 1 & 0 \end{pmatrix} \quad \text{then} \quad \mathbf{A}^n = \begin{pmatrix} F_{n+1} & F_n \\ F_n & F_{n-1} \end{pmatrix}. \qquad (28.2)$$

No solution is given for this exercise. Matrix formulæ (including matrix × matrix) may be found on p.485. Hints: For the base case, check (28.2) with n set to a small number; then for the inductive step, write (28.2) in j instead of n, assume it is true, and evaluate $\mathbf{A}\mathbf{A}^j$.

Chapter 29

Superlattices and metamaterials

New ways of thinking

A (surprisingly quiet) revolution is taking place in physics and materials science that will soon touch many aspects of our lives. In order to position ourselves to understand the role of ϕ and the Fibonacci recurrence in these new developments, in this chapter we shall *very briefly* review key principles and applications. And while some of the subjects have long, even very long histories, they are now benefitting from new theoretical insights—'their time has come'. These new ways of thinking have sparked a frenzy of research and development, and certain new areas have suddenly generated a vast literature. As a result, we must skim to a great extent and take a simplified view—in order to move on to the next chapter. There, we shall see how the direct application of Fibonacci- and ϕ-based structures and techniques adds a host of extra 'levers to pull and buttons to press' in this new family of technologies.

Here, we shall major on the latest light-based advances, but we shall also note how some of the new concepts have important applications in heat and sound engineering, silicon electronics, local control of ocean waves, and even the protection of buildings from earthquakes.

Resonance

We met a basic example of resonance in the *Art* chapter—that of bathwater sloshing back and forth in the bathtub. This serious scientific experiment very simply illustrates the key features of resonance. By accelerating the water forwards we give it 'kinetic energy'.[1] And when the moving water meets the end of the tub, it heaps up—accumulating 'potential energy'[2] and slows to a standstill. But, because there is nothing to hold it (as in a cartoon animation), it then starts to fall and rushes back towards the other end of the tub, only to repeat the same sequence. Similarly, with a garden swing, we see just the same cyclic interchange between kinetic and potential energy. As another very basic demonstration, we might 'twang' a rule (aka ruler) at the edge of a desk. In this case the energy interchange is between kinetic and the stored elastic energy which is at its maximum as the end of the rule slows to a stop just before it reverses direction.

Only an analogy

Some systems include more than two forms of energy, but we shall concentrate on those with just two—probably the most common. The other essential attributes of resonance are (1) an easy way for one form of energy to convert to the other and back, and (2) a particular time taken for a full cycle of the to-and-fro process. In contrast, for the Fibonacci Resonance, *the resonance is only an analogy*—a way of visualizing the quantized distances or abacus beads as standing waves in sound, comparing them with, for example, the pattern made by a violin string sounding a musical note (Fig. 220).

Figure 220: **Standing wave resonance, 2 wavelengths (= 4 halves).**

---

[1] The energy it has due to its mass and speed.

[2] The energy it has due to its mass having been lifted—work done against the force of gravity.

Maxwell, the speed of light, ϵ and μ

Figure 221: **James Clerk Maxwell—published a rich set of 20 equations.** Image: Popular Science Monthly, 1880 [997].

The three scientists that Einstein most admired were Newton, Faraday, and Maxwell. Einstein kept a picture of each on his wall in Princeton [974]. Isaac Newton had encapsulated the whole of classical mechanics in just three laws of motion.[3] Michael Faraday was one of the greatest experimentalists and visual thinkers—a pioneer in electromagnetism and the father of field theory [18]. And James Clerk Maxwell published 20 equations (in 1865), summarizing all he knew about electricity and magnetism. He wrote these equations in quaternions[4] and cartesian coordinates [639].

---

[3] When asked how he had made his discoveries, Newton said: 'Truth is the offspring of silence and unbroken meditation.' [116]—compare this with Patañjali [729].

[4] A number system discovered in 1843 by Irish physicist, astronomer, and mathematician William Hamilton. A quaternion has 4 elements $a + bi + cj + dk$, a, b, c, d in $\mathbb{R}$ and where $i^2 = j^2 = k^2 = ijk = -1$. Trying to find this consistent relation had tantalized Hamilton for some time, and upon its discovery (in his elation) Hamilton immediately carved this i, j, k relationship into Brougham ('Broom') Bridge in Dublin [966, 301]. A quaternion has a scalar part a and a vector part (b, c, d). Quaternions were recommended by 'Sage'—John Desmond Bernal [53] and Alan Mackay (p.300) for expressing the relative positions of molecules (distance and rotation) [613]. They also eliminate the 'gimbal lock' problem in computer graphics [16].

However, these equations were then significantly rationalized and restated in vector form—as just 4 elegant equations—by Oliver Heaviside in 1884.[5] It is Heaviside's four equations that are in near-universal use today, and it has long been fashionable not to distinguish, and to call these 'Maxwell's equations' [398]. In practical terms, Heaviside's was a great achievement—providing an easier-to-use basis for the development of electrodynamics.[6] One of the most interesting things about Maxwell's equations was that they predicted the speed of propagation of electromagnetic waves, and this matched the speed of light (now standardized as 299,792,458 metres per second, which is about 186,000 miles per second). It was in this way that the true nature of light was discovered and understood—a great triumph for mathematical physics. This speed simply depended on just two properties of the medium (or empty space): the electrical response[7] and the magnetic response[8] (and *nothing else*[9]). These values are represented by Greek letters epsilon (ϵ) and mu (μ), and each of these may be positive or negative; but, in nature, ϵ and μ are never found both negative at the same time.

IoR—Index of refraction

In Fig. 222 we see the familiar bending of a ray of light when it impinges on a transparent material such as a slab of glass or a tank of water—a material with a positive index of refraction. When moving from air to glass (for example), light is slowed down. If a light ray arrives at a glancing angle as in Fig. 222, then the beam is bent downwards. The exact amount of bending is predictable according to Snell's law (aka *la loi de Descartes*)—a law that was first discovered in about 984 AD by Ibn Sahl, a Muslim Persian mathematician [470].[10] The 'index of refraction' (IoR) measures an optical material's ability to slow down light waves. Therefore, the bigger the index of refraction, the greater the bending effect.

---

[5] Heaviside did his work in parallel with that of Gibbs and Hertz.

[6] But, more and more people believe that Heaviside oversimplified Maxwell's work—particularly with his deliberate elimination of Maxwell's scalar and vector potentials [367, 199]. Yakir Aharonov & David Bohm for example, discussed how these potentials could be shown to have independent significance [11] and the 'Aharonov-Bohm effect' has since been confirmed experimentally [745].

[7] The permittivity.

[8] The permeability.

[9] Hence special relativity.

[10] Abu Sa'd al-'Ala' ibn Sahl.

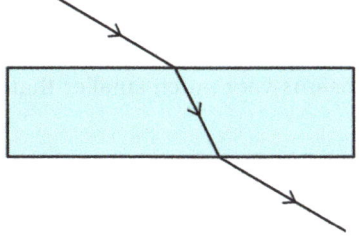

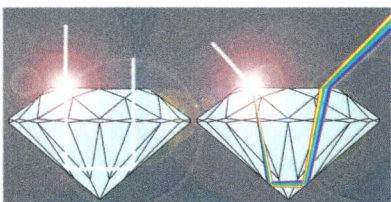

Figure 222: **Refraction—positive index material.**

Figure 223: **'Brilliance' and 'fire' in diamonds.** Principles only, angles approx.

Water has an IoR of 1.33 compared with air at 1.00003 and it is this contrast in indices that causes a straight straw to appear kinked as it is put into water. The index of refraction 'n' depends on ϵ and μ only: $n=\sqrt{\epsilon\mu}$. This is not a coincidence—as we noted, ϵ and μ determine the speed of propagation of the light. Diamond has the unusually high IoR of 2.4, giving a high contrast between its index and that of air. This in turn makes internal reflection very efficient—and also, high refraction combined with an ideal cut[11] results in double reflection (internally) that routes incoming light back out again towards the viewer, giving the diamond its characteristic sparkle—its 'brilliance' (Fig. 223, left). Also the high refractive index produces dispersive refraction, which splits white light into flashes of colour—the 'fire' in the gem (Fig. 223, right).

Thinking sub-wavelength

Until recently, designers of optical equipment have had to accept that (using the best theory available), their ability to focus light would be limited. The barrier they faced is known as 'the diffraction limit'. This was studied in depth and experimentally confirmed by Ernst Abbe in 1873.[12] Abbe became Carl Zeiss's partner at his Optical Works (now Carl Zeiss AG). Together they produced state-of-the-art microscopes based on Abbe's advanced optical knowledge. According to the limiting theory, the smallest dot into which light could be focused would be about half a wavelength wide.

---

[11] With the pavilion (conical rear body) not too shallow, not too deep.

[12] The so-called 'Abbe diffraction limit formula' itself was noted a year later by Helmholtz, but he credited it to Lagrange who had died 61 years earlier [332].

However, as we shall see shortly, with new ways of thinking and modern nano-scale fabrication techniques:

- Light may be confined to dimensions very much smaller than half a wavelength.

- Deeply sub-wavelength structures may be made that nevertheless interact strongly with electromagnetic waves to produce exotic results.

- In particular, non-magnetic metals can produce a significant magnetic response (using novel *geometry* at a suitable scaling).

Photonics—advanced light technologies

Photonics is 'the new optics'. It takes into account the particle/ quantum aspects of light, but not exclusively so—much of the work still relies on classical wave theory. Its name is intended to underline an analogy with electronics. Some semiconductor concepts cross over as is, and some are extended and generalized. It is increasingly connected with information, communication, and computing. As a practical field of research, it effectively began with the invention of the laser by Theodore Maiman in 1960 [615], though the basic 'stimulated emission' laser theory had been given by Einstein in 1916 [200]. Photonics is now a substantial hi-tech industry concerned with the generation and manipulation of 'light'—here understood as electromagnetic radiation, visible or otherwise (e.g. infrared) in:

- lasers, light emitting diodes (LED's)
- optics, imaging, microscopy, spectroscopy
- fibre optic telecommunications: emission, transmission, modulation, signal processing, switching, amplification, storage, and detection of light[13]
- silicon photonics, photonic integrated circuits (PICs)
- holography, data recording, photonic/quantum computing
- sensors and detectors, optical coatings, test and measurement, camera components, night vision
- medical diagnosis and treatment
- light harvesting ('green' conversion of light to electricity)

---

[13] Reducing or eliminating slow, physically large, and energy-hungry electronic elements—moving towards 'greener', ultra-fast, ultra-high-bandwidth, 'all-optical' systems.

Band gaps

Let's compare three materials. Our first is a slab of iron—simply opaque to light. Our second is window glass—clear and translucent. But the third is much more complex—and all the more interesting for that—a filter attachment for a camera lens. Such a filter will typically let some colours through but not others; it may even have an extremely intricate pattern of behaviour, custom-designed for some specific use. When a light filter stops only certain ranges of wavelengths while allowing the rest to pass—it is said to have 'band gaps'. This basic concept has now been applied across a wide range of phenomena. In general, a gap is a set of energy states that cannot exist—ones not allowed as a result of the properties of the system being studied. 'Band gaps are important', and although the concept originated in condensed matter physics—particularly in semiconductor theory—it is now also applied to electromagnetic responses (e.g. to light) and to elastic vibrations such as sound and heat. Band gap phenomena are usually associated with, or may be entirely dependent on, *resonance* effects. A further point though, is that band-gap properties may depend on the direction of the incoming energy (photon, vibration wave bundle, or the like) as it approaches the material. The response in one direction may be very different to that in another. As a result, the design goals for a band-gap material often include trying to get a single even response, regardless of incoming direction. Such a response is called 'isotropic'.

Nature already has examples of visible spectrum band gaps—for example the colours seen on butterfly wings and bird plumage are often not just due to pigmentation. Finely structured surface details produce colour by light interference. And as the band gaps formed are not complete in 3D terms, they result in iridescence. The *Morpho peleides* butterfly has been closely studied, and its nano-scale surface structure has been successfully replicated in aluminium oxide. Such mineral copies still exhibit the waveguide and beam splitting effects that were present on the original wing [422, 178]. Similarly, the brilliant iridescence of opal stones arises from sub-micron sized silica spheres (SiO_2) arranged in a magnified crystal structure (called a 'superlattice'—more on this shortly), again, the effect of incomplete band gaps. Remarkably, the inverse structure—with a high index of refraction 'fill' and with air-hole spheres in place of the silica spheres—*can* have a complete band gap [424].

Scaling structures for use with visible light

As we saw in the *Quasicrystals* chapter, the combination of the ordered structure of a crystal along with layer-to-layer distances of 'the right size' causes X-rays to be Bragg diffracted (p.327). Here 'the right size' means that the wavelength of the X-rays is comparable with the distance between the layers of atoms. Adjacent layers then cause rays to interfere with each other—constructively or destructively. However, this effect is utterly dependent on wavelength; and if, for example, we were to try the same experiment again using say green light instead of X-rays, we should not see any diffraction pattern. This is because visible light wavelengths are of the order of 1000× longer than those of X-rays. So, to get the adding and subtracting effect using green light, we shall need reflecting surfaces that are 1000× further apart—that is, we must match the scale of the structure to the wavelength of the light. (This recalls the situation we saw at the start of the *Is it fractal?* chapter, where long measuring rods could not 'interact' with the finer detail of the coastline—p.230). In practice, to position multiple reflecting surfaces correctly, two different materials are combined in alternating layers. And as this constitutes a new and higher level of 'lattice geometry' (each layer having its own atomic lattice), the term 'superlattice' is adopted.

Superlattices and photonic crystals

An important application of this superlattice principle is in the artificial construction of photonic crystals in one, two, and three dimensions. These typically consist of alternating layers of high and low dielectric constant (which is related to refractive index). Simple devices are periodic—that is, they have the same regular layer pattern repeated, where the layer thickness is typically one quarter or one half the wavelength of the light to be used. When light crosses a boundary from one index of refraction to another, it may be partly reflected, partly refracted and partly absorbed—and according to the specific situation, the amounts of each may be predicted by well known laws of physics. So, what will happen when we build up a periodic set of partially reflecting layers by regularly alternating layers of two kinds of material differing in index of refraction—'ABABABA… ', such as the leftmost example in Fig. 224? In 1887 Lord Rayleigh found that as long as the wavelength of the structure was close to half the wavelength of the light, then even with a very low contrast in the index of refraction layer to layer, he could obtain a *complete reflection*

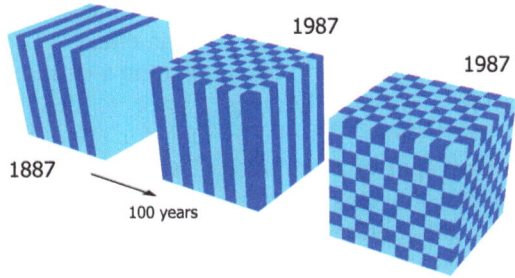

Figure 224: **Basic concept—1D, 2D, and 3D photonic crystal structures.**
1887: Lord Rayleigh, 1987: Eli Yablonovitch, Sajeev John.

effect [777]. However, 100 years then had to pass before Eli Yablonovitch (and independently and near simultaneously, Sajeev John) devised 2D and 3D versions of Rayleigh's 1D photonic crystal. Yablonovitch recalls using a chess-board analogy to envisage the structure (Fig. 224) [994]. This work built on that of both Rayleigh and C G Darwin (grandson of biologist Charles Darwin) [777, 161], and it put the photonic band gap on a level footing with the well understood semiconductor electron band gap. Rayleigh and Darwin had worked with low-contrast layers. Both Yablonovitch and John proposed combining 3D with high-contrast layering, in order to obtain the raft of phenomena already developed in semiconductor physics during the 1930s for electrons—but this time round—for light—for photons. Here was the birth of a new technology, delivered complete with an off-the-shelf analogue formalism 'to go'—a quantum leap indeed.

What are metamaterials?

Alongside the physical development of composite materials such as the photonic crystal, there has been a fundamental shift in thinking. This has been away from materials themselves (with characteristics fixed by physics and chemistry) and onto ways to shape and configure them—ways based on theory and intuitive 'what if?' experiments, ways limited only by imagination. The focus moved to abstraction, to fine geometry and its mathematical consequences—in order to alter bulk material properties significantly. There are various definitions of the term 'metamaterial'; but mostly it refers to artificial materials that achieve properties far exceeding those found in nature (*meta*: Greek, 'go beyond'). Typically they are designed according to a new 'sub-wavelength' paradigm: carefully fabricated micro- and nanostructures may be included within materials that significantly alter bulk material properties and produce exotic effects. The 'paradigm' word is important here, as apart from meaning a pattern or a model, it also connotes a way of thinking. Prior to this technological revolution, it was believed absolutely that, for example, light could not be focused to a spot smaller than half its wavelength, and that (for non-cephalopods at least) invisibility cloaking was strictly science-fiction. The metamaterial approach has turned out to have an unexpectedly wide scope—applicable in many 'wave' contexts, even including sound, heat, seismic, and ocean waves. The waves of greatest interest are the electromagnetic. These exhibit a vast range of vibration frequencies: from extra low frequency radio waves, through radio, microwave, terahertz, infrared, visible, ultra-violet, X-rays, and gamma rays (overall, about 14 orders of magnitude). Metamaterials may now be designed and constructed to slow light down, and to steer the path it takes precisely. This is done using exceptional indices of refraction. Metamaterials also offer the possibility of building invisibility cloaks and flat lenses. To appreciate basic metamaterial principles, we shall first consider how materials interact with light. As we noted earlier, the interaction depends only on ϵ and μ (the electrical and the magnetic responses), and in nature, ϵ and μ can each have a positive or a negative value, but they are never found both negative at the same time.

Victor Veselago

Russian physicist Victor Veselago (Fig. 225) was the first to consider: 'What if a material *could* have both negative ϵ and negative μ together? What properties would such a material have?' Veselago concluded that double-negative materials were indeed theoretically possible, and as he analysed, he discovered certain strange electro-magnetic consequences. He published these in his landmark paper of 1968 in which he called these hypothetical media 'left-handed materials' [927]. He noted that although substances with negative ϵ and negative μ together had never been observed, he regarded them as being: *'of undoubted interest'.* His paper includes the first description and diagram of a flat lens—simply a plate of left-handed material. In Fig. 226 we summarize material properties for each of the four cases of ϵ and μ, positive and negative. Many of the materials showing in the single-negative quadrants only do so in particular frequency ranges [841].

Figure 225: **Victor Veselago.** Image: Nicolas Guérin [287, 144].

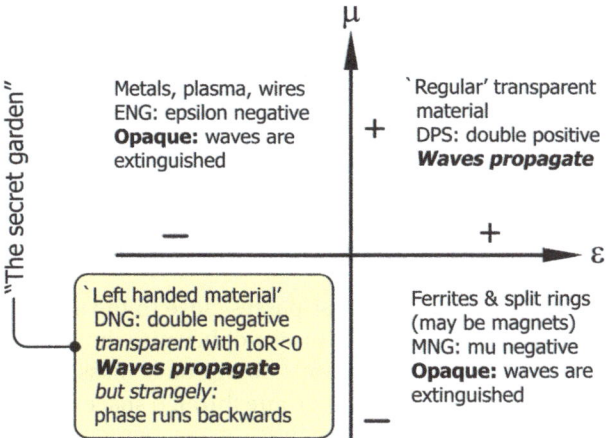

Figure 226: **4 regions—positive & negative values of ϵ & μ.**
(Sources: Soukoulis [851], Smith [841], Chen [113], and others).

Negative index of refraction

One application of being able to choose ϵ and μ, is to create a material that has a negative index of refraction. Such a material will have remarkable properties, and will facilitate the building of flat lenses, and super lenses that break the diffraction barrier.

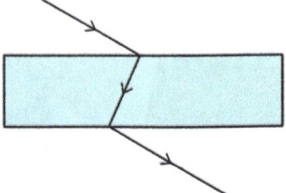

Figure 227: **Refraction—negative-index material.**

A signature property of negative index of refraction materials is that when light travels through them, the wave ripples travel 'backwards'.[14] We might imagine throwing a pebble into a lake, only to see the patch of disturbance expanding, while inside it, a succession of rings form on the circular edge and these rapidly shrink inwards to the centre as if time-reversed. A consequence of this kind of behaviour is that in 'left-handed materials' the Doppler effect is *reversed*.[15]

John Pendry

Given the absence of actual materials upon which to experiment, Veselago's work was largely forgotten. Though not forever—physicist John Pendry (Fig. 228) later described Veselago's special double-negative combination of ϵ and μ as: 'the secret garden of electro-magnetics—a dream that had to wait 30 years' [850, 733, 736].

Pendry's analogue gravity—transformation optics

When the light from a distant star passes close to the sun on its way to us, its path appears to bend. This is not because the light has mass and 'falls' towards the sun—photons do not have mass. Instead, it is because the light takes the shortest route through space—space which has been *distorted* by the sun's gravity.

---

[14]   The phase velocity is anti-parallel to the direction of energy flow.

[15]   An example of the normal effect would be the apparent increase in pitch of an approaching train whistle, and the decrease as it passes and speeds away.

Figure 228: **John Pendry.** Image courtesy of John Pendry, Imperial College London.

Einstein not only demonstrated this effect (during the eclipse of 1919)—he also gave an exact formula for the effective index of refraction of the curved space (at any point). Pendry's breakthrough was to realize how this transformation mathematics (from general relativity) could be directly applied in the design of metamaterials —to seize control of the routing of light. Pendry's transformation technique uses the space-bending-by-gravity analogy to provide a straightforward two-stage process which gives a recipe for how ϵ and μ should change in order to produce a required path distortion exactly [938, 808, 736].

Alongside this theoretical leap there were also major technological advances in fabricating metamaterials suitable for its exploitation. As a result it became possible to design and implement continuous media through which ϵ and μ changed smoothly according to a precise specification (as in Figs. 229, 230, and 231).

Split rings and rods

Once we know the recipe for the right values of ϵ and μ in the right places, then how can we apply it? One popular technique is that of 'split rings and rods'. These are tiny (sub-wavelength) electrically conducting shapes that interact with the electromagnetic field. When the 'I' rod is parallel to the electric field it couples; and similarly when the normal passing through the centre of the 'C' split ring is parallel to the magnetic field, it couples magnetically.[16] It is the detailed geometry of these shapes that delivers accurately predictable values for ϵ and μ. So in Figs. 229 and 231, we see theory then practice, where the rod and ring elements are efficiently combined. It is a great benefit to designers that the electric and magnetic parameters may be independently set. A defining feature of such configurations is the scale of the functional elements compared with the wavelength of the light. Whereas in photonic crystals we saw layer thicknesses comparable with the wavelength—metamaterials depend on elements that are markedly sub-wavelength.

Metamaterial superlenses and antennas

One of Veselago's important findings was the possibility of flat lenses with extraordinary properties, and in 2000, John Pendry published a paper on such 'superlensing' extending Veselago's work [734]. Reminiscent of the reaction that followed Dan Shechtman's discovery of quasicrystals, the announcement of 'left-handed materials' (later known as negative-index materials), was met with a wave of disbelief,

---

[16] As the magnetic component of electromagnetic radiation passes back and forth through the ring, it induces a current in the conductor which charges the capacitance formed by the gap. The combined inductance and capacitance form a resonant circuit with energy swinging back and forth between stored electrostatic and stored magnetic states. For a certain range of frequencies of operation, this arrangement will produce a significant effect on the permeability (mu) of the material. The smaller the gap the greater the μ. Also this magnetic interaction is achieved without the use of iron or any other such magnetic material. In a very similar way, an 'I' shaped conductor can interact with the electric field. In this case, the taller the 'I' the greater the ϵ (Fig. 229).

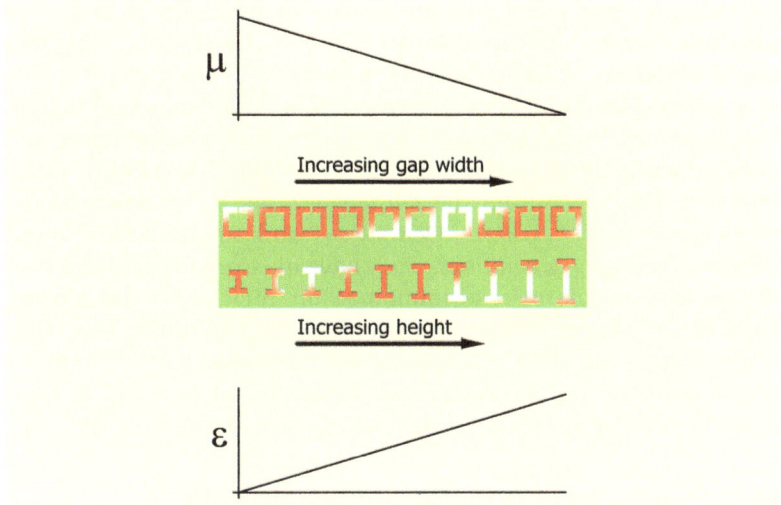

Figure 229: **Gradient metamaterials.** Diagram based on David R Smith presentation slide [843] with kind permission.

criticism and rejection. Some well known experts were initially confused by the possibility of negative index [842]. Then in 2003 Pendry wrote an article—*Positively negative*—which to a great extent settled the debate in favour of acceptance. Since then, Veselago and Pendry's predictions have been confirmed experimentally; and Pendry particularly credits Nicholas Fang, working at the time with Xiang Zhang at the University of California, Berkeley [208]. Resolutions of better than 1/20th of a wavelength are now possible—so there is no doubt that the diffraction limit has been well and truly smashed. However, Pendry notes that superlenses are a challenge to manufacture—being extremely sensitive to defects [735, 809]. In contrast, the application of metamaterials to satellite antenna design is well under way. First products are being hailed as disruptive technology—'game changers'—being flat, low weight, low cost, and ultra low power (a reduction of 3 orders of magnitude). With no moving parts, they are a step advance on traditional mechanically steered solutions—ideal for fixing to the tops of trains and boats and planes, even to cars—aiming for 'always connected' broadband [520].

Invisibility cloaking in nature

We shall presently take a look at an artificial invisibility cloak for use with microwaves. But first, let us consider the cloaking examples found in nature. A basic example is that of an insect whose colour exactly matches the leaf it is feeding on; so it is hard for a bird (say) to recognize it and 'meet for lunch'. A step up from this is the striped and dappled camouflage of tigers and giraffes—allowing them to blend into their habitat. But for the *masters of disguise,* we return to the cephalopod family (p.81). Octopusses and squids have evolved *advanced cloaking capabilities.* They have the extraordinary ability to display a view of their surroundings on their skin—the shapes, colours, brightnesses, even the textures. And by doing this, they *simply disappear* [233]. These are truly extraordinary creatures— whose 3 hearts pump copper-based blue blood (our red is iron-based)—and that get about using their own jet propulsion [989].

Invisibility cloaking with metamaterials

By gaining unprecedented power to control the path of the light with mathematical precision, the science fiction 'invisibility cloaking' (for humans) might one day become science fact. As with all such major goals, it is best to 'start simple' and build from there. In the world of metamaterials, this means starting with relatively large structures and studying their effect on correspondingly large wavelengths—such as microwaves (in the centimetres) [807]. The goal then is to scale down successively: into the infra-red (microns), and eventually into visible light wavelengths (hundreds of nanometres). Ideally a cloak should bend light (or initially, microwaves) round an object so that to the viewer, the object is no longer blocking light rays from behind—so it disappears. A similar re-routing happens in a mirage— when the road is cloaked in hot air, rays from above are bent towards us and we see the sky instead, giving a watery effect. With a cloaked object, we look in the direction of the object, but we only see the image directly behind it. The trick then, is not just to take in the light at the rear surface, steer it round and output it again—in addition, the light must be *accurately* radiated towards us just as if the obstruction was absent (as in Fig. 230). John Pendry refers to this as: 'making light flow like water around the hidden object'. Pendry expected some attention when in 2006 he started talking about invisibility cloaking. But he was not prepared for the vast publicity that he received—the idea really gripped people's imaginations [979].

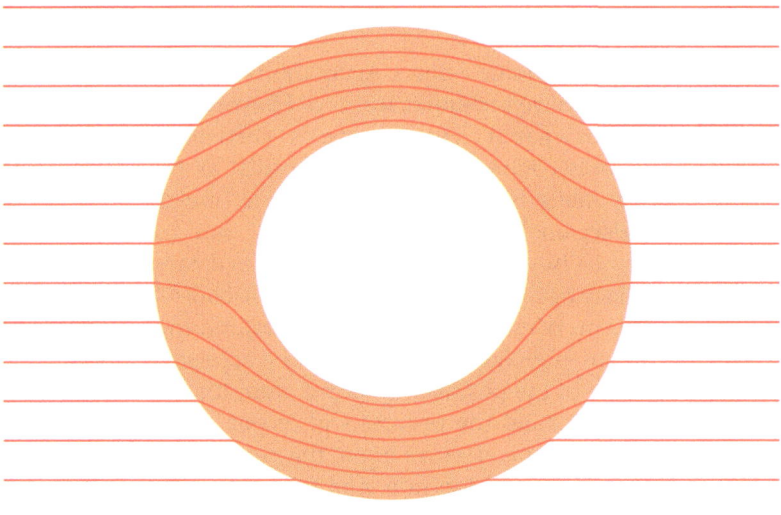

Figure 230: **Invisibility cloaking.** Illustrating the main concept using ray-trace formula given by Schurig et al. [808].

Figure 231: **Invisibility cloak** designed for use with microwaves [807]. Picture courtesy of Duke University. Also, Narayana & Sato have since demonstrated DC magnetic field cloaking using a stacked superconducting metamaterial [667]. There is now significant interest in terahertz-metamaterial applications including quantum cascade lasers ('QCLs') [831, 887] and in optical metasurfaces: sub-wavelength-thickness arrays of sub-wavelength antennas or resonators—a new class of flat and compact optical components for 'wave-front engineering', e.g. lenses, polarizers, and filters [99, 256].

Plasmonics—when light sticks to metal

Next we turn to 'plasmonic' effects. Again this is in preparation for the final chapter where we see applications of Fibonacci and ϕ. We may think of plasma as a gas of charged particles. This can result from very high temperatures—stars such as our sun exist as balls of plasma. However, our main interest here will be in metals, where the conduction electrons can move around.[17]   But as an electron separates from its host atom, an electric force is created that tries to accelerate it back again. Just as in the resonant-bath-water example, we have two forms of energy (this time electrostatic and kinetic), plus a way for them to interchange. As the electron accelerates back in the general direction of the positively charged atom (i.e. ion), then it will typically overshoot, and the process will repeat. As a result of the average separations, the electron mass, the charges and so on, there will be a particular oscillation rate, and this is known as the plasma frequency—typically in the ultra-violet for metals. A plasmon is a single quantum of plasma oscillation in the same way that a photon is a quantum of electromagnetic radiation (e.g. light).

Plasmonics is largely (but not exclusively) about the way that photons interact with metals—'metal optics'.   Although plasmonic resonance has been theoretically understood only in recent times, its practical application has been known since at least Roman times for colouring glass.   When light interacts with sparse nanometre sized particles of noble metals that have been mixed in during glass manufacture, interesting colour effects may be obtained.[18] The Romans were able to make glass that included (approximately 70nm) nano-particles of noble metals gold and silver well dispersed ('in colloidal form') throughout the product [30]. This may give the glass the unusual 'dichroic' property that when lit from the front it is pea-green, but when back lit, it is deep wine-red (Fig. 232).   The particles that produce this effect are so small that they cannot be seen using an optical microscope: they are a number of times smaller than the wavelength of blue light. Also the size of the metal particles is critical for the required interaction between light and electrons in the surface.   And as a similar example: for a long time it was a mystery how the medieval stained glass makers achieved the exceptional purity and brilliance of colour that can be seen in say, the windows of Notre Dame cathedral (Fig. 233).

---

[17]   As there is a high density of free electrons, it is probably best (in this case) to think of them as a liquid rather than gas.

[18]   *Nature Photonics* published a review issue, focused on plasmonics [672].

Figure 232: **Roman plasmonics—the Lycurgus Cup.** Images © The Trustees of the British Museum, all rights reserved [87].

Figure 233: **Medieval plasmonics and Ori(2^n) geometry—the Rayonnant north rose window of Notre Dame Cathedral, Paris.** Derived from image by Krzysztof Mizera [653, 144].

We now know that this was achieved using the very same plasmonic technology of evenly dispersing nano-particles of gold throughout the glass. In addition to its religious significance, the rose window design is a celebration of the powers of two, with the fifth power—32—being quoted twice. We see one centre element surrounded by 8 elements, then 16 inner radials, and 32 outer radials, with these all surrounded by an outer ring of a further 32 elements. We could argue that these special glasses (the Roman and medieval examples) count as metamaterials. This is not because they rely on ingenious geometric shapes for their unusual properties—rather it is because they use simple, but deeply sub-wavelength inclusions to alter the effective electromagnetic properties noticeably.

Now, when a photon with the right energy collides with the surface of bulk metal from a certain angle, it can (in a manner of speaking) 'stick to the surface', then proceed along it for a while in a specially constrained way. It is a little like a raindrop pelting against a glass window. Once it has hit, the drop moves down much more slowly along the surface. The photon collision initiates a ripple of charges in the surface called a 'surface plasmon polariton'. Further, the ripple wavelength is very much smaller than the incoming wavelength although the vibration frequency stays the same. This effect is tremendously important as it breaks through the 'diffraction limit' —that is, Abbe's law that light cannot be focused by normal means to a dot much smaller than half a wavelength. In Fig. 234 we see an incoming light photon coupling to a metal surface to create a 'surface plasmon polariton' (SPP). The crests of the wave motion of the free electrons travel along in the surface of the metal in lock-step with the now confined light wave in the air above. The resulting SPP wavelength is surprisingly short—10 to 100 times smaller than that of the incident light ray. To incoming light (say visible) it appears that the metal surface has a very high index of refraction, giving a 'slow light' effect, and very short wavelengths—for example those normal for X-rays. The plasmonic resonance is extremely sensitive to changes in the surrounding environment, and this property can be used in advanced sensing applications, for example in biosciences—registering binding events in real time, from ions to viruses [376]. Plasmonics is also providing unprecedented advances in spectroscopic analysis [556]. In computing and fibre optic telecommunications there has been a scaling mismatch between optical elements (hundreds of nm up to 1500 nm) and silicon chip fabrication (with its tens of nm scale).

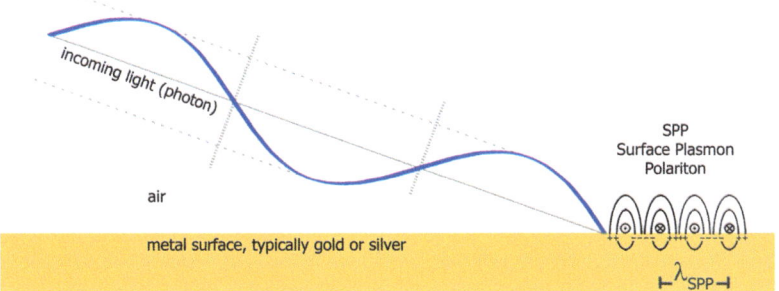

Figure 234: **Metal optics—the creation of a surface plasmon polariton—an 'SPP'.** λ_{SPP} is SPP wavelength.   Incorporates sine wave by Omegatron [713, 144].

Now, using the plasmonic coupling effect—Fig. 234, light can be confined into much smaller geometry. This breakthrough will provide new ways to get information from different parts of a silicon chip to where it is needed, and very quickly too. Graphene is emerging as a very attractive material for use in combinations of plasmonic and nano-optic applications. Graphene is a flat crystal one atom thick, with carbon atoms arranged in a honeycomb lattice (we mentioned this structure on p.332). Used in a nanotechnology context, it is tunable, ultra-fast, durable, flexible, ultra-thin, and suitable for large-scale processing with nanometre-scale patterning [505].

Extending beyond E-M

Until recently, metamaterials have been mostly associated with electromagnetism—light, infra-red, microwaves, and so on. However, the concept has quite general application as its purpose is to control and manipulate wave and wave-like interactions (using sub-wavelength and deeply sub-wavelength inclusions).[19] In addition to electromagnetic waves, the concepts may be applied to heat, sound, fluid dynamics, mechanical materials, seismic waves, and ocean waves, and even to artificial liquids ('pentamode'/meta-liquids).

---

[19]   Electron behaviour in a crystal is characterized by the Schrödinger equation, photon behaviour in a photonic crystal by Maxwell's equations, and mechanical waves in a phononic crystal by the elastic wave equation [871].

Phononics and acoustics—
advanced mechanical wave technologies

Green light and X-rays are very different things. But, looked at from the point of view of electromagnetic radiation of photons, they are essentially the same thing but with different energies, different rates of e-m field vibration. Similarly, we think of sound and heat as being very different, but these also may be considered to be basically the same—in that each is a mechanical vibration. Again the big difference is in the rate of vibration. Sound (as all who are serious about 'hi-fi' will know) ranges from around 20 Hertz (vibrations per second) up to 20,000 Hz. Slower vibrations are referred to as 'subsonic' or 'infrasound'. At the other extreme, heat vibrations are measured in the terahertz (0.1–100×10^{12} Hz). Between the sound and heat regimes are 'ultrasound' (vibrations at around 1MHz) and 'hypersound' (in the gigahertz). And just as light may be thought of in terms of particles (photons), both heat and sound are carried by bundles of vibration energy called 'phonons'. Hence the associated field of study and technology is known as 'phononics'.

However, because phonons have no mass or charge, they are difficult to control. (We continue to build our list of 'onics'—candidates for enhancement using Fibonacci and ϕ in the next and final chapter: *Quasicrystal '-onics'.*) Sigalas & Economou began studies on phononic crystals in 1992 [833, 834], and in 1993, Kushwaha et al. made full band-structure calculations [519]. By analogy with photonic crystals, phononic crystals are made by periodically varying the acoustic properties—for example by alternating layers which differ in elasticity and mass density.[20] This can give rise to a useful band gap, preventing a range of frequencies or wavelengths being propagated through the material. The band gaps occur for much the same reasons that we saw in photonic crystals—that is, the faces where the different materials meet cause reflection of the waves, and these (as we saw in Bragg diffraction Fig. 201 on page 327) can add and subtract. Again it is the spacing of the layers and the wavelength and direction of the incoming waves that determines the interference effect—destructive where waves are cancelled, or constructive where they are reinforced. Phononic crystals will provide acoustic engineers with the kind of control and flexibility provided by lenses and mirrors in optics.

---

[20] In acoustic metamaterials, the bulk modulus and the mass density provide the equivalents of permittivity and permeability in e-m metamaterials.

And just as with photonic crystals, the scale of the geometry needs to be comparable with the wavelength of the vibration. For sound and ultrasound applications, structural periodicity of 10 cm to 1 mm is appropriate. But for high-frequency phonons (10^9–10^{12} Hz), sub-micron scaling is needed [619]. Gaining control over phonons allows the design and fabrication of novel structures that can function as sonic filters, waveguides, sonic 'diodes' (one-way transmission channels), and resonant cavities, vibration filtering and control, even acoustic cloaking devices (analogous to Pendry & Smith's microwave cloaking device)—e.g. García-Meca et al. [236, 235]. Further, thermal metamaterials are being developed for the management of heat conduction and the generation of electricity from waste heat (thermoelectricity) [620]. An important application of phononic crystals is the conversion of acoustic and vibration energy into electricity, with the potential to power wearable and portable gadgets, and small autonomous sensor devices [113].

Seismic cloaking

Following the success of metamaterial research on small scales, teams started to look at scaling up designs to interact with much larger waves such as seismic and ocean. Perhaps seismic metamaterial structures could route earthquake waves around a building, saving it from destruction? The problem with seismic wave control is that the Earth's surface is highly irregular in terms of its density and elastic properties. However, a French team led by Stéphane Brûlé performed a successful 'proof of concept' experiment near the ski-city of Grenoble in August 2012. They drilled 3 lines of empty boreholes 5 metres deep, then vibrated the ground on one side and measured the attenuation on the other [91]. Later in the same year (December 2012), Torres-Silva and Cabezas published their study based on an analogy between Maxwell's equations and elastic theory equations, to suggest a practical method for absorbing seismic waves, suitable for protecting buildings [901]. Their approach would be to build a metamaterial ring around a building, which then requires available space. Hence this technique is not applicable directly to say, a building in a city, but it *is* suited to isolated sites—such as a dam or power station, airport, and so on. The main advantage is in not having to change the structure that is being protected in any way.

Taking a different tack, Kim and Das have developed a method that produces a seismic shadow using acoustic metamaterials. This shadow protects not just those buildings immediately behind, but ones behind those too. The technique relies on burying huge empty boxes that have a few side holes—designed to resonate with the seismic waves and dissipate seismic energy as sound and heat [472].

Ocean waves and tsunamis

By using the cloaking approach, it may be possible to shield vulnerable strips of coastline from dangerous ocean waves (at least to some useful extent)—ideally while harvesting wave energy. In 2008 an Anglo-French team (Farhat, Enoch, et al.) applied Pendry's coordinate transformation technique, and proposed a design applicable to tsunami protection—using concentric rings of peg/towers [209]. Xinhua Hu and team took another approach—that of using arrays of resonant cylinders to get metamaterial effects. Although their scheme offers the possibility of protection and energy production from ordinary and large waves—John Pendry has pointed out that given the very low frequency and enormous energy of tsunami waves, Hu's approach would be insufficient [97]. Again applying Pendry's technique, Berraquero and colleagues have shown how water waves may be controlled in a bent waveguide [55].

Bio-medical metamaterials

As reviewed by Rosaline & Raghavan, the use of the metamaterial approach has allowed step improvements in a number of medical areas. Magnetic Resonance Imaging (MRI) is widely used for visualizing structures within the body—with greater precision than with X-rays. With a large and strong magnet surrounding the patient, an oscillating magnetic field is added causing protons (hydrogen nuclei in water in the body) to resonate and transmit. Techniques for rapid analysis of the received signals allow remarkably detailed images to be produced in minutes. Various metamaterial geometries are used to increase resolution and sensitivity including: 'Swiss rolls', parallel wires, loaded rings, split rings, and even an array of spirals (typically used to produce chiral effects). Metamaterials are also used in wireless endoscopy, organic tissue analysis, and cancer heat treatment (hyperthermia), and negative-index lenses have been proposed for better detection of breast tumours [788].

Chapter 30

Quasicrystal '-onics'

Having reviewed basic photonics, plasmonics, phononics, plus a couple of related fields—we now ask: 'What benefits can quasicrystal theory add?' Well, the most useful is in providing a much more even response to waves, regardless of their incoming direction—ideally an isotropic response. As Steurer & Sutter-Widmer note, quasicrystals offer perfectly ordered structure combined with arbitrarily high rotational symmetry—prerequisites for isotropic band-gap composites [871, 515, 516]. Even in 1D layer designs, the use of Fibonacci and other non-periodic arrangements can offer unprecedented flexibility for tailoring and tuning transmission spectra.

Merlin's Fibonacci superlattice

As discussed by Alan MacDonald, interest in quasi-periodic systems predates Shechtman's 1982 discovery of quasicrystals.[1] But Mac-Donald also notes how the field was 're-invigorated' by the event [603, 829]. We mentioned earlier the theoretical work of Levine and Steinhardt on quasicrystals (p.333) [548]. Part of this included a 1D quasicrystal model which Roberto Merlin et al. set about fabricating in the form of a multi-layer semiconductor.[2] This was pioneering work indeed—in 1985 they were the first to start exploring the rich properties of non-periodic (yet highly ordered) systems—in particular using Fibonacci-word-based layer separations. They built and studied a 'Fibonacci superlattice' [641, 604].[3]

---

[1]   E.g. the 'almost periodic Schrödinger equation' in maths [835] and physics [846].

[2]   Using AlAs and GaAs layer pairs as building blocks—'A': 17Å, 42Å; and 'B': 17Å, 20Å.

[3]   Using X-ray and Raman scattering techniques.

Kohmoto's Fibonacci photonics

Shechtman's quasicrystal discovery also inspired a new research topic in photonics—that of photonic quasicrystals. Mahito Kohmoto, working with Bill Sutherland and K Iguchi wrote the first paper on the subject in 1987—in which they proposed an experiment to probe the quasi-localization of photons in a photonic crystal where the layers were sized according to the Fibonacci word. We took a brief first look at the Fibonacci word in the *Is it Fractal?* chapter (p.238); but now we shall review it in detail.

The Fibonacci word—the Golden String

The Fibonacci word (also known as the Fibonacci chain, the Golden String or the Fibonacci string—Sloane A003849), provides a simple and very well studied scheme for setting up a quasicrystal structure in one dimension—that is, a set of layers with the all-important properties of non-periodicity and long-range order. It is very surprising that this word can be generated in three apparently quite different ways—all producing exactly the same result (Figs. 235, 236, and 237). We see the simplest case first in Fig. 235. As with the numeric Fibonacci recurrence, we start with two seeds—here they are $G_0 = A$ and $G_1 = AB$. We then progress by externally combining the two previous generations to get the next; and we repeat this pattern to build and build without limit. (Again, '⌢' means concatenate):

$$G_{n+2} = G_{n+1} \, ^\frown G_n. \tag{30.1}$$

This gives us

ABAABABAABAABABAABABA...

Yet (remarkably), this same result may also be achieved by making one-by-one 'internal' substitutions. This way, to produce the next generation, each letter is transformed according to the 'rabbits' substitution rule: $A \rightarrow AB$ and $B \rightarrow A$ (Fig. 236).[4] To understand this diagram fully, we note that the generations that fan out below the A on the right (marked 'R'), are the same as those below the A on the left ('L'); just that those on the right lag by one generation. Indeed the consequences of every A are the same—only the start position changes. Hence, if this tree is extended indefinitely, we know it will contain an infinite number of copies of itself.

---

[4] For Fibonacci's (1202) rabbit rule, see the footnote on p.284.

$$A$$
$$AB$$
$$AB\frown A = ABA$$
$$ABA\frown AB = ABAAB$$
$$ABAAB\frown ABA = ABAABABA$$
$$ABAABABA\frown ABAAB = ABAABABAABAAB$$
$$ABAABABAABAAB\frown ABAABABA = ABAABABAABAABABAABABA$$

Figure 235: **Fibonacci word: 'External' growth—by block concatenation.**

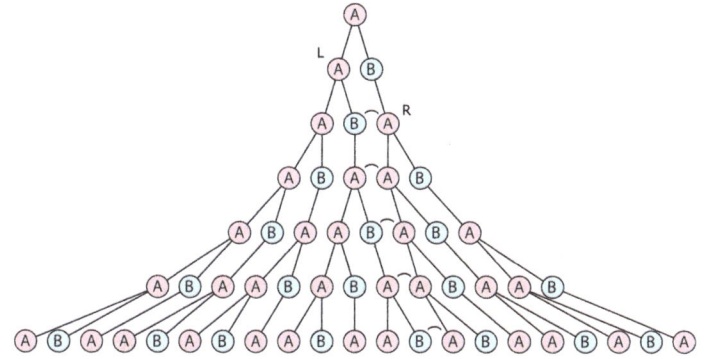

Figure 236: **Fibonacci word: 'Internal' growth—by atomic substitution.**

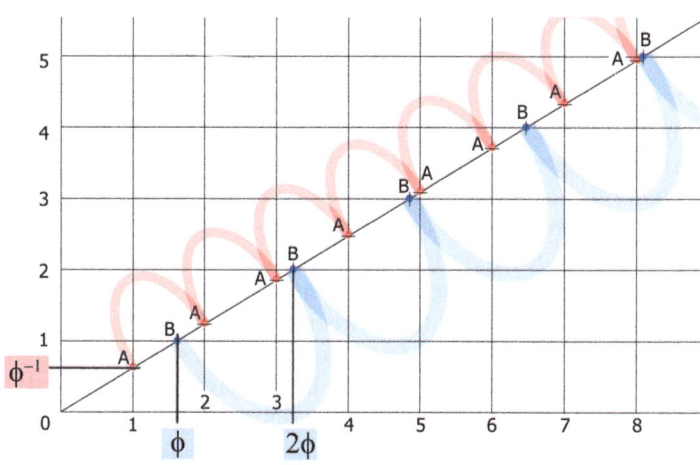

Figure 237: **Fibonacci word: Higher-dimensional 'cutting sequence'.**

In both cases, the word lengths increase according to the Fibonacci sequence: 1, 2, 3, 5, 8, 13, 21... Also, we see the first two letters are always AB, and the last two letters alternate between AB and BA. And if we remove the last two letters, we are left with a palindrome. (To allow immediate comparison, the concatenation positions '⌒' from Fig. 235 are added to Fig. 236.)

Initial pattern soon breaks

In Fig. 235 on page 385, if we count the A's and B's in successive generations—say in the form (a, b)—then we shall get

$$(1, 0), \quad (1, 1), \quad (2, 1), \quad (3, 2), \quad (5, 3), \quad (8, 5), \quad (13, 8), \quad \ldots$$

Immediately, we recognize Fibonacci numbers—paired as they are in the convergents of the Golden Ratio; hence we know that in the limit, the ratio of A's to B's will tend to ϕ. So, if and when we spot a pattern at the beginning of the Fibonacci word, we know it cannot continue. The irrationality of ϕ, is (of course) fundamentally inconsistent with periodicity. In Fig. 238 we visualize an initial pattern using the gate shapes from page 11. These are shown as Sun ('$\phi 1 \phi$') in gold and red; and Moon ('$1 \phi 1$') in blue, with column widths $A = \phi$, $B = 1$. The pale-green/blue combined represents a Sun Gate scaled $\times \phi$—and in this scaling we again see the 'rabbits' atomic substitution rule, with pillars $A \rightarrow AB$; and door $B \rightarrow A$ (p.284). But checking Figs. 235 and 236, we see that the very last 2 elements shown, viz. ... BA, break the opening pattern: (ABA) A (BAB) A; (ABA) A (BAB) A; (ABA) B̶A̶ ...

The higher-dimensional cut

Fig. 237, p.385 shows yet a third way to obtain the Fibonacci word—by taking a 'higher-dimensional' cut. In this case, in order to obtain a non-periodic 1-dimensional result, we only need cut through a regular, periodic, 2-dimensional lattice. Here it is the irrational slope gradient of ϕ^{-1} that delivers the non-periodic result. We mark A where this slope cuts a vertical grid line, and B for an intersection with a horizontal. Hence the A's are periodic (red arcs)—and so are the B's (blue arcs)—but the ratio between their periods is not rational. This projects (onto the slope line) a 1D quasicrystal structure from the 'higher-dimensional' (i.e. 2D) lattice. In the same way, 2D and 3D structures may be obtained from higher dimensions too. As we saw earlier, aperiodic tiling research initially had a very local focus—that is, on individual tiles and their matching rules.

Figure 238: **Initial periodicity of Fibonacci word—Sun & Moon gates**

But Robert Ammann then took a global view with his bars, and later, in 1981 Nicolaas Govert de Bruijn devised two powerful global construction techniques [89]. They are the 'cut-and-project' and the 'pentagrid' methods. These are discussed by Au-Yang & Park who note how N G de Bruijn used his irrational-cut method to produce 2D Penrose tilings. They also note a direct relationship between Conway worms, pentagrids, and Ammann bars [21, 22]. Marjorie Senechal identified de Bruijn as one of the first to see Penrose tiling in terms of a projection from a higher dimension. She recalls that when de Bruijn heard about quasicrystals he remarked: 'This is a gold mine for mathematics.' [821]. Ammann originally drew lines that were parallel equispaced, but he then found by moving them to a short and long distancing (with ratio ϕ) he could get the same pattern to appear on all the thin rhombs and similarly for all the thick rhombs (p.322, Fig. 197 on the right) [818]. And because the sequences of distancing LSLLSLSL etc. turn out to be substrings (subsections from some position) in the Fibonacci word, it is then possible to use the cut-and-project approach to find bar-crossing vertices from which tile placements are fixed. De Bruijn started with a 5D lattice and projected onto a cutting plane all vertices within a threshold distance [89]. Then in 1982, Frans Beenker built on de Bruijn's algebraic approach to project the Silver-Ratio-based Ammann 'A5' tiling which we saw on page 337 [47]. And two years later, Kramer and Neri projected icosahedral (3D) quasicrystals from a regular 6D lattice [511]. However, this technique was not seen as the ideal way forward by all crystallographers. N David Mermin published a paper in 1992 comparing the 'resort to higher dimensions' with the use of epicycles in pre-Copernican times. Instead, he proposed a new classification which did not turn on periodicity [642].[5]

---

[5] We also recall Mermin's famous remark about quantum mechanics—p.297.

Nevertheless, the irrational-cut approach has become very well established. As we mentioned back on page 334, in 2010 Alexandra Ledermann published her work (done along with Martin Wegener and Georg von Freymann) which showed the possibility of other 3D-quasicrystal structures in addition to the existing icosahedral. She started with a regular 12D lattice and made an irrational cut based on the Silver Ratio—in order to project rhombicuboctahedra in 3D. These have a local 8-fold rotational symmetry [527].

Never AAA or BB

Looking back at Fig. 237 on page 385, we see that along the slope, the instances of A appear regularly—once for each crossing of a vertical grid line. Similarly the B instances occur once for every crossing of a horizontal. Because of the less than unity slope (actually ϕ^{-1}), there are more A instances than B. For every interval between B's there is at least one A, and sometimes another, but there is no room for a third. This is why we never see groupings: AAA, $AAAA$, or more in the Fibonacci word. And in the same way, we never see: BB, or BBB, or more, no matter how far the sequence is extended.

Scaling the word

It is also interesting to look at the scaling properties of the Fibonacci word (Fig. 239). If we let $A=1$ and $B=\phi^{-1}$ (upper section), and scale these by ϕ, then we shall get $A'=\phi$ and $B'=1$. This graphically demonstrates to us the Fibonacci substitution rule, because $A'=\phi A=A+B$, and $B'=\phi B=A$, so $A\rightarrow AB$ and $B\rightarrow A$. The dotted lines show where the new boundary positions correspond to positions in the original. Although this is a property of the Fibonacci word, it is not true in general—for example if the top row started $ABBB$, then the expanded B's ($B'=1$ each) would not all line up. Again we come back to the fundamental point—that although the Fibonacci sequence is not periodic; it *is* very well ordered.

Fibonacci favourite

Because the Fibonacci word combines *long-range order without periodicity* and *extreme simplicity of generation*, it has become the one-dimensional quasicrystal scheme of choice in widely different fields of research.

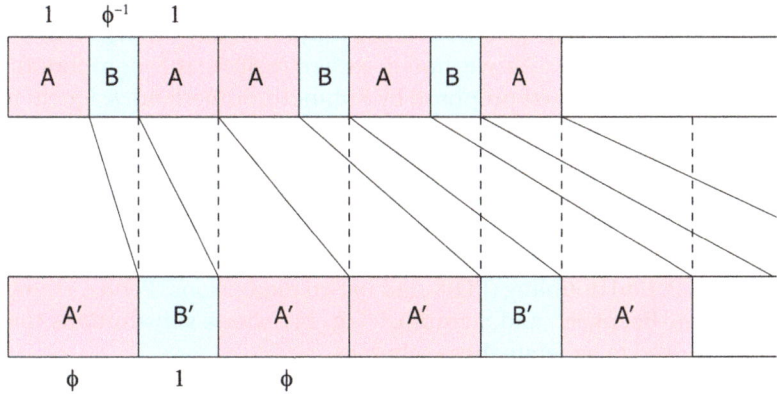

Figure 239: **Scaling the Fibonacci word by ϕ.**

In Fig. 240 we compare two 1D photonic crystals—the simple ABABABAB periodic and the ABAABABAAB Fibonacci word quasicrystal (A blue, B cyan). In an example implementation given by Vardeny et al. the A layer is titanium dioxide (TiO_2) and the B layer, silicon dioxide (SiO_2) [919]. These have indices of refraction of 2.30 and 1.45 respectively, and in the periodic crystal, the thickness of each layer is chosen to contain one half wavelength.

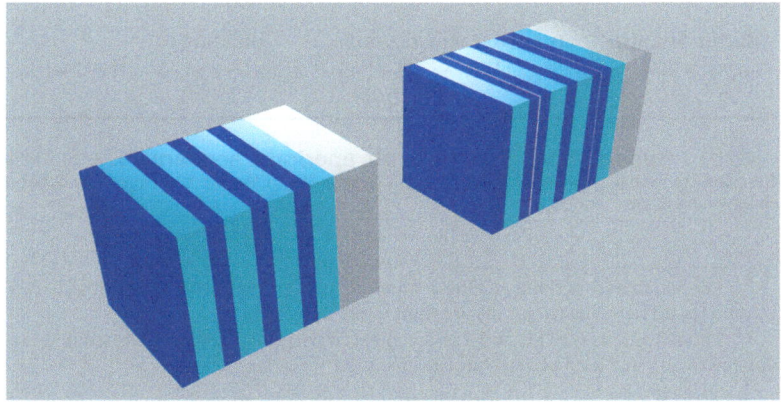

Figure 240: **Photonic crystals.** Left: Periodic, ABABABAB. Right: 1D Fibonacci word quasicrystal, ABAABABAAB. End blocks: glass substrate. Simplified version of Vardeny et al. [919], after Kohmoto [503], image created in *Blender* [787].

Because light is slowed more in the TiO_2, the wavelength is compressed, the blue layers are made proportionally thinner than the cyan (SiO_2) ones. As we noted earlier (p.384), the non-periodic configuration was first proposed by Kohmoto, Sutherland, & Iguchi in 1987 (not that long after Shechtman's quasicrystal discovery p.328). They used an elegant transfer matrix method to show that the resulting transmission spectrum of the quasicrystal was multi-fractal with considerable self-similarity [503]. Other aperiodic[6] 1D schemes include: Cantor Triadic, Thue-Morse (aka PTM, Prouhet-Thue-Morse), Period doubling (PD)[7], and in two dimensions: Penrose tiling, Ammann-Beenker,[8] and Stampfli.[9] Fig. 241 shows substitutions that may be used to generate these schemes.

Sequence	Substitutions		Initial generations
Fibonacci A003849 [839]	$(A \to AB,$	$B \to A)$	$b,\ A,\ AB,\ ABA,\ ABAAB, \dots$
—*AB* swapped A005614	$(B \to BA,$	$A \to B)$	$A,\ B,\ BA,\ BAB,\ BABBA, \dots$
Cantor Triadic A088917 [623, 31]	$(A \to ABA,$	$B \to BBB)$	$A,\ ABA,\ ABABBBABA, \dots$
Thue-Morse A010060	$(A \to AB,$	$B \to BA)$	$A,\ AB,\ ABBA,\ ABBABAAB, \dots$
Double-period A096268	$(A \to AB,$	$B \to AA)$	$A,\ AB,\ ABAA,\ ABAAABAB, \dots$
Rudin-Shapiro A020985 [85]	pair-wise substitutions (e.g. +1 +1 $\to$ +1 +1 +1 -1)		building to +1 +1 +1 -1 +1 +1 -1 +1 ...

Figure 241: **Substitutions that generate aperiodic sequences.** F, CT, TM, & DP rules are from the review by Albuquerque & Cottam [14]. Element '*b*' indicates an optional *B* beginning to the Fibonacci.

---

[6] Aperiodic/non-periodic: As the distinction made when considering tilings (p.309) is not relevant here, the two terms are used interchangeably.

[7] Though CT, TM/PTM, and PD are not quasicrystal, they are also studied and applied to obtain a level of deterministic disorder.

[8] See page 337.

[9] This distinctive pattern is built from squares and equilateral triangles (all sides equal) to make a dodecagonal quasiperiodic lattice. It inflates in a similar way to Penrose tilings but with a scale factor of $\sqrt{2+\sqrt{3}}$. Stampfli relates this pattern to small quasicrystal nickel chromium particles [862].

We note that starting the Fibonacci generations with *B* instead of *A* immediately takes us back to the canonical *A*-start version we have seen so far. But, in this *B*-start word, if we swap letters *B* and *A* we get an equivalent Fibonacci word sequence which is used by some authors (e.g. Au-Yang & Perk [21]). This word still obeys the same concatenation rule, viz. (30.1) p.384.

In their paper about Fibonacci chains of metal nanoparticles, Dal Negro and Feng give references to Fibonacci-based studies of wave-like excitations in electronic, optical, mechanical, spin waves (magnetic), mixed waves, and polariton waves [157]. We shall only have space to look into some of these in superficial detail; but the main point here is the vast range of applicability of this new 'quasicrystal' way of thinking, yet further extended by compatibility with metamaterial techniques too…

Quasicrystal photonics

In their 2013 review of photonic quasicrystals, Vardeny, Nahata, and Agrawal observed that for applications involving light interference— *periodic* structures were in overwhelming use. They concluded that the reason for this was that the benefits of using aperiodic structures —such as quasicrystals—were still not widely recognized. They pointed out that when compared with periodic structures, the large variety of aperiodic structures available could add substantial flexibility and richness when engineering optical responses [920]. They went on to differentiate between quasicrystal-aperiodic structures (such as the Fibonacci superlattice) and 'deterministic-aperiodic' structures (such as the Thue-Morse and period-doubling which we saw in Fig. 241, p.390). While the quasicrystals are characterized by their sharp Bragg peaks, the second group show far more complex Fourier properties.

Photonic quasicrystals may be designed and built in 2D and 3D, and one of the most common in 2D is the Penrose P3 rhomb tiling (which we reviewed on p.317). We have already mentioned (at the beginning of this chapter) the importance of quasicrystal arrangements when designing for evenness of directional response (isotropy). For 2D *periodic* photonic crystals, rotational symmetry is limited to 6-fold; yet for 2D photonic quasicrystals, 10-fold (Penrose) and 12-fold (dodecahedral) symmetries are common [636]. Using the 2D Penrose structure, Notomi et al. were the first to demonstrate lasing in a photonic quasicrystal. They used a silicon substrate (SiO_2 on Si) and fabricated a hole pattern where each hole was placed at the centre of a Penrose P3 rhomb (be it thick or thin) [687].

Quasicrystal phononics and acoustics

As we have noted, the quasicrystal approach can add more ways to tune more parameters and thereby open up new areas of application. This is especially true in the case of phononics and acoustics. Chen et al. compared two-component periodic phononic crystals and two-component Fibonacci phononic quasicrystals in configurations very much the same as we saw in Fig. 240 (photonic). They then did example calculations for the Fibonacci case with layers made of lead (A) and epoxy (B). In one instance, to make a (potentially infinite) calculation finite, they chose to work to the 20th generation of the Fibonacci sequence (10,946 layers). In other exercises, they compared the band structures of ABA, ABAAB, and ABAABABA superlattices (generations 3, 4, and 5) [112]. In their article in Physics World 2008, Lei Wang and Baowen Li report on early work in phononics [532]. They discuss a thermal 'diode', the equivalent of the electronic diode, but one that lets heat flow in only one direction [891]. They point out that such a device would provide tremendous advantages over air conditioning systems in tropical countries—allowing cooling outward heat flow from a building at night, but reducing heat inflow during the day. They go on to discuss phononic switching and aim to build phononic equivalents of electronic logic gates and transistors.

Quasicrystal metamaterials

One of the first to consider combining the strengths of the quasicrystal and metamaterial approaches was Luca Maini in his 2010 PhD thesis [616]. He looked at applications of 'metamaterials based on photonic quasicrystals: from superlensing to new photonic devices.' As mentioned on page 365, a frequent goal in the design of band-gap materials is that the response be even for a wide range of incidence angles. Kruk et al. realized how quasicrystal geometry could help them achieve this. In metamaterials, typically the sub-wavelength inclusions ('meta-atoms') are simply included in a periodic arrangement. This can result in widely uneven response as a function of incidence angle. However, in their paper, Kruk and team discuss meta-atom inclusions located in a quasicrystal arrangement. They show how this can provide isotropic optical properties while still preserving (desired) pronounced resonances [515, 516].

Quasicrystal plasmonics

We discussed in the last chapter how plasmonics was identified as a possible bridge between the photonic scale and that of silicon-chip electronics. Dal Negro and Feng note how this led to a revival in fundamental studies of surface plasmons in arrays and chains of metal nano-particles [157]. They then took the next step of considering the properties of metal nano-particles arranged in Fibonacci-quasicrystal chain structures. Their results confirmed the presence of plasmonic band gaps (among other findings) [158]. Their expectation that this would have a large impact in nano-photonic device design proved correct.[10] Ricciardi, Crescitelli et al. studied combined metal and dielectric nano-scale quasicrystal structures, and found combined plasmonic and photonic resonances. In comparing their results with those from periodic structures, they reported a richer spectrum of resonances, well suited to the development of high-performance optical devices for use in communications, energy, and sensing [778].

Quasicrystal nanophotonics and light harvesting

Nanophotonics is all about thinking sub-wavelength and controlling light—using waveguides and phenomena such as Anderson localization. Working down to single-atom scales, the effects of the quantization of light must increasingly be taken into account. This new field owes its existence to dramatic practical progress in advanced fabrication techniques, along with considerable theoretical work regarding the confinement of light—using novel geometry. A major area of research is energy efficient data communications (using 10×, 100×, even 1000× less power) to meet the challenge of exponentially increasing internet use. This encompasses data centres, high-performance computers, personal computers, and broadband fibre optic to homes. The goal is to use integrated optical devices and circuits in 'all-optical' networks with petabit/second data rates. Already, work is being done on single-photon processing (using 'qubits': quantum information processing). Although photons do not interact in a vacuum, they do interact with atomic electron clouds, and this distortion then affects other photons. Such non-linear effects can be used to switch light in the *femtosecond* regime (10^{-15} s) [198].

---

[10] This and following studies have resulted in remarkable efficiency increases in surface enhanced Raman spectroscopy and photoluminescence along with increased biosensing sensitivity—Vardeny [919] p.183.

PV (photo-voltaic) solar cells convert light directly into electricity, and given the sun's *up to 180°* change in incidence angle, isotropy is a key focus in PV design. In a 2012 paper, Bauer et al. describe their study of 2D quasiperiodic plasmonic crystals—modelling and measuring their interaction with incident light at normal and oblique angles [44]. And in 2014, Xavier et al. reported the design and fabrication of cells that offered significant advances by using an ultra-thin structure based on a nanophotonic (10-fold) quasicrystal. They coated a glass substrate with a gel, then pressed this using a block having a quasicrystal arrangement of holes. This formed vertical rod stubs, and these were then used as the basis for the rest of the fabrication process.[11] As the main expense of traditional PV cells is their bulk silicon content, we should expect major cost savings by using very thin films. Significant improvements in light trapping are also achieved, which more than double the available current output density (compared with unstructured planar films) [991].

Quasicrystal magnonics

(In a relationship similar to that between photons and light waves), magnons are the quasi-particles associated with spin waves, and magnonic crystals may be built in much the same way as photonic crystals, but with alternating layers of different magnetic permeability (μ). In this way, useful 'magnonic band gaps' may be achieved, and again the quasicrystal approach can significantly increase the flexibility and richness of the response spectra. The study by Costa et al. (2012) is particularly interesting from our point of view as it considers not only the Fibonacci/Golden Ratio case, but also several other 'PQ' instances: the Silver, Bronze, and Nickel Ratios, (see pages 267 and 429–430) [137, 493, 854].

Overview of quasicrystal R & D

At first it may seem strange that things as different as a fragment of a meteorite and the Fibonacci word 'ABAABABA...' should equally be referred to as *quasicrystal*. But we find that quasicrystal research and development is taking place on all of three different levels of abstraction at the same time. Lewis Carroll's Cheshire cat can help us compare and contrast (Fig. 242 and Table in Fig. 243).

---

[11] The quasicrystal pattern used for this 'nano-imprint lithography' (NIL) included rosettes reminiscent of the screen pattern we saw in Fig. 196 on p.321.

'I said pig,' replied Alice; 'and I wish you wouldn't keep appearing and vanishing so suddenly: you make one quite giddy.'

'All right,' said the Cat; and this time it vanished quite slowly, beginning with the end of the tail, and ending with the grin, which remained some time after the rest of it had gone.

'Well! I've often seen a cat without a grin,' thought Alice; 'but a grin without a cat! It's the most curious thing I ever saw in my life!'

Figure 242: **John Tenniel (1820–1914), Alice speaking to the Cheshire cat, 1866.** From Lewis Carroll's *Alice's Adventures in Wonderland* [100, 889].

Level	Examples	Applications
Molecular	Shechtman's metal alloys	Surgeons' instruments Aviation Razor blades Non-stick coatings Protective coatings Hydrogen storage Catalysis.
Superlattice	Photonic	LEDs, Lasers, Displays Solar Cells (PV) Optical communications Optical computing
	Plasmonic	Spectroscopic sensors Bio-trace (quantum dots)
	Phononic	Control of heat and sound.
Abstract	Fibonacci word (1D) Irrational cut (esp. ϕ) Penrose tilings (ϕ) Ammann-Beenker (δ) Stampfli (dodecagonal)	Quasicrystal design. Design goals may include tuned transmission spectra and isotropic response.

Figure 243: **Quasicrystal R & D—the 3 levels of abstraction.**

Molecular ('Cat with grin')

In *molecular* quasicrystals—such as the metal alloy phases discovered by Shechtman et al.—it is the atomic level structure of the physical material itself that constitutes the non-periodic pattern with its long-range order.

Superlattice ('Grin becoming separate from cat')

In *superlattice* quasicrystal realizations, designers stack components of different materials (which might include metamaterials)—for example by layering. In this case, it is the *layer boundaries* that determine the non-periodic long-range order.

Abstract ('Just the grin')

And finally, in the case of purely *abstract* quasicrystals—such as the Fibonacci word and Penrose tilings—it is the mathematical structure alone that is studied.

The future

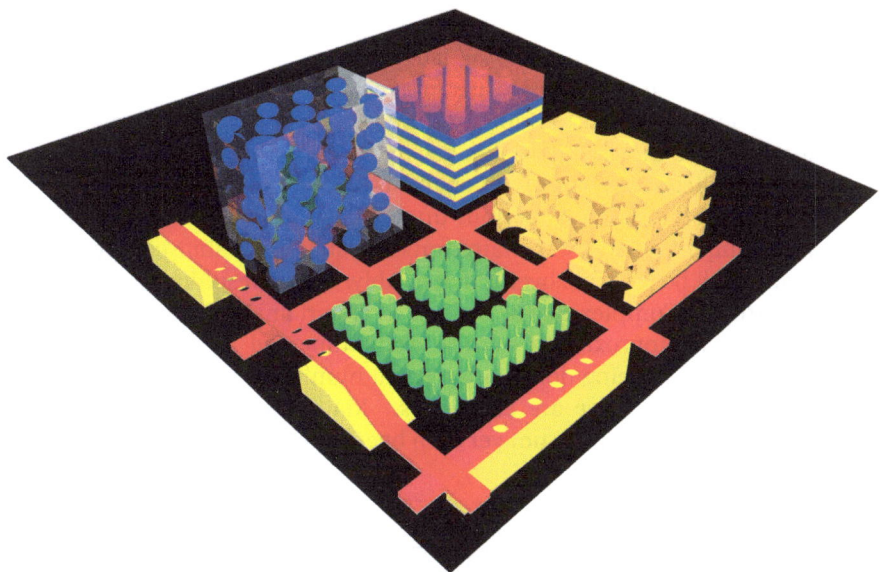

Figure 244: **Molding the flow of light.** Front cover illustration from *Photonic Crystals*, by Joannopoulos et al. © 2008 Princeton University Press [423].

The interdisciplinary approach (which we mentioned earlier regarding phyllotaxis and quasicrystals), is already well established in these '-onics' technologies. We are seeing diverse collaborations of researchers from: optics, crystallography, mathematics, physics, materials science, and advanced silicon fabrication technology, electronic engineering, computing science, optical computing, telecommunications, seismology, acoustics, thermal engineering, green engineering (energy harvesting), and more. Yablonovitch [860], Engheta [204], Sun [878], and others, all predict that effects will increasingly be analysed in terms of lumped circuit elements. And in practice, circuits will become componentized and standardized for reuse, just as now happens in electronics (with its resistors, capacitors, inductors, transistors, antennas, waveguides, and so on). Fig. 244 illustrates the concept of an integrated optical circuit. The various photonic components included are detailed and analysed in the 2008 book by Joannopoulos et al. *Photonic crystals, molding the flow of light* [423]. For example, on the left, the suspended red strip with holes is a 'waveguide-cavity-waveguide filter'. The two groups of

4 holes each form a periodic sequence while the extra spacing between the groups forms a 'defect'. The hole spacing is approximately 500 nanometres [425]. Similarly (the more recent) Fig. 245 shows 3 all-optical components fabricated on one chip. We should also expect an increasing use of metamaterials in superlattice configurations—both periodic and quasicrystal, in 1D, 2D, and 3D. And with the increasing challenges in nano-fabrication (top-down), some physicists are looking to chemists and molecular biologists for self-assembly solutions (bottom-up). With a greater understanding (for example) of how soft-matter quasicrystals are stabilised, greater control will be possible over self assembly, ideally resulting in perfect formations. Complexity will grow quickly and analytic solutions will give way to approximate models and numerical methods. Most current work is being done with a small number of types of material configured, but these are early days. Materials will soon become commonplace that have not yet even been invented.

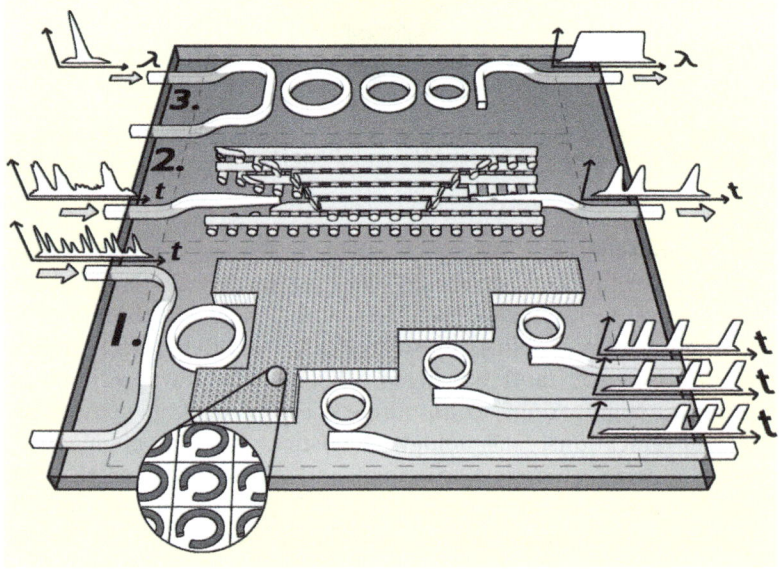

Figure 245: **All-optical processing.** Image © 2015 Ben Eggleton, Cudos [198].
1—Demux: Metamaterial-based photonic circuitry comprising metallic split-ring resonators (as on p.372), producing artificial magnetic properties that can be harnessed for efficient information processing (here separating out three different colour components from a multiplexed signal).
2—Filter: Three-dimensional 'woodpile' structure, here conditioning/cleaning a noisy signal.
3—Frequency converter: Coupled ring resonators are used to widen frequency range. The large ring resonators operate in 'whispering gallery modes'—discovered by Lord Rayleigh in his late 19th century acoustic studies (we saw his 1887 photonic superlattice on p.367) [776].

Part VI

APPENDICES

Appendix A—Ori32 trigonometry

The MIK and trigonometry—recap

The MIK is the basic 'structural' unit of angle in Ori32 geometry. It is $(2\pi)/32$ radians. Sailors and navigators will recognize it as a turning angle of one point. We saw it first on page 162 where we adopted a box notation, with $\boxed{m}$ denoting m MIK:

$$
\begin{aligned}
1\ \text{MIK} &= \boxed{1} &=& \quad (2\pi)/32 &=& \quad 11.25° \\
2\ \text{MIK} &= \boxed{2} &=& \quad (2\pi)/16 &=& \quad 22.50° \\
3\ \text{MIK} &= \boxed{3} &=& \quad (2\pi)\cdot 3/32 &=& \quad 33.75° \\
4\ \text{MIK} &= \boxed{4} &=& \quad (2\pi)/8 &=& \quad 45.00°.
\end{aligned}
$$

On page 405, the Table (Fig. 248) lists conversions from MIK to degrees and radians. In Ori32 we follow the maths standard and measure increasing angle anticlockwise, (not the seafaring and air-navigation clockwise).

The trigonometry functions in terms of their complements, where m is integer number of MIK and $0 \le m \le 8$ are:

$$
\begin{aligned}
\sin \boxed{m} &= \cos \boxed{8-m} & \qquad \tan \boxed{m} &= \cot \boxed{8-m} \\
\cos \boxed{m} &= \sin \boxed{8-m} & \qquad \cot \boxed{m} &= \tan \boxed{8-m}.
\end{aligned}
$$

Analytic Tables for: sin, cos, tan, and cot now follow...

Ori32 sines and cosines

As can be seen in Fig. 246, the sin ( ) and cos ( ) values depend entirely on combinations of 2 and $\sqrt{2}$, and roots of combinations and so on [970]. According to circumstance, either the κ, λ (half-angle derived) or the α, β (sum, difference) short forms may prove more useful. For sample derivations see page 404.

Letting $\mu = \sqrt{2 + \sqrt{2}},$ $\quad v = \sqrt{2 - \sqrt{2}}$ $\qquad \Longrightarrow \quad \mu v = \sqrt{2}$

and $\quad \alpha = \sqrt{2 + \mu},$ $\quad \beta = \sqrt{2 - \mu}$ $\qquad \Longrightarrow \quad \alpha\beta = v$

and $\quad \kappa = \sqrt{2 + v},$ $\quad \lambda = \sqrt{2 - v}$ $\qquad \Longrightarrow \quad \kappa\lambda = \mu$

$\sin \boxed{0} \quad = \qquad\qquad\qquad\qquad 0 \qquad\qquad = \quad \cos \boxed{8}$

$\sin \boxed{1} \quad = \qquad \dfrac{\beta}{2} \qquad = \quad \tfrac{1}{2}\sqrt{2 - \sqrt{2 + \sqrt{2}}} \quad = \quad \cos \boxed{7}$

$\sin \boxed{2} \quad = \qquad \dfrac{v}{2} \qquad = \quad \tfrac{1}{2}\sqrt{2 - \sqrt{2}} \qquad = \quad \cos \boxed{6}$

$\sin \boxed{3} \quad = \quad \dfrac{\lambda}{2} = \dfrac{\alpha - \beta}{2\sqrt{2}} \quad = \quad \tfrac{1}{2}\sqrt{2 - \sqrt{2 - \sqrt{2}}} \quad = \quad \cos \boxed{5}$

$\sin \boxed{4} \quad = \qquad\qquad\qquad \tfrac{1}{2}\sqrt{2} \qquad\qquad = \quad \cos \boxed{4}$

$\sin \boxed{5} \quad = \quad \dfrac{\kappa}{2} = \dfrac{\alpha + \beta}{2\sqrt{2}} \quad = \quad \tfrac{1}{2}\sqrt{2 + \sqrt{2 - \sqrt{2}}} \quad = \quad \cos \boxed{3}$

$\sin \boxed{6} \quad = \qquad \dfrac{\mu}{2} \qquad = \quad \tfrac{1}{2}\sqrt{2 + \sqrt{2}} \qquad = \quad \cos \boxed{2}$

$\sin \boxed{7} \quad = \qquad \dfrac{\alpha}{2} \qquad = \quad \tfrac{1}{2}\sqrt{2 + \sqrt{2 + \sqrt{2}}} \quad = \quad \cos \boxed{1}$

$\sin \boxed{8} \quad = \qquad\qquad\qquad\qquad 1 \qquad\qquad = \quad \cos \boxed{0}$

Figure 246: **Ori32 sines and cosines.**

Ori32 tangents and cotangents

Tangents and their reciprocals (aka cotangents) [970] may be derived from the Ori32 sines and cosines chart (Fig. 246), using the relation

$$\tan \boxed{\text{m}} \ = \ \sin \boxed{\text{m}} \ / \ \cos \boxed{\text{m}}.$$

Again the expressions are almost entirely in terms of 2 and $\sqrt{2}$. We also note the multiple appearance of the Silver Ratio $\delta = (1 + \sqrt{2})$, and its conjugate, $-\delta^{-1} = (1 - \sqrt{2})$, (p.268).

Letting $\quad \mu = \sqrt{2 + \sqrt{2}}, \qquad \nu = \sqrt{2 - \sqrt{2}} \qquad\qquad \Longrightarrow \quad \mu\nu = \sqrt{2}$

and $\qquad \alpha = \sqrt{2 + \mu}, \qquad \beta = \sqrt{2 - \mu} \qquad\qquad \Longrightarrow \quad \alpha\beta = \nu$

and $\qquad \kappa = \sqrt{2 + \nu}, \qquad \lambda = \sqrt{2 - \nu} \qquad\qquad \Longrightarrow \quad \kappa\lambda = \mu$

$\tan \boxed{0} \quad = \qquad\qquad\qquad\qquad 0 \qquad\qquad\qquad\qquad = \quad \cot \boxed{8}$

$\tan \boxed{1} \quad = \quad \dfrac{\beta}{\alpha} \quad = \quad \sqrt{2}\sqrt{2+\sqrt{2}} - (1 + \sqrt{2}) \quad = \quad \cot \boxed{7}$

$\tan \boxed{2} \quad = \qquad\qquad\qquad -(1 - \sqrt{2}) \qquad\qquad = \quad \cot \boxed{6}$

$\tan \boxed{3} \quad = \quad \dfrac{\lambda}{\kappa} \quad = \quad \sqrt{2}\sqrt{2-\sqrt{2}} + (1 - \sqrt{2}) \quad = \quad \cot \boxed{5}$

$\tan \boxed{4} \quad = \qquad\qquad\qquad\qquad 1 \qquad\qquad\qquad\qquad = \quad \cot \boxed{4}$

$\tan \boxed{5} \quad = \quad \dfrac{\kappa}{\lambda} \quad = \quad \sqrt{2}\sqrt{2-\sqrt{2}} - (1 - \sqrt{2}) \quad = \quad \cot \boxed{3}$

$\tan \boxed{6} \quad = \qquad\qquad\qquad (1 + \sqrt{2}) \qquad\qquad\qquad = \quad \cot \boxed{2}$

$\tan \boxed{7} \quad = \quad \dfrac{\alpha}{\beta} \quad = \quad \sqrt{2}\sqrt{2+\sqrt{2}} + (1 + \sqrt{2}) \quad = \quad \cot \boxed{1}$

$\tan \boxed{8} \quad = \qquad\qquad\qquad\qquad \infty \qquad\qquad\qquad\qquad = \quad \cot \boxed{0}$

Figure 247: **Ori32 tangents and cotangents.**

Example derivations of Ori32 short forms

Expressions in terms of $\alpha, \beta, \kappa, \lambda, \mu, \nu$—as used in the Tables on p.402 and p.403—may be derived using standard trigonometric identities. For example, by using the half angle formula:

$$\sin\left(\frac{A}{2}\right) = \sqrt{\frac{1-\cos(A)}{2}}.$$ [953]

So, to obtain $\sin \boxed{3}$ from $\cos \boxed{6}$,

$$\sin \boxed{3} \;=\; \sin \frac{\boxed{6}}{2} \;=\; \sqrt{\frac{1-\cos \boxed{6}}{2}}$$

$$= \sqrt{\frac{1-(\nu/2)}{2}}$$

$$= \frac{1}{2}\sqrt{2-\nu}$$

$$= \frac{\lambda}{2}.$$

Or alternatively: We may apply the sine difference formula,

$$\sin(A-B) = \sin(A)\cos(B) - \sin(B)\cos(A).$$ [969]

Hence letting $A = \boxed{4}$ and $B = \boxed{1}$

$$\sin \boxed{3} \;=\; \sin\left(\boxed{4}-\boxed{1}\right)$$

$$= \sin \boxed{4}\cos \boxed{1} \;-\; \sin \boxed{1}\cos \boxed{4}$$

$$= \frac{\sqrt{2}}{2}\cdot\frac{\alpha}{2} \;-\; \frac{\beta}{2}\cdot\frac{\sqrt{2}}{2}$$

$$= \frac{\alpha-\beta}{2\sqrt{2}}.$$

MIK	degrees	radians	MIK	degrees	radians
0	**0**	0	16	**180**	$(2\pi)\cdot 1/2$
1	11.25	$(2\pi)\cdot 1/32$	17	191.25	$(2\pi)\cdot 17/32$
2	22.50	$(2\pi)\cdot 1/16$	18	202.50	$(2\pi)\cdot 9/16$
3	33.75	$(2\pi)\cdot 3/32$	19	213.75	$(2\pi)\cdot 19/32$
4	45	$(2\pi)\cdot 1/8$	20	225	$(2\pi)\cdot 5/8$
5	56.25	$(2\pi)\cdot 5/32$	21	236.25	$(2\pi)\cdot 21/32$
6	67.50	$(2\pi)\cdot 3/16$	22	247.50	$(2\pi)\cdot 11/16$
7	78.75	$(2\pi)\cdot 7/32$	23	258.75	$(2\pi)\cdot 23/32$
8	**90**	$(2\pi)\cdot 1/4$	24	**270**	$(2\pi)\cdot 3/4$
9	101.25	$(2\pi)\cdot 9/32$	25	281.25	$(2\pi)\cdot 25/32$
10	112.50	$(2\pi)\cdot 5/16$	26	292.50	$(2\pi)\cdot 13/16$
11	123.75	$(2\pi)\cdot 11/32$	27	303.75	$(2\pi)\cdot 27/32$
12	135	$(2\pi)\cdot 3/8$	28	315	$(2\pi)\cdot 7/8$
13	146.25	$(2\pi)\cdot 13/32$	29	326.25	$(2\pi)\cdot 29/32$
14	157.50	$(2\pi)\cdot 7/16$	30	337.50	$(2\pi)\cdot 15/16$
15	168.75	$(2\pi)\cdot 15/32$	31	348.75	$(2\pi)\cdot 31/32$

Figure 248: **Ori32 angle conversion chart.**

Ailles extended

As a further simple example of Ori32, the arrangement shown in Fig. 249 is based on the 'Ailles rectangle'—which (in original) permits analysis of a 15-75-90° triangle [917]. Here the Ailles is realized in Ori32 geometry (tinted area) and then extended to form a square ABCD.

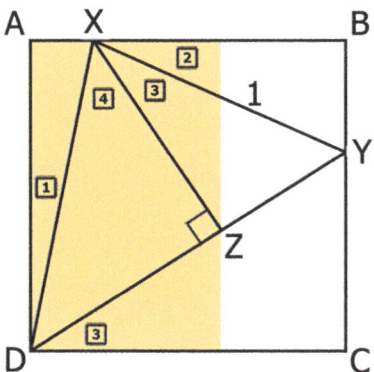

Figure 249: **Ori32—An arrangement of Fibons.**

Example solution—Exercise 1, verticals, p.166

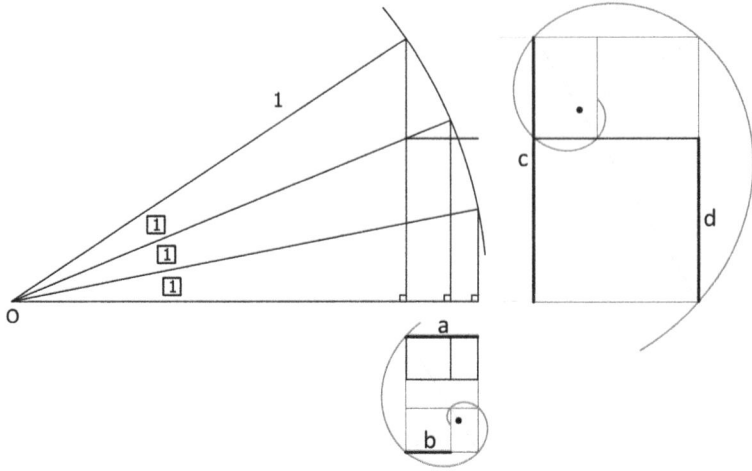

Figure 250: **Twice near-Golden.**

In Fig. 250 first a and b are the AC and AB we saw in Fig. 121 on page 164. And second we find the very-near-Golden relationship of $c : d$. (The two Golden Spirals are decorative and not part of the solution.) After trying a few likely candidate ratios, we find that of c and d (Fig. 250):

$$c = \sin \boxed{3}$$

$$d = \cos \boxed{3} \tan \boxed{2}$$

$$\frac{c}{d} = \frac{\tan \boxed{3}}{\tan \boxed{2}}$$

$$= 1.613\ldots = \phi - 0.3\%.$$

This we compare with the earlier 'horizontal' find, (p.164)

$$\frac{a}{b} = 1.616\ldots = \phi - 0.14\%.$$

How do these approximate values compare with the perfect Golden Ratio? Well, actually, these golds are better than 23.9 karat—that is, by taking the measure of fully pure as 24 karat and applying the percentage differences from the exact Golden Ratio $\phi = 1.618\ldots$

Example solution—Exercise 2, proving '*XYZ* is Fibon4' using angles   p.166

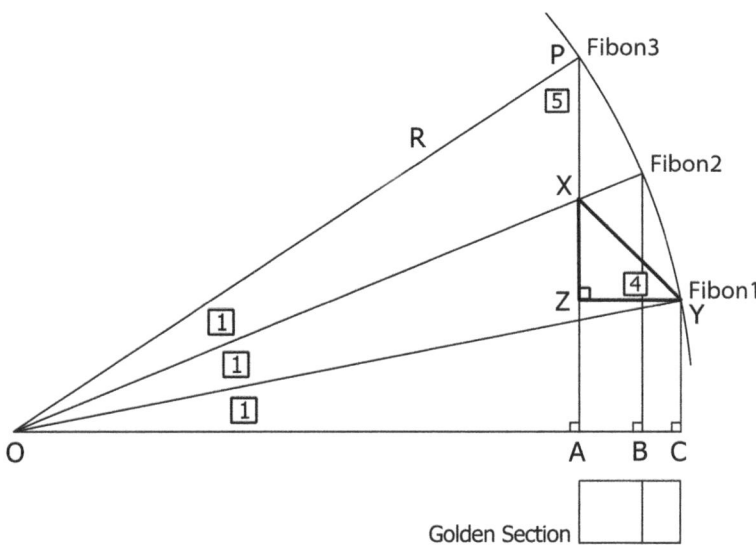

Figure 251: **On △ *XYZ* being Fibon4.**

In Fig. 250 we begin by identifying the Fibon3 and locating its $\boxed{5}$ angle at the top: vertex P. We confirm this as

$$\angle PAO = \boxed{8} \text{ and } \angle AOP = \boxed{3} \implies \angle OPA = \boxed{5}.$$

Therefore, by reflection symmetry about OX,

$$\angle OYX = \boxed{5}. \tag{A.1}$$

Now, as ZY is parallel to OA, and $\angle YOA = \boxed{1}$, then

$$\angle OYZ = \boxed{1}.$$

We subtract this from $\angle OYX$, which we showed in (A.1) is $\boxed{5}$

$$\angle XYZ = \boxed{5} - \boxed{1} = \boxed{4}.$$

Which means that $\triangle XYZ$ is Fibon4, and therefore $ZX = ZY$. ∎

Example solution—Exercise 3, proving *XYZ* is Fibon4 using sines and cosines   p.166

Again using Fig. 250 (p.406), first we let $R = 1$. It then follows that

$$OA = \cos \boxed{3}$$

$$AX = \cos \boxed{3} \tan \boxed{2}$$

$$CY = \sin \boxed{1}$$

$$XZ = AX - AZ = AX - CY$$

$$= \cos \boxed{3} \tan \boxed{2} - \sin \boxed{1}. \tag{A.2}$$

$$ZY = AC = \cos \boxed{1} - \cos \boxed{3}. \tag{A.3}$$

If $\triangle XYZ$ is Fibon4, then $XZ = ZY$;
and as $ZY = AC$, then $XZ - AC = 0$.

To show this is the case, we restate the Ori32 trig. expressions using the short-form analytic values (in α and β) from the reference Tables on page 402. So using (A.2) and (A.3),

$$XZ - AC = \left(\cos \boxed{3} \tan \boxed{2} - \sin \boxed{1}\right) - \left(\cos \boxed{1} - \cos \boxed{3}\right)$$

$$= \frac{\alpha + \beta}{2\sqrt{2}} \cdot (\sqrt{2} - 1) \quad - \frac{\beta}{2} \quad - \frac{\alpha}{2} \quad + \frac{\alpha + \beta}{2\sqrt{2}}.$$

Now we let $\gamma = \alpha + \beta$, so

$$XZ - AC = \frac{\gamma}{2\sqrt{2}} \cdot (\sqrt{2} - 1) - \frac{\gamma}{2} + \frac{\gamma}{2\sqrt{2}}$$

and finally multiply all through by $2\sqrt{2}$ to simplify,

$$2\sqrt{2}(XZ - AC) = \gamma(\sqrt{2} - 1) - \gamma\sqrt{2} + \gamma \quad = 0.$$

Appendix B—Fibonacci hexads modulo 32

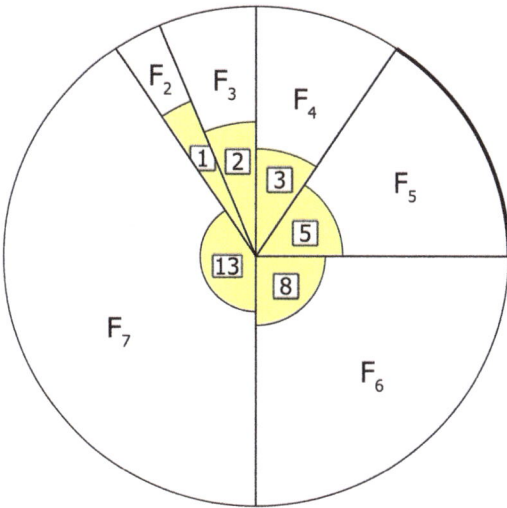

Figure 252: **Ori32 Fibonacci circle** (p.168).

Abstract

We prove that for the 'hexad' of 6 Fibonacci numbers $F_2 = 1$ through $F_7 = 13$ (Fig. 252), and for each successive hexad (and for each preceding too); the sum of its elements is a multiple of 32. Also modulus 32 residue ones of the partial sums of F_n are considered. Next, (mod 32) residue twos are found with period 6 in the Fibonacci sequence and Tables are given for these. Further, successive triads of even-index terms centred on F_{6n} are shown to be multiples of 32. Residue distribution within the 8 hexads of a Pisano period of $\langle F_n \rangle$ is explored, along with frequencies of particular sets of residues in a Pisano period of the cumulative sequence (i.e. partial sums of F_n, whose residues also have period 48), and these are compared with Elliot T Jacobson's results for the Fibonacci sequence itself.

409

Theorem:

$$\sum_{n=2+6k}^{7+6k} F_n \equiv 0 \pmod{32}, \qquad \text{for all } k \text{ in } \mathbb{Z} \tag{B.1}$$

Proof by Induction:

Base case:

We show that (B.1) is true when $k = 0$,

$$\sum_{n=2}^{7} F_n = 1 + 2 + 3 + 5 + 8 + 13 = 32, \qquad 32 \equiv 0 \pmod{32}.$$

Inductive step for $k \geq 0$:

If we assume for any integer $j \geq 0$ it is true that

$$A = \sum_{n=2+6j}^{7+6j} F_n \equiv 0 \pmod{32} \tag{B.2}$$

then we must show that (say) $B \equiv 0 \pmod{32}$, where

$$B = \sum_{n=2+6(j+1)}^{7+6(j+1)} F_n = \sum_{n=8+6j}^{13+6j} F_n. \tag{B.3}$$

We start by simplifying the bounds in (B.2) and (B.3)—letting

$$m = 2 + 6j \tag{B.4}$$

so that (B.2)'s summation becomes

$$A = \sum_{n=m}^{m+5} F_n \tag{B.5}$$

and similarly (B.3)'s becomes

$$B = \sum_{n=m+6}^{m+11} F_n. \tag{B.6}$$

Simplifying range sums down to two terms

Now let us express the numbers being summed in these (B.5) and (B.6) as the sums of successively smaller Fibonacci numbers. We do this by repeated application of the recurrence relation just as we did on page 196. For example, to begin decomposing the elements of (B.5):

$$F_{m+5} = F_{m+4} + F_{m+3}, \qquad \text{then} \quad F_{m+4} = F_{m+3} + F_{m+2}, \qquad \text{and so on...}$$

We halt this process when we have got down to terms in F_{m+1} and F_m

$$A = \sum_{n=m}^{m+5} F_n \qquad = \quad 12F_{m+1} + 8F_m \qquad\qquad \text{(B.7)}$$

and similarly treating (B.6),

$$B = \sum_{n=m+6}^{m+11} F_n \qquad = \quad 220F_{m+1} + 136F_m. \qquad\qquad \text{(B.8)}$$

Now, as we shall be interested in the factors of the coefficients, we will see advantage in analysing the difference $B - A$. Let

$$C = B - A \qquad = \quad 208F_{m+1} + 128F_m. \qquad\qquad \text{(B.9)}$$

Justification for using C

Our target is to show that $B \equiv 0 \pmod{32}$, and in (B.2) we assumed that $A \equiv 0 \pmod{32}$, therefore

$$C = B - A \equiv 0 \pmod{32} \qquad \text{implies} \qquad B \equiv 0 \pmod{32}. \qquad \text{(B.10)}$$

Proof that $C \equiv 0$ (mod 32), part 1/2:   Analysing the '208' term

We will consider the $208F_{m+1}$ term in (B.9). But first, we must note the odd/even parity in the Fibonacci sequence—it is well known that this repeats with period three: $\underline{0}, 1, 1;$ $\underline{2}, 3, 5;$ $\underline{8}, 13, 21;$ $\ldots$

$$\langle \textit{even, odd, odd} \rangle, \quad \langle \underline{F_0}, F_1, F_2 \rangle, \quad \langle \underline{F_3}, F_4, F_5 \rangle, \quad \langle \underline{F_6}, F_7, F_8 \rangle, \quad \ldots$$

We set an index i, where

$$i = m + 1, \quad \text{and applying} \quad m = 2 + 6j \quad \text{from (B.4), this becomes}$$
$$= 3 + 6j, \quad j \leq 0. \qquad\qquad \text{(B.11)}$$

This (B.11) shows that for *any* integer j our index i will always be a multiple of 3; so $F_i = F_{m+1}$ is always *even* in this context. It then follows that F_{m+1} can only have residues (mod 32) of $0, 2, 4, \ldots 30$. And as $208 = 13 \times 16$, then the even F_{m+1} residue will multiply the 16 factor to produce some multiple of 32, hence

$$208F_{m+1} \equiv 0 \pmod{32}. \qquad\qquad \text{(B.12)}$$

Proof that $C \equiv 0$ (mod 32), part 2/2:   Analysing the '128' term

In (B.9), we notice that $128 = 4 \times 32$. So, clearly,

$$128 F_n \equiv 0 \quad (\text{mod } 32) \quad \text{for all } F_n \qquad \text{(B.13)}$$

and using (B.12) and (B.13) in (B.9)

$$C \equiv 0 \quad (\text{mod } 32)$$

so from (B.10), we fulfill the requirement of (B.3), namely that
$$B \equiv 0 \quad (\text{mod } 32).$$

This completes the proof of the theorem for $k \geq 0$. Now, to address k negative, we shall make the inductive step *backwards*—this technique is called 'reverse induction'.

Reverse inductive step for $k < 0$:

If we again start from (B.2), and now assume that for *any* integer j it is true that

$$A = \sum_{n=2+6j}^{7+6j} F_n \equiv 0 \quad (\text{mod } 32) \qquad \text{(B.2) copy}$$

then we must show that (say) $W \equiv 0$ (mod 32), where

$$W = \sum_{n=2+6(j-1)}^{7+6(j-1)} F_n = \sum_{n=6j-4}^{1+6j} F_n. \qquad \text{(B.14)}$$

As before we simplify the bounds, this time letting $m = 6j - 4$.
So from (B.2), and then reusing the form of (B.8)

$$A = \sum_{n=m+6}^{m+11} F_n = 220 F_{m+1} + 136 F_m \qquad \text{(B.15)}$$

and similarly from (B.14) and reusing the form of (B.7)

$$W = \sum_{n=m}^{m+5} F_n = 12 F_{m+1} + 8 F_m. \qquad \text{(B.16)}$$

Again we choose to work with a difference, this time $A - W$. Let

$$C = A - W = 208 F_{m+1} + 128 F_m.$$

Our target now is to show $W \equiv 0$ (mod 32), and as we initially assumed that $A \equiv 0$ (mod 32), then $C = A - W \equiv 0$ (mod 32) implies $W \equiv 0$ (mod 32). The benefit to us is that C has the same definition in m here as it did for the $k \geq 0$ case (B.9). And although m is different here, the only consequence of this comes at the i index stage (B.11). Nevertheless, as $m = 6j - 4$; $i = m + 1 = 6j - 3$, then 3 divides i for all integer j, and the original outcome is maintained—so F_{m+1} is again always even. Hence (B.12) and (B.13) both hold, and we have therefore shown that $W \equiv 0$ (mod 32), which proves theorem (B.1) for negative k and thus completes its proof for all integer k. ∎

Corollary: Same result for any number of hexads

We have just proved that the total of any hexad—whose start is offset according to (B.1)—will be congruent with 0 (mod 32). Therefore by adding any number of such hexads together, we shall still get zero residue (mod 32), and it follows without further proof that

$$\sum_{n=2+6m}^{7+6k} F_n \equiv 0 \quad (\text{mod } 32), \quad \text{for all } k, m \text{ in } \mathbb{Z}, \ m \leq k \qquad \text{(B.17)}$$

∎

Corollary: Summation starting from F_0

From this last result, we may quickly obtain the expression for the cumulative summation starting with F_0 which is

$$\sum_{n=0}^{1+6k} F_n \equiv 1 \quad (\text{mod } 32), \quad k = 0, 1, 2, 3, \ldots \qquad \text{(B.18)}$$

Proof:

In (B.17) we set $m = 0$ and rename k as j to get

$$\sum_{n=2}^{7+6j} F_n \equiv 0 \quad (\text{mod } 32), \quad j = 0, 1, 2, 3, \ldots.$$

We then add $F_0 + F_1 = 1$ to both sides to start the summation from F_0

$$\sum_{n=0}^{7+6j} F_n \equiv 1 \quad (\text{mod } 32),$$

and then let $j = k - 1$ so that $6j = 6k - 6$ and $7 + 6j$ becomes $1 + 6k$, giving us the result

$$\sum_{n=0}^{1+6k} F_n \equiv 1 \quad (\text{mod } 32), \quad k = 0, 1, 2, 3, \ldots.$$

∎

$$\sum_{n=0}^{1+6k} F_n \equiv 1 \pmod{32}, \qquad k = 0, 1, 2, 3, \dots$$

n	F_n	Cumulative starting F_0	Residue (mod 32)	k
0	0	0	0	
1	1	1	**1**	0
2	1	2	2	
3	2	4	4	
4	3	7	7	
5	5	12	12	
6	8	20	20	
7	13	33	**1**	1
8	21	54	22	
9	34	88	24	
10	55	143	15	
11	89	232	8	
12	144	376	24	
13	233	609	**1**	2
14	377	986	26	
15	610	1596	28	
16	987	2583	23	
17	1597	4180	20	
18	2584	6764	12	
19	4181	10945	**1**	3
20	6765	17710	14	
21	10946	28656	16	
22	17711	46367	31	
23	28657	75024	16	
24	46368	121392	16	
25	75025	196417	**1**	4
26	121393	317810	18	
27	196418	514228	20	
28	317811	832039	7	
(continued...)				

Figure 253: **Residue ones of cumulative F_n (mod 32) from F_0.** Each entry in the k column marks a residue 1. Also we note how each cumulative (i.e. partial sum) is a Fibonacci number less 1 (Sloane A000071) according to the identity $\Sigma F_n = F_{n+2} - 1$ [490]. (This Table is continued in Fig. 254).

n	F_n	Cumulative starting F_0	Residue (mod 32)	k
29	514229	1346268	28	
30	832040	2178308	4	
31	1346269	3524577	**1**	5
32	2178309	5702886	6	
33	3524578	9227464	8	
34	5702887	14930351	15	
35	9227465	24157816	24	
36	14930352	39088168	8	
37	24157817	63245985	**1**	6
38	39088169	102334154	10	
39	63245986	165580140	12	
40	102334155	267914295	23	
41	165580141	433494436	4	
42	267914296	701408732	28	
43	433494437	1134903169	**1**	7
44	701408733	1836311902	30	
45	1134903170	2971215072	0	
46	1836311903	4807526975	31	
47	2971215073	7778742048	0	
48	4807526976	12586269024	0	
49	7778742049	20365011073	**1**	8
50	12586269025	32951280098	2	
51	20365011074	53316291172	4	
52	32951280099	86267571271	7	
53	53316291173	139583862444	12	
54	86267571272	225851433716	20	
55	139583862445	365435296161	**1**	9
56	225851433717	591286729878	22	
57	365435296162	956722026040	24	
58	591286729879	1548008755919	15	
59	956722026041	2504730781960	8	
60	1548008755920	4052739537880	24	
...				

Figure 254: **Residue ones of cumulative F_n (mod 32) from F_0, continued.**

415

Corollary: Residue twos, period 6

On page 248 we considered F_n (mod 10) and saw it had period 60. As for the sequence of residues of F_n (mod 32), this repeats in a cycle of 48 residues. Such cycles are called 'Pisano periods' (after Leonardo Pisano: Fibonacci). *Ohio mathematics professor (and gaming specialist) Eliot Jacobson, in his much more general study of this topic, states that for modulus 32 there are 8 instances of residue 2 per (Pisano) period of 48* [406, 407].[1] Here we show that

$$\boxed{F_{3+6k} \equiv 2 \pmod{32}, \quad k = 0, 1, 2, 3, \ldots}$$ (B.19)

Proof:

The identity for partial sums of $\langle F_n \rangle$ [490] (that is, cumulative Fibonacci) states

$$\sum_{n=0}^{m} F_n = F_{m+2} - 1.$$

We rearrange this as

$$F_{m+2} = 1 + \sum_{n=0}^{m} F_n.$$

We then let $m = 1 + 6k$, to give

$$F_{3+6k} = 1 + \sum_{n=0}^{1+6k} F_n.$$ (B.20)

Now, from page 413, we already have the congruence

$$\sum_{n=0}^{1+6k} F_n \equiv 1 \pmod{32}, \quad k = 0, 1, 2, 3, \ldots$$ (B.18) copy

We therefore combine this (B.18) with (B.20) to obtain the result

$$F_{3+6k} \equiv 2 \pmod{32}, \quad k = 0, 1, 2, 3, \ldots$$

The effect of (B.19) is illustrated by the Tables in Figs. 255 and 256.

■

---

[1] The periodic behaviour of Fibonacci modulo residues was studied in detail by Lagrange and more recently by Donald Dines Wall [931].

$$F_{3+6k} \equiv 2 \pmod{32}, \quad k = 0, 1, 2, 3, \ldots$$

n	F_n	Residue (mod 32)	k	Pisano start
0	0	0		
1	1	1		
2	1	1		$\leftarrow F_2$
3	2	2	0	
4	3	3		
5	5	5		
6	8	8		
7	13	13		
8	21	21		
9	34	2	1	
10	55	23		
11	89	25		
12	144	16		
13	233	9		
14	377	25		
15	610	2	2	
16	987	27		
17	1597	29		
18	2584	24		
19	4181	21		
20	6765	13		
21	10946	2	3	
22	17711	15		
23	28657	17		
24	46368	0		
25	75025	17		
26	121393	17		
27	196418	2	4	
28	317811	19		
(continued...)				

Figure 255: **Residue twos of F_n (mod 32).** Each entry in the k column marks a residue 2—page 171. Here we choose a starting point for Pisano cycles at $n = 2$ to match the sequence on page 419, and this pattern of residues then repeats with its next start at $n = 50$, then $n = 98$, and so on. (This Table is continued in Fig. 256).

n	F_n	Residue (mod 32)	k	Pisano start
29	514229	21		
30	832040	8		
31	1346269	29		
32	2178309	5		
33	3524578	**2**	5	
34	5702887	7		
35	9227465	9		
36	14930352	16		
37	24157817	25		
38	39088169	9		
39	63245986	**2**	6	
40	102334155	11		
41	165580141	13		
42	267914296	24		
43	433494437	5		
44	701408733	29		
45	1134903170	**2**	7	
46	1836311903	31		
47	2971215073	1		
48	4807526976	0		
49	7778742049	1		
50	12586269025	1		← F_{50}
51	20365011074	**2**	8	
52	32951280099	3		
53	53316291173	5		
54	86267571272	8		
55	139583862445	13		
56	225851433717	21		
57	365435296162	**2**	9	
58	591286729879	23		
59	956722026041	25		
60	1548008755920	16		

...

Figure 256: **Residue twos of F_n (mod 32), continued.** Again we confirm the period as 6, set in the context of a Pisano period of 48. The Pisano residues we saw in Fig. 255 started from $n = 2$ as $\langle 1, 2, 3, 5, 8, 13, 21, 2, \ldots \rangle$, and here from $n = 50$, we see the same sequence.

Corollary: Even-Indexed Triads

$$F_{6n+2} + F_{6n} + F_{6n-2} \equiv 0 \quad (\text{mod } 32), \qquad n = 0, 1, 2, 3, \ldots$$

$$(\text{B.21})$$

Proof:

Theorem (B.21) is clearly true for $n=0$, as $F_2 + F_0 + F_{-2}=0$. Also, a consequence of (B.19), 416 is that

$$F_{3+6m} - F_{3+6k} \equiv 0 \quad (\text{mod } 32), \qquad m, k \text{ in } \mathbb{Z}, \ 0 \le k, \ k \le m$$

and looking at this with $m = k+1$, and expanding the first term (and resulting odd-index terms) into constituent pairs of preceding Fibonacci numbers,

$$F_{3+6(k+1)} - F_{3+6k} \equiv 0 \quad (\text{mod } 32)$$

$$F_{9+6k} - F_{3+6k} \equiv 0 \quad (\text{mod } 32)$$

$$[F_{8+6k} + \{F_{6+6k} + (F_{4+6k} + F_{3+6k})\}] - F_{3+6k} \equiv 0 \quad (\text{mod } 32)$$

$$F_{8+6k} + F_{6+6k} + F_{4+6k} \equiv 0 \quad (\text{mod } 32).$$

Putting $6n = 6 + 6k$, we are left with a triad of successive even-index terms, centred on F_{6n}

$$F_{6n+2} + F_{6n} + F_{6n-2} \equiv 0 \quad (\text{mod } 32). \qquad \blacksquare$$

Residue distribution within a Pisano period

Finally, we shall examine the distribution of Fibonacci residues. One Pisano period of the Fibonacci sequence (mod 32) has length 48. Starting from F_2 it consists of: $\langle 1, \mathbf{2}, 3, 5, 8, 13, 21, \mathbf{2}, 23, 25, 16, 9, 25, \mathbf{2}, 27, 29, 24, 21, 13, \mathbf{2}, 15, 17, 0, 17, 17, \mathbf{2}, 19, 21, 8, 29, 5, \mathbf{2}, 7, 9, 16, 25, 9, \mathbf{2}, 11, 13, 24, 5, 29, \mathbf{2}, 31, 1, 0, 1 \rangle$. The residue twos here repeat with period 6 —Figs. 255 and 256, starting on page 417.

Jacobson explores the residue frequencies of $\langle F_n \rangle$ for various moduli [406]. But instead, we shall focus only on modulus 32 residues, and these only in the sequence of partial sums of $\langle F_n \rangle$ starting from F_2. This cumulative sequence also exhibits a (mod 32) Pisano period of 48. Starting from the residue for F_2 we shall list all 48 residues in 8 groups of 6 (i.e. 8 hexads), so that each group ends with a zero: $\langle 1, 3, 6, 11, 19, \mathbf{0}, 21, 23, 14, 7, 23, \mathbf{0}, 25, 27, 22, 19, 11, \mathbf{0}, 13, 15, 30, 15, 15, \mathbf{0}, 17, 19, 6, 27, 3, \mathbf{0}, 5, 7, 14, 23, 7, \mathbf{0}, 9, 11, 22, 3, 27, \mathbf{0}, 29, 31, 30, 31, 31, \mathbf{0}. \rangle$ As F_0 and F_1 are omitted, each of these elements is one less (mod 32) than the corresponding residue in the cumulative sequence from F_0 (Figs. 253 and 254, starting on page 414).

Now we will check the first residue of each hexad, and then check the second for each, and so on. To do this, we write the residues for each hexad into successive columns, which then allows us to read off all the first terms as the top row, all the 2nd terms as the next row and so on:

1st:	1,	21,	25,	13,	17,	5,	9,	29,
2nd:	3,	23,	27,	15,	19,	7,	11,	31,
3rd:	6,	14,	22,	30,	6,	14,	22,	30,
4th:	11,	7,	19,	15,	27,	23,	3,	31,
5th:	19,	23,	11,	15,	3,	7,	27,	31,
6th:	0,	0,	0,	0,	0,	0,	0,	0.

We observe per Pisano period (in order of greatest frequency):

- **8** instances of 0—when every 6th term is added to the sum, the resulting residue (mod 32) will always be 0, as we saw earlier in the main theorem (p.168),[2]

- **3** instances of each of: 3, 7, 11, 15, 19, 23, 27, 31—the 2nd, 4th, and 5th adds all produce this same set of odd residues, differently ordered,[3]

- **2** instances of each of: 6, 14, 22, 30—the 3rd add produces only these even residues,[4]

- **1** instance of each of: 1, 5, 9, 13, 17, 21, 25, 29—the 1st term added produces these odd residues—the 'in betweens' of the '3 instance' odd residues,[5]

- **0** instances of: 2, 4, 8, 10, 12, 16, 18, 20, 24, 26, 28—these (even) residues are *never* produced.

This set of **5** residue frequencies: 8, 3, 2, 1, and 0, are the same set found by Jacobson for the (non-cumulative) sequence $\langle F_n \rangle$—but with different residues, e.g. his 8 instances are of residue 2 (mod 32)—which we saw demonstrated in Fig. 255 on page 417 [406]. It would be hard not to notice that each of these frequencies and even their count (i.e. 5 frequencies listed) is a Fibonacci number.

---

[2] For the 8 instances case, each zero residue is congruent with '8 (mod 4)'.
[3] For the 3 instances case, each residue is congruent with 3 (mod 4).
[4] For the 2 instances case, each residue is congruent with 2 (mod 4).
[5] For the 1 instance case, each residue is congruent with 1 (mod 4).

Appendix C—
Continued fractions

Continued fraction for ϕ

We already know that if we take the reciprocal of 1.618... we get 0.618... ($1/\phi$), which earlier we called q_1. Working in the opposite direction, we have $1/q_1 = 1 + q_1$ ((15.7) page 210). From this we shall now see how a simple recursive definition provides us with continued fraction representations for both ϕ and $1/\phi$. We start with Eqn. (1.3) from p.2—written in its positive root ϕ

$$\phi^2 - \phi - 1 = 0.$$

Rearranging we get

$$\phi = 1 + \frac{1}{\phi}. \tag{C.1}$$

Now, targetting the ϕ in the denominator on the right of (C.1), we substitute the whole of (C.1) in again

$$\phi = 1 + \cfrac{1}{1 + \cfrac{1}{\phi}} \tag{C.2}$$

and we repeat this process, ('recursively substituting'), to give

$$\phi = 1 + \cfrac{1}{1 + \cfrac{1}{1 + \cfrac{1}{1 + \ddots}}}. \tag{C.3}$$

But as this notation can become rather unwieldy...

a more elegant alternative has been devised, whereby

$$a_0 + \cfrac{1}{a_1 + \cfrac{1}{a_2 + \cfrac{1}{a_3 + \cdots}}} \qquad \text{is written as} \qquad [\, a_0;\ a_1,\ a_2,\ a_3,\ \ldots\,].$$

So, using this notation, (C.3) becomes

$$\phi = [\,1;\ 1,\ 1,\ 1,\ 1,\ 1,\ \ldots\,]. \qquad\qquad (C.4)$$

Also as $\phi^{-1} = \phi - 1$, then

$$\phi^{-1} = [\,0;\ 1,\ 1,\ 1,\ 1,\ 1,\ \ldots\,]. \qquad\qquad (C.5)$$

Now, because the continued fraction for ϕ has 'all ones', it is the slowest to converge (compared to fractions with larger denominators whose successive adjustments dwindle more rapidly). Hence, ϕ is sometimes called 'the most irrational number' (but see footnote p.48). To make the elegant notation even more so, there is a convention that repeating digits (or groups of digits) are indicated by an overline (vinculum), thus $\phi = [1;\overline{1}]$. We shall see further examples of this shortly.[1]

Fibonacci fractions—convergents of ϕ

We have just seen that the continued fraction for ϕ is $[1; 1, 1, 1, 1, \ldots]$. Let's evaluate this stage by stage, to obtain successive rational approximations. These increasingly precise results are called 'the convergents' of ϕ. We start with $\phi \approx 1$. Then for each stage, we take the reciprocal of the last result and add one, and this gives us our next approximation:

$$\phi \approx 1, \qquad \phi \approx 1 + \frac{1}{1} = 2, \qquad \phi \approx 1 + \cfrac{1}{1 + \cfrac{1}{1}} = \frac{3}{2},$$

$$\phi \approx 1 + \cfrac{1}{1 + \cfrac{1}{1 + \cfrac{1}{1}}} = \frac{5}{3}, \qquad\qquad \phi \approx 1 + \cfrac{1}{1 + \cfrac{1}{1 + \cfrac{1}{1 + \cfrac{1}{1}}}} = \frac{8}{5},$$

$$\text{thence} \qquad \phi \approx \frac{13}{8},\ \frac{21}{13},\ \frac{34}{21},\ \frac{55}{34},\ \ldots \ \frac{F_{n+1}}{F_n}. \qquad (C.6)$$

---

[1] Surprisingly, it is almost always true (*though clearly not for ϕ*), that as $n \to \infty$, then $\sqrt[n]{a_1 a_2 a_3 \ldots a_n} \to K_0$, where Khinchin's constant, $K_0 = 2.68545\ldots$ Sloane A002210.

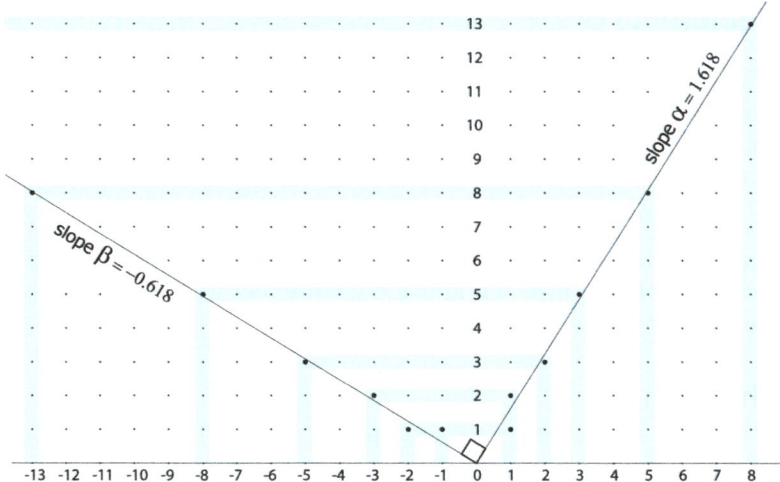

Figure 257: **Alpha and beta as slopes.**
Diagram developed from ϕ line diagram by Ron Knott with kind permission [495].

We recall that Kepler had a hunch that the ratio of successive pairs of Fibonacci numbers tend closer and closer to ϕ [467]. In (C.6) we see these same pairs emerging naturally as we evaluate the continued fraction for ϕ, stage by stage. In Fig. 257, the larger dots mark coordinates that are successive pairs of Fibonacci numbers—$(1, 1)$ $(1, 2)$ $(2, 3)$ $(3, 5)$ etc. These give us a way of visualizing the convergents of the Golden Ratio. Representing the denominators and numerators of the convergents, the dots at each (x, y) get closer and closer to the ϕ sloped line—they zigzag either side of it.[2] We know that a slope gradient m will have a normal with gradient $-1/m$. Here we have a slope of ϕ (which we often call α), so its normal will have slope $-1/\phi$ (which we often call β). This diagram shows very clearly the nature of irrationality. Both slope lines extend infinitely, and although they get arbitrarily close to integer grid points, they *never* pass through them—if a slope did pass through a single grid point, then that would mean its slope could be expressed as one integer divided by another, and it would be rational. These slopes are irrational. In terms of the decimal expansion expansion of an irrational, its digits must continue indefinitely and never fall into a repeating pattern—otherwise (again) the number would have a corresponding slope line that passed through some integer grid point—it would be rational.

---

[2]   In the diagram, the slope lines make angles with the x-axis of $arctan\,(1/\phi)$=31.72° and $arctan\,(\phi)$=58.28°.

Reciprocal of q_5

When we first saw ϕ^{-5} (Equation (13.3), p.195) we called it just q. Now, for consistency with the generalization we made (for many spirals, each with its own associated quantum size), we shall refer to it as q_5. Now let's consider its reciprocal

$$q_5 = 0.0901699437495\ldots, \quad \text{so} \quad q_5^{-1} = 11.0901699437495\ldots$$

Magic? Perhaps. On the face of it, we appear to have found that the reciprocal of q_5 is just: '11, and add the number we first thought of.' We shall now prove this.

Theorem:

$$x = q_5 = \phi^{-5} \quad \text{is a solution of} \quad x^{-1} = 11 + x. \quad (C.7)$$

Proof:

Rearranging (C.7) we get $x^2 + 11x - 1 = 0$. Therefore, using the quadratic formula and taking the positive root, we get (in terms of $\sqrt{5}$)

$$x = \frac{-11 + 5\sqrt{5}}{2}.$$

Next we subtract 5ϕ from each side to eliminate $\sqrt{5}$ in favour of ϕ (and recall that $\phi^{-5} = 5\phi - 8$, (13.4), p.195):

$$x - 5\phi = \frac{-11 + 5\sqrt{5}}{2} - \frac{(5 + 5\sqrt{5})}{2} = -8$$

$$x = 5\phi - 8 = \phi^{-5} = q_5. \quad \blacksquare$$

Silver means—the meaning of

Again to avoid confusion, we need to note that while the terms Golden Ratio and Golden mean are used interchangeably, the same is not true for Silver. We know the Silver Ratio very well as $1 + \sqrt{2}$, and it is specific. On the other hand, the term 'Silver mean' is used as a generalization. As an example, in (C.7), $1/x = \phi^5$, so we may restate the equation as $\phi^5 = 11 + 1/\phi^5$—and we see that ϕ^5 is 'a number that is a whole number greater than its reciprocal'. All such numbers are called 'Silver means' [493]. The distinction here echoes the one we met earlier, between the specific term 'Lucas number sequence' and the general term 'Lucas sequences'. In this book we shall call the specific cases 'ratios' and (in this context) endeavour to reserve the term 'mean' solely for generalizations. We will return to this subject shortly.

Rapid convergence in CF for ϕ^{-5}

If we apply the recursive principle given in Eqn. (C.2) on page 421 to (C.7) (p.424), then it is easy to see that the CF for q_5 is [0; 11, 11, 11, 11, 11, ...]. We may quickly check how fast this CF converges with a hand calculator (Table, Fig. 258). In fact, only 3 simple iterations are sufficient to give the value of q_5 correct to 7 places of decimals (results are bolded for comparison). And as $q_5 = \phi^{-5} = 5\phi - 8$, if we now 'add 8 and divide by 5', we shall get the Golden Ratio to the same precision. Another way to approximate ϕ is to take two large adjacent Fibonacci numbers and divide the greater by the lesser.

hand calculator		displayed result
clear		0.0
add 11	*then* $\frac{1}{x}$	0.09090909
add 11	*then* $\frac{1}{x}$	0.09016393
add 11	*then* $\frac{1}{x}$	**0.09016999**
compare with q_5		**0.09016994**

Figure 258: **Rapid convergence of continued fraction for ϕ^{-5}.**

Continued fractions for sundry powers of ϕ

Checking the CF's for several Golden Powers (Fig. 259 below), we may be struck by the repeated appearance of Lucas numbers (2, 1, 3, 4, 7, 11, 18, 29, 47, ...), or 'one or two less'. This is not a coincidence, and we shall prove an existing theorem as confirmation. (Elements with an overline form a block that repeats indefinitely.)

Odd ϕ powers	Even ϕ powers
$\phi^1 = [1;\overline{1}]$	$\phi^2 = [2;\overline{1,1}]$
$\phi^3 = [4;\overline{4}]$	$\phi^4 = [6;\overline{1,5}]$
$\phi^5 = [11;\overline{11}]$	$\phi^6 = [17;\overline{1,16}]$
$\phi^7 = [29;\overline{29}]$	$\phi^8 = [46;\overline{1,45}]$
...	...
$\phi^n = [L_n;\ \overline{L_n}]$	$\phi^n = [(L_n-1);\ \overline{1,(L_n-2)}]$

Figure 259: **Continued fractions for odd and even powers of ϕ** [857, 494].

Theorem:

$$\phi^n = \begin{cases} [\, L_n;\ \overline{L_n}\,], & \text{if } n \text{ is odd,} \\ [\, (L_n - 1);\ \overline{1,\ (L_n - 2)}\,], & \text{if } n \text{ is even.} \end{cases} \qquad \text{(C.8)}$$

Proof:

This proof is offered for the above already known theorem and is based on the initial approach of Kalia [457], and uses identities from Vajda and Ruggles. (Another proof is given by de Spinadel [857].)

When n is odd,

$$\text{let } x = [\, L_n;\ \overline{L_n}\,] \;=\; L_n + [0;\ \overline{L_n}\,].$$

Therefore,

$$x = L_n + \frac{1}{x} \quad \Longrightarrow \quad x^2 - L_n x - 1 = 0. \qquad \text{(C.9)}$$

When n is even,

$$\text{let } x = [\, (L_n - 1);\ \overline{1,\ (L_n - 2)}\,] \quad \text{and} \qquad \text{(C.10)}$$
$$\text{let } y = (L_n - 2). \qquad \text{(C.11)}$$

Therefore, subtracting 1 from each side in (C.10)

$$x - 1 = [\, (L_n - 2);\ \overline{1,\ (L_n - 2)}\,]$$

and substituting $(L_n - 2)$ with y (from (C.11))

$$= [\, y;\ \overline{1;\ y}\,]$$

$$= y + \cfrac{1}{1 + \cfrac{1}{y + \cfrac{1}{1 + \cfrac{1}{y + \cfrac{1}{1 + \ddots}}}}}. \qquad \text{(C.12)}$$

But as we meet $y + 1/$etc. again in the continued fraction (C.12), we realize we can substitute the whole equation itself at that point—that is, with $x - 1$

$$x - 1 = y + \cfrac{1}{1 + \cfrac{1}{(x-1)}} = y + \frac{x-1}{x}.$$

And using $y = L_n - 2$ (C.11) we get

$$x^2 - L_n x + 1 = 0. \qquad (C.13)$$

We may now combine odd case (C.9) and even case (C.13) by introducing $(-1)^n$, which suits both,

$$x^2 - L_n x + (-1)^n = 0. \qquad (C.14)$$

Now, the discriminant for the combined quadratic is[3]

$$\Delta = L_n^2 - 4(-1)^n. \qquad (C.15)$$

To simplify this, we use two identities given by Vajda:

$$L_n^2 = L_{2n} + 2(-1)^n \qquad \text{[915] and}$$

$$L_{2n} = 5F_n^2 + 2(-1)^n \qquad \text{[916].}$$

So, the (C.15) discriminant Δ reduces to

$$\Delta = 5F_n^2.$$

Applying the quadratic formula to (C.14) using this Δ value, we get x, which is the Ruggles identity [794] for ϕ^n, thus proving (C.8)

$$x = \frac{L_n + F_n\sqrt{5}}{2} = \phi^n.$$

■

Although this last identity may at first look mysterious, we see it is found simply by adding the F_n and L_n Binet forms and dividing by 2:

$$\sqrt{5}F_n = \alpha^n - \beta^n \qquad \text{from (1.8), page 25, and}$$
$$L_n = \alpha^n + \beta^n \qquad \text{from (1.10), page 28.}$$
$$\sqrt{5}F_n + L_n = 2\alpha^n.$$

---

[3] In the quadratic formula, the value $b^2 - 4ac$ (p.2, footnote 4).

ϕ^{-3} and $\sqrt{5}$ CF convergents

In the *Art—Mondrian* chapter (p.102), we saw (twice) how the gnomon width ϕ^{-3} is related to the Lucas number 4. And to find the continued fraction for ϕ^{-3} we use (C.8) (p.426) to find ϕ^3 and take the reciprocal:

$$\phi^3 \;=\; [\,L_3;\; \overline{L_3}\,] \;=\; [\,4;\; \overline{4}\,]$$

$$\phi^{-3} = [\,0;\; \overline{4}\,] \;=\; \cfrac{1}{4 + \cfrac{1}{4 + \cfrac{1}{4 + \ddots}}}.$$
We evaluate this in Fig. 260:

hand calculator		displayed result
clear		0.0
add 4	*then* $\frac{1}{x}$	0.25000000
add 4	*then* $\frac{1}{x}$	0.23529412
add 4	*then* $\frac{1}{x}$	0.23611111
add 4	*then* $\frac{1}{x}$	0.23606557
add 4	*then* $\frac{1}{x}$	**0.23606811**

compare with ϕ^{-3}	**0.23606797**.

Figure 260: **Convergence of continued fraction for ϕ^{-3}.**

As expected, convergence is slower for 4 than it was with 11 (p.425). Now, does this 0.236067 look familiar? Well, yes, it is the decimal part of $\sqrt{5} = 2.236067\ldots$ In fact, $\sqrt{5} = 2 + \phi^{-3}$. Therefore we only need add 2 to the continued fraction for ϕ^{-3} in order to get the CF for $\sqrt{5}$—that is, $[\,2;\; \overline{4}\,]$. Then, by evaluating the successive steps of refinement (just as with the Fibonacci fractions p.422), we get the convergents of $\sqrt{5}$:

$$\frac{2}{1}, \;\; \frac{9}{4}, \;\; \frac{38}{17}, \;\; \frac{161}{72}, \;\; \frac{682}{305}, \;\; \cdots$$

The sequences of numerators (A001077) and denominators (A001076) share the same PQ-Lucas-Sequence recurrence viz. $a_n = 4a_{n-1} + a_{n-2}$, ($P=4$, $Q=-1$). Here, the seeds for the numerators are: $\langle 1, 2 \rangle$, and those for the denominators are: $\langle 0, 1 \rangle$; while the $(4, -1)$ companion sequence is the even Lucas numbers (L_{3n}, pp.267 & 274, A014448), with seeds: $\langle 2, 4 \rangle$ (that is, twice the numerator seeds). Hence the numerators are half even Lucas numbers $\langle L_{3n}/2 \rangle$. (Similarly, numerators of $\sqrt{2}$ convergents are half companion Pell numbers— p.480.)

428

Silver means—as part of the Metallic Means Family

As we have seen, the continued fraction for the Golden Ratio is 'all ones': $\phi = [1; 1, 1, 1, \dots]$. For the Silver Ratio $\delta = 1 + \sqrt{2}$, it is 'all twos': $\delta = [2; 2, 2, 2, \dots]$ [944]. And for the Bronze Ratio $b = (3 + \sqrt{13})/2$, it is 'all threes': $b = [3; 3, 3, 3, \dots]$ [855].

$$\phi = 1 + \cfrac{1}{1 + \cfrac{1}{1 + \ddots}}, \qquad \delta = 2 + \cfrac{1}{2 + \cfrac{1}{2 + \ddots}}, \qquad b = 3 + \cfrac{1}{3 + \cfrac{1}{3 + \ddots}}.$$

This is illustrated in Fig. 261, where in each rectangle we see the integer part as a number of unit squares, and a downscaled rectangle is included at its end in the same proportion as the overall rectangle. As this is 'the right shape', it may be further partitioned in the same way indefinitely. Each stage corresponds to a partial evaluation of the continued fraction (as on p.422). The three associated spirals are also shown (Golden, Silver, and Bronze)—their centres being located by the shared intersection of successive rectangle diagonals. The Silver means have their basis in the PQ-generalized recurrences studied by Lucas (p.251). Also, in 1997 Vera W de Spinadel introduced her 'MMF'—Metallic Means Family where each mean is a real positive root of the PQ characteristic quadratic (20.4) on page 252 [853, 854]. As such the Silver means are a subset of the MMF, and every MMF member will be characterized by its unique pair of values—P and Q. (Note that de Spinadel does not negate q, e.g. her $p=1, q=2$ corresponds to Lucas' $P=1, Q=-2$.) We mentioned the Copper Ratio (=2) and the Nickel Ratio (=$(1 + \sqrt{13})/2$) earlier (p.267). Now we see these as examples of Metallic means which are not Silver means. This distinction is further clarified in the Table, Fig. 262.

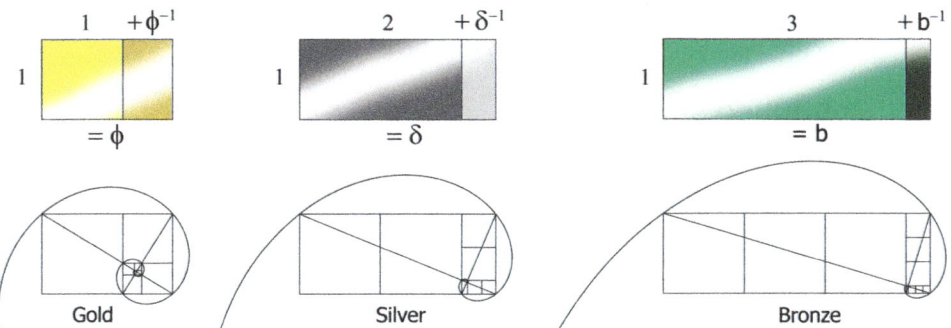

Figure 261: **The Golden, Silver, and Bronze Ratios—these are all 'Silver means'.**

Metallic Means summary

Ratio	(P,Q)	Formula	'Silver mean'	CF	Decimal
ϕ Golden	$(1,-1)$	$(1+\sqrt{5})/2$	✓	$[1;\overline{1}]$	1.61803...
δ Silver	$(2,-1)$	$(2+2\sqrt{2})/2$	✓	$[2;\overline{2}]$	2.41421...
b Bronze	$(3,-1)$	$(3+\sqrt{13})/2$	✓	$[3;\overline{3}]$	3.30278...
ϕ^3	$(4,-1)$	$(4+2\sqrt{5})/2$	✓	$[4;\overline{4}]$	4.23607...
ϕ^5	$(11,-1)$	$(11+5\sqrt{5})/2$	✓	$[11;\overline{11}]$	11.09017...
δ^3	$(14,-1)$	$(14+10\sqrt{2})/2$	✓	$[14;\overline{14}]$	14.07107...
2 Copper	$(1,-2)$	$(1+\sqrt{9})/2$	×	$[2;\overline{0}]$	2.00000...
Nickel	$(1,-3)$	$(1+\sqrt{13})/2$	×	$[2;\overline{3}]$	2.30278...
ϕ^2	$(3,1)$	$(3+\sqrt{5})/2$	×	$[2;\overline{1}]$	2.61803...
δ^2	$(6,1)$	$(6+4\sqrt{2})/2$	×	$[5;\overline{1,4}]$	5.82843...

Figure 262: **MMF—Metallic Means Family—sample members.**

Formulæ are the 'alphas' found by applying the (P,Q) value pair in Equations (20.10) and (20.11) on page 254. We think of them as the underlying growth rates of the corresponding PQ Lucas Sequences (both primary and companion). They are here all expressed in halves to highlight correspondence with the integer part of the CF in the 'Silver means'. Now, while we associate the Golden Ratio ϕ with the pentagon, and the Silver Ratio δ with the octagon—'Could there be a regular polygon associated with the Bronze Ratio?' Researcher and maths teacher Antonia Redondo Buitrago has already checked—she finds (through a set of inequalities) that there is not [93].

Continued square root

Lastly, as $\phi^2 = 1+\phi$, then in the positive square root, $\phi = \sqrt{1+\phi}$, we may recursively substitute for the final ϕ—just as we did with the continued fraction for ϕ—(C.1) on p.421—to create the nested radical

$$\phi = \sqrt{1+\sqrt{1+\sqrt{1+\ldots}}}$$

On a calculator: Take seed 1 (say), then repeat 'add 1 and square root', 'add 1 and square root', ... for better and better approximations to ϕ.

Appendix D—Powers of ϕ: Golden Powers

Root Golden [RG], $1 : \sqrt{\phi}$ —neglected?

Apparently, both Seurat (p.85) and Mondrian (p.99) used the [RG] Root Golden Rectangle—with the $\sqrt{\phi} : 1$ being located top left in each of the two cases that we studied. Also we noted that the basic Root-Golden geometry was known to Kepler (p.7). It also turns out that the Sun Gate (Chapter 1, p.11) outer shape (not counting the shallow arc top) has approximately Root-Golden proportion. We recall that to start the *Art—Seurat* section, we discussed 2 as an 'artistic approximation' of $\phi\sqrt{\phi}$, and considered the Root Golden Rectangle and its approximants (p.73). From this point of view, the Sun Gate rectangle has an aspect ratio of $\phi^3 : 2\phi$—that is, $\phi^2 : 2$. We recognize this ratio as [RG+]—the greater of the pair of first-order Root Golden approximants. Perhaps because of its overall rarity though, Wolfgang von Wersin did not include the Root Golden as one of his 'orthogons' (p.69).[1] But, although it is little discussed nowadays, the Root Golden Rectangle does have some further interesting geometry—in addition to the Root Golden Spiral construction we made on p.33—which in turn links it to complex Binet spirals (p.137) and to spiral proportional musical analysis (Lendvai and Howat, pp.141 and 155).

(Note from page 180.)[2]

---

[1] In his *Greek Vase* book [300], Hambidge (p.68) discusses Golden Rectangles and simple root rectangles at length, but never the Root Golden—perhaps he was not aware of it.

[2] By successive approximation, we find parabola crossing spiral at $x = 0.685927751$ (referred to the $y = x^2$ parabola cartesian coordinates). The spiral's radial vector to the crossing point makes an angle with the x-axis of $\gamma = \arctan(x^2/x) = 34.44730751°$. The slope of the parabola at x is given by the arctan of $dy/dx = 2x$—that is, $\theta = 53.91020543°$. Also we know the equiangular value for the Golden Spiral which is $\sigma = 72.9676089°$ from (2.9), p.41. It therefore follows that the intersection is at an angle of $\theta - \gamma + \sigma = 92.43...°$

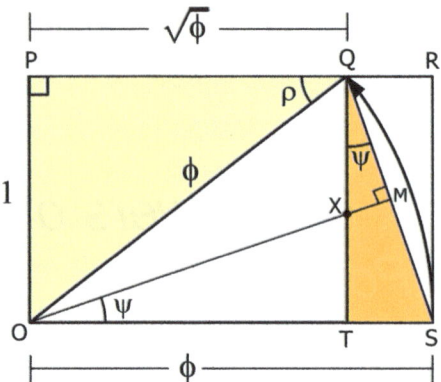

Figure 263: **Root-Golden geometry** $\rho = 2\psi$.

Let's revisit the simple Kepler triangle construction we made on p.33, and consider the rectangle that remains after the Root Golden is removed from the Golden Rectangle (that is, $QRST$ in Fig. 263). We note first that the angle $\rho = arctan(1/\sqrt{\phi})$ is almost two-thirds of a radian (actually 1.999 thirds), and second that angle ψ is exactly half this (i.e. almost one-third radian)—angle ρ=38.17°, angle ψ=19.09°. It is easy to see this 'factor 2' angle relationship via chord QS with midpoint M. Right triangles QMX and OTX are clearly similar—hence $\angle$XOT=ψ. And as $\triangle$OQS is isosceles, then $\angle$QOX=$\angle$XOT=ψ, so $\rho = 2\psi$. But, to explore this relationship in terms of ϕ and $\sqrt{\phi}$, we shall compare the tangents of angles ρ and ψ. We expect to be able to demonstrate that $tan(\rho) = tan(2\psi)$. Now

$$OP/PQ = tan(\rho) = 1/\sqrt{\phi}, \qquad \text{and} \qquad (D.1)$$

$$TS/QT = tan(\psi) = \phi - \sqrt{\phi}. \qquad (D.2)$$

So using (D.2) in the double angle formula,
$tan(2\psi) = 2tan(\psi)/[1 - tan^2(\psi)]$ [948],
and (in the expanded denominator) noting $\phi^2 = 1 + \phi$,

$$tan(2\psi) = \frac{2(\phi - \sqrt{\phi})}{1 - (\phi - \sqrt{\phi})^2} = \frac{2(\phi - \sqrt{\phi})}{-2\phi + 2\phi\sqrt{\phi}} = \frac{\phi - \sqrt{\phi}}{\phi(\sqrt{\phi} - 1)}$$

$$= \frac{\phi - \sqrt{\phi}}{\sqrt{\phi}\sqrt{\phi}(\sqrt{\phi} - 1)} = \frac{\phi - \sqrt{\phi}}{\sqrt{\phi}(\phi - \sqrt{\phi})} = \frac{1}{\sqrt{\phi}} \qquad (D.3)$$

thus showing the equality of (D.3) and (D.1). ∎

Summing powers of ϕ in terms of ϕ— additive approach

In the *5 Golden Powers* chapter, we added the powers explicitly one by one. Here we shall derive general formulæ. We shall take two different approaches to this and then check they match. The first way will give us an appreciation of what is involved along the way, and the second will be more direct. To begin we use the identity $\phi^n = \phi F_n + F_{n-1}$ [491], and tabulate the stages as follows:

$$
\begin{aligned}
\phi^0 &= \phi F_0 + F_{-1} &= 0 + 1 &= 1 \\
\phi^1 &= \phi F_1 + F_0 &= \phi + 0 &= \phi \\
\phi^2 &= \phi F_2 + F_1 &= \phi + 1 \\
\phi^3 &= \phi F_3 + F_2 &= 2\phi + 1 \\
\phi^4 &= \phi F_4 + F_3 &= 3\phi + 2 \\
&\cdots \\
\phi^k &= \phi F_k + F_{k-1}
\end{aligned}
$$

$$
\begin{aligned}
\sum_{n=0}^{k} \phi^n &= \phi \sum_{n=0}^{k} F_n + \sum_{n=-1}^{k-1} F_n \\
&= \phi \sum_{n=0}^{k} F_n + \sum_{n=0}^{k} F_n - F_k + F_{-1}.
\end{aligned}
$$

Now using $\sum_{n=0}^{k} F_n = (F_{k+2} - 1)$ [490], and also $F_{-1} = +1$, we get

$$
\sum_{n=0}^{k} \phi^n = \phi(F_{k+2} - 1) + (F_{k+2} - \cancel{1}) - F_k + \cancel{1}
$$

and applying the Fibonacci recurrence $F_{k+2} - F_k = F_{k+1}$ gives

$$
\boxed{\sum_{n=0}^{k} \phi^n = \phi(F_{k+2} - 1) + F_{k+1}} \tag{D.4}
$$

If we wish to start the summation from $n = -2$ (in order to synchronize with Golden Power pentad starts as on page 177), then we shall need to include ϕ^{-1} and ϕ^{-2}, which sum to 1. That is, (D.4) right-hand side will have 1 added. We shall next check this result, proving it by induction.

Proof by induction

$$\sum_{n=0}^{k} \phi^n = \phi(F_{k+2} - 1) + F_{k+1}. \tag{D.5}$$

Base case: We show that (D.5) is true when $k = 0$:

$$\phi^0 = \phi \cdot (0) + 1.$$

Inductive step: If we assume for any integer j it is true that

$$\sum_{n=0}^{j} \phi^n = \phi(F_{j+2} - 1) + F_{j+1} \tag{D.6}$$

then we must show that

$$\sum_{n=0}^{j+1} \phi^n = \phi(F_{j+3} - 1) + F_{j+2}. \tag{D.7}$$

We start by adding the next higher power ϕ^{j+1} to each side of (D.6),

$$\sum_{n=0}^{j} \phi^n + \phi^{j+1} = \phi(F_{j+2} - 1) + F_{j+1} + \phi^{j+1}$$

and then apply the identity $\phi^n = \phi F_n + F_{n-1}$ [491], to the ϕ^{j+1} term

$$\sum_{n=0}^{j+1} \phi^n = \phi(F_{j+2} - 1) + F_{j+1} + (\phi F_{j+1} + F_j)$$

$$= \phi(F_{j+2} + F_{j+1} - 1) + F_{j+1} + F_j. \tag{D.8}$$

Now twice using the Fibonacci recurrence relation in (D.8), both as $F_{j+2} + F_{j+1} = F_{j+3}$ and as $F_{j+1} + F_j = F_{j+2}$ we are left with

$$\sum_{n=0}^{j+1} \phi^n = \phi(F_{j+3} - 1) + F_{j+2}$$

demonstrating (D.7) as required.

Summing powers of ϕ in terms of ϕ— geometric series approach

A more direct route to this sum of powers is to note that it is a geometric series. Naming the sum as S,

$$\sum_{n=0}^{k} \phi^n = S = 1 + \phi + \phi^2 + \phi^3 + \ldots \phi^k \tag{D.9}$$

$$\phi S = \phi + \phi^2 + \phi^3 + \ldots \phi^k + \phi^{k+1}. \tag{D.10}$$

Subtracting (D.9) from (D.10), we get

$$S(\phi - 1) = \phi^{k+1} - 1.$$

Now dividing by $(\phi - 1)$ and noting that $1/(\phi - 1) = \phi$, we get

$$S = \phi^{k+2} - \phi \tag{D.11}$$

and as $\phi^n = \phi F_n + F_{n-1}$ [491] we have

$$S = \phi F_{k+2} + F_{k+1} - \phi.$$

Then by gathering ϕ terms we match our 'additive' result (D.4), p.433:

$$\boxed{\sum_{n=0}^{k} \phi^n = \phi (F_{k+2} - 1) + F_{k+1}} \tag{D.12}$$

Also, we compare the intermediate result (D.11) with its Fibonacci equivalent [490]—putting $1 = F_1$:

$$\boxed{\sum_{n=0}^{k} \phi^n = \phi^{k+2} - \phi^1, \qquad \sum_{n=0}^{k} F_n = F_{k+2} - F_1} \tag{D.13}$$

Further, as powers of the conjugate root (i.e. β^n) also satisfy the Fibonacci recurrence relation, we should expect to be able to replace ϕ in (D.13) with $\beta = (-\phi)^{-1}$:

$$\sum_{n=0}^{k} (-\phi)^{-n} = (-\phi)^{-(k+2)} - (-\phi)^{-1} \qquad \text{(D.14)}$$

But to reassure ourselves, let's prove that too by induction.

Proof by induction

We start with (D.14) in its β form

$$\sum_{n=0}^{k} \beta^n = \beta^{k+2} - \beta. \qquad \text{(D.15)}$$

Base case: We show that (D.15) is true when $k = 0$.

By putting $k = 0$, (D.15) immediately reduces to the basic Golden Ratio quadratic—(1.3) on page 2—which we already know is satisfied by β:

$$\beta^0 = \beta^2 - \beta.$$

Inductive step: If we assume for any integer j it is true that

$$\sum_{n=0}^{j} \beta^n = \beta^{j+2} - \beta \qquad \text{(D.16)}$$

then we must show that

$$\sum_{n=0}^{j+1} \beta^n = \beta^{j+3} - \beta. \qquad \text{(D.17)}$$

We add β^{j+1} to each side of (D.16),

$$\sum_{n=0}^{j} \beta^n + \beta^{j+1} = \beta^{j+2} + \beta^{j+1} - \beta \qquad \text{(D.18)}$$

and (on the right) as $\beta^{j+2} + \beta^{j+1} = \beta^{j+3}$ from the recurrence relation, then (D.18) simplifies to become (D.17). ∎

Summing powers of ϕ—in terms of $\sqrt{5}$

Next we look at expressing (D.12) in terms of $\sqrt{5}$. We have

$$\sum_{n=0}^{k} \phi^n = S = \phi(F_{k+2} - 1) + F_{k+1}.$$

And as $\phi = (\sqrt{5} + 1)/2$, we double up to get

$$2S = (\sqrt{5} + 1)(F_{k+2} - 1) + 2F_{k+1}$$
$$= (F_{k+2} - 1)\sqrt{5} + F_{k+2} - 1 + 2F_{k+1}.$$

We split the end F_{k+1} pair and apply one of these in $F_{k+2} + F_{k+1} = F_{k+3}$, giving

$$2S = (F_{k+2} - 1)\sqrt{5} + F_{k+3} + F_{k+1} - 1.$$

But, as $F_{n+2} + F_n = L_{n+1}$ [488], then $F_{k+3} + F_{k+1} = L_{k+2}$, so

$$2S = (F_{k+2} - 1)\sqrt{5} + L_{k+2} - 1$$

which we halve to get our result:

$$\boxed{\sum_{n=0}^{k} \phi^n = \frac{1}{2}\left\{(F_{k+2} - 1)\sqrt{5} + (L_{k+2} - 1)\right\}} \qquad (D.19)$$

As we noted just earlier, to sum from $n = -2$ we must add in $\phi^{-2} + \phi^{-1} = 1$, and taking this as two halves, the final -1 in (D.19) will become $+1$,

$$\boxed{\sum_{n=-2}^{k} \phi^n = \frac{1}{2}\left\{(F_{k+2} - 1)\sqrt{5} + (L_{k+2} + 1)\right\}} \qquad (D.20)$$

And from (D.20) it is easy now to express the cumulative sum (starting F_{-2} in terms of the number of pentads P—by substituting $k = 5P - 3$,

$$\sum_{n=-2}^{5P-3} \phi^n = \frac{1}{2}\left\{(F_{5P-1} - 1)\sqrt{5} + (L_{5P-1} + 1)\right\}. \qquad (D.21)$$

ϕ higher powers—the 4th and 5th pentads

In the 'Five Golden Powers' chapter we went as far as the third pentad (page 186). If we check the next cycle (powers 13–17), and the next (powers 18–22), then we find the results wander well away from whole numbers of turns: $1487.8 \cdot (2\pi)$, $16501.4 \cdot (2\pi)$, ... This contrasts with the integer Fibonacci Circle where the hexads always cycled back to the exact starting point (page 170).

Summary

Our conclusion therefore is that the first two pentads are the most interesting, but to summarize all the cases we have considered, we shall make a Table of results, and to do this, we write

$$\sum_{n=-2}^{5P-3} \phi^n + \epsilon = T \cdot (2\pi) \tag{D.22}$$

where P is the number of pentads (groups of 5 consecutive powers of ϕ), and ϵ is the angular shortfall—i.e. the positive (or negative) angle in radians that must be added to the sum to bring it to a whole number of turns (T). This residual is then shared amongst all the sectors giving the gap angle (Fig. 264). But, unless we have some target value for T, then by just taking 'the nearest whole turn', we shall never be further than 180° away from it. So, although the exercise may be justified for the initial pentads where ϵ is small, it makes less and less sense as we continue.

P	Sum of P pentads	Page	T turns	ϵ radians	ϵ°	Sectors	Gap
1	$4 + \sqrt{5}$	177	1	0.047	2.7°	5	0.54°
2	$11 \cdot (7 + \sqrt{5} \cdot 3)/2$	182	12	0.0031	0.18°	10	0.018°
3	$422 + 188\sqrt{5}$	186	134	−0.434	−25°	15	−1.66°
4	$4675 + 2090\sqrt{5}$	438	1488	0.998	57.2°	20	2.86°
5	$(103683 + 46367\sqrt{5})/2$	438	16501	−2.54	−145.6°	25	−5.8°

Figure 264: **Summary of pentads P=1, 2, 3, 4, 5. A negative gap is an overlap.**
Formulæ for cumulative pentads up to 4 and 5 are found using Equation (D.21), p.437. (The first pentad P=1 starts with F_{-2}.)

Appendix E—Binet from a generating function

$$F_n = \frac{\phi^n - (-\phi)^{-n}}{\sqrt{5}}.$$

We have already seen the great power of the Binet formula—allowing potentially difficult proofs to be reduced to just a few lines. So, if we are placing such a great dependence on it, then it makes sense for us to understand how it may be derived. In this section, we shall combine the approaches of both Donald Knuth [498] and Ivan Galkin [232], and for clarity we shall express our working in 'alphas and betas' [592].

Power series as generating functions for recurrence sequences

We start by imagining an infinite power series whose every coefficient is a Fibonacci number, in ascending sequence,

$$G(x) = F_0 + F_1 x + F_2 x^2 + F_3 x^3 + \cdots \tag{E.1}$$

Ideally we should like to encapsulate this infinite series in some simple equivalent expression—in fact, one which will allow us to find a formula for a single coefficient in the series—that is, an analytic expression for the nth Fibonacci number. To start, we 'build in' the Fibonacci recurrence relation by multiplying $G(x)$ by x, and then again separately by x^2, to give us equations (E.3) and (E.2).

$$xG(x) = F_0 x + F_1 x^2 + F_2 x^3 + F_3 x^4 + \cdots \tag{E.2}$$

$$x^2 G(x) = F_0 x^2 + F_1 x^3 + F_2 x^4 + F_3 x^5 + \cdots \tag{E.3}$$

We then subtract both (E.3) and (E.2) from the original (E.1):

$$(1 - x - x^2)G(x) = F_0 + (F_1 - F_0)x + (F_2 - F_1 - F_0)x^2 + (F_3 - F_2 - F_1)x^3 + \cdots \tag{E.4}$$

$n,$	$\big[G(x)$	$-xG(x)$	$-x^2 G(x) \big]$	$=$	$(1 - x - x^2)G(x)$		
0	F_0			$=$	F_0	$=$	0
1	$F_1 x$	$-F_0 x$		$=$	$(F_1 - F_0)x$	$=$	x
2	$F_2 x^2$	$-F_1 x^2$	$-F_0 x^2$	$=$	$(F_2 - F_1 - F_0)x^2$	$=$	0
3	$F_3 x^3$	$-F_2 x^3$	$-F_1 x^3$	$=$	$(F_3 - F_2 - F_1)x^3$	$=$	0
4	$F_4 x^4$	$-F_3 x^4$	$-F_2 x^4$	$=$	$(F_4 - F_3 - F_2)x^4$	$=$	0
...							

$x.$

Figure 265: **Simplifying the generating function.**

Just how much we achieve in (E.4) becomes more obvious in the Table Fig. 265, where we write our 3 formulæ in 3 columns, with each power of x (that is, x^n) having its own row and horizontal summation. We consider (row by row) the coefficients of successive x^n. For power $n = 0$ we just have F_0 which is zero. For power $n = 1$ we have $(1 - 0)x$— that is, just x. Then for all the higher-power terms, the coefficients *vanish* owing to the recurrence relations, $F_2 = F_1 + F_0$, $F_3 =$, and so on. Hence, all three infinite series have combined to produce one simple expression[1] $(1 - x - x^2) G(x) = x$

$$G(x) = \frac{x}{(1 - x - x^2)}$$
$$= \frac{-x}{(x^2 + x - 1)}. \tag{E.5}$$

In order to factorize the denominator of (E.5), we find the roots of $x^2 + x - 1 = 0$ which are

$$x_1 = \frac{-(1 + \sqrt{5})}{2} \quad \text{and} \quad x_2 = \frac{-(1 - \sqrt{5})}{2}.$$

So (E.5) becomes

$$G(x) = \frac{-x}{(x - x_1)(x - x_2)}. \tag{E.6}$$

---

[1] Decimal place value depends on powers of 10. Hence if we set x to some negative power of 10, then the right-hand side of (E.1) will 'place' Fibonacci numbers in a never ending decimal. For example, by setting $x = 1/100$, the right-hand side of (E.1) will become 0.01 01 02 03 05 08 13 21... Similarly, by setting $x = 1/10$ in both (E.5) and (E.1) and dividing all through by 10, we find that $1/89$ comprises $0.0 + 0.01 + 0.001 + 0.0002 + 0.00003 + 0.000005 + 0.0000008 + 0.00000013 + \ldots$

We recognize these roots as being very similar to those we saw earlier for the Golden Section (p.3). In that case the quadratic was

$$x^2 - x - 1 \quad \text{with roots} \quad \alpha = (1 + \sqrt{5})\big/2 \quad \text{and} \quad \beta = (1 - \sqrt{5})\big/2.$$

Here we have negated versions of these $(x_1 = -\alpha,\ x_2 = -\beta)$, so we may substitute,

$$G(x) = \frac{-x}{(x + \alpha)(x + \beta)}. \tag{E.7}$$

But, for reasons soon to become plain, we seek denominator terms of the form $(1 - kx)$. Therefore, we use the fact that $\alpha\beta = -1$, and multiply top and bottom by $(-\alpha)(-\beta)$,

$$G(x) = \frac{-x(-\alpha)(-\beta)}{(x + \alpha)(x + \beta)(-\alpha)(-\beta)}.$$

We apply $(-\alpha)(-\beta)$ to the denominator in stages, giving

$$-\beta(x + \alpha) = (1 - \beta x), \quad \text{and}$$
$$-\alpha(x + \beta) = (1 - \alpha x) \quad \text{with which, (E.7) becomes}$$
$$G(x) = \frac{x}{(1 - \alpha x)(1 - \beta x)}. \tag{E.8}$$

Next, we expand this in partial fractions with unknown numerators p and q that we must find

$$G(x) = \frac{x}{(1 - \alpha x)(1 - \beta x)} = \frac{p}{(1 - \alpha x)} + \frac{q}{(1 - \beta x)}. \tag{E.9}$$

In (E.9), multiplying through by $(1 - \alpha x)(1 - \beta x)$, we get

$$x = p(1 - \beta x) + q(1 - \alpha x). \tag{E.10}$$

This Eqn. (E.10) should be true for any x. For example when $x = 0$

$$0 = p + q. \tag{E.11}$$

Hence (E.10) reduces to $x = -p\beta x - q\alpha x$. And so when $x = 1$

$$1 = -p\beta - q\alpha. \tag{E.12}$$

Using $-q = p$ from (E.11) in (E.12), $1 = -p\beta + p\alpha,$ so,

$$p = \frac{1}{\alpha - \beta} \quad \text{and from (E.11),} \quad q = \frac{-1}{\alpha - \beta}.$$

Applying these back in (E.9), we get

$$G(x) = \frac{x}{(1 - \alpha x)(1 - \beta x)} = \frac{1}{\alpha - \beta}\left[\frac{1}{(1 - \alpha x)} - \frac{1}{(1 - \beta x)}\right]. \quad \text{(E.13)}$$

Now we come to the reason for wanting 'denominator terms of the form $(1 - kx)$'. As an example, let us study $1/(1 - \alpha x)$ in detail. If we expand this using polynomial long division, we obtain a remarkably simple power series,

$$
\begin{array}{r}
1 \quad +\alpha x \quad +\alpha^2 x^2 \quad +\alpha^3 x^3 \quad + \cdots \\
\hline
(1 - \alpha x) \;\big|\; 1 \qquad\qquad\qquad\qquad\qquad\quad \\
\underline{1 \quad -\alpha x} \qquad\qquad\qquad\qquad \\
\alpha x \qquad\qquad\qquad\qquad \\
\underline{\alpha x \quad -\alpha^2 x^2} \qquad\qquad \\
\alpha^2 x^2 \qquad\qquad \\
\underline{\alpha^2 x^2 \quad -\alpha^3 x^3} \quad \\
\alpha^3 x^3 \\
\cdots
\end{array}
$$

which we summarize as...

$$\frac{1}{(1 - \alpha x)} = \sum_{k=0}^{\infty} \alpha^k x^k. \quad \text{(E.14)}$$

We now apply (E.14) twice in (E.13) to give us

$$G(x) = \frac{1}{\alpha - \beta}\left[\sum_{k=0}^{\infty} \alpha^k x^k - \sum_{k=0}^{\infty} \beta^k x^k\right]$$

$$= \frac{1}{\alpha - \beta}\sum_{k=0}^{\infty} (\alpha^k - \beta^k)x^k. \quad \text{(E.15)}$$

Back at the start in (E.1) we set things up so the nth coefficient in $G(x)$ be equal to the nth Fibonacci number. Therefore we may finally assert

$$F_n = \frac{\alpha^n - \beta^n}{\alpha - \beta} \quad \text{or in terms of } \phi, \quad F_n = \frac{\phi^n - (-\phi)^{-n}}{\sqrt{5}}.$$

Such formulæ, which immediately give the value of the nth member of a sequence—(i.e. that 'solve' a given recurrence)—are known as 'closed forms'.

Appendix F—Inductive Binet, instant Binet

Appendix E may have given the impression that it necessarily takes three or four pages to prove the Binet formula $F_n = (\alpha^n - \beta^n)/\sqrt{5}$, but here we shall use a quicker method (induction), followed by an almost 'by observation' method. While induction confirms that the Binet formula 'works', the second method has the advantage of showing why it works—that is, just why it takes this form.

Binet inductive proof

$$F_n = \left(\alpha^n - \beta^n\right)\Big/\sqrt{5}. \tag{F.1}$$

Base case: We show that (F.1) is true when $n = 0$:

$$F_0 = \left(\alpha^0 - \beta^0\right)\Big/\sqrt{5} \;=\; 0.$$

Inductive step: If we assume for any integer j it is true that

$$F_j = \left(\alpha^j - \beta^j\right)\Big/\sqrt{5} \tag{F.2}$$

then we must show that

$$F_{j+1} = \left(\alpha^{j+1} - \beta^{j+1}\right)\Big/\sqrt{5}. \tag{F.3}$$

To do this, we take the Fibonacci recurrence and scale it by $\sqrt{5}$

$$\sqrt{5}F_{j+1} = \sqrt{5}F_j + \sqrt{5}F_{j-1}. \tag{F.4}$$

We then apply (F.2) twice in the scaled form (F.4) to get

$$\sqrt{5}F_{j+1} = \alpha^j - \beta^j + \alpha^{j-1} - \beta^{j-1}. \tag{F.5}$$

Now, in Chapter 1, page 31, we saw how powers of both α and β must obey the Fibonacci recurrence too—therefore

$$\alpha^j + \alpha^{j-1} = \alpha^{j+1} \quad \text{and} \tag{F.6}$$

$$\beta^j + \beta^{j-1} = \beta^{j+1}. \tag{F.7}$$

So, using (F.6) and (F.7) in (F.5), we get

$$\sqrt{5}F_{j+1} = \alpha^{j+1} - \beta^{j+1}$$

which rearranged, becomes (F.3). ∎

(Almost) instant proof

This next derivation is based on a very elegant outline given by Keith Tognetti in his article: *Fibonacci—his rabbits and his numbers and Kepler,* [899]. Back on page 181 we noted that by successively applying $\phi^2 = \phi + 1$, any integral power of ϕ, say ϕ^n may be reduced to a linear form: $F_n\phi + F_{n-1}$. That is

$$\phi^2 = \phi + 1 \quad \Longrightarrow \quad \phi^n = F_n\phi + F_{n-1}.$$

But, after all, $\phi^2 = \phi + 1$ is the equation that not only defines the Golden Ratio ϕ (that we have called α)—it also gives us the conjugate solution $\beta = -\phi^{-1}$. We see these both directly in the parabola diagram Fig. 266 which demonstrates exactly the quadratic relation, (repeated here from p.5). Therefore we may write equally for each root

$$\alpha^2 = \alpha + 1 \quad \Longrightarrow \quad \alpha^n = F_n\alpha + F_{n-1} \quad \text{and} \tag{F.8}$$

$$\beta^2 = \beta + 1 \quad \Longrightarrow \quad \beta^n = F_n\beta + F_{n-1}. \tag{F.9}$$

It is now sufficient to subtract (F.9) from (F.8)

$$\alpha^n - \beta^n = F_n(\alpha - \beta)$$

and rearrange:

$$F_n = \frac{\alpha^n - \beta^n}{\alpha - \beta} = \frac{\phi^n - (-\phi)^{-n}}{\sqrt{5}} \tag{F.10}$$

∎

From this, we may also quickly obtain a complementary expression for the Lucas numbers, using the identity (F.11), [956]

$$L_n = F_{n+1} + F_{n-1}. \tag{F.11}$$

In (F.11), we deploy equation (F.10) twice to give us

$$(\alpha - \beta)L_n = (\alpha^{n+1} - \beta^{n+1}) + (\alpha^{n-1} - \beta^{n-1})$$

$$= \alpha^n(\alpha + \alpha^{-1}) - \beta^n(\beta + \beta^{-1}).$$

Next, as $\alpha\beta = -1$, we substitute $\alpha^{-1} = -\beta$, and $\beta^{-1} = -\alpha$, giving

$$(\alpha - \beta)L_n = \alpha^n(\alpha - \beta) - \beta^n(\beta - \alpha)$$

which simplifies to the result

$$L_n = \alpha^n + \beta^n = \phi^n + (-\phi)^{-n}$$

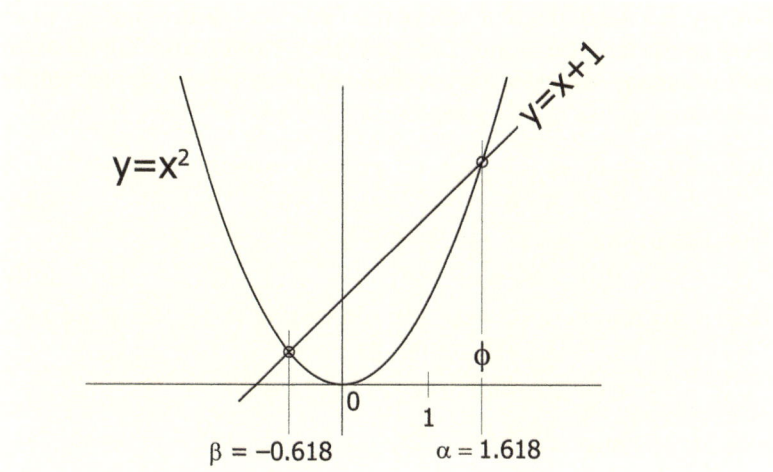

Figure 266: **Fundamental quadratic relation for the Golden Ratio.**

'Any linear combination'

Any linear combination of the powers of the roots of the characteristic equation will satisfy a given linear recurrence rule [533].[1] It is the choice of seeds (or equivalently the choice of coefficients for the power terms in the Binet formula) that determine a specific case.[2] For example, we know that both the Fibonacci numbers and the Lucas numbers share the same recurrence rule: $U_n = U_{n-1} + U_{n-2}$. For the Fibonacci number sequence $\langle 0, 1, 1, 2, 3, 5, 8, \dots \rangle$ with seeds $(0, 1)$, the closed form is

$$F_n = \frac{\alpha^n - \beta^n}{\sqrt{5}} \qquad \text{where the coefficients are} \quad \frac{1}{\sqrt{5}} \quad \text{and} \quad \frac{-1}{\sqrt{5}}.$$

And for the Lucas number sequence $\langle 2, 1, 3, 4, 7, 11, 18, \dots \rangle$ with seeds $(2, 1)$, the closed form is

$$L_n = \alpha^n + \beta^n \qquad \text{where the coefficients are} \quad 1 \quad \text{and} \quad 1.$$

---

[1] This is well worth thinking about. 'Add the two predecessors to get the next' works for any power of α—for example $\alpha^{n-63} + \alpha^{n-62} = \alpha^{n-61}$, and this is also true if all members are scaled by the same amount; and the same applies to all powers of β (p.31).

[2] We have already noted that the 'growth converging to ϕ' property of the Fibonacci recurrence does not depend on particular starting seeds—any two numbers will do. We explored the accuracy of successive convergents in a practical example using seeds 10 and 12 (the 'ten twelve' sequence, p.187) [525]—and we shall return to that case shortly.

But, given a sequence that obeys the Fibonacci recurrence, how can these coefficients be found? Let's call them A and B and find them for the Lucas sequence. Our linear combination of powers of roots will be

$$L_n = A\alpha^n + B\beta^n.$$

Hence for $n = 0$, $L_0 = 2$ and $\alpha^0 = \beta^0 = 1$, leaving

$$2 = A + B \quad \Longrightarrow \quad B = (2 - A). \tag{F.12}$$

And similarly for $n = 1$, $L_1 = 1$ giving

$$1 = A\alpha + B\beta. \tag{F.13}$$

Then using B from (F.12) in (F.13), and writing $(1 - \alpha)$ for β we get

$$1 = A\alpha + (2 - A)(1 - \alpha)$$

$$0 = A(\alpha + \alpha - 1) + (-1 + 2 - 2\alpha)$$

$$A = \frac{-(-1 + 2 - 2\alpha)}{(\alpha + \alpha - 1)} = 1.$$

$$B = 2 - A = 1.$$

Now using the same approach, we shall derive a general formula[3] for Fibonacci recurrences $\langle G_n \rangle$ with arbitrary seed values G_0 and G_1, and then apply it to the 'ten twelve' sequence. As previously,

$$G_n = A\alpha^n + B\beta^n.$$

Hence for $n = 0$

$$G_0 = A + B \quad \Longrightarrow \quad B = (G_0 - A). \tag{F.14}$$

And similarly for $n = 1$

$$G_1 = A\alpha + B\beta. \tag{F.15}$$

So, using B from (F.14) in (F.15), and again writing $(1 - \alpha)$ for β

$$G_1 = A\alpha + (G_0 - A)(1 - \alpha)$$

$$0 = A(\alpha + \alpha - 1) + (-G_1 + G_0 - G_0\alpha)$$

$$A = \frac{G_0\alpha + G_1 - G_0}{2\alpha - 1} = \frac{G_0\phi + G_1 - G_0}{\sqrt{5}}, \quad B = G_0 - A. \tag{F.16}$$

Let's apply (F.16) to $\langle K_n \rangle$, the 'ten twelve' sequence (p.187). For this $K_0 = 10$ and $K_1 = 12$, (with $\alpha = \phi = (1 + \sqrt{5})/2$, and $\beta = 1 - \phi$),

$$A = (2 + 10\alpha)/\sqrt{5} = 5 + \left(7/\sqrt{5}\right).$$

$$B = 10 - A = 5 - \left(7/\sqrt{5}\right).$$

$$K_n = \left[5 + \left(7/\sqrt{5}\right)\right]\alpha^n + \left[5 - \left(7/\sqrt{5}\right)\right]\beta^n. \tag{F.17}$$

---

[3] A standard identity (e.g. 56 and 57, Vajda [911], p.180) which has symmetric form $\sqrt{5}A = G_1 - G_0\beta$ and $\sqrt{5}B = -G_1 + G_0\alpha$. (Whereas (F.16) is slightly simpler to apply.)

Appendix G—Mersenne primes, Hilbert's 10th

To round off, we shall briefly review two examples of how studies of: '*add two numbers to get the next,* plus various generalizations, and many interesting consequences', have contributed to achievements in more advanced mathematics.

Hunting Mersenne primes: $\boxed{2^p - 1 \ \textit{may be prime}}$

Marin Mersenne (1588–1648) lived more than half of his life in Paris. He became a monk in 'The Order of the Minims' in the same year that Kepler wrote his *Snowflake* essay (1611).[1] He is known for discovering the cycloid curve; and after initially rejecting the works of Galileo, he later became a significant and influential promoter of them. But today, he is best remembered for his study of numbers of the form $2^n - 1$, which are now called 'Mersenne numbers' [703].[2] When prime, these numbers are written 'M_n'.[3] A theorem (sometimes called the Cataldi-Fermat theorem) states that: 'If for some $n > 0$, $2^n - 1$ is prime, then n too is prime.' There exist short proofs for this, for example Shanks [826], Caldwell [96]. Note however, that the converse cannot be relied upon—that is, n being prime is not sufficient to ensure that $2^n - 1$ is prime.[4]

The $2^n - 1$ primes: 3, 7, 31, and 127 (now called M_2, M_3, M_5, and M_7) were all known to the ancient Greeks c. 300 BC, as was the intimate connection between these and the (then known) four perfect numbers [329].

---

[1] Which was also the year (many say), that William Shakespeare wrote his last play, *The Tempest*.
[2] Sloane A000225 Mersenne numbers $2^n - 1$.
[3] Sloane A001348 Mersenne primes $2^p - 1$, p prime.
[4] E.g. $2^{11} - 1 = 2047$ is composite: 23×89. Hence earlier, U_{11} was not perfect (p.281).

Édouard Lucas developed a powerful technique for testing the primality of $2^n - 1$ numbers. It still required a vast amount of manual calculation, but using it, in 1876 he was able (remarkably) to present a proof of the primality of $M_{127} = 2^{127} - 1$, a 39 digit number [593, 168]. The previous largest known Mersenne prime was due to Euler: $M_{31} = 2^{31} - 1$. Lucas' method was simplified in 1930 by Derrick Henry Lehmer to become the 'Lucas-Lehmer' test (LL or LLT) [702]. The full details of the LLT are beyond the scope of this book. However, as a result of the generalization analysis we did starting on page 251, we *are* in a position to appreciate the fundamental role of Lucas Sequences in the test. The LLT method is based on the recurrence

$$S_{n+1} = S_n^2 - 2, \qquad \text{with } S_0 = 4 \tag{G.1}$$

which generates the Lucas-Lehmer sequence [5]

$$\langle S_n \rangle = \langle 4, 14, 194, 37634, 1416317954, \dots \rangle. \tag{G.2}$$

To test $2^p - 1$ for primality, the recurrence is evaluated $p-2$ times; then a check is made whether $2^p - 1$ divides S_{p-2}. If it does, then $2^p - 1$ is prime; otherwise $2^p - 1$ is composite [529, 530].[6]

We mentioned in Chapter 21, *Pell, Jacobsthal, and Mersenne* (p.284), that: 'Steven Vajda relates the case of $P=4$, $Q=1$ to the study of Mersenne numbers [913].' Here we shall explore his very brief (LLT) discussion in expanded detail. For these coefficients $P=4$ and $Q=1$, we have $\Delta=(P^2 - 4Q)=12$, hence $\sqrt{\Delta}=2\sqrt{3}$. And as before (p.254), the roots α and β are given by $(P \pm \sqrt{\Delta})/2$, and $\alpha\beta=Q$. Hence

$$\alpha = 2 + \sqrt{3} \qquad \beta = 2 - \sqrt{3} \qquad \alpha\beta = Q = 1.$$

Now using these values, we consider the companion sequence of the recurrence, $V_n = 4V_{n-1} - V_{n-2}$, with seeds 2 and 4 (=P), which therefore starts [7]

$$\langle V_n \rangle = \langle 2, \underline{4}, \underline{14}, 52, \underline{194}, 724, 2702, 10084, \underline{37634}, 140452, \dots \rangle.$$

But from it, we pick only the members whose index is a power of 2— those underlined members at $1, 2, 4, 8, \dots$ This forms a new sequence

$$\langle V_{2^n} \rangle = \langle 4, 14, 194, 37634, 1416317954, \dots \rangle$$

—which we identify with (G.2), the LL sequence—i.e. $\langle S_n \rangle \equiv \langle V_{2^n} \rangle$.

---

[5] Sloane A003010.

[6] Note that some authors seed $\langle S_n \rangle$ as $S_1 = 4$ or even $S_2 = 4$ leading to corresponding variations in the statement of this test.

[7] Sloane A003500.

To see how the recurrence expression for S_{n+1} (G.1) is derived, we first examine V_{2^n} squared using the (companion) Binet form, and again recall that $\alpha\beta = Q$, with $Q=1$ in this case,

$$(V_{2^n})^2 = \left(\alpha^{2^n} + \beta^{2^n}\right)^2$$

$$= \left(\alpha^{2^n}\right)^2 + \left(\beta^{2^n}\right)^2 + 2\alpha^{2^n}\beta^{2^n}$$

$$= \alpha^{2 \cdot 2^n} + \beta^{2 \cdot 2^n} + 2(\alpha\beta)^{2^n}$$

$$= \alpha^{2^{n+1}} + \beta^{2^{n+1}} + 2$$

$$= \left(V_{2^{n+1}}\right) + 2.$$

Rearranging, we obtain the recurrence relationship

$$\left(V_{2^{n+1}}\right) = (V_{2^n})^2 - 2. \tag{G.3}$$

Or writing (G.3) in terms of S, we get the Lucas-Lehmer recurrence:

$$\boxed{S_{n+1} = S_n^2 - 2 \quad \text{with } S_0 = 4}$$

We could just use this formula as is to find the sequence $\langle S_n \rangle$, but with the successive squarings, the values grow very rapidly and become unnecessarily cumbersome. 'Unnecessarily', as we may discard integer multiples of M_p along the way, because in the end, we are only interested in the *residue* (mod M_p). In other words, taking the congruence modulo M_p at every stage will not affect the final test result. To work a simple example, we shall choose (as does Haghighi [293]) M_7—that is, $2^7 - 1 = 127$. And as $p = 7$ this will require us make only $7 - 2 = 5$ iterations. We show a value s_n, followed on the same line by its use in the calculation of s_{n+1}:

$s_0 = 4,$ $(4^2 - 2) = 14,$ $14 \pmod{127} \equiv 14$

$s_1 = 14,$ $(14^2 - 2) = 194,$ $194 \pmod{127} \equiv 67$

$s_2 = 67,$ $(67^2 - 2) = 4487,$ $4487 \pmod{127} \equiv 42$

$s_3 = 42,$ $(42^2 - 2) = 1762,$ $1762 \pmod{127} \equiv 111$

$s_4 = 111,$ $(111^2 - 2) = 12319,$ $12319 \pmod{127} \equiv 0$

$s_5 = 0,$ confirming M_7 is prime.

449

The Lucas-Lehmer test is still in use today. But since Lucas, with his many years of manual calculation, all larger Mersenne primes have been found using computers and the LLT. In 1996 George Woltman and Scott Kurowski founded a community project they called 'GIMPS'—*The Great Internet Mersenne Prime Search,* dedicated to finding ever bigger Mersenne primes. In January 2013, Curtis Cooper, using GIMPS software, discovered the 48th known Mersenne prime: $M_{57885161} = (2^{57,885,161} - 1)$, a 17,425,170 digit prime number [987]. To recap, we test because 'p prime' is insufficient (p.447)—the GIMPS logo is absolutely clear on this, declaring: '$2^p - 1$ *may be prime !*' [986].   Yet back in 1811, in his book *The Theory of Numbers,* Peter Barlow noted that not only was M_{31} the greatest Mersenne prime known at the time—but also its associated perfect number would probably be the greatest that would ever be discovered: 'For, as they are merely curious without being useful, it is not likely that any person will attempt to find one beyond it.' [32]. A recent application of Mersenne primes is in (pseudo-) random number generation in computer programs. The 'Mersenne Twister' algorithm is currently the most popular 'PRNG'—with implementations in all the major computer programming languages [637].

Hilbert's 10th problem

German mathematician David Hilbert (1862–1943) made tremendous contributions both to mathematics and to mathematical physics. At the International Congress of Mathematicians (ICM) in Paris in 1900, Hilbert gave a historic presentation in which he challenged the mathematical community with a list of unsolved problems. By the time he published his list, it had grown to 23 items. One of these problems, his tenth [401], was solved 70 years later by Russian mathematician and computer scientist Yuri Matiyasevich (then 22 years old) [634]. It concerns Diophantine equations. Although these are algebraic—interest is focused solely on integer solutions. In general these are hard to solve—Fermat's last theorem is a famous example (that $a^n + b^n = c^n$ has no integer solutions for $n>2$). Hilbert's 10th asks: 'Is there a method to determine if a Diophantine equation (with any number of unknown quantities and with rational numerical coefficients) has an integral solution?' Matiyasevich proved the negative—he proved that such a method cannot exist. In an interview, Julia Bowman Robinson (Fig. 267)—who, with Martin Davis, had long worked towards the solution—summarized how this final proof depended on Fibonacci numbers [13, 707].

Glossary

This glossary contains mostly terms coined during the Fibonacci Resonance work, with the addition of a few conventional terms. No attempt is made to cover all the scientific terms included in the 'ϕ Science' chapters. A page number is given to locate the term's introduction in the main text. Cross references between entries are indicated by *emphasis*.

3 5 8 triangle	A right triangle having vertex angles of $3 \times 11.25°$, $5 \times 11.25°$, and $8 \times 11.25°$, which (in *Ori32 geometry*) is also known as the 'Fibon3'—page 163.
5 Golden Powers	A geometric sequence of five powers of ϕ, initially identified as the powers ϕ^{-2} through ϕ^{+2}—page 175.
Adjustment	In the polar/spiral visualization scheme: The distance along a radial axis between a number point and a spiral crossing point. For the chosen set of spirals, adjustments are consistently quantized in units that are powers of ϕ, here called *Golden Quanta*—page 47.
AH	Islamic year, *Anno Hegiræ*—in the year of the Hijra—page 345.

Aliquot divisor
The aliquot divisors of n are the divisors of n that are less than n, including 1—page 280.

Anchor
Defines a particular *Golden Spiral* by requiring that it intersect a quadrant axis at a distance from the origin that is a given Fibonacci number. The particular axis to be used for an anchoring is determined in the visualization scheme—page 209.

Backbone growth
The underlying growth rate shared by primary and companion sequences, e.g. ϕ per step for both Fibonacci and Lucas numbers—page 207.

Brute-force search
Also known as an 'exhaustive search'. All possible solutions are checked one by one, with no attempt to find time-saving shortcuts—page 197.

Companion axis
Provides the line of measurement from a companion number point (for example a Lucas number) to the intersection with the relevant spiral—page 213.

Deflation
(In connection with Penrose tiling.) A partitioning and rebuilding of tiles in a pattern so as to make a new Penrose tiling pattern with edge size ϕ times smaller [243]. Across the literature readers may encounter inconsistencies in the use of this term, especially when it is extended to substitution sequences—page 314.

Eigenvalue/eigenvector   (In connection with Islamic tiling patterns.) When a vector is transformed by a matrix, it typically changes direction. However, a special vector may retain its direction, and only be scaled: $Ax=\lambda x$. Such a vector x is called an eigenvector and the scaling factor λ is called an eigenvalue. The subdivision of tiling shapes may be described in this way with each shape from a set giving rise to a combination of multiple shapes from the same tiling set. In this context, irrational eigenvalues are: 'a signature of quasiperiodicity' [588] —page 348.

Fibon   (Pronounced 'Fibbon'.)
In *Ori32 geometry*, one of four right triangles in which each vertex angle is a multiple of 11.25°—page 163.

Fibonacci circle   Circle divided into six sectors whose angles are each a multiple of $1/32$nd of a turn, such that the counts form a Fibonacci sequence. The sequence is: $1/32$nd of a turn, $2/32$nds, $3/32$, $5/32$, $8/32$, and $13/32$, giving the full-turn total of $32/32$—page 168.

Fibonacci Resonance   A visualization of the relationship between ϕ, the Fibonacci numbers, the Lucas numbers, and a chosen infinite set of *Golden Spirals*. Each spiral is *anchored* to its own Fibonacci number, and it has an associated *quantum* size which is a power of ϕ. This quantum may be imagined as an abacus bead or a standing wave (half wavelength) in an acoustic resonance, e.g. a violin string—page 221.

Gnomonic expansion	Where an old form is contained within a larger new form, while the shape remains the same. The shape that is added to achieve the growth may be called a gnomon—page 103.
Golden Noise	Sound analogy—an audio comb spectrum comprising discrete frequency peaks regularly spaced by a factor of ϕ—that is, the musical interval of $8\frac{1}{3}$ semitones—page 221.
Golden Power	Simply a power of ϕ, the *Golden Ratio*, for example ϕ^7—page 175.
Golden Quantum	A whole unit of *adjustment*, being a power of ϕ. *Quanta* are never subdivided and may be thought of rather like abacus beads counting out the distance along a radial axis between a number point and a spiral crossing—page 208.
Golden Ratio	Also known as ϕ ('phi'), the 'Golden Number' and many other names. In this book we define it as $\phi = (1 + \sqrt{5})/2$—page 2.
Golden Rectangle	Has a long side that is ϕ times its short side—page 15.
Golden Sector	In a circle, a sector whose angle in radians is an integer power of ϕ —page 179.
Golden Spiral	Logarithmic spiral that grows by ϕ per quadrant—page 42.
Golden Triangle	Isosceles triangle with angles 36°, 72°, and 72° —page 16.

Hexad

6 consecutive Fibonacci numbers with a start-position constraint. The reference hexad starts on F_2 and all hexads that follow form a contiguous sequence (i.e. one without gaps). For instance, the next hexad after the reference starts on F_8 —page 168.

Inflation

(In connection with Penrose tiling.) A partitioning and rebuilding of tiles in a pattern so as to make a new Penrose tiling pattern with edge size ϕ times larger [243]. As with *deflation*, across the literature readers may encounter inconsistencies in the use of this term, especially when it is extended to substitution sequences—page 314.

Kepler's triangle

A right triangle having sides in the proportion $1 : \sqrt{\phi} : \phi$ —page 7.

MIK

(Pronounced 'Mick'), the name of the basic unit of angle in *Ori32 geometry*: $11.25°$, being one thirty-second of a full turn—page 162.

Modular arithmetic

Consider a 12 hour clock. If the clock shows 12 o'clock and 16 hours pass, what will the clock then show? To know, we divide the 16 by 12 and thereby find the remainder 4—which is our result. In modular arithmetic, the divisor (here 12) is called the 'modulus', and the remainder (here 4) is called the 'residue'. So we say that 16 is 'congruent' to 4 (mod 12) —page 169.

Nine and phi

A near-integer construction in *Ori32 geometry* originally based on the observation that $10\cos\boxed{3} \approx 9\cos\boxed{2}$ —page 165.

Offset angle	In the context of a log spiral, the rotation of the *companion axes* that is equivalent to a given growth (dilation). For example in the spiral visualization scheme, Lucas numbers 'lead in growth' by $\sqrt{5}$ compared with the Fibonacci numbers, and this can be represented by an equivalent 'lead in rotation' of 150.505°—page 213.
Offset axes	Same as *companion axes*. These axes are rotated by the *offset angle* —page 213.
Ori32 geometry	A constrained Euclidean plane geometry concerned with lines and angles. There are no curves. Given a starting reference line, all other lines are required to be at an angle to it that is an integer multiple of 11.25° —page 161.
Parastichies	In *phyllotaxis* (botanic & generalized), e.g. in a sunflower head, the arrange-ment of florets in conspicuous spirals —page 52.
Pentad	A geometric sequence of 5 powers of ϕ —page 177.
Phi	*Golden Ratio*—page 2.
Phyllotaxis	Initially referring to leaf arrangements on plants, this topic now encompasses a particular set of spiral patterns such as those seen in flower heads and pine cone scales. Although the numbers of spirals are typically found to be Fibonacci numbers, they may instead be Lucas numbers —page 52.

Pisot number   A Pisot number is a positive algebraic integer greater than 1, whose conjugate elements all have absolute value less than 1 [964]—page 72.

Polar Fibonacci   A scheme for placing Fibonacci numbers in polar coordinates (r, θ) where the points increasingly approach a chosen Golden Spiral. Number points are placed at $r = F_n$ from the polar origin and at an angle θ which is n quadrants anticlockwise —page 45.

Proportion   When two ratios are equal they are said to be 'in proportion'. For example, in the Golden Proportion, the ratio of the greater to the lesser is equal to the ratio of the whole to the greater—page 2.

Quantum/quanta   Term borrowed from physics. The concept is of an 'atomic' unit of measurement, one that is not subdivided. When we say here that distances in a set are quantized, we mean that they share a common measure—that each distance is an integer multiple of a basic unit length —page 208.

Rational number   A number that can be expressed as one integer divided by another integer (the latter being non-zero), for example $3/4$—page 23.

Residual angle   The angle in radians that must be added to a sum of *Golden Powers* to bring it to a whole number of turns —page 178.

Residue   The remainder in *modular arithmetic* —page 170.

Root Golden growth	Growth at a rate of $\sqrt{\phi}$ per step. However, in practice, an approximation using the Golden Section of 2 may be encountered. This deviates from $\sqrt{\phi}$ by 2.9% —page 73.

Root Golden Rectangle	Rectangle having sides in the ratio of $1 : \sqrt{\phi}$. The diagonal divides the figure into two Kepler's triangles, joined along their hypotenuses —page 99.

Root Golden Spiral	Logarithmic spiral that grows by $\sqrt{\phi}$ per quadrant—that is, ϕ per half-turn —page 137.

Silver Ratio	This is defined as $\delta = 1 + \sqrt{2}$ —page 268.

Silver Spiral	Logarithmic spiral that grows by the *Silver Ratio* per quadrant—page 270.

Sloane A123456	Such are references into the *Sloane Online Encyclopedia of Integer Sequences* oeis.org [839]—page 34.

Spiral number 's'	Key parameter that both identifies a spiral and directly determines its *anchor* point. For example, the *Golden Spiral* identified by $s = 4$ will be anchored with a radial vector—*'spoke'*—of $F_4 = 3$ units —page 209.

Spoke	In the context of a spiral: A radial vector—page 36.

Zero-q point	Another name for the *anchor* point. While either side of this point we see quantized *adjustments*—here, by definition, we see zero *quanta* of *adjustment*, hence 'zero-q' —page 210.

Symbols used

Most of the symbols used in this book already have conventional meanings, and those are kept. For example, the Lucas Sequence coefficients P and Q date back to Lucas' original works of the 1870s.

Sequence notation: Angle brackets $\langle\ \rangle$ are used to denote sequences—rather than curly set brackets { }—so as to emphasize ordering. For example the Fibonacci number sequence is written as
$$\langle F_n \rangle = \langle 0,1,1,2,3,5,8,13,\dots\ \rangle.$$

Logarithms: Where natural logarithms are required 'ln( )' is used. But when interest is limited to the *ratio* of two logarithms, then the base is not relevant (so long as it is the same for both) and the generic 'log( )' is used. For example on page 212, the use of both base 10, or both base e logs will give the same result.

Fractals: The symbols N, r, and D are often used in the context of fractals; these are adopted in the *Is it fractal?* chapter—p.232.

Ori32 Geometry: In the introduction to Ori32 and the reference Tables—the symbols α, β, κ, λ, μ, and ν are used for roots of combinations of 2 and $\sqrt{2}$ (p.402). It is important to note that the meanings of α and β in Ori32 have no relation to ϕ and its conjugate—again this is a matter of context.

Inverse trigonometry: 'Arc' notation, e.g. $\arcsin(x)$ is used in place of $\sin^{-1}(x)$ as the latter is inconsistent with expressions such as $\sin^2(x)$.

Fibonacci Resonance: The Table that follows provides full details of the symbols used in the development of the Fibonacci Resonance. The four symbols specific to this book (ω, q_s, s, and t) are marked as 'new'. Other symbols—neither new nor checked as conventional—have been seen used, but not widely enough to warrant being referred to as standard or conventional.

Greek Alphabet: This Table also includes references to usages in the text. For convenience, several such references are duplicated from the Fibonacci Resonance Table.

Symbol	Conven-tional	Name	Explanation	See page
⌢	✓		Concatenate.	384
(2π)		circle constant	Angle in radians of one full turn, *also* circumference of the unit circle.	33 / 179
a			Coefficient parameter in search.	197
α	✓	alpha	Root of $x^2 - Px + Q = 0$ $\alpha = (P + \sqrt{\Delta})/2.$	254
β	✓	beta	Root of $x^2 - Px + Q = 0$ $\beta = (P - \sqrt{\Delta})/2.$	254
$\mathbb{C}$	✓	black-board C	The complex numbers.	134
δ		delta	Silver Ratio $\delta = 1 + \sqrt{2}.$	268
Δ	✓	Delta	Discriminant of $x^2 - P + Q = 0$ $\Delta = P^2 - 4Q.$	254

continued...

460

Symbol	Conven-tional	Name	Explanation	See page
$\langle F_n \rangle$	✓	sequence F-sub-n	Fibonacci numbers $F_n = F_{n-1} + F_{n-2}$.	1
$\langle f_n \rangle$			$2^n - 1$ sequence 'Mersenne' $f_n = 3f_{n-1} - 2f_{n-2}$.	277
$\langle g_n \rangle$			$2^n + 1$ sequence 'Fermat' $g_n = 3g_{n-1} - 2g_{n-2}$.	277
$\langle J_n \rangle$	✓		Jacobsthal sequence $J_n = J_{n-1} + 2J_{n-2}$.	275
$\langle j_n \rangle$			Jacobsthal-Lucas sequence $j_n = j_{n-1} + 2j_{n-2}$.	275
λ	✓	lambda	Wavelength.	223
$\langle L_n \rangle$	✓		Lucas numbers $L_n = L_{n-1} + L_{n-2}$.	27
(mod 32)	✓	modulo 32	Remainder after ÷32.	171

(contd.) appears at top.

continued...

Symbol	Conven- tional	Name	Explanation	See page
(contd.)				
$\tilde{\mathbb{N}}$	✓	N-tilde	The set of non-negative integers: $\{0, 1, 2, 3, \dots\}$	256
n	✓		Primary index into sequences e.g. $\langle F_n \rangle$.	
ω	new	omega	Companion axes offset angle (in radians).	213
Ω	✓	Omega	Divergence angle.	52
P	✓		Lucas Sequence recurrence coefficient.	252
$\langle P_n \rangle$	✓		Pell sequence $P_n = 2P_{n-1} + P_{n-2}$.	268
ϕ	✓	phi	The Golden Ratio $\phi = (1 + \sqrt{5})/2$ a root of $x^2 - x - 1 = 0$.	14
Q	✓		Lucas Sequence recurrence coefficient.	252
$|Q|$	✓	modulus of Q	The absolute value of Q.	261

continued…

Symbol	Conventional	Name	Explanation	See page
(contd.)				
$\langle Q_n \rangle$	✓		Pell-Lucas sequence $Q_n = 2Q_{n-1} + Q_{n-2}$.	268
$\mathbb{R}$	✓	black-board R	The real numbers.	134
q_s	new	q-sub-s	Quantum associated with spiral s, $\quad q_s = \phi^{-s}$.	208
r	✓		Radial polar coordinate.	35
s	new		Spiral identification number (= index offset).	209
t	new		Sign parameter, value $+1$ or -1.	262
θ	✓	theta	Angular polar coordinate.	35
$\langle U_n \rangle$	✓		PQ Lucas Sequence $U_n = P \cdot U_{n-1} - Q \cdot U_{n-2}$.	252
$\langle V_n \rangle$	✓		Companion Lucas Sequence $V_n = P \cdot V_{n-1} - Q \cdot V_{n-2}$.	252
$\mathbb{Z}$	✓	black-board Z	the set of all integers: $\{\ldots, -2, -1, 0, 1, 2, \ldots\}$ (G.) Zählen: numbers/counting.	199

The Greek alphabet

Lower case	Name	Example usage	Upper case
α	alpha	Root of characteristic Eqn. p.3, Ori32 p.402.	A
β	beta	—ditto—	B
γ	gamma	Angle p.431.	Γ
δ	delta	Tiny angle $\delta\theta$ p.40, Silver Ratio p.125, difference Δ p.194, discriminant Δ p.254.	Δ
ϵ	epsilon	Electric p.362, small angular shortfall p.438.	E
ζ	zeta		Z
η	eta		H
θ	theta	General angle p.35.	Θ
ι	iota		I
κ	kappa	Ori32 p.402.	K
λ	lambda	Wavelength p.223, eigenvalue p.348, Ori32 p.402.	Λ
μ	mu	Magnetic p.362, Ori32 p.402.	M
ν	nu	Ori32 p.402.	N
ξ	xi		Ξ
o	omicron		O
π	pi	Conventional p.7, but (2π) preferred p.33.	Π
ρ	rho	Root Golden approximants p.74.	P
σ	sigma	Angle p.40, summation Σ p.169.	Σ
τ	tau	New circle constant p.33 (discussed but not used).	T
υ	upsilon		Y
ϕ	phi	Golden Ratio p.3, and throughout.	Φ
χ	chi		X
ψ	psi	Angle p.432, plastic number Ψ p.94 (footnote).	Ψ
ω	omega	Offset angle p.214, divergence angle Ω p.293.	Ω

Collected formulæ

Golden Spiral 'with walls' ....

Shallow angle between quadrant axes and radial vectors through tangent points,

$$90 - 72.9676089 = \mathbf{17.0\,32\,39\,11}°$$ (p.43)

Ori32 Fibonacci circle ....

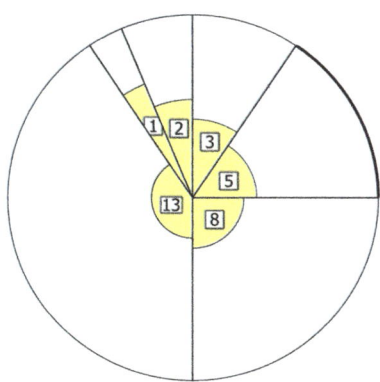

Residue 0:   (allowing $k < 0$ in Eqn.(11.3), p.169)

$$\sum_{n=2+6k}^{7+6k} F_n \equiv 0 \pmod{32}, \quad \text{for all } k \text{ in } \mathbb{Z} \quad \text{(Eqn.(B.1), p.410)}$$

Residue 1:   (Eqn.(11.4), p.170)

$$\sum_{n=0}^{1+6k} F_n \equiv 1 \pmod{32}, \quad k = 1, 2, 3, \dots \quad \text{(Eqn.(B.18), p.413)}$$

465

Residue 2: (allowing $k < 0$ in Eqn.(11.7), p.171)

$$F_{3+6k} \equiv 2 \pmod{32}, \quad \text{for all } k \text{ in } \mathbb{Z} \quad \text{(Eqn.(B.19), p.416)}$$

Even-index triads, residue 0: (allowing $k < 0$ in Eqn.(11.10), p.172)

$$F_{6k-2} + F_{6k} + F_{6k+2} \equiv 0 \pmod{32}, \quad \text{for all } k \text{ in } \mathbb{Z}$$
$$\text{(Eqn.(B.21), p.419)}$$

Particular results:

Fibonacci indices,

$$L_5 + F_8 = 32 \qquad\qquad \text{(Eqn.(11.8), p.172)}$$

Lucas indices,

$$L_7 + F_4 = 32. \qquad\qquad \text{(Eqn.(11.9), p.172)}$$

Generalized relationship summary (in 3 equivalent expressions):

$$\left.\begin{array}{l} F_{6k-2} + F_{6k} + F_{6k+2} \equiv 0 \pmod{32} \\ L_{6k-1} + F_{6k+2} \equiv 0 \pmod{32} \\ L_{6k+1} + F_{6k-2} \equiv 0 \pmod{32} \end{array}\right\} \quad \text{(Eqn.(11.13), p.173)}$$

from which it follows that

$$8F_{6k} \equiv 0 \pmod{32}. \qquad\qquad \text{(Eqn.(11.14), p.173)}$$

Five Golden Powers ....

First pentad:

$$\sum_{n=-2}^{2} \phi^n = 2\phi + 3 = 4 + \sqrt{5} \approx (2\pi). \quad \text{(Eqn.(12.2), p.178)}$$

Sum of the first and second pentads—incorporating the first 5 primes:

$$\sum_{n=-2}^{7} \phi^n = 11 \cdot \left(7 + \sqrt{5} \cdot 3\right)\big/2. \qquad\qquad \text{(Eqn.(12.3), p.181)}$$

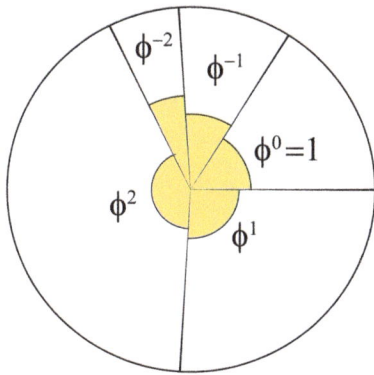

Hence the expression incorporating the first 4 primes:

$$\phi^4 \;=\; \left(7 + \sqrt{5}\cdot 3\right)\big/ 2. \qquad \text{(Eqn.(12.5), p.182)}$$

Observation leading to Fibonacci numbers 55 and 144:

$$\phi^4 \cdot \frac{11}{24} \;=\; \pi \times 0.9999589\ldots \qquad \text{(Eqn.(12.8), p.185)}$$

Sum of the first three pentads:

$$\sum_{n=-2}^{12} \phi^n \;=\; 422 + 188\sqrt{5}. \qquad \text{(Eqn.(12.9), p.186)}$$

The Fibonacci Resonance . . . .

Case of quantum $= \phi^{-5}$, index offset 5:

$$F_n = 5\phi^{n-5} - F_{n-5}\,q \qquad \text{where} \quad q = \phi^{-5}. \qquad \text{(Eqn.(13.6), p.196)}$$

Generalization of Fibonacci and Lucas expressions with index offset s:

$$\left.\begin{aligned} F_n &= & F_s\,\alpha^{n-s} &+ F_{n-s}\,\beta^s \\ L_n &= & \sqrt{5}\,F_s\,\alpha^{n-s} &+ L_{n-s}\,\beta^s \end{aligned}\right\} \quad \text{for all } n, s \text{ in } \mathbb{Z}.$$

$$\text{(Eqn.(13.16), p.200)}$$

$$\left.\begin{aligned} F_n &= & F_s\,\phi^{n-s} &+ F_{n-s}\,(-\phi)^{-s} \\ L_n &= & \sqrt{5}\,F_s\,\phi^{n-s} &+ L_{n-s}\,(-\phi)^{-s} \end{aligned}\right\} \quad \text{for all } n, s \text{ in } \mathbb{Z}.$$

$$\text{(Eqn.(13.17), p.200)}$$

Visualization—Fibonacci and Lucas .... (1, −1)

Quantized form for Fibonacci and Lucas numbers:

$$
\left.\begin{aligned}
q_s &= \phi^{-s} \\
F_n &= F_s\,\phi^{n-s} + (-1)^s\,F_{n-s}\,q_s \\
L_n &= \sqrt{5}\,F_s\,\phi^{n-s} + (-1)^s\,L_{n-s}\,q_s
\end{aligned}\right\} \quad \text{for all } n, s \text{ in } \mathbb{Z}.
$$

(Eqn.(15.2), p.208)

Golden Spiral anchored on F_s:

$$
r = F_s\,\phi^{\frac{4\theta}{(2\pi)} - s}.
$$

(Eqn.(15.5), p.209)

Equivalence of dilation and rotation, for Golden Spiral 5× scaling:

$$
\frac{log(5)}{4\,log(\phi)} \cdot (2\pi) = 5.2536098\ldots \quad \text{radians} \qquad \text{(Eqn.(15.8), p.212)}
$$

$$
\equiv 301.0096689\ldots^{\circ} = 360 - \mathbf{58.99003311\ldots}^{\circ}
$$

Companion quadrant axes offset angle (Fibonacci to Lucas):

$$
\omega = \frac{log(\sqrt{5})}{4\,log(\phi)} \cdot (2\pi) = 2.62680\ldots \equiv \mathbf{150.505\ldots}^{\circ} \quad \text{(Eqn.(15.9), p.213)}
$$

A numeric coincidence involving the diagonal of a Golden Rectangle:

$$
tan^{-1}\left(\frac{1}{\phi}\right) - (180^{\circ} - 150.505^{\circ}) = 2.222^{\circ} \qquad \text{(Eqn.(15.11), p.215)}
$$

Polar coordinate placements for Fibonacci and Lucas numbers:

$$
\left.\begin{aligned}
\langle F_n\rangle \text{ are placed at } (r,\theta) &= \left(F_n,\ n\cdot\frac{(2\pi)}{4}\right) \\
\langle L_n\rangle \text{ are placed at } (r,\theta) &= \left(L_n,\ n\cdot\frac{(2\pi)}{4} + \omega\right)
\end{aligned}\right\}
$$

(Eqn.(15.10), p.214)

Generalization—Lucas Sequences ....(P, Q)

Generalized formulæ for Lucas Sequences U_n and V_n:

$$\left.\begin{array}{l} U_n = \quad\quad U_s\alpha^{n-s} + U_{n-s}\beta^s \\ V_n = (\alpha - \beta)\,U_s\alpha^{n-s} + V_{n-s}\beta^s \end{array}\right\} \quad \left\{\begin{array}{l}\text{for all } n, s \text{ in } \mathbb{Z}, \text{ if } |Q| = 1, \\ \text{for all } s \le n, \text{ and } n, s \text{ in } \tilde{\mathbb{N}}, \text{ if } |Q| \ne 1 \end{array}\right.$$

(Eqn.(20.34), p.261)

where $\tilde{\mathbb{N}}$ (N-tilde) is the set of non-negative integers $\{0,1,2,3, \dots\}$ (p.256)

Quantized form for U_n and V_n with sign parameter t:

$$\left.\begin{array}{l} t \;=\; \begin{cases} -1, & \text{if } \beta < 0 \text{ and } s \text{ is odd,} \\ +1, & \text{otherwise.} \end{cases} \\[2em] q_s \;=\; |\beta^s| \\ U_n = \quad\quad U_s\alpha^{n-s} + t\,U_{n-s}\,q_s \\ V_n = (\alpha - \beta)\,U_s\alpha^{n-s} + t\,V_{n-s}\,q_s \end{array}\right\} \quad \left\{\begin{array}{l}\text{for all } n, s \text{ in } \mathbb{Z}, \text{ if } |Q| = 1, \\ \text{for all } s \le n, \text{ and } n, s \text{ in } \tilde{\mathbb{N}}, \\ \quad\quad \text{if } |Q| \ne 1. \end{array}\right.$$

(Eqn.(20.37), p.262)

Log spiral with α per quadrant growth, anchored on U_s:

$$r = U_s\,\alpha^{\frac{4\theta}{(2\pi)} - s}.$$

(Eqn.(20.40), p.263)

Companion quadrant axes offset angle in terms of α and β:

$$\omega = \frac{log(\alpha - \beta)}{4\,log(\alpha)} \cdot (2\pi).$$

(Eqn.(20.41), p.263)

Companion quadrant axes offset angle in terms of P and Q:

$$\omega = \frac{log(P^2 - 4Q)}{8\left[log\left(P + \sqrt{P^2 - 4Q}\right) - log(2)\right]} \cdot (2\pi).$$

(Eqn.(20.42), p.264)

Polar coordinate placements for U_n and V_n:

$$\left.\begin{array}{l} \langle U_n \rangle \text{ are placed at } (r,\theta) \;=\; \left(U_n,\; n \cdot \dfrac{(2\pi)}{4}\right) \\[1.5em] \langle V_n \rangle \text{ are placed at } (r,\theta) \;=\; \left(V_n,\; n \cdot \dfrac{(2\pi)}{4} + \omega\right) \end{array}\right\}$$

(Eqn.(20.43), p.264)

Pell (Silver Ratio) .... (2, −1)

Quantized form for Pell and Pell-Lucas numbers, with $\delta = 1 + \sqrt{2}$ (the Silver Ratio—ϕ and δ are examples of 'metallic means' [493, 854]):

$$
\left.
\begin{aligned}
q_s &= \delta^{-s} \quad \Longrightarrow \text{ fractal-like} \\[4pt]
P_n &= \phantom{2\sqrt{2}\,} P_s \delta^{n-s} + (-1)^s P_{n-s}\, q_s \\[4pt]
Q_n &= 2\sqrt{2}\, P_s \delta^{n-s} + (-1)^s Q_{n-s}\, q_s
\end{aligned}
\right\}
\quad \text{for all } n, s \text{ in } \mathbb{Z}.
$$

(Eqn.(21.12), p.269)

Silver Spiral anchored on P_s:

$$
r = P_s \delta^{\frac{4\theta}{(2\pi)} - s}.
$$

(Eqn.(21.13), p.270)

Companion quadrant axes offset angle:

$$
\omega = \frac{log(2\sqrt{2})}{4\, log(\delta)} \cdot (2\pi) \;=\; 1.853005\ldots \;\equiv\; \mathbf{106.169\ldots}^{\circ}
$$

(Eqn.(21.14), p.270)

Polar coordinate placements for Pell and Pell-Lucas numbers:

$$
\left.
\begin{aligned}
\langle P_n \rangle \text{ are placed at} \quad (r, \theta) &= \left(P_n, \; n \cdot \frac{(2\pi)}{4} \right) \\[8pt]
\langle Q_n \rangle \text{ are placed at} \quad (r, \theta) &= \left(Q_n, \; n \cdot \frac{(2\pi)}{4} + \omega \right)
\end{aligned}
\right\}
$$

(Eqn.(21.15), p.270)

'Bronze Fibonacci numbers' (Bronze Ratio) (3, −1)

In passing, we note another metallic mean, the Bronze Ratio, with value $(3 + \sqrt{13})/2$. The associated Lucas Sequence is A006190 with companion A006497. The companion axes offset $\omega = 96.61°$ [493, 854]. (See also p.429.)

$\sqrt{5}$ Denominators / even Lucas numbers .... (4, −1)

$$
\alpha = 2 + \sqrt{5} = \phi^3
$$

(Eqn.(21.22), p.274)

$$
\beta = 2 - \sqrt{5} = (-\phi)^{-3}.
$$

(Eqn.(21.23), p.274)

Companion quadrant axes offset angle:

$$\omega = \frac{log(20)}{8\left(log(4+\sqrt{20}) - log(2)\right)} \cdot (2\pi) \;=\; 1.629803829\ldots$$

$$\equiv \; \mathbf{93.3\,808\,808\,58\ldots}^{\circ} \qquad\qquad \text{(Eqn.(21.24), p.274)}$$

The companion sequence lists the even Lucas numbers:

$$V_n = L_{3n}. \qquad\qquad \text{(Eqn.(21.25), p.274)}$$

Jacobsthal (Copper Ratio) ....(1, −2)

Quantized form for Jacobsthal and Jacobsthal-Lucas numbers:

$$\left.\begin{aligned}
q_s &= |\beta^s| = 1, \;\;\Longrightarrow\; \text{not fractal-like} \\[4pt]
J_n &= \;\; J_s 2^{n-s} + (-1)^s J_{n-s} q_s \\[4pt]
j_n &= 3 J_s 2^{n-s} + (-1)^s j_{n-s} q_s
\end{aligned}\right\} \quad \text{for all } s \leq n, \text{ and } n, s \text{ in } \tilde{\mathbb{N}}.$$

$$\text{(Eqn.(21.29), p.276)}$$

Log spiral with double per quadrant growth ($\alpha=2$, the Copper Ratio), anchored on J_s:

$$r = J_s 2^{\frac{4\theta}{(2\pi)} - s}. \qquad\qquad \text{(Eqn.(21.30), p.276)}$$

Companion quadrant axes offset angle:

$$\omega = \frac{log(3)}{4\,log(2)} \cdot (2\pi) \;=\; 2.48965327\ldots$$

$$\equiv \; \mathbf{142.646625\ldots}^{\circ} \qquad\qquad \text{(Eqn.(21.31), p.276)}$$

Polar coordinate placements for Jacobsthal and Jacobsthal-Lucas:

$$\left.\begin{aligned}
\langle J_n\rangle \text{ are placed at } (r,\theta) &= \left(J_n, \; n\cdot\frac{(2\pi)}{4} \right) \\[10pt]
\langle j_n\rangle \text{ are placed at } (r,\theta) &= \left(j_n, \; n\cdot\frac{(2\pi)}{4} + \omega \right)
\end{aligned}\right\}$$

$$\text{(Eqn.(21.32), p.277)}$$

'Nickel Fibonacci numbers' (Nickel Ratio) .. (1, −3)

Again in passing, we note a further metallic mean, the Nickel Ratio, with value $(1+\sqrt{13})/2$. The associated Lucas Sequence starts with zero and continues as A006130. The companion sequence is A075118, and the companion axes offset is $\omega = 138.38°$ [493, 854]. (See also p.429.)

Mersenne and Fermat .... (3, 2)

Quantized form for 'Mersenne' and 'Fermat' numbers:

$$q_s = |\beta^s| = 1, \implies \text{not fractal-like}$$
$$f_n = f_s 2^{n-s} + f_{n-s} q_s$$
$$g_n = f_s 2^{n-s} + g_{n-s} q_s$$

for all $s \leq n$, and n, s in $\tilde{\mathbb{N}}$.

(Eqn.(21.36), p.278)

Log spiral with double per quadrant growth, anchored on f_s:

$$r = f_s 2^{\frac{4\theta}{(2\pi)} - s}.$$

(Eqn.(21.37), p.278)

Offset angle of companion axes:

$$\omega = \frac{log(1)}{4\,log(2)} \cdot (2\pi)$$
$$= 0° \quad \textbf{(zero offset).}$$

(Eqn.(21.38), p.278)

Polar coordinate placements for 'Fermat' and 'Mersenne' numbers:

$$\langle f_n \rangle \text{ are placed at } (r,\theta) = \left(f_n, \ n \cdot \frac{(2\pi)}{4} \right)$$
$$\langle g_n \rangle \text{ are placed at } (r,\theta) = \left(g_n, \ n \cdot \frac{(2\pi)}{4} \right)$$

(Eqn.(21.39), p.278)

Perfect host .... (6, 8)

Quantized form:

$$q_s = |\beta^s| = 2^s, \implies \text{fractal-like}$$
$$U_n = U_s 4^{n-s} + U_{n-s} q_s$$
$$V_n = 2 U_s 4^{n-s} + V_{n-s} q_s$$

for all $s \leq n$, and n, s in $\tilde{\mathbb{N}}$.

(Eqn.(21.45), p.281)

Perfect host, Binet form:

$$U_n = 2^{n-1}(2^n - 1). \qquad \text{(Eqn.(21.48), p.282)}$$

Log spiral with 'quadruple per quadrant' growth, anchored on U_s:

$$r = U_s\, 4^{\frac{4\theta}{(2\pi)} - s}. \qquad \text{(Eqn.(21.46), p.281)}$$

Companion axes offset angle (special case):

$$\omega = \frac{log(2)}{4\cdot 2\cdot log(2)}\cdot(2\pi) \;=\; \frac{(2\pi)}{8} \;=\; 0.785398\ldots$$

$$\equiv \; \mathbf{45°} \quad \textbf{(special case).} \qquad \text{(Eqn.(21.41), p.280)}$$

Polar coordinate placements:

$$\langle U_n\rangle \text{ are placed at } (r,\theta) \;=\; \left(U_n,\; n\cdot\frac{(2\pi)}{4} \right) \Bigg\}$$

$$\langle V_n\rangle \text{ are placed at } (r,\theta) \;=\; \left(V_n,\; n\cdot\frac{(2\pi)}{4} \;+\; \frac{(2\pi)}{8} \right)$$

(Eqn.(21.47), p.282)

Offset $\omega = 0$ ....

$$P^2 \;=\; 4Q+1 \;\implies\; \omega = 0 \implies$$

$$(P, Q) \;=\; (3, 2),\; (5, 6),\; (7, 12),\; (9, 20),\; (11, 30),\; \ldots$$

(Eqn.(21.40), p.279)

Fibonacci recurrences $\langle G_n\rangle$, $\langle K_n\rangle$ closed forms.

For any seed pair G_0 and G_1, with $\alpha = \phi$ and $\beta = -\phi^{-1}$:

$$G_n \;=\; G_{n-1} + G_{n-2} \;=\; A\alpha^n + B\beta^n.$$
$$A \;=\; (G_0\alpha + G_1 - G_0)\big/\sqrt{5},$$
$$B \;=\; G_0 - A \;=\; (G_0\alpha - G_1)\big/\sqrt{5}.$$

(Eqn.(F.16), p.446)

Example—'ten twelve' sequence (p.187) where $K_0 = 10$, and $K_1 = 12$:

$$K_n \;=\; \left[5 + \left(7/\sqrt{5}\right)\right]\alpha^n \;+\; \left[5 - \left(7/\sqrt{5}\right)\right]\beta^n. \qquad \text{(Eqn.(F.17), p.446)}$$

Root Golden $\cdots\cdots\cdots\cdots\cdots\cdots\cdots\cdots\cdots\cdots\cdots$

[RG−] $= 2/\phi$ (Golden Section of 2) $\approx \sqrt{\phi}$. Also [RG+] $= \phi^2/2$. (p.73)

Successive approximations of $\sqrt{\phi}$, using Babylonian method: [171]

$$\rho_1 = 2/\phi \qquad (\sqrt{\phi} \text{ within 3\%}) \qquad \text{(Eqn.(6.1), p.74)}$$

$$\rho_2 = \frac{5\phi - 3}{4} \qquad (\sqrt{\phi} \text{ within 0.05\%}) \qquad \text{(Eqn.(6.2), p.74)}$$

$$\rho_3 = \frac{11\phi + 34}{8(5\phi - 3)} \qquad (\sqrt{\phi} \text{ within 0.00001\%}). \qquad \text{(Footnote, p.74)}$$

$$r = \left(\sqrt{\phi}\right)^{\frac{4\theta}{(2\pi)}} \qquad \text{Root Golden Spiral.} \qquad \text{(Eqn.(9.2), p.137)}$$

Powers of ϕ $\cdots\cdots\cdots\cdots\cdots\cdots\cdots\cdots\cdots\cdots\cdots$

Summing powers of ϕ in terms of ϕ:

$$\sum_{n=0}^{k} \phi^n = \phi(F_{k+2} - 1) + F_{k+1}. \qquad \text{(Eqn.(D.12), p.435)}$$

'Umbral' comparison (correspondence between powers and indices):

$$\sum_{n=0}^{k} \phi^n = \phi^{k+2} - \phi^1, \qquad \sum_{n=0}^{k} F_n = F_{k+2} - F_1. \qquad \text{(Eqn.(D.13), p.435)}$$

And similarly for the conjugate root $\beta = (-\phi)^{-1}$:

$$\sum_{n=0}^{k} (-\phi)^{-n} = (-\phi)^{-(k+2)} - (-\phi)^{-1}. \qquad \text{(Eqn.(D.14), p.436)}$$

Summing powers of ϕ in terms of $\sqrt{5}$, starting with ϕ^0:

$$\sum_{n=0}^{k} \phi^n = \frac{1}{2}\left\{(F_{k+2} - 1)\sqrt{5} + (L_{k+2} - 1)\right\}. \qquad \text{(Eqn.(D.19), p.437)}$$

Summing powers of ϕ in terms of $\sqrt{5}$, starting with ϕ^{-2}:

$$\sum_{n=-2}^{k} \phi^n = \frac{1}{2}\left\{(F_{k+2} - 1)\sqrt{5} + (L_{k+2} + 1)\right\}. \qquad \text{(Eqn.(D.20), p.437)}$$

Proportional music structure, ϕ^2 based ....

Using the layout of the negative-index Fibonacci numbers along the real number line (Fig. 268) (page 145)

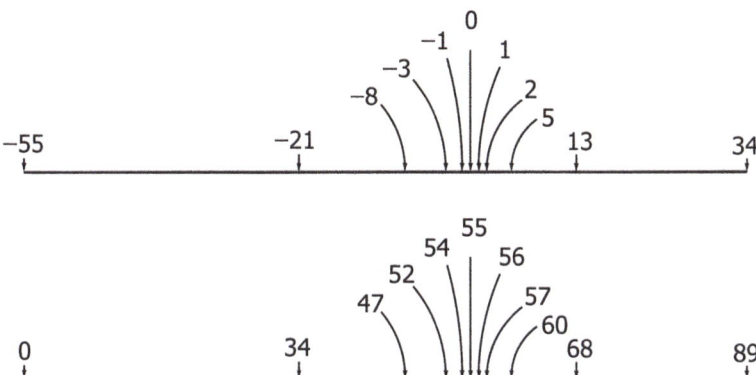

Figure 268: **'Volcano' of negative-index Fibonacci numbers (p.145) with offset version showing example bar numbers along the number line.**

Simple Golden Section ....

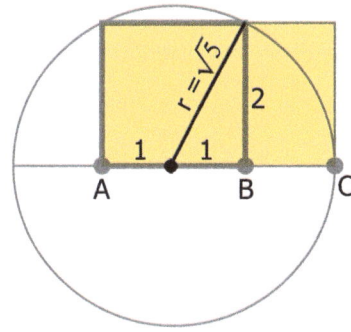

Figure 269: **ABC Golden Section**

Fig. 269 was introduced in Fig. 1 (p.3). We obtain the radius of the circle using Pythagoras. That is, because $1^2 + 2^2 = 5$, then $r = \sqrt{5}$, and as

$$AB = 2$$
$$AC = 1 + \sqrt{5}, \quad \text{then}$$

$$\frac{AC}{AB} = \frac{1 + \sqrt{5}}{2} = \phi.$$

For an example Fibonacci integer approximation for this construction, see p.122.

Ori32 trigonometry ·······························

Formulæ for Ori32 trigonometry are shown in Appendix A, (page 401).

Ori8 radials and angles are a subset of those of Ori32, comprising only the cardinal and ordinal orientations (i.e. spaced in units of 45°). The main and companion axes for the $(6, 8)$ perfect host sequence (whose members include the perfect numbers) share this geometry.

(Fig. 178, page 283)

Skip-Pell: Make Pell-Lucas ························

In the same form as the skip-Fibonacci identity $L_n = F_{n+1} + F_{n-1}$ [488], we have the skip-Pell summation $P_{n+1} + P_{n-1}$ which produces the Pell-Lucas (companion sequence) member Q_n :

$$Q_n = P_{n+1} + P_{n-1}.$$ (Eqn.(21.16), p.272)

Skip generalization ·····························

Generalization of skip-Fibonacci and skip-Pell identities.

For the subset of Lucas Sequences with recurrence coefficient $Q = -1$ (here showing this value of Q applied)

$$\begin{cases} U_n & = P{\cdot}U_{n-1} + U_{n-2}, & U_0 = 0,\ U_1 = 1, & P \text{ in } \mathbb{Z} \\ V_n & = P{\cdot}V_{n-1} + V_{n-2}, & V_0 = 2,\ V_1 = P, \\ V_n & = U_{n+1} + U_{n-1}. \end{cases}$$

(Eqn.(21.21), p.273)

PQ Lucas Sequences Table ·····················

(See page 267, which lists six pairs of Lucas Sequences along with their ω offset angles, their initial elements, and Sloane references.)

Golden Ratio—Fibonacci, Lucas, and ϕ^n ....

n	F_n	L_n	ϕ^n	=	$F_n\phi + F_{n-1}$		
-10	-55	123	ϕ^{-10}	=	$-55\phi + 89$	=	0.008
-9	34	-76	ϕ^{-9}	=	$34\phi - 55$	=	0.013
-8	-21	47	ϕ^{-8}	=	$-21\phi + 34$	=	0.021
-7	13	-29	ϕ^{-7}	=	$13\phi - 21$	=	0.034
-6	-8	18	ϕ^{-6}	=	$-8\phi + 13$	=	0.056
-5	5	-11	ϕ^{-5}	=	$5\phi - 8$	=	0.090
-4	-3	7	ϕ^{-4}	=	$-3\phi + 5$	=	0.146
-3	2	-4	ϕ^{-3}	=	$2\phi - 3$	=	0.236
-2	-1	3	ϕ^{-2}	=	$-\phi + 2$	=	0.382
-1	1	-1	ϕ^{-1}	=	$\phi - 1$	=	0.618
0	0	2	ϕ^0	=	1	=	1.000
1	1	1	ϕ^1	=	ϕ	=	1.618
2	1	3	ϕ^2	=	$\phi + 1$	=	2.618
3	2	4	ϕ^3	=	$2\phi + 1$	=	4.236
4	3	7	ϕ^4	=	$3\phi + 2$	=	6.854
5	5	11	ϕ^5	=	$5\phi + 3$	=	11.090
6	8	18	ϕ^6	=	$8\phi + 5$	=	17.944
7	13	29	ϕ^7	=	$13\phi + 8$	=	29.034
8	21	47	ϕ^8	=	$21\phi + 13$	=	46.979
9	34	76	ϕ^9	=	$34\phi + 21$	=	76.013
10	55	123	ϕ^{10}	=	$55\phi + 34$	=	122.992
11	89	199	ϕ^{11}	=	$89\phi + 55$	=	199.005
12	144	322	ϕ^{12}	=	$144\phi + 89$	=	321.997
13	233	521	ϕ^{13}	=	$233\phi + 144$	=	521.002
14	377	843	ϕ^{14}	=	$377\phi + 233$	=	842.999
15	610	1364	ϕ^{15}	=	$610\phi + 377$	=	1364.001

Figure 270: **Fibonacci, Lucas, and ϕ powers Table,** including ϕ powers in their linear form $j\phi + k$, with numeric values rounded to 3 decimal places [491]. The Golden Ratio $\phi = (1 + \sqrt{5})/2$.

Golden Ratio—ϕ^n in terms of $\sqrt{5}$ ....

n	ϕ^n	=	$(L_n + F_n\sqrt{5})/2$		
-10	ϕ^{-10}	=	$(123 - 55\sqrt{5})/2$		
-9	ϕ^{-9}	=	$(-76 + 34\sqrt{5})/2$	=	$-38 + 17\sqrt{5}$
-8	ϕ^{-8}	=	$(47 - 21\sqrt{5})/2$		
-7	ϕ^{-7}	=	$(-29 + 13\sqrt{5})/2$		
-6	ϕ^{-6}	=	$(18 - 8\sqrt{5})/2$	=	$9 - 4\sqrt{5}$
-5	ϕ^{-5}	=	$(-11 + 5\sqrt{5})/2$		
-4	ϕ^{-4}	=	$(7 - 3\sqrt{5})/2$		
-3	ϕ^{-3}	=	$(-4 + 2\sqrt{5})/2$	=	$-2 + \sqrt{5}$
-2	ϕ^{-2}	=	$(3 - \sqrt{5})/2$		
-1	ϕ^{-1}	=	$(-1 + \sqrt{5})/2$		
0	ϕ^0	=	1		
1	ϕ^1	=	$(1 + \sqrt{5})/2$		
2	ϕ^2	=	$(3 + \sqrt{5})/2$		
3	ϕ^3	=	$(4 + 2\sqrt{5})/2$	=	$2 + \sqrt{5}$
4	ϕ^4	=	$(7 + 3\sqrt{5})/2$		
5	ϕ^5	=	$(11 + 5\sqrt{5})/2$		
6	ϕ^6	=	$(18 + 8\sqrt{5})/2$	=	$9 + 4\sqrt{5}$
7	ϕ^7	=	$(29 + 13\sqrt{5})/2$		
8	ϕ^8	=	$(47 + 21\sqrt{5})/2$		
9	ϕ^9	=	$(76 + 34\sqrt{5})/2$	=	$38 + 17\sqrt{5}$
10	ϕ^{10}	=	$(123 + 55\sqrt{5})/2$		
11	ϕ^{11}	=	$(199 + 89\sqrt{5})/2$		
12	ϕ^{12}	=	$(322 + 144\sqrt{5})/2$	=	$161 + 72\sqrt{5}$
13	ϕ^{13}	=	$(521 + 233\sqrt{5})/2$		
14	ϕ^{14}	=	$(843 + 377\sqrt{5})/2$		
15	ϕ^{15}	=	$(1364 + 610\sqrt{5})/2$	=	$682 + 305\sqrt{5}$

Figure 271: ϕ **powers in terms of $\sqrt{5}$,** using $\phi^n = (L_n + F_n\sqrt{5})/2$ [794].

Silver Ratio—Pell, Pell-Lucas, and δ^n . . . .

n	P_n	Q_n	δ^n	$=$	$P_n\delta + P_{n-1}$		
-10	-2378	6726	δ^{-10}	$=$	$-2378\delta + 5741$	$=$	0.0001
-9	985	-2786	δ^{-9}	$=$	$985\delta - 2378$	$=$	0.0004
-8	-408	1154	δ^{-8}	$=$	$-408\delta + 985$	$=$	0.0009
-7	169	-478	δ^{-7}	$=$	$169\delta - 408$	$=$	0.0021
-6	-70	198	δ^{-6}	$=$	$-70\delta + 169$	$=$	0.0051
-5	29	-82	δ^{-5}	$=$	$29\delta - 70$	$=$	0.0122
-4	-12	34	δ^{-4}	$=$	$-12\delta + 29$	$=$	0.0294
-3	5	-14	δ^{-3}	$=$	$5\delta - 12$	$=$	0.0711
-2	-2	6	δ^{-2}	$=$	$-2\delta + 5$	$=$	0.1716
-1	1	-2	δ^{-1}	$=$	$\delta - 2$	$=$	0.4142
0	0	2	δ^0	$=$	1	$=$	1.0000
1	1	2	δ^1	$=$	δ	$=$	2.4142
2	2	6	δ^2	$=$	$2\delta + 1$	$=$	5.828
3	5	14	δ^3	$=$	$5\delta + 2$	$=$	14.07
4	12	34	δ^4	$=$	$12\delta + 5$	$=$	33.97
5	29	82	δ^5	$=$	$29\delta + 12$	$=$	82.01
6	70	198	δ^6	$=$	$70\delta + 29$	$=$	197.99
7	169	478	δ^7	$=$	$169\delta + 70$	$=$	478
8	408	1154	δ^8	$=$	$408\delta + 169$	$=$	1154
9	985	2786	δ^9	$=$	$985\delta + 408$	$=$	2786
10	2378	6726	δ^{10}	$=$	$2378\delta + 985$	$=$	6726
11	5741	16238	δ^{11}	$=$	$5741\delta + 2378$	$=$	16238
12	13860	39202	δ^{12}	$=$	$13860\delta + 5741$	$=$	39202
13	33461	94642	δ^{13}	$=$	$33461\delta + 13860$	$=$	94642
14	80782	228486	δ^{14}	$=$	$80782\delta + 33461$	$=$	228486
15	195025	551614	δ^{15}	$=$	$195025\delta + 80782$	$=$	551614

Figure 272: **Pell, Pell-Lucas, and δ powers Table,** This includes δ powers in their linear form $j\delta + k$, with numeric values rounded. The Silver Ratio $\delta = (1 + \sqrt{2})$. For other 'metallic means', see Knott [493], and also de Spinadel [854].

Silver Ratio—δ^n in terms of $\sqrt{2}$ ·················

n	δ^n	$=$	$H_n + P_n\sqrt{2}$
-10	δ^{-10}	$=$	$3363 - 2378\sqrt{2}$
-9	δ^{-9}	$=$	$-1393 + 985\sqrt{2}$
-8	δ^{-8}	$=$	$577 - 408\sqrt{2}$
-7	δ^{-7}	$=$	$-239 + 169\sqrt{2}$
-6	δ^{-6}	$=$	$99 - 70\sqrt{2}$
-5	δ^{-5}	$=$	$-41 + 29\sqrt{2}$
-4	δ^{-4}	$=$	$17 - 12\sqrt{2}$
-3	δ^{-3}	$=$	$-7 + 5\sqrt{2}$
-2	δ^{-2}	$=$	$3 - 2\sqrt{2}$
-1	δ^{-1}	$=$	$-1 + \sqrt{2}$
0	δ^0	$=$	1
1	δ^1	$=$	$1 + \sqrt{2}$
2	δ^2	$=$	$3 + 2\sqrt{2}$
3	δ^3	$=$	$7 + 5\sqrt{2}$
4	δ^4	$=$	$17 + 12\sqrt{2}$
5	δ^5	$=$	$41 + 29\sqrt{2}$
6	δ^6	$=$	$99 + 70\sqrt{2}$
7	δ^7	$=$	$239 + 169\sqrt{2}$
8	δ^8	$=$	$577 + 408\sqrt{2}$
9	δ^9	$=$	$1393 + 985\sqrt{2}$
10	δ^{10}	$=$	$3363 + 2378\sqrt{2}$
11	δ^{11}	$=$	$8119 + 5741\sqrt{2}$
12	δ^{12}	$=$	$19601 + 13860\sqrt{2}$
13	δ^{13}	$=$	$47321 + 33461\sqrt{2}$
14	δ^{14}	$=$	$114243 + 80782\sqrt{2}$
15	δ^{15}	$=$	$275807 + 195025\sqrt{2}$

Figure 273: **Table of δ powers in terms of $\sqrt{2}$,** using $\delta^n = (H_n + P_n\sqrt{2})$, where H_n is the 'half companion Pell number' (aka 'modified Pell number q_n' [162]) viz. $Q_n/2$. Also $\langle H_{n \geq 0} \rangle$ is the sequence of numerators of $\sqrt{2}$ convergents (p.268, Sloane A001333).

Standard formulæ used ....

For a list of page numbers where a formula has been used, follow the associated bibliographic reference, e.g. [483].

Fibonacci and Lucas recurrences

$$F_n = F_{n-1} + F_{n-2} \text{ with } F_0 = 0, F_1 = 1 \qquad \text{(Eqn.(1.1), p.1)}$$

$$L_n = L_{n-1} + L_{n-2} \text{ with } L_0 = 2, L_1 = 1. \qquad \text{(Eqn.(1.9), p.27)}$$

The Golden Ratio defining quadratic

$$x^2 - x - 1 = 0. \qquad \text{(Eqn.(1.3), p.2)}$$

Both its roots

$$x = \frac{(1 \pm \sqrt{5})}{2}. \qquad \text{(Eqn.(1.4), p.2)}$$

Their values

$$\alpha = \frac{1 + \sqrt{5}}{2} \quad = +\phi^{+1} \quad = +1.618\ldots \qquad \text{(Eqn.(1.5), p.3)}$$

$$\beta = \frac{1 - \sqrt{5}}{2} \quad = -\phi^{-1} \quad = -0.618\ldots$$

with combinations and comparisons [275]

$$\alpha + \beta = 1 \qquad \alpha^2 + \beta^2 = 3 \qquad \alpha\beta = -1 \qquad \alpha^2 = \alpha + 1$$

$$\alpha - \beta = \sqrt{5} \qquad \alpha^2 - \beta^2 = \sqrt{5} \qquad 1 - 2\beta = \sqrt{5} \qquad \beta^2 = \beta + 1.$$

Cassini [489]

$$F_{n+1}F_{n-1} - F_n^2 = (-1)^n.$$

Binet forms [654, 54, 206]

$$F_n = \frac{\alpha^n - \beta^n}{\sqrt{5}} = \frac{\phi^n - (-\phi)^{-n}}{\sqrt{5}} = \frac{(1 + \sqrt{5})^n - (1 - \sqrt{5})^n}{2^n \sqrt{5}}$$

$$\text{(Eqn.(1.8), p.25)}$$

$$L_n = \alpha^n + \beta^n = \phi^n + (-\phi)^{-n} = \frac{(1 + \sqrt{5})^n + (1 - \sqrt{5})^n}{2^n}.$$

$$\text{(Eqn.(1.10), p.28)}$$

Negative indices

$$F_{-n} = (-1)^{n+1} F_n \qquad \text{[482]}$$

$$L_{-n} = (-1)^n L_n. \qquad \text{[483]}$$

Sums

$$3F_n = F_{n+2} + F_{n-2} \qquad \text{[485]}$$

$$5F_n = L_{n+1} + L_{n-1} \qquad \text{[486]}$$

$$L_n = F_n + 2F_{n-1} \qquad \text{[487]}$$

$$L_{n+1} = F_n + F_{n+2} \qquad \text{[488]}$$

$$\sum_{i=0}^{m} F_i = F_{m+2} - 1. \qquad \text{[490]}$$

Powers of ϕ

$$\phi^n = F_n \phi + F_{n-1} \qquad \text{[491]}$$

$$= (L_n + F_n \sqrt{5})/2. \qquad \text{[794]}$$

Golden Angle (p.53)

$$360 / \phi^2 = 137.50776405\ldots^\circ$$

Lucas Angle (p.58)

$$360 \cdot \left(3 + \frac{1}{\phi}\right)^{-1} = \frac{360}{3 + \phi^{-1}} = \frac{360}{2 + \phi} = \frac{360}{1 + \phi^2} = 99.50155^\circ.$$

Vogel's sunflower formula (p.56)

$$\begin{cases} r &= scale \times \sqrt{n} \\ \theta &= n \times 137.508^\circ, \quad n = 0, 1, 2, 3, \ldots \end{cases} \qquad \text{[930]}$$

Macro for Ori32 MIK boxes (p.162)

Expanding on the note on page 162—the simplest LaTeX code \fbox{3} produces the large $\boxed{3}$. Alternatively, the two part coding below will give a more compact version.

1. Define this macro in the preamble before the \begin{document}:

```
\newcommand{\mikbox}[1]{\hspace{0.1em}\raisebox{0.05em}%
{\scalebox{0.65}{\fbox{\scalebox{1.5}{#1}}}}}
```

2. Use the above macro as required: \mikbox{3} produces $\boxed{3}$.

Continued fractions for Golden Powers [457]

$$\phi^n = \begin{cases} [\,L_n;\ \overline{L_n}\,], & \text{if } n \text{ is odd,} \\ [\,(L_n-1);\ \overline{1,\,(L_n-2)}\,], & \text{if } n \text{ is even.} \end{cases}$$

(Eqn.(C.8), p.426)

Fractal dimension

$$N = r^D.$$

$$\log(N) = D\log(r)$$

$$D = \frac{\log(N)}{\log(r)}.$$

(Eqn.(18.1), p.232)

Trigonometry: Half angle formula

$$\sin\left(\frac{A}{2}\right) = \sqrt{\frac{1-\cos(A)}{2}}.$$

[953]

Double angle formulæ

$$\sin(2A) = 2\sin(A)\cos(A)$$

[948]

$$\tan(2A) = \frac{2\tan(A)}{1-\tan^2(A)}.$$

[948]

Sine difference formula

$$\sin(A-B) = \sin(A)\cos(B) - \sin(B)\cos(A).$$

[969]

Cosecant formula for 18°

$$\csc\left(\frac{2\pi}{20}\right) = 2\phi.$$

[945]

Generating function and related [498, 232, 592]

$$G(x) = F_0 + F_1 x + F_2 x^2 + F_3 x^3 + \cdots$$

(Eqn.(E.1), p.439)

$$G(x) = \frac{-x}{(x^2+x-1)}.$$

(Eqn.(E.5), p.440)

Summation form

$$\frac{1}{(1-\alpha x)} = \sum_{k=0}^{\infty} \alpha^k x^k.$$

(Eqn.(E.14), p.442)

Lucas Sequences—PQ generalization [592]

Recurrences

$$U_n = P{\cdot}U_{n-1} - Q{\cdot}U_{n-2} \qquad U_0 = 0,\ U_1 = 1 \qquad \text{(Eqn.(20.1), p.252)}$$

$$V_n = P{\cdot}V_{n-1} - Q{\cdot}V_{n-2} \qquad V_0 = 2,\ V_1 = P. \qquad \text{(Eqn.(20.2), p.252)}$$

PQ-generalized quadratic

$$x^n = Px^{n-1} - Qx^{n-2}$$
$$x^2 - Px + Q = 0. \qquad \text{(Eqn.(20.4), p.252)}$$

This has discriminant Δ; $\Delta > 0$ ensures roots α and β are real and distinct.

$$\Delta = P^2 - 4Q. \qquad \text{(Eqn.(20.10), p.254)}$$

The roots are

$$\alpha = \frac{P + \sqrt{\Delta}}{2} \qquad \text{(Eqn.(20.11), p.254)}$$

$$\beta = \frac{P - \sqrt{\Delta}}{2}. \qquad \text{(Eqn.(20.12), p.254)}$$

Their combinations are

$$\alpha + \beta = P \qquad \text{(Eqn.(20.13), p.254)}$$

$$\alpha\beta = Q \qquad \text{(Eqn.(20.14), p.254)}$$

$$\alpha - \beta = \sqrt{\Delta}. \qquad \text{(Eqn.(20.15), p.254)}$$

The Binet forms of the recurrences are [592]

$$U_n = \frac{\alpha^n - \beta^n}{\alpha - \beta} \qquad \text{(Eqn.(20.24), p.255)}$$

$$V_n = \alpha^n + \beta^n. \qquad \text{(Eqn.(20.25), p.255)}$$

The negative-index identities are

$$U_{-n} = -Q^{-n}U_n \qquad \text{(Eqn.(20.26), p.256)}$$

$$V_{-n} = Q^{-n}V_n. \qquad \text{(Eqn.(20.27), p.256)}$$

Matrix multiplication (p.348)

(Transforming a vector using a transformation matrix.)
Principle: for a column vector (x, y), the matrix transformed value of x is the scalar (dot) product of the zeroth row of the matrix with the vector. The transformed value of y is the scalar product of row 1 with the vector, and so on similarly for higher-order matrices which transform correspondingly longer column vectors.

For 2×2,

$$\begin{pmatrix} a_{00} & a_{01} \\ a_{10} & a_{11} \end{pmatrix} \begin{pmatrix} x \\ y \end{pmatrix} = \begin{pmatrix} a_{00}x + a_{01}y \\ a_{10}x + a_{11}y \end{pmatrix}.$$

And for 3×3,

$$\begin{pmatrix} a_{00} & a_{01} & a_{02} \\ a_{10} & a_{11} & a_{12} \\ a_{20} & a_{21} & a_{22} \end{pmatrix} \begin{pmatrix} x \\ y \\ z \end{pmatrix} = \begin{pmatrix} a_{00}x + a_{01}y + a_{02}z \\ a_{10}x + a_{11}y + a_{12}z \\ a_{20}x + a_{21}y + a_{22}z \end{pmatrix}. \qquad [957]$$

And for matrix × matrix (easier to see when not subscripted)—this rule was discovered by Binet in 1812 [507]

$$\begin{pmatrix} a & b \\ c & d \end{pmatrix} \begin{pmatrix} E & F \\ G & H \end{pmatrix} = \begin{pmatrix} aE + bG & aF + bH \\ cE + dG & cF + dH \end{pmatrix}.$$

Determinants (p.348)

For a 2×2 matrix $\mathbf{T}$, where

$$\mathbf{T} = \begin{pmatrix} a_{00} & a_{01} \\ a_{10} & a_{11} \end{pmatrix},$$

the determinant of $\mathbf{T}$ is defined as

$$det(\mathbf{T}) = \begin{vmatrix} a_{00} & a_{01} \\ a_{10} & a_{11} \end{vmatrix} = a_{00}a_{11} - a_{01}a_{10}. \qquad [946]$$

And for a 3×3 matrix $\mathbf{U}$, where

$$\mathbf{U} = \begin{pmatrix} a_{00} & a_{01} & a_{02} \\ a_{10} & a_{11} & a_{12} \\ a_{20} & a_{21} & a_{22} \end{pmatrix},$$

$$det(\mathbf{U}) = \begin{vmatrix} a_{00} & a_{01} & a_{02} \\ a_{10} & a_{11} & a_{12} \\ a_{20} & a_{21} & a_{22} \end{vmatrix}.$$

This may be expanded using 2×2 'minors'

$$det(\mathbf{U}) = a_{00} \begin{vmatrix} a_{11} & a_{12} \\ a_{21} & a_{22} \end{vmatrix} - a_{01} \begin{vmatrix} a_{10} & a_{12} \\ a_{20} & a_{22} \end{vmatrix} + a_{02} \begin{vmatrix} a_{10} & a_{11} \\ a_{20} & a_{21} \end{vmatrix}. \qquad [947]$$

Modulor as a precision tool

Should we wish to use The Modulor (p.127) as a precision design tool and avoid any inaccuracies introduced by practical values, accumulated rounding errors, conversions, and so on; we need only remember Le Corbusier's 'policeman' decision—that is, to take 72.000 inches as the primary reference length for the whole system (let's call it h). As a result, the infinite sets of sequence members for the red and the blue series—$\langle R_n \rangle$ and $\langle B_n \rangle$, are given (in inches) by

$$\left. \begin{aligned} h &= 72 \\ R_n &= \phi^n h \\ B_n &= 2\phi^n h \end{aligned} \right\} \quad \text{where n= } \ldots, -2, -1, 0, 1, 2, \ldots$$

This means that after choosing a suitable range of n values, we may then use the above formulæ to derive consistent lists of dimensions to whatever precision we require.[1] Further, if we choose our own value for h (to give a 'non-policeman' Modulor), then we may apply the system to any self-contained design project (surface or spatial), and it will still deliver the main Modulor benefits of:

- high orderliness—selected combinations of incommensurate dimensions still 'line up' exactly

- consistent harmony of proportion—*Golden of course,* and

- reduced or zero periodicity—at our choice

---

[1] But—a small point regarding metric specification—the Modulor system predates the standardization of the inch to 2.54 cm—which happened in 1959 [670]. Le Corbusier used the conversion of 2.539 cm/inch. But this need not present any problem—so long as a single reference is chosen and adhered to.

References

(DOIs are persistent digital object identifiers. See www.doi.org for details.)

[1] Abengoa Solar S.A. (2007) *PS10 Tower Plant.* Solar electricity generation, 11 megawatts, Solúcar Complex, Sanlúcar la Mayor, Seville. http://www.abengoasolar.com/ accessed 30Oct2014, (cited on p.296).

[2] Adler I (1994) *A model of contact pressure in phyllotaxis.* J. Theor. Biol. 45:1, pp.1–79, doi:10.1016/0022-5193(74)90043-5 (cited on p.293).

[3] ———. (2012) *Solving the riddle of phyllotaxis, why the Fibonacci numbers and the Golden Ratio occur on plants.* Compilation of papers over 20 years. Singapore: World Scientific Publ., p.171, http://www.worldscientific.com/worldscibooks/10.1142/8500 accessed 18Jun2013, (cited on pp.52, 52, 494, 558).

[4] Adler I, Barabe D, & Jean RV (1997) *A history of the study of phyllotaxis.* Annals of Botany 80, *crystallography* p.231, doi:10.1006/anbo.1997.0422 (cited on p.54).

[5] —Ibid. *Kepler,* p.233, (cited on pp.23, 52).

[6] —Ibid. *Bravais brothers,* p.234, (cited on p.52).

[7] —Ibid. [*Rejecting mathematics*], p.236, (cited on pp.52, 557).

[8] Ael2 (2007) *Decagone de Gummelt.* Public domain png file, SVG version by SharkD and Geometry Guy (2009), http://en.wikipedia.org/wiki/File:Gummelt_decagon.svg accessed 6Nov2013, (cited on pp.336, 587).

[9] Agostinho LCL, Barobosa CMBM, Nascimento L, & Rodbari JR (2013) *Catalytic dehydration of methanol to dimethyl ether (DME) using the* $Al_{62,2}Cu_{25,3}Fe_{12,5}$ *quasicrystalline alloy.* J. Chem. Eng. Process Technol. 4:5, doi:10.4172/2157-7048.1000164 (cited on p.341).

[10] Agrawala VS (1969) *Pāṇinikālīna Bhāratavarṣa.* (Hn.), Varanasi-1: Chowkhamba Vidyabhawan, cited by Singh [838], (cited on p.12).

[11] Aharonov Y & Bohm D (1959) *Significance of electromagnetic potentials in the quantum theory.* Phys. Rev. 115:3, 1Aug1959, pp.485–491, doi:10.1103/PhysRev.115.485 (cited on p.362).

[12] Ahuja A (2013) *A profile of Professor John Pendry, pioneer of metamaterials.* Website article, Imperial College London, http://www3.imperial.ac.uk/newsandeventspggrp/imperialcollege/newssummary/news_22-5-2013-11-25-22 accessed 28Nov2013, (cited on p.xi).

[13] Albers DJ, Alexanderson GL, & Reid C eds. (1990) *More Mathematical People.* Boston, MA: Harcourt Brace Jovanovich, *Julia Robinson / Hilbert's tenth,* cited by O'Connor & Robertson [707], (cited on p.450).

[14] Albuquerque EL & Cottam MG (2003) *Theory of elementary excitations in quasiperiodic structures.* Physics Reports 376:4–5, Mar2003, p.230, doi:10.1016/S0370-1573(02)00559-8 (cited on p.390).

[15] —Ibid. *cold spinning wheel,* p.226, (cited on p.328).

[16] Akenine-Möller T, Haines E & Hoffman N (2008) *Real-Time Rendering, Third Edition.* Natick MA: A K Peters/CRC Press, pp.54–55, p.67, pp.71–79, pp.84–85, (cited on p.361).

[17] Archibald RC (1920) *Notes on logarithmic spirals, the Golden Section & Fibonacci numbers.* In Hambidge *Dynamic symmetry: the Greek vase.* [300], New Haven & New York: Yale Univ. Press, V–XIV, pp.146–161, https://archive.org/details/cu31924019526882 accessed 29Apr2014, https://ia600407.us.archive.org/9/items/cu31924019526882/cu31924019526882.pdf accessed 29Apr2014, (cited on pp.34, 39, 40, 41).

[18] Arnheim R (1996) *Visual Thinking.* Berkeley and Los Angeles: University of California Press, p.286, (cited on p.361).

[19] Ascani E (2011) *Epic Conway's Game of Life.* Video 00h:06m:32s, http://www.youtube.com/watch?v=C2vglCfQawE accessed 8Apr2015, (cited on p.302).

[20] Assul, Ayub Farabi (2009) *Photograph of Gunbad-i-Quābūd, Marāgha, Iran.* Wikimedia Commons file, http://commons.wikimedia.org/wiki/File:Gonbad_Kabud4.jpg accessed 1Aug2014, (cited on pp.345, 587).

[21] Au-Yang H & Perk JHH (2013) *Quasicrystals—the impact of NG de Bruijn.* Oklahoma State Univ. http://arxiv.org/abs/1306.6698v3 accessed 5Dec2013, (cited on pp.387, 391, 495, 552, 552).

[22] —Ibid. *Penrose tiling, Conway worms, Ammann bars, pentagrids,* p.8, (cited on p.387).

[23] Bacaër N (2011) *A Short History of Mathematical Population Dynamics.* London: Springer Verlag, p.2, doi:10.1007/978-0-85729-115-8 (cited on pp.25, 492).

[24] Baggot J (1990) *Krypton atoms cling together in 'shells'.* New Scientist, 3Mar1990, p.31, (cited on pp.303, 535).

[25] Bainbridge K, Guyomarc'h S, Bayer E, et al. (2008) *Auxin influx carriers stabilize phyllotactic patterning.* Genes Dev. 2008 22, pp.810–823, doi:10.1101/gad.462608 http://genesdev.cshlp.org/content/22/6/810.full.pdf+html accessed 7Jul2013, (cited on p.289).

[26] Balashov Y (1994) *Should Plato's Line be divided in the Mean and Extreme Ratio?* Dept. Phil. Duquesne Univ., Ancient Philosophy Jnl., Pittsburgh: Mathesis Publications 14, doi:10.5840/ancientphil19941423 https://yuri.myweb.uga.edu/Papers/divided_line.pdf accessed 18Jun2013, (cited on p.16).

[27] Balius Juli R (1983) *Picasso and sport.* Olympic Review, Los Angeles: LA84 Foundation, *boxing and boxers,* p.300–301, http://library.la84.org/OlympicInformationCenter/OlympicReview/1983/ore187/ORE187s.pdf accessed 15Apr2014, (cited on pp.112, 504).

[28] Bamford CH, Brown L, Elliott A, et al. (1954) *Alpha- and beta-forms of poly-l-alanine.* Nature, 173, 2Jan1954, p.27 cited by Frey-Wyssling [228], doi:10.1038/173027a0 (cited on p.290).

[29] Bancel PA, Heiney PA, Horn PM, & Steinhardt P (1989) *Comment on a paper by Linus Pauling.* Proc. Natl. Acad. Sci. USA, Nov1989 86:22, pp.8600–8601, *extraordinarily large unit cell … 10,000 atoms … [potentially] 425,000 atoms ,* doi:10.1073/pnas.86.22.8600 (cited on p.332).

[30] Barber DJ & Freestone IC (1990) *An investigation of the origin of the colour of the Lycurgus Cup by analytical transmission electron microscopy.* Archaeometry 32:1, pp.33–45, doi:10.1111/j.1475-4754.1990.tb01079.x Cited by Freestone et al. *The Lycurgus Cup—a Roman nanotechnology,* http://master-mc.u-strasbg.fr/IMG/pdf/lycurgus.pdf accessed 4Dec2013, (cited on p.376).

[31] Barile M & Weisstein EW (1999–2014) *Cantor Set.* From MathWorld—a Wolfram web resource. http://mathworld.wolfram.com/CantorSet.html accessed 8Jul2014, (cited on p.390).

[32] Barlow P (1811) *An elementary investigation of the Theory of Numbers.* London: J Johnson and Co., St. Paul's Church-Yard, p.43, cited by O'Connor & Robertson [706], (cited on p.450).

[33] Baron JH (1998) *Intimate music: a history of the idea of chamber music.* New York: Pendragon Press, p.384, (cited on p.140).

[34] Barthel-Calvet AS (2002) *Biography of Iannis Xenakis,* in *Portrait(s) de Iannis Xenakis.* Ed. François-Bernard Mâche, Paris: Bibliothèque nationale de France, http://www.iannis-xenakis. org/xen/bio/bio.html accessed 27Jan2001, (cited on pp.71, 130, 151).

[35] Bartók B (undated) *Manuscript 80FSS1.* Sketchbook including Turkish folk song transcriptions. The New York Bartók Archive, recto pp.1–2, (facsimile, Howat [392], Ex.2, p.86), (cited on p.65).

[36] ———. (1911) *(Duke) Bluebeard's Castle, Sz. 48, BB 62, (A kékszakállú herceg Vára).* One act opera, opus 11. Symbolist libretto by Béla Balázs, English version by Chester Kallman, London: Universal Edition; USA: Boosey & Hawkes, (cited on pp.140, 151).

[37] ———. (1921) *The relation of folk music to the development of the art and music of our time.* Essay in The Sackbut 2:1, pp.5–11, republished as *At the sources of folk music.* Muzyka, 2:6, Jun1925, pp.230–233, and 4:6, Jun1927, pp.256-259, cited by Lendvai [535] & by Antokoletz E (1990) in *The Music of Béla Bartók: A Study of Tonality and Progression in Twentieth Century music.* Berkeley: University of California Press, p.3 note 9, (cited on p.153).

[38] ———. (1928) *String Quartet No.4, Sz.91, BB 95.* Publ. (1929) Vienna: Universal Edition, reissued (1939) London: Boosey & Hawkes, (cited on pp.140, 151).

[39] ———. (1934) *String Quartet No.5, Sz.102, BB 110.* Publ. (1936) Vienna: Universal Edition, reissued (1939) London: Boosey & Hawkes, (cited on pp.140, 151).

[40] ———. (1936) *Music for string instruments, percussion & celesta, Sz 106, BB 114.* 'MFSPC'. Philharmonia Score, Vienna: Universal Edition, (cited on pp.62, 64, 139, 142, 142, 151, 582).

[41] ———. (1937) *Sonata for two pianos and percussion, Sz 110, BB 115.* 1942, London: Hawkes & Son, (cited on pp.142, 151).

[42] Basin SL (1963) *The Fibonacci sequence as it appears in nature.* Fib. Qtrly, 1:1, pp.53–57, cited by Scott & Marketos [812], http://www.fq.math.ca/Scanned/1-1/basin.pdf accessed 24apr2014, (cited on pp.284, 559).

[43] Battersby S (2005) *Saturn's moon reveals bulging equator.* New Scientist article 10Jan2005, available online, http://www.newscientist.com/article/dn6860-saturns-moon-reveals-bulging-equator.html accessed 4Oct2014, (cited on p.24).

[44] Bauer C, Kobiela G, & Giessen H (2012) *2D quasiperiodic plasmonic crystals.* Scientific Reports 2:681, doi:10.1038/srep00681 (cited on p.394).

[45] Beardsley MB (2012) *Potential use of quasicrystalline materials as thermal barrier coatings for diesel engine components.* Paperback, ISBN 1248991133, Proquest, Umi Dissertation Publishing, (cited on p.340).

[46] Béatrix A-L (2012) *The Galerie du Temps at the Louvre-Lens.* Press release, Musée du Louvre, *Statuette of Lady Tuya, matron of the hareem of Min.* New Kingdom, 18th Dynasty, reign of Amenophis III (1391–1353 BC, E10655.) pp.2,4, (cited on p.75).

[47] Beenker FPM (1982) *Algebraic theory of non-periodic tilings of the plane by two simple building blocks: a square and a rhombus.* (82-WSK04), Technical report, Eindhoven University of Technology, (cited on p.387).

[48] Bell JS (1964) *On the Einstein Podolsky Rosen Paradox.* Physics (Long Island City, NY) 1, pp.195–200, reprinted in Bell (1987) *Speakable and Unspeakable in Quantum Mechanics.* Cambridge: Cambridge UP, (cited on p.298).

[49] Bellhouse DR & Genest C (2007) *Maty's biography of Abraham de Moivre, translated, annotated and augmented.* Inst. of Math. Stats., Statistical Science, 22:1, p.122, doi:10.1214/088342306000000268 (cited on p.25).

[50] Benjamin AT & Quinn JJ (2003) *Proofs that really count.* Mathematical Association of America. Cited by Knott [480], with note: B&Q use f_n for F_{n-1}, (cited on pp.529, 529, 530, 530).

[51] Berg JM, Tymoczko JL, & Stryer L (2002) *Biochemistry, 5th edition.* New York: WH Freeman, § 3.2. (cited on p.290).

[52] Berger R (1966) *The undecidability of the domino problem.* Memoirs of the AMS 66, doi:10.1090/memo/0066 (cited on p.310).

[53] Bernal JD (1923) *The analytic theory of point systems.* Prize Essay for Emmanuel College, Cambridge, published (1981) London: Birkbeck College, http://www.iucr.org/__data/assets/pdf_file/ 0008/25559/Bernal_monograph.pdf accessed 16Jul2014, http://www.iucr.org/ education/teaching-resources/bernal-essay accessed 16Jul2014, (cited on p.361).

[54] Bernoulli D (1728) *Observationes de seriebus ...* Comment. Acad. Sci. Imp. Petropolitanæ (1728/1732), 3, pp.85–100. Reproduced in: *Die Werke von Daniel Bernoulli.* Band 2, Birkhäuser, Basel, (1982), pp.49–64, (cited by Bacaër [23] and by Thompson [893] p.923), (cited on pp.25, 481).

[55] Berraquero CP, Maurel A, Petitjeans P, & Pagneux V (2013) *Experimental realization of a water-wave metamaterial shifter.* Phys. Rev. E, Stat. Nonlin. Soft Matter Phys. 19Nov2013 88:5, 051002, doi:10.1103/PhysRevE.88.051002 (cited on p.382).

[56] Besant A & Leadbeater CW (1901) *Thought-Forms.* London: Theosophical Publishing House, ref. TF21–24, cited by Taylor [888], (cited on p.112).

[57] Bicknell M & Hoggatt VE (1973) *Proofs of $F_n \mid F_{nk}$.* In *A primer for the Fibonacci numbers.* The Fibonacci Association. *1. Binet, 2. induction, 3. generating functions and polynomials, & 4. general case,* Part IX, p.111, http://www.fq.math.ca/Books/Primer/bicknell6. pdf accessed 2Sep2014, (cited on p.174).

[58] Bier C (2012) *The decagonal tomb tower at Maragha and its architectural context: lines of mathematical thought.* Nexus Network Journal 14:2, *Smith MB (1937) decagonal plan.* Fig.2, p.251, doi:10.1007/s00004-012-0108-6 (cited on p.345).

[59] Blavatsky HP (1879) *The Theosophist.* Bombay: Theosophical Soc., 1:1, § 1, *What is Theosophy?,* (cited on p.90).

[60] Blotkamp C (1994) *Mondrian: the art of destruction.* London: Reaktion Books, quoting Mondrian's letter to HP Bremmer, Paris 29Jan1914, p.81, (also see Holtzman & James [375]), also see Joosten [429], (cited on pp.91, 493).

[61] —Ibid. *strategic move ... to silence ... critics ... inclined to see his work as cerebral,* p.11, (cited on p.113).

[62] —Ibid. ... *work of Picasso, whom I greatly admire,* from letter to HP Bremmer, 29Jan1914 (same letter as referred to in Blotkamp [60]) p.81, (cited on p.112).

[63] —Ibid. ... *(for Mondrian, evolution closely related with destruction ... destruction not negative concept for him ... essential to make way for the new),* p.15, (cited on p.110).

[64] —Ibid. *influence of Schoenmaekers* v. *'got everything from The Secret Doctrine',* p.111, (cited on p.106).

[65] —Ibid. *Rosenberg boxing and fencing,* p.182, (cited on p.112).

[66] —Ibid. *argument over diagonals,* p.192, (cited on p.96).

[67] Blumenfeld D (2011) *2011 Nobel Laureate in Chemistry: Dan Shechtman.* Photograph © 2011 Nobel Media, http://www.nobelprize.org/press/nobelmedia/#/image/view/ press-image-nobel-laureate-2011-dan-shechtman-104697 accessed 2Oct2013, (cited on pp.329, 587).

[68] Bois YA (1990) *Painting as model.* Massachusetts: MIT Press. Cites retrospective 1927 article by van Doesburg about the 3 aspects of *De Stijl: Journal ... group ... idea,* p.101, (cited on p.89).

[69] Bogousslavsky J, Walusinski B, & Veyrunes D (2009) *Crime, hysteria and Belle Époque hypnotism: the path traced by Jean-Martin Charcot and Georges Gilles de la Tourette.* Eur. Neurol. 62, *Hypnotism and crime,* p.196, doi:10.1159/000228252 (cited on p.83).

[70] —Ibid. *automatons,* p.196, (cited on p.83).

[71] Bohlen H (2012) *An 833 cents scale.* Website article, http://www. huygens-fokker.org/bpsite/833cent.html accessed 23Apr2014, (cited on p.224).

[72] Bohm D (1980) *Wholeness and the implicate order.* London, Boston: Routledge & Kegan Paul, p.188, (cited on p.95).

[73] Bonduelle M & Gelfand T (1999) *Hysteria behind the scenes: Jane Avril at the Salpêtrière.* Journal of the History of the Neurosciences: Basic and Clinical Perspectives 8:1, pp.35–42, doi:10.1076/jhin.8.1.35.1778 (cited on p.83).

[74] Bonnet C (1754) *Recherches sur l'usage des feuilles dans les plantes, et sur quelques autres sujets relatifs à l'histoire*

de la végétation. Chez Elie Luzac, fils. imp.-lib., cited in Atela P & Golé C (2003) *Phyllotaxis; a brief history of phyllotaxis.* Smith College, http://www.math.smith.edu/phyllo/OldFiles/ History/historynoroll.html accessed 2Jun2013, (cited on p.52).

[75] Bouleau C (1963) *Charpentes, la géométrie secrète des peintres.* Éditions de Seuil, (cited on p.494).

[76] ———. (1963) *The painter's secret geometry—a study of composition in art.* Transl. from the French [75] by Griffin J, New York: Harcourt, Brace & World Inc., reissued (1980) New York: Hacker Art Books, *Mondrian: Painting I,* p.248. The black bars have thicknesses in the proportion 3:4:5; three of the four corners of the 'black bars' square are cropped by the 45° lozenge canvas edges, (cited on p.92).

[77] —Ibid. *Pacioli's Divina Proportione,* pp.74–75, (cited on p.21).

[78] Bourgoin J (1879) *Les éléments de l'art arabe: le trait des entrelacs.* Paris: Firmin-Didot. Plates reprinted (1973) in *Arabic geometrical patterns and designs.* New York: Dover Publications, cited by Grünbaum et al. [278], (cited on p.343).

[79] Bragg M, Sautoy M du, Stedall J, & Knott R (2007) *In Our Time: the Fibonacci sequence.* BBC Radio4 programme 29Nov2007, http://www.bbc.co.uk/programmes/b008ct2j accessed 22Apr2013, (cited on p.34).

[80] Brandmüller J (1992) *Fivefold symmetry in mathematics, physics, chemistry and beyond.* In (1992) *Fivefold symmetry.* Ed. Hargittai I, Singapore: World Scientific, *quoting from Kepler's Harmonices Mundi: 'Das Rechteck aus 1 und 3 erzeugt ein Weibchen, ... '—'The rectangle from 1 and 3 produces a female, ... less by unity than the square of two', followed by examples: 2, 5 with 3^2—an excess of 1 and therefore male; 3, 8 with 5^2 female; and 5, 13 with 8^2 male; and so on,* p.16, (cited on p.24).

[81] Bravais L & A (1837) *Essai sur la disposition sur des feuilles curvisériées.* (Cited by Adler [3]), Annals Sci. Nat. Bot. 2, pp.42–110, (cited on p.52).

[82] Brent RP, Cohen GL, Riele te HJJ (1991) *Improved techniques for lower bounds for odd perfect numbers.* Math. Comput. 57:196, pp.857–868, cited by Greathouse & Weisstein [270], doi:10.1090/S0025-5718-1991-1094940-3 http://www.ams.org/journals/mcom/ 1991-57-196/S0025-5718-1991-1094940-3/home.html accessed 18Aug2014, (cited on p.284).

[83] Brezinski C (1991) *History of continued fractions and Padé approximants.* Springer Series in Computational Mathematics, New York: Springer-Verlag, p.96, (cited on p.24).

[84] Briggs R (1985) *Knowledge representation in Sanskrit and artificial intelligence.* NASA Ames Research Center, California. AI Mag. 6:1, p.32, doi:10.1609/aimag.v6i1.466 (cited on p.12).

[85] Brillhart J & Morton P (1996) *A Case Study in Mathematical Research: The Golay-Rudin-Shapiro Sequence.* The American Mathematical Monthly, 103, p.854, (cited on pp.159, 390).

[86] Briscoe JR (1999) *Timbre, voice-leading and the musical arabesque in Debussy's piano music.* Essay in *Debussy in performance.* Ed. James Briscoe, New Haven: Yale University Press, p.228, (cited on pp.65, 130, 151).

[87] British Museum (4th-century AD) *The Lycurgus Cup—a dichroic glass cup with a mythological scene.* Height 16.5 cm (with modern metal mounts), diameter 13.2 cm. BM Reg. 1958,1202.1, (cited on pp.377, 589).

[88] Brouillet A (1887) *Une leçon clinique à la Salpêtrière.* 290 × 430 cm (figures nearly life size). Painting located in corridor of Descartes University in Paris 6th arr, near the *Musée d'Histoire de la Médecine* entrance, http://commons.wikimedia.org/wiki/File: Une_le%C3%A7on_clinique_%C3%A0_la_Salp%C3%AAtri%C3%A8re.jpg accessed 14Dec2014, (cited on pp.82, 83, 580).

[89] Bruijn NG de (1981) *Algebraic theory of Penrose's non-periodic tilings of the plane, Parts I and II.* Koninklijke Nederlandse Akademie van Wetenschappen (KNAW), Proc. Ser. A 84:1, 20Mar1981, Part I pp.38–52, Part II pp.53–66. (= Indagationes Mathematicæ 43:1, pp.39–66), cited by Au-Yang & Perk [21], (cited on pp.322, 387, 387).

[90] Bruijn PJ de (1974) *An extension of Fibonacci's sequence.* Fibonacci Quarterly 12:3, pp.251–258, cited by Knott [481], http://www.fq.math.ca/Scanned/12-3/debruijn.pdf accessed 21Nov2013, (cited on p.136).

[91] Brûlé S, Jauvelaud EH, Enoch S, & Guenneau S (2014) *Experiments on seismic metamaterials: molding surface waves.* Phys. Rev. Lett. 31Mar2014 112:133901, doi:10.1103/PhysRevLett.112.133901 (cited on p.381).

495

[92] Buckley OE (2010) *Oliver E Buckley Condensed Matter Prize.* American Physical Society, http://www.aps.org/programs/honors/prizes/ buckley.cfm accessed 3Oct2013, (cited on p.333).

[93] Buitrago AR (2007) *Polygons, diagonals, and the Bronze mean.* Nexus Network Journal, 9, pp.321–326, (cited on p.430).

[94] Burkov SE (1991) *Structure model of the Al-Cu-Co decagonal quasicrystal.* Phys. Rev. Lett. 67:5, pp.614–617, doi:10.1103/PhysRevLett.67.614 (cited on p.335).

[95] Cahn RW & Haasen P (1996) *Physical metallurgy, Volume 1.* Amsterdam: Elsevier Science, § 2.1, p.375, (cited on pp.318, 319).

[96] Caldwell CK (undated) *A proof that if $2^n - 1$ is prime, then so is n.* Dedicated website: 'The Prime Pages'. Univ. Tennessee at Martin, http://primes.utm.edu/notes/proofs/Theorem2.html accessed 3Sep2014, (cited on pp.447, 533, 534).

[97] Campbell M (2011) *Wave 'invisibility cloak' could shield coastlines.* New Scientist, 2815, p.14, (cited on p.382).

[98] Caparrini S (2006) *On the common origin of some of the works on the geometrical interpretation of complex numbers.* Essay in *Two cultures: essays in honour of David Speiser.* Ed. Williams K, Basel: Birkhäuser Verlag (Springer), p.139, (cited on p.134).

[99] Capasso F (2014) *New frontiers in optics and photonics with designer electronic and optical materials.* Bagwell Lecture, Purdue Univ. 14May2014. https://nanohub.org/resources/21372/download/ 2014.05.14-Capasso-BAGWELL.pdf accessed 17Feb2015, (cited on p.375).

[100] Carroll L (1865) *Alice's adventures in Wonderland.* with (1871) *Through the looking-glass and what Alice found there.* Intro. Green RL; illust. Tenniel J, 1982, 1971. Oxford & New York: Oxford University Press, ch. 6, (cited on pp.395, 566).

[101] Casselman W (2004) *The difficulties of kissing in three dimensions.* University of British Columbia. Notices of the AMS, 51:8, p.884, http://www.ams.org/notices/200408/comm-cass.pdf accessed 7Mar2014, (cited on p.300).

[102] Castro, LN de (2007) *Fundamentals of natural computing: basic concepts, algorithms, and applications.* Boca Raton FL: CRC Press, Chapman & Hall, § 7, pp.327–387, (cited on p.304).

[103] Cb89: Wikimedia contributor (2007) *Fleur de prunus cerasifera.* Photograph, http://commons.wikimedia.org/wiki/File: 2007-03-fleur_prunus_cerasifera.jpg accessed 15Aug2014, (cited on pp.51, 579).

[104] Cdang (2004) *Loi de Bragg.* Illustration, SVG by Gregors, http: //en.wikipedia.org/wiki/File:Braggs_Law.svg accessed 7Sep2013, (cited on pp.327, 586).

[105] Chaplin MF (2000) *A proposal for the structuring of water.* Biophysical Chemistry 83, pp.211–221, doi:10.1016/S0301-4622(99)00142-8 (cited on pp.306, 306, 519, 519, 585).

[106] ———. (2000–) *Water structure and science.* Dedicated website hosted by London South Bank University, http://www.lsbu.ac.uk/ water/sitemap.html accessed 8Jun2013, (cited on p.306).

[107] —Ibid. *Platonic solids, water and the Golden Ratio,* http://www. lsbu.ac.uk/water/platonic.html accessed 23Sep2013, (cited on p.308).

[108] —Ibid. *Anomalous properties of water,* http://www.lsbu.ac.uk/water/ anmlies.html accessed 19Jun2013, (cited on p.306).

[109] —Ibid. *Icosahedral $(H_2O)_{280}$ water clusters,* http://www1.lsbu.ac. uk/water/icosahedra.html accessed 2Feb2015, (cited on p.306).

[110] —Ibid. *Platonic solids, water and the Golden Ratio,* http://www. lsbu.ac.uk/water/platonic.html accessed 8Jun2013, (cited on p.306).

[111] Chapman RE (1951) *The fifth quartet of Béla Bartók.* The Music Review 12, pp.296–303, cited by Parker ME in (2005) *String quartets: a research and information guide.* Routledge Music Bibliographies, New York: Routledge Taylor & Francis Group, note 379, p.92, (cited on p.140).

[112] Chen A-Li, Wang Yue-Sheng, Guo Ya-Fang, & Wang Zheng-Dao (2008) *Band structures of Fibonacci phononic quasicrystals.* Solid State Communications 145:3, pp.103–108, doi:10.1016/j.ssc.2007.10.023 (cited on pp.392, 533).

[113] Chen Zhongsheng, Bin Guo, Yongmin Yang, Congcong Cheng (2014) *Metamaterials-based enhanced energy harvesting: a review.* Physica B: Condensed Matter 1Apr2014, 438, pp.1–8, doi:10.1016/j.physb.2013.12.040 (cited on pp.369, 381).

[114] Chippindale C (1983) *Stonehenge complete.* London: Thames and Hudson, p.48, (cited on p.120).

[115] Chris 73 (2004) *Nautilus cutaway logarithmic spiral.* Wikimedia Commons image, http://commons.wikimedia.org/wiki/File: NautilusCutawayLogarithmicSpiral.jpg accessed 4May2013, (cited on pp.49, 579).

[116] Christianson GE (2005) *Isaac Newton.* Oxford: Oxford University Press, p.21, (cited on p.361).

[117] Clarke J (2012) *Iannis Xenakis and the Philips Pavilion.* The Journal of Architecture 17:2, p.213 & p.220 doi:10.1080/13602365.2012.678641 (cited on p.132).

[118] —Ibid. *likeness should not be overemphasised,* p222, (cited on p.132).

[119] Clement RT & Houzé A (1999) *Neo-impressionist Painters.* Westport, Connecticut & London: Greenwood Publishing, *Chevreul & Blanc,* note 47, p.14, (cited on p.75).

[120] Coldea R, Tennant DA, et al. (2010) *Quantum criticality in an Ising chain: experimental evidence for emergent E_8 symmetry.* Helmholtz Assoc. of German Research Centres, Science 327, 8Jan2010, pp.177–180 doi:10.1126/science.1180085 (cited on pp.298, 517).

[121] Cole JH (1925) *Determination of the exact size and orientation of the Great Pyramid of Giza.* Cairo: Government Press, Survey of Egypt Paper No. 39, cited by Herz-Fischler [349], *mean side ('b')=9069.4 inches (0.6 inches more than Petrie measurement 9068.8* [746] *),* p.7, http://www.ronaldbirdsall.com/gizeh/ errata/Cole%20Survey.pdf accessed 18Sep2014, (cited on p.7).

[122] Conway JH (1968) *A perfect group of order 8, 315, 553, 613, 086, 720, 000 and the sporadic simple groups.* Proceedings of the National Academy of Sciences of the United States of America 61:2, pp.398–400, doi:10.1073/pnas.61.2.398 (cited on p.304).

[123] Conway JH & Sloane NJA (1993) *Sphere packings, lattices and groups.* 3rd edn. (1999) Series: Grundlehren der mathematischen Wissenschaften (Book 290), New York: Springer-Verlag, § 2.1, p.21, (cited on p.300).

[124] Cook TA (1914) *The curves of life: being an account of spiral formations and their application to growth in nature, to science and to art: with the special reference to the manuscripts of Leonardo da Vinci.* London: Constable, reprinted (1979) New York: Dover Pubs., p.420, http://archive.org/details/cu31924028937179 accessed 18Jun2013, (cited on p.28).

[125] Corbusier, Le & Jeanneret P (1931) *Villa Savoye*. Modernist villa. Address: 82 Rue de Villiers, 78300 Poissy, France, http://villa-savoye. monuments-nationaux.fr/ accessed 29Sep2014, (cited on pp.125, 151).

[126] —Ibid. *Villa Savoye front elevation proportions.* Deduced from photo by Paul Koslowski ©FLC/ADAGP. Photo not reproduced: used only as an indicator of relative dimensioning, http://www.fondationlecorbusier.fr/CorbuCache/900x720_2049_2897.jpg accessed 4Jul2014, (cited on p.124).

[127] Corbusier, Le (1954) *The Modulor: A harmonious measure to the human scale universally applicable to architecture and mechanics.* First English edn. transl. Francia P de & Bostock A, (from 1948 French edn. *Le Modulor*), London: Faber & Faber, (cited on p.124).

[128] —Ibid. *Einstein quotation*: 'Making the bad difficult, and the good easy', in 1946 letter from Einstein to Le Corbusier, p.5, (cited on pp.128, 156).

[129] —Ibid. *sound 'divided into sections and measured',* p.16, (cited on p.124).

[130] —Ibid. *'mathematics of the human body … source of … harmony … beauty',* p.19, (cited on p.124).

[131] —Ibid. *1945 work on 'proportioning grid'—precursor to the Modulor,* p.43, (cited on p.151).

[132] —Ibid. *the unit, the double unit, and ϕ,* p.50, (cited on p.126).

[133] —Ibid. *'… in English detective novels, good looking men such as policemen, are always 6 feet tall',* p.56, (cited on p.126).

[134] —Ibid. *red and blue series measurements,* p.82, (cited on pp.126, 127).

[135] —Ibid. *the panel exercise/game,* pp.90–101, (cited on pp.128, 156).

[136] Corbusier, Le & Xenakis I (1959) *Sainte Marie de La Tourette.* Modernist Dominican priory near Lyon, 69210 Eveux, France, http://www.couventdelatourette.fr/ accessed 21Feb2015, (cited on p.132).

[137] Costa CHO, Vasconcelos MS, Barbosa PHR, & Barbosa Filho FF (2012) *Fractal spectra in generalized Fibonacci one-dimensional magnonic quasicrystals.* Jnl. of Magnetism & Magnetic Materials, 324, p.2315, doi:10.1016/j.jmmm.2012.02.123 (cited on p.394).

[138] Courtauld Gallery (2011) *Toulouse-Lautrec and Jane Avril beyond the Moulin Rouge*. Web article in support of exhibition 16Jun–18Sep2011, http://www.courtauld.ac.uk/gallery/exhibitions/2011/Lautrec2.shtml accessed 15Dec2014, (cited on p.83, 83).

[139] Coxeter, HSM (1961) *Introduction to geometry*. New York: Wiley, *phyllotaxis ... a prevalent tendency'*, p.172, (cited on p.58).

[140] Creative Commons (2001–) *Attribution 2.0 Generic (CC BY 2.0)*. License, http://creativecommons.org/licenses/by/2.0/deed.en accessed 7May2014, (cited on pp.9, 11, 119, 121, 273, 577, 581).

[141] Creative Commons (2001–) *Attribution-ShareAlike 2.0 France (CC BY-SA 2.0 FR)*. License, http://creativecommons.org/licenses/by-sa/2.0/fr/deed.en accessed 5Nov2013, (cited on pp.229, 584, 590).

[142] Creative Commons (2001–) *Creative Commons Attribution-ShareAlike 2.5 Generic (CC BY-SA 2.5)*. License, http://creativecommons.org/licenses/by-sa/2.5/deed.en accessed 13May2014, (cited on pp.107, 581, 590).

[143] Creative Commons (2001–) *Attribution 3.0 Unported (CC BY 3.0)*. License, http://creativecommons.org/licenses/by/3.0/deed.en accessed 3Nov2013, (cited on pp.iv, 16, 311, 337, 345, 577, 585, 587, 587).

[144] Creative Commons (2001–) *Attribution-ShareAlike 3.0 Unported (CC BY-SA 3.0)*. License, http://creativecommons.org/licenses/by-sa/3.0/deed.en accessed 28May2013, (cited on pp.iv, iv, 27, 49, 51, 131, 143, 313, 315, 315, 317, 319, 327, 330, 346, 355, 355, 369, 377, 379, 578, 579, 579, 582, 582, 585, 586, 586, 586, 586, 586, 587, 587, 588, 588, 588, 588, 589, 589, 590).

[145] Crick FHC (1953) *The packing of α-helices: simple coiled coils*. Acta Cryst. 6, p.689, doi:10.1107/S0365110X53001964 (cited on p.295).

[146] Cromwell PR (2009) *The search for quasi-periodicity in Islamic 5-fold ornament*. The Mathematical Intelligencer, Jan2009, 31:1, *Topkapı scroll MS.H.1956 library of Topkapı Palace*, p.37, doi:10.1007/s00283-008-9018-6 (cited on pp.346, 546).

[147] —Ibid. *expanded girih tile set*, Fig.6, p.41, (cited on p.354).

[148] —Ibid. *Gunbad-i-Quābūd*, Fig.9, p.43, (cited on p.354).

[149] —Ibid. *subdivide and enlarge: inflation*, p.46, (cited on p.354).

[150] —Ibid. *largest eigenvalue is the square of the scale factor*, p.46, (cited on p.354).

500

[151] —Ibid. *eigenvector with the largest eigenvalue contains relative frequencies,* p.46, (cited on p.354).

[152] —Ibid. *designs of different scales superimposed,* p.47, (cited on p.347).

[153] —Ibid. *(Isfahan, Darb-i Imam) scale factor* $4 + 2\sqrt{5}$ [$= 2\phi^3$], Fig.21, p.43, (cited on p.354).

[154] —Ibid. *(Topkapı scroll) scale factor of* $3 + \sqrt{5}$ [$= 2\phi^2$], Fig.15, p.48, (cited on p.354).

[155] —Ibid. *Spain & Morocco... local 8-fold symmetry,* p.55, (cited on p.356).

[156] Cronholm144 (2007) *Girih tiles.* Illustration, http://en.m.wikipedia.org/wiki/File:Girih_tiles.svg accessed 9Jun2014, (cited on pp.346, 588).

[157] Dal Negro L & Feng NN (2007) *Spectral gaps and mode localization in Fibonacci chains of metal nanoparticles.* Optical Express 15:22, p.14397, doi:10.1364/OE.15.014396 (cited on pp.391, 393).

[158] —Ibid. *impact in the design ... nanophotonic devices,* p.14402, (cited on p.393).

[159] Dalí S (1983) *Untitled. Swallow's tail and cellos (Catastrophes Series).* 73.2 × 92.2 cm, oil on canvas. Dalí's last painting, which concerns the theories of René Thom. Fundació Gala-Salvador Dalí, Figueras: Dalí Theatre-Museum, http://www.salvador-dali.org/museus/teatre-museu-dali/la-collection/114/sans-titre-queue-daronde-et-violoncelles-serie-des-catastrophes accessed 26Aug2014, (cited on p.71).

[160] Darvas G (2007) *Symmetry: cultural-historical and ontological aspects of science-arts relations; the natural and man-made world in an interdisciplinary approach.* Springer Science Basel: Birkhäuser Verlag, p.208, (cited on p.322).

[161] Darwin CG (1914) *The theory of X-ray reflexion, Part II.* Phil. Mag. Series 6, 27:160, pp.675–690, cited by Yablonovitch [994] doi:10.1080/14786440408635139 (cited on p.367).

[162] Dasdemir A (2011) *On the Pell, Pell-Lucas and modified Pell numbers by matrix method.* Applied Mathematical Sciences, 5:64, pp.3173–3181, (cited on p.480).

501

[163] Debnath L (2009) *The Legacy of Leonhard Euler.* Singapore: World Scientific, p.62, (cited on p.282).

[164] Debussy C (1905) *La mer, trois esquisses symphoniques pour orchestre, L.109.* 'The sea, three symphonic sketches for orchestra', (1909) Paris: Durand et Fils, (cited on pp.142, 150).

[165] ———. (1888) *«Aquarelles II. Spleen»: Les roses étaient toutes rouges.* Song 6 in *Ariettes oubliées (Forgotten songs) L.60.* Six songs for voice and piano. Paris: Veuve Girod, reissue (1913) Paris: E Fromont, reprinted (1981) as *Claude Debussy: Songs, 1880–1904.* New York: Dover Publications, (cited on pp.146, 150).

[166] ———. (1913) *Taste.* S.I.M. 15Feb1913, reprinted in Debussy, *Debussy on Music.* [545], *intrinsic mystery of music ... threatened by over-scrupulous analysis ... beauty of a work ... always remain mysterious ... one can never find out exactly 'how it is done',* (cited on p.65).

[167] Decaillot-Laulagnet A-M (1999) *Édouard Lucas (1842–1891) : le parcours original d'un scientifique français dans la deuxième moitié du XIXe siècle.* Doctoral thesis 17Dec1999, University of René Descartes, Paris V, Centre Universitaire des Saints-Pères, UFR Biomedicale. *Announcement by Charles-Ange Laisant at the congrès de l'AFAS de 1879,* Ch.11 p.143, *'L'édition préparée avec C Henry comprendra... ',* p.145, *Lucas et Henry... envoyés... en Italie, [Florence],* p.146, *rupture,* p.150, http://www.worldcat.org/oclc/490673160 accessed 2Feb2014, (cited on p.149).

[168] —Ibid. *Les nombres de Mersenne,* pp.110-112, (cited on p.448).

[169] Deicher S (2004) *Mondrian.* Cologne: Taschen, p.67, (cited on p.96).

[170] Delaunay R, Le Fauconnier H, Gleizes A, Léger F, Metzinger J, (plus listed collaborators) (1912) *La Section d'Or.* Numéro spécial, 9Oct1912, ... *qu'ils se rattachent à la grande tradition [d'art],* p.1, http://en.wikipedia.org/wiki/File:La_Section_d%27Or,_numero_special,_9_Octobre_1912.jpg accessed 20May2014, (cited on p.70).

[171] Dellajustina FJ & Martins LC (2014) *The hidden geometry of the Babylonian square root method.* Appl. Maths. 6Nov2014, 5:19, Eqn.1.1, p.2983, doi:10.4236/am.2014.519284 (cited on pp.74, 474).

[172] Devlin K (2011) *The man of numbers: Fibonacci's arithmetic revolution.* London: Bloomsbury Publishing, p.3, (cited on p.19).

[173] ———. (2012) *The Golden Ratio & Fibonacci Numbers: Fact versus Fiction.* Stanford University Continuing Studies Program, video 1h:43m:17s, 11Dec2012, http://www.youtube.com/watch?v= 4oyyXC5lzEE accessed 3Nov2013,

[174] ———. (2007) *The myth that will not go away.* Archived article, Mathematical Association of America. http://www.maa.org/external_ archive/devlin/devlin_05_07.html accessed 3Nov2013, (cited on p.70).

[175] Dickson LE (1919) *History of the theory of numbers, Vol.I, Divisibility & primality.* Carnegie Institute of Washington publication 256, reprinted 1992 by American Mathematical Society. New York: AMS Chelsea Publishing, p.393 et seq., (cited on pp.252, 539, 539).

[176] —Ibid. *'Euler, in a posthumous paper, proved that every even perfect number is of Euclid's type', & proof,* p.19, (cited on p.282).

[177] —Ibid. *relatively prime integers,* p.396, (cited on p.252).

[178] Yong Ding, Sheng Xu, & Zhong Lin Wang (2009) *Structural colors from Morpho peleides butterfly wing scales.* J. Appl. Phys. 106:074702, doi:10.1063/1.3239513 (cited on p.365).

[179] Doig A (2009) *(About the artist) Theo van Doesburg (Christian Emil Marie Küpper).* Source: Oxford University Press, quoted by MoMA, New York, http://www.moma.org/collection/artist.php?artist_id=6076 accessed 30Apr2014, (cited on p.90).

[180] Donnay JDH & G (1976) *A graphical derivation of the crystallographic rotation axes.* Canadian Mineralogist, 1Nov1976, 14:4, note 1, p.567, (cited on p.323).

[181] Donnay M (1926) *Autour du Chat Noir.* Paris: Grasset, bocks p.19, *mépris* p.28, (cited on pp.149, 152).

[182] Douady S & Couder Y (1992) *Phyllotaxis as a physical self-organized growth process.* Physical Review Letters 68, pp.2098–2101, cited in [227], doi:10.1103/PhysRevLett.68.2098 (cited on p.293).

[183] ———. (1992) *Douady & Couder phyllotaxis experiment, film DouadyCouderExp5.9MB.mov* http://cs.smith.edu/~phyllo/Assets/ Movies/DouadyCouderExp5.9MB.mov accessed 31Aug2013, (cited on p.293).

[184] ———. (1996) *Phyllotaxis as a dynamical self-organizing process (Part I, II, III).* J. Theor. Biol. 178:3, pp.255–273, pp.275–294, pp.295–312, cited in [227], doi:10.1006/jtbi.1996.0024 (cited on p.293).

[185] Doyle, Arthur Conan (1892) *Silver Blaze: ... the curious incident of the dog in the night time*—(the dog that did not bark) in *The adventures of Sherlock Holmes.* The Strand Magazine Vol.IV, p.645, (cited on p.138).

[186] Düchting H (2000) *Seurat.* Cologne: Taschen, *regularly attended his lectures,* p.59, (cited on p.149).

[187] Duncan DD (1957) *Photograph of Picasso sparring with his son Claude at the Villa La Californie.* Cited by R Balius Juli [27], http://www.hrc.utexas.edu/exhibitions/web/ddd/gallery/picasso/036.html accessed 15Apr2014, (cited on p.112).

[188] Dunlap RA (1997) *The Golden Ratio and Fibonacci numbers.* Singapore: World Scientific Pub. Co. Inc. Eqns. 1.3, 1.4, p.5, (cited on pp.184, 529, 529, 529, 530, 530).

[189] —Ibid. *Additive & geometric,* p.8, (cited on p.31).

[190] Dürer A (1525) *Underweysung der Messung mit dem Zirkel und Richtscheyt.* 'The Four Books of Measurement with Compass and Ruler', Nuremberg, cited by Lück [600]. 1538 edn., pentagons: 'spread 115', http://www.rarebookroom.org/Control/duruwm/index.html accessed 28Feb2014, English transl. (1977) *The painter's manual.* New York: Abaris, cited by Grünbaum & Shephard [279], (cited on p.23).

[191] Dvivedi KD & Singh SL (2008) *The prosody of Piṅgala.* Varanasi: Vishwavidyalaya Prakashan, (reference to Sadguruśiṣya in Foreword), (cited on p.12).

[192] —Ibid. *Piṅgala & Pāṇini,* Foreword, (cited on p.12).

[193] —Ibid. *systematizing an already established body of knowledge,* in Foreword, (cited on p.13).

[194] Dyson F (2009) *Birds and frogs.* AMS Notices, 56:2, p.215, http://www.ams.org/notices/200902/rtx090200212p.pdf accessed 12Sep2013, (cited on p.334).

[195] Eckermann JP (1836) *Conversations with Goethe in the last years of his life.* Transl. from the German by Fuller SM (1839), Boston: Hilliard Gray, & Co., p.282, Univ. California, California Digital Library, https://archive.org/details/conversationswit00goetiala accessed 31Dec2013, (cited on p.130).

[196] Edwards S (2000–2004) *Aperiodic tiling: Penrose tilings.* Website, Southern Polytechnic State Univ., Marietta GA, http://fac-web.spsu. edu/math/tile/aperiodic/ accessed 8Jun2013, (cited on pp.313, 317, 585, 586).

[197] Egan G (2013) *Gummelt (an applet) draws Petra Gummelt's quasiperiodic tiling of the plane with overlapping decagons.* Refs: [289] & Voss D (1999) *Smash the system.* New Scientist 2175, 27Feb1999, p.44, http://www.gregegan.net/APPLETS/06/06.html accessed 18Sep2013, (cited on pp.336, 587).

[198] Eggleton B (2013) *Photonic circuits for the new information age: Part 2.* ISS2013 Nanoscience lecture video. Cudos all-optical processing chip: image at 0h:14m:58s, https://www.youtube.com/ watch?v=9gZ1qFYGSa4 accessed 23Jan2013, (cited on pp.393, 398, 589).

[199] Ehrenberg W & Siday RE (1949) *The refractive index in electron optics and the principles of dynamics.* Proc. Phys. Soc. B 62, p.8, cited by Hiley [367], doi:10.1088/0370-1301/62/1/303 (cited on p.362).

[200] Einstein A (1916) *Strahlungs-Emission und Absorption nach der Quantentheorie.* (Emission and absorption of radiation in quantum theory.) Deutsche Physikalische Gesellschaft, Verhandlungen 18, pp.318–323, http://www.ulp.ethz.ch/education/ quantenelektronik/Paper_Einstein1.pdf accessed 17Jun2014, (cited on p.364).

[201] Elder J (2006) *Different like me: my book of autism heroes.* London: Jessica Kingsley, p.20, (cited on p.106).

[202] Elstein C (1980–1985) *The Spath Road Workshop.* Teaching drawing, painting and sculpture, http://www.cecileelstein.com/ Biography/index.html accessed 8Apr2014, (cited on p.94).

[203] Eneström G (1913) *Die Schriften Eulers chronologisch nach den Jahren geordnet, in denen sie verfasst worden sind.* (Euler's writings, by year), Jahresbericht de Deutschen Mathematiker-Vereinigung (Annual Report of the German Mathematical Society), (cited on p.25).

[204] Engheta N (2012) *Of light, electroncs and metamaterials.* ECE Lecture 15Feb2012, Univ. of Delaware, https://www.youtube.com/ watch?v=dchwRsaY2-E accessed 19Jan2015, (cited on p.397).

[205] Espenak F (2012) *NASA Eclipse website.* NASA Goddard Space Flight Center, Greenbelt, MD: NASA Heliophysics Science Division, http://eclipse.gsfc.nasa.gov/SEsaros/SEsaros.html accessed 8May2014, (cited on p.119).

[206] Euler L (1767) *Observationes analyticæ—E326*. Novi commentarii academiæ scientiarum Petropolitanæ, 11, 1765, pp.124–143, reprinted in Opera Omnia Series I 15, pp.50–69, facsimile: Euler Archive [207], (cited on pp.25, 481).

[207] Euler Archive (c.2003–) *The Euler archive, a digital library dedicated to the work and life of Leonhard Euler.* Website hosted by the MAA. http://eulerarchive.maa.org/docs/originals/E326.pdf accessed 23Apr2013, (Eneström E326 Latin facsimile), (cited on pp.25, 506).

[208] Fang N et al. (2003–) *[... a series of publications regarding superlenses...]* Zhang Lab, Dept. Mechanical Engineering, Univ. California Berkeley, *Journal publication list and pdf files*, refs: 019, 020, 027, 040, 046, 059, 070, http://xlab.me.berkeley.edu/pub_full. html accessed 7Jan2015, (cited on p.373).

[209] Farhat M, Enoch S, Guenneau S, & Movchan AB (2008) *Broadband cylindrical acoustic cloak for linear surface waves in a fluid.* Phys. Rev. Lett. 25Sep2008, 101:134501, doi:10.1103/PhysRevLett.101.134501 (cited on p.382).

[210] Fathauer R (2005) *Extending Escher's recognizable motif tilings.* In *M C Escher's legacy: a Centennial celebration.* Eds. Schattschneider & Emmer, for Rome conference 1998. Berlin: Springer-Verlag, p.155, (cited on p.316).

[211] Fechner GT (1876) *Vorschule der Aesthetik [Experimental aesthetics]*. Leipzig, Germany: Breitkopf & Haertel, cited by McManus [640], (cited on pp.67, 69, 149).

[212] Fedorov ES (1891) *Symmetry of regular systems of figures (in Russian)*. Notices of the Imperial St. Petersburg Mineralogical Society, 28, pp.1–146, English transl. by Harker D & K, ACA Monograph No.7, New York: American Crystallographic Association, (cited on p.343).

[213] Feng Duan & Jin Guojun (2005) *Introduction to condensed matter physics: Part 1.* Singapore: World Scientific Publishing, pp.56–57, (cited on p.300).

[214] Fernando L & Molina JE (2011) *Islamic Patterns research project.* Nomad Inception—specialist website, http://www.nomadinception. com/gallery-arabic-patterns-islamic-patterns-research.aspx accessed 23Jan2015, (cited on pp.356, 356, 588).

[215] Fernholm A (2011) *Crystals of Golden proportions.* Nobel Prize in Chemistry 2011. Science Editors: Lidin S & Thelander L, the Nobel Committee for Chemistry, illustrations by Jarnestad J, Stockholm: The Royal Swedish Academy of Sciences, Fig.4, p.5, http://www.nobelprize.org/nobel_prizes/chemistry/laureates/2011/popular-chemistryprize2011.pdf accessed 21Mar2014, (cited on pp.357, 588).

[216] Ferretti-Bocquillon M, et al. (2001) *Signac 1863–1935.* New York: Met. Museum of Art, Yale Univ. Press, (17Jan1890—tenuous connection), p.303, (cited on p.71).

[217] Feynman RP (1959) *There's plenty of room at the bottom.* Lecture at meeting of the American Physical Society 29Dec1959. Caltech Engineering & Science, 23:5, (1960), http://www.zyvex.com/nanotech/feynman.html accessed 18Jan2015, (cited on p.304).

[218] ———. (1979) *Photons: corpuscles of light.* Douglas Robb Memorial Lecture, University of Auckland, video 1h:17m:57s, *remarkably accurate agreement of theory and experiment,* 0h:12m:00s, http://www.youtube.com/watch?v=xdZMXWmlp9g accessed 11Jul2014, (cited on p.297).

[219] —Ibid. *(not) understanding quantum mechanics—I don't understand it—nobody understands it—it's the way nature works—[if] you don't like it: go somewhere else.* 0h:20m:56s–0h:30m:00s, (cited on p.297).

[220] Fibonacci Association, (1963–) *The Fibonacci Quarterly ('FQ').* Journal, with website hosted by the Department of Mathematics & Statistics at Dalhousie University in Halifax, Nova Scotia, Canada, http://www.fq.math.ca/ accessed 12Mar2013, http://www.mscs.dal.ca/Fibonacci/index.html accessed 12Mar2013, (cited on pp.34, 520).

[221] Fibonacci L (1202, rev. 1228) *Liber Abaci (Manuscript).* Re-publ. as *Scritti di Leonardo Pisano* in 2 Vols. Rome: Boncompagni, B, 1857–1862. Vol.I, pp.283–285, re-publ. as *Fibonacci's Liber Abaci: a translation into modern English of Leonardo Pisano's Book of Calculation,* transl. Sigler LE (2002) New York: Springer, p.404, (cited on pp.19, 22, 284, 284, 577).

[222] Fisher IR, Cheon KO, Panchula AF, Canfield PC, Chernikov M, Ott HR, & Dennis K (1999) *Magnetic and transport properties of single-grain R-Mg-Zn icosahedral quasicrystals [R=Y, $(Y_{1-x}Gd_x)$, $(Y_{1-x}Tb_x)$, Tb, Dy, Ho, and Er].* Phys. Rev. B, 59:1, pp.308–321, doi:10.1103/PhysRevB.59.308 (cited on pp.333, 587).

[223] Ford H (1922) *My life and work.* New York: Garden City Publishing. *The idea came in a general way ... [from] Chicago packers,* p.55, (cited on p.81).

[224] Forestry Commission UK (2014) *Scots pine—pinus sylvestris.* Website article, http://www.forestry.gov.uk/forestry/INFD-5NLFAP accessed 20Jan2014, (cited on p.56).

[225] Frank FC (1952) *Supercooling of liquids.* Proc. R. Soc. London Ser. A 215, p.43, cited by Kelton et al. [463], doi:10.1098/rspa.1952.0194 (cited on p.307).

[226] Frank FC & Kasper JS (1958) *Complex alloy structures regarded as sphere packings.* Acta Cryst. 11, p.184, doi:10.1107/S0365110X58000487 (cited on p.300).

[227] Fratini RM (2012) *PhiTaxis: Fibonacci digital simulation of spiral phyllotaxis.* Website http://www.sciteneg.com/PhiTaxis/PHYLLOTAXIS.htm accessed 3Jun2013, (cited on pp.293, 503, 503).

[228] Frey-Wyssling A (1954) *Divergence in helical polypeptide chains and in phyllotaxis.* Nature, 173, 27Mar1954, p.596, doi:10.1038/173596b0 (cited on pp.290, 489, 551).

[229] Frings M (2002) *The Golden Section in architectural theory.* Nexus Network Journal, 4:1, pp.9–32, *Neufert disciple of Gropius, works: Bauentwurfslehre 1936, Bauordnungslehre 1943,* p.20, doi:10.1007/s00004-001-0002-0 http://www.marcus-frings.de/text-nnj.htm accessed 26Nov2014, (cited on pp.132, 519).

[230] —Ibid. *One should not underestimate Zeising's impact.* And further, (from his historical discussion) Frings concludes that: *Zeising 'discovered' [φ] for architecture and the pictorial arts,* p.19, (cited on pp.67, 550).

[231] Galenson D (2008) *The rise and (partial) fall of abstract painting in the Twentieth Century.* NBER Working Paper 13744, p.5, (cites Holty [374]), http://www.nber.org/papers/w13744 accessed 10Jul2014, (cited on pp.520, 560).

[232] Galkin I (undated) *Fibonacci numbers spelled out.* Univ. Massachusetts Lowell, http://ulcar.uml.edu/~iag/CS/Fibonacci.html accessed 30Apr2013, (cited on pp.439, 483).

[233] Gallo D (2007) *Underwater astonishments (highlighting work of Edith Widder & Roger Hanlon & others)*. TED Talk video 0h:05m:21s, *dramatic footage of neurally controlled invisibility cloaking by octopus,* at 0h:04m:15s, http://www.ted.com/talks/david_ gallo_shows_underwater_astonishments accessed 7Jan2015, (cited on p.374).

[234] Garcia MQ, Pfänder M, Gertig C, & Gonzales-Aguilar O (2010) *Method for distributing heliostats in a tower plant.* Abengoa Solar New Technologies, SA, Seville. US Patent Application: 13/513,181, US 2013/0092156 Al, (cited on p.296).

[235] García-Meca C, Carloni S, Barceló C, Jannes G, Sánchez-Dehesa J, & Martínez A (2014) *Analogue transformation acoustics.* Photonics and Nanostructures— Fundamentals and Applications, Aug2014 12:4, pp.312–318, doi:10.1016/j.photonics.2014.05.001 (cited on p.381).

[236] ———. (2014) *Transformational acoustic metamaterials based on pressure gradients.* Phys. Rev. B, 90, 024310, doi:10.1103/PhysRevB.90.024310 (cited on p.381).

[237] Gardner M (1970) *Mathematical Games: The fantastic combinations of John Conway's new solitaire game 'Life'.* Scientific American, Oct1970, 223, pp.120–123, (cited on pp.302, 304, 304).

[238] ———. (1977) *Mathematical Games: Extraordinary nonperiodic tiling that enriches the theory of tiles.* Scientific American, Jan1977, 236:1, pp.110–121, reprinted as Gardner trapdoor [239] & Gardner revised [240], (cited on p.310).

[239] ———. (1989) *Penrose tiles to trapdoor ciphers ...and the return of Dr. Matrix.* The Mathematical Association of America, Spectrum Series. New York: WH Freeman. Chapters 1 & 2 available online: Gardner revised [240], (cited on pp.310, 509).

[240] ———. (1997) *Penrose tiles to trapdoor ciphers ... and the return of Dr. Matrix.* New York: WH Freeman. Revised edn. (1997) The Mathematical Association of America, Chapters 1 & 2, http: //www.maa.org/sites/default/files/pdf/pubs/focus/Gardner_PenroseTilings1-1977.pdf accessed 5Dec2013, (cited on p.509, 509).

[241] —Ibid. *1973 Penrose [P1 tiling],* p.6, (cited on p.312, 312).

[242] —Ibid. *1.618 [times] as many kites as darts,* p.7, (cited on p.314).

[243] —Ibid. [*inflation & deflation*] *discovered by Penrose,* [*named by*] *Conway,* p.8, (cited on pp.314, 452, 455).

[244] —Ibid. *structure is fractal,* p.8, (cited on pp.314, 318).

[245] —Ibid. *Conway's theorem: repetition density,* $d\phi^3/2$, pp.9–10, (cited on p.318).

[246] —Ibid. *you can never determine which tiling you are on,* p.10, (cited on p.319).

[247] —Ibid. *uncountable infinity of... tilings... differ... in infinitely many ways,* p.10, (cited on p.319).

[248] —Ibid. *'empire' (area commanded by vertex configuration),* p.11, (cited on p.314).

[249] —Ibid. *The cartwheel pattern,* Fig.11, p.13, (cited on p.346).

[250] —Ibid. *Ammann ... independently discovered ... rhomb tiles in 1976,* p.19, (cited on p.310).

[251] —Ibid. *(nets for) obtuse & acute Golden Rhombohedra.* All faces are 'P3 thick' rhombs, 72° & 108°, Fig. 17, p.23, (cited on pp.322, 342).

[252] —Ibid. *quoting Coxeter on Golden Rhombohedra—there are only two kinds: acute & obtuse, and both were studied by Kepler,* p.24, (cited on pp.322, 342, 557).

[253] —Ibid. *Penrose 1976 letter: quasi-periodic 'crystals',* p.24, (cited on p.324).

[254] —Ibid. *only one centre for 5-fold symmetry,* Barlow (1862)/Conway, p.27, (cited on p.319).

[255] Gaspard JP, Hodges CH, & Gordon MB (1977) *Structural stability of transition and noble metal clusters.* J. Phys. Colloques, C2, supplément au no. 7, tome 38, Jul1977, p.C2 63–67, doi:10.1051/jphyscol:1977213 (cited on p.303).

[256] Genevet P & Capasso F (2015) *Holographic optical metasurfaces: a review of current progress.* Rep. Prog. Phys. 78, 024401 (19pp), (cited on p.375).

[257] Geometry Guy (2009) *Kite Dart: Penrose kite and dart with two sorts of matching rule.* Illustration, http://en.wikipedia.org/wiki/File: Kite_Dart.svg accessed 8Jun2013, (cited on pp.313, 317, 585, 586).

[258] ———. (2010) *Kite-Dart star and corresponding Penrose tiling by rhombs.* Illustration. http://en.wikipedia.org/wiki/File:Penrose_tilings_ P2_and_P3.svg accessed 1Jun2014, (cited on pp.319, 586).

[259] Gerkan A von & Müller-Wiener W (1961) *Das Theater von Epidauros.* Stuttgart: Kohlhammer Verlag, cited by Scott & Marketos [811], *Golden triangles near centre of Theatre of Epidaurus,* (cited on pp.16, 559).

[260] Ghyka MC (1927) *Esthétique des proportions dans la nature et dans les arts.* Paris: Gallimard, (cited on pp.71, 72, 151, 518).

[261] ———. (1931) *Le Nombre d'Or; rites et rythmes pythagoriciens dans le développement de la civilisation occidentale.* 2 volumes, Paris: Gallimard, (cited on pp.71, 72).

[262] ———. (1946) *The geometry of art and life.* Franklin, WI: Sheed & Ward, (cited on p.72).

[263] Gillmor AM (1988) *Erik Satie.* London: Macmillan Press, *Golden Section & diagrams,* pp.87–88, (cited on p.150).

[264] —Ibid. *(use of Golden Section) 'more than coincidental ... same year',* p.87, (cited on p.150).

[265] Goldman AI, Anderegg JW, et al. (1996) *Quasicrystalline materials.* American Scientist May/Jun1996 84:3, pp.230–241, (cited on p.339).

[266] Goodhew P, Green A, et al. (1993–) *MATTER.* Materials science CBL consortium led by Univ. of Liverpool, http://www.matter.org.uk/ diffraction/electron/electron_diffraction.htm accessed 18Feb2015, (cited on p.327).

[267] Gotlieb M (2002) *The Painter's Secret: Invention and Rivalry from Vasari to Balzac.* The Art Bulletin, Sep2002 84:3, re artists guarding their secrets—*Seurat's paranoia regarding the dissemination of his method ... disinclined to exhibit his pictures ... charged his friends with theft and betrayal,* p.469, (cited on p.85).

[268] Gray H & Lewis WH (1918) *Anatomy of the human body.* 20th edn. (aka *Gray's anatomy.*) Philadelphia: Lea & Febiger. *Median sagittal section of brain: Figure 720,* (cited on pp.154, 582).

[269] Gray MW (1987) *Sophie Germain (1776–1831)* chapter in Campbell PJ & Grinstein LS (1987) *Women of Mathematics: A Bio-Bibliographic Sourcebook.* Westport Connecticut: Greenwood Publ., pp.47–56, (cited on p.26).

[270] Greathouse C & Weisstein EW (1999–2014) *Odd Perfect Number.* From MathWorld, a Wolfram web resource, http://mathworld. wolfram.com/OddPerfectNumber.html accessed 7Feb2014, (cited on pp.284, 494, 549).

[271] Green CD (1995) *All that glitters: a review of psychological research on the aesthetics of the Golden Section.* Perception 24:8, pp.937–968, doi:10.1068/p240937 (cited on p.113).

[272] —Ibid. *Table 1. Summary of Golden Section research,* p.962, (cited on p.113).

[273] Griffiths J (2012) *Carom activity 1–5: kites and darts.* Microsoft PowerPoint® slide set, slide 14 of 20, components by Tovstra [902], http://www.s253053503.websitehome.co.uk/carom/carom-final/carom-1-5.ppt accessed 30May2014, (cited on pp.315, 586).

[274] Grimaldi RP (2012) *Fibonacci and Catalan numbers, an introduction.* Rose Hulman Institute of Technology, Hoboken: Wiley, p.54, (cited on p.2).

[275] —Ibid. *Properties of α & β,* p.55, (cited on pp.4, 481).

[276] Grimes P (1997) *Jung: the Dialectic and Plato's divided line.* Lecture DVD V075, The Noetic Society, *divided line based on Golden Section,* at 0h:06m:05s, http://www.openingmind.com/ accessed 5Feb2015, http://noeticsociety.org/ accessed 5Feb2015, (cited on p.16).

[277] Gringer (2008) *Major chord root and inversions.* Public domain svg file, Wikimedia Commons, http://commons.wikimedia.org/wiki/File:Major_chord_root_and_inversions.svg accessed 13May2014, (cited on pp.iv, 107, 581).

[278] Grünbaum B, Grünbaum Z & Shephard GC (1986) *Symmetry in Moorish and other ornaments.* Comp. & Maths. with Appls. (Pergamon), 12B:3/4, pp.641–653, doi:10.1016/0898-1221(86)90416-5 (cited on pp.494, 554).

[279] Grünbaum B & Shephard GC (1987) *Tilings and patterns.* New York: WH Freeman, *Penrose tiling classification* § 10.3, P1: p.531, P2: p.535, P3: p.542, (note: the abridged version—*T&P an introduction*—excludes Penrose tiling), (cited on pp.312, 504).

[280] —Ibid. *Seven kinds of vertex neighbourhood (kites & darts),* Fig.10.5.3, p.531, (cited on p.314).

[281] —Ibid. *equivalence of P1 and P2 tilings,* Figs. 10.3.21 & 10.3.22, p.546, (cited on p.316).

[282] —Ibid. *P2 / P3 equivalence,* Fig. 10.3.19, p.543, (cited on p.316).

[283] —Ibid. *uniqueness of composition and decomposition,* p.558, (cited on p.318).

[284] —Ibid. *congruent to infinitely many patches in every tiling,* p.562, (cited on p.318).

[285] —Ibid. *musical sequences/Ammann bars,* p.571, (cited on p.321).

[286] —Ibid. *systems of bars (e.g. Ammann) are fundamental,* p.581, (cited on p.321).

[287] Guérin N (2006) *Photo du Professeur Veselago.* http://en.wikipedia. org/wiki/File:Veselago.jpg accessed 3Oct2013, (cited on pp.369, 588).

[288] Gullberg J (1997) *Mathematics from the birth of numbers.* New York: WW Norton, p.287, (cited on p.24).

[289] Gummelt P (1996) *Penrose tilings as coverings of congruent decagons.* Geometriae Dedicata 62:1, 1Aug1996, pp.1–17, doi:10.1007/BF00239998 (cited on pp.335, 505).

[290] Hagens W (1958) *Expo 1958 paviljoen van Philips (Expo 1958 Philips Pavilion).* Photograph, http://en.wikipedia.org/wiki/File:Expo58_ building_Philips.jpg accessed 12Sep2014, (cited on pp.131, 582).

[291] Hagerty DJ (2010) *The Art of Maynard Dixon.* Layton Utah: Gibbs Smith, *Hambidge dynamic symmetry, ... prominent in 1920s ... based on Golden Section,* p.112, (cited on p.69).

[292] —Ibid. *Albert Barrows: mathematician, painter, writer, lecturer,* p.113, (cited on p.69).

[293] Haghighi M (1992) *Computation of Mersenne primes using Cray X-MP.* Internat. J. Computer Math. 41:3–4, M_7 *example,* Illustrn. 1.2, p.251, doi:10.1080/00207169208804044 (cited on p.449).

[294] Hales TC (2000) *Cannonballs and honeycombs.* University of Michigan at Ann Arbor. Notices of the AMS, 47:4, p.441, http://www.ams.org/notices/200004/fea-hales.pdf accessed 7Mar2014, also see 'Project Flyspec' completion, https://code.google.com/p/flyspeck/wiki/ AnnouncingCompletion accessed 19Sep2015, (cited on p.299).

[295] Hall GA (2013) *Comments on The Modulor.* The Liverpool Architectural Society. Personal communication quoted with kind permission, (cited on p.130).

[296] Hambidge J (1920) *The Diagonal.* New Haven & New York: Yale University Press, (the original work), (cited on pp.68, 514).

[297] ———. (1967) *The elements of dynamic symmetry.* Unaltered version published 1926 by Brentano's Inc. and reprinted in 1948 by Yale University Press, being content from Hambidge *The Diagonal* [296], *static symmetry,* p.xiii, (cited on p.68).

[298] —Ibid. *Greeks learnt from Egyptians,* p.xv, (cited on p.68).

[299] —Ibid. *life and movement,* p.xv, (cited on p.68).

[300] ———. (1920) *Dynamic symmetry: the Greek vase.* New Haven & New York: Yale Univ. Press, *Root rectangles,* Fig. 10, p.24, https://archive.org/details/cu31924019526882 accessed 29Apr2014, https://ia600407.us.archive.org/9/items/cu31924019526882/cu31924019526882.pdf accessed 29Apr2014, (cited on pp.68, 68, 431, 488, 580).

[301] Hamilton WR (1843) *Letter to John T Graves.* Philosophical Magazine & Journal of Science. Publ. London, Edinburgh, & Dublin, XXV (1844), pp.489–495, http://www.maths.tcd.ie/pub/HistMath/People/Hamilton/QLetter/QLetter.pdf accessed 8Jan2015, (cited on p.361).

[302] Hammond C (2009) *The basics of crystallography and diffraction.* Oxford: Oxford University Press, p.1, (cited on p.342).

[303] Hardy GH (1919) *A problem of diophantine approximation.* Journal Ind. Math. Soc. 11, pp.205–243, (cited on p.72).

[304] Hardy L (1993) *Nonlocality for two particles without inequalities for almost all entangled states.* Phys. Rev. Lett. 71:11, pp.1665–1668, doi:10.1103/PhysRevLett.71.1665 (cited on p.298).

[305] Hargittai I (1997) *Quasicrystal discovery: a personal account.* (Interviews with Mackay, Shechtman, Cahn, Levine, Steinhardt, & Senechal.) The Chemical Intelligencer, Oct1997. New York: Springer Verlag, *first electron microscope at Technion 1967 ... Shechtman one of the first to use it,* p.34, (cited on pp.300, 331).

[306] —Ibid. *laws that were artificially constructed,* p.31, cited by Lifshitz [559], (cited on p.342).

[307] ———. (2002) *Alan L Mackay—crystallographer, universalist, humanist.* Struct. Chemistry, Aug2002, (special issue dedicated to ALM), 13:3–4, p.213, doi:10.1023/A:1015899107760 (cited on p.300).

[308] ———. (2010) *Structures beyond crystals.* J. Mol. Struct. 976:1, *Mackay diffraction from Penrose pattern,* p.81, doi:10.1016/j.molstruc.2010.02.009 (cited on p.324).

[309] ———. (2013) *Buried Glory: Portraits of Soviet Scientists.* Aleksandr Kitaigorodskii: Soviet Maverick, p.251, (cited on p.100).

[310] Hargittai I & Hargittai M (2000) *In our own image: personal symmetry in discovery.* Springer, *Penrose's concern—aperiodicity, Mackay's—hierarchic structures,* p.153, (cited on p.324).

[311] —Ibid. *Escher: Pólya & Coxeter,* pp.136–137, (cited on p.69).

[312] —Ibid. *Bernal: icosahedral structure of water,* p.152 & p.156, (cited on p.306).

[313] —Ibid. *Greeks: icosahedral structure of water,* note 25, p.179, (cited on p.306).

[314] —Ibid. *noting Pauling's icosahedral results,* p.152, (cited on p.330).

[315] —Ibid. *Mackay shows simulated diffraction pattern in 1981,* p.152, (cited on p.324).

[316] —Ibid. *Shechtman & Blech paper bounced back from JAP,* p.162, (cited on p.328).

[317] —Ibid. *Ammann bars to Ammann planes,* p.170, (cited on p.333).

[318] Harriss E & Frettlöh D (undated) *Tilings encyclopedia—in rhombs, and wedges, and half-moons, and wings.* Website hosted by Dept. Maths. University of Bielefeld, http://tilings.math. uni-bielefeld.de/glossary/ammann_bars accessed 12Mar2014, (cited on pp.321, 322, 586).

[319] —Ibid. *Robinson triangle, substitutions,* http://tilings.math. uni-bielefeld.de/substitution_rules/robinson_triangle accessed 6Jun2014, (cited on p.322).

[320] Hart G (2012) *The Golden Ratio nautilus: what would a nautilus look like if it actually was a Golden Spiral?* Video 0h:2m:21s, http://youtu.be/_gxC8OjoQkQ accessed 17Jun2013, http://georgehart.com accessed 17Jun2013, (cited on pp.50, 579).

[321] Hartl M (2010) *The tau manifesto: 'No, really, pi is wrong...'* http://tauday.com/tau-manifesto accessed 12Mar2013, (cited on p.33).

[322] —Ibid. *Golden Ratio,* Note 10, http://tauday.com/tau-manifesto#fn-0_10 accessed 20Jan2014, (cited on p.33).

[323] Hansen RT (1972) *Generating identities for Fibonacci and Lucas triples.* Fib. Qtrly. Dec1972 10:6, theorem 2, p.574, cited by Knott [492], http://www.fq.math.ca/Scanned/10-6/hansen.pdf accessed 15Jul2014, (cited on pp.196, 530).

[324] Hausdorff F (1919) *Dimension und äußeres Maß. (Dimension and outer measure).* Mathematische Annalen 79, pp.157–179, http://www.digizeitschriften.de/dms/resolveppn/?PPN=GDZPPN002266989 accessed 5Jul2013, (cited on p.242).

[325] Haüy R-J (1784) *Essai d'une théory sur la structure des crystaux, appliquée à plusieurs genres de substances crystallisées.* Paris: Gogué & Née de la Rochelle, (cited on p.323).

[326] Heath TL (1908) *The thirteen books of Euclid's Elements.* Transl. from text of Heiberg JL by Heath TL. Cambridge: The University Press, *'To cut a given finite straight line in extreme and mean ratio',* Vol.II, Book VI, Prop. 30, p.267, (cited on p.17).

[327] —Ibid. *'The five so-called Platonic figures',* Vol.III, Book XIII, p.438 et seq., (cited on p.17).

[328] —Ibid. *'Historical note—origins of the Platonic solids',* Vol.III, Book XIII, p.438, (cited on p.14, 14).

[329] —Ibid. *'if … the product will be perfect',* Vol.II, Book IX, Proposition 36, p.421, 'If as many numbers as we please beginning from an unit be set out continuously in double proportion, until the sum of all becomes prime, and if the sum multiplied into the last make some number, the product will be perfect.', (cited on pp.282, 447).

[330] Heidelberg Printing Machines AG (2008) *An introduction to screening technology.* Available online as *Expert Guide 'Screening Technology'.* Heidelberg: Heidelberger Druckmaschinen AG, *standard set of screen angles,* p.9, http://www.heidelberg.com/www/html/en/binaries/files/prinect/expert_guide_screening_tech_pdf accessed 13Mar2014, (cited on p.320).

[331] —Ibid. *Rational and irrational screening,* p11, (cited on p.320).

[332] Helmholtz H von, Fripp H (1876) *On the limits of the optical capacity of the microscope.* Monthly Microscopical Jnl. July 16:1, pp.15–39, doi:10.1111/j.1365-2818.1876.tb05606.x (cited on p.363).

[333] Helmholtz-Zentrum (2010) *Golden Ratio discovered in a quantum world.* Press release re Coldea et al. [120], Berlin: Helmholtz-Zentrum Berlin für Materialien und Energie (HZB), 7Jan2010, http://www.helmholtz-berlin.de/pubbin/news_datei?did=4061 accessed 11May2013, (cited on p.298).

[334] Henry C (1885) *Introduction à une esthétique scientifique.* Paris, also in *La Revue Contemporaine.* II, May–Aug1885, pp.441–469, cited by Howat [385], Bibliothèque nationale de France: http://gallica.bnf.fr/ark:/12148/bpt6k329774 accessed 27Jan2014, (cited on pp.146, 149).

[335] Herbert RL (1968) *Neo-Impressionism.* Catalogue for exhibition held Feb–Apr1968. New York: Solomon R Guggenheim Foundation. On Seurat's *Parade—'little drawing for the picture ... cites Henry's words ... reproduces [Henry's] diagram defining the Golden Section,* p.22. *BUT*: Note that Herbert was later persuaded to change his mind about Seurat and the Golden Section [338]—amongst his reasons, he cites arguments by Herz-Fischler 1983 [363] and Neveux 1990, (cited on pp.70, 71, 150).

[336] —Ibid. *'abstract' ... did not mean 'devoid of reference to the real world', as we now use the term,* p.23, (cited on p.94).

[337] ———. (1991) *Seurat.* Publication to accompany exhibition: 'Seurat 1859–1891', Paris Apr-Aug1991, New York Sep1991–Jan1992. New York: Metropolitan Museum of Art, *The Lady Tuya,* p.174, (cited on p.75).

[338] —Ibid. *'Foremost among the myths... ' [the one that Seurat used] 'geometric formulas such as the golden section.',* p.4; also see Note 4, p.8, and discussion p.392, (cited on p.517).

[339] —Ibid. *Seurat ... met Henry ... in 1886,* p.392, (cited on pp.79, 150).

[340] Hermán P, Kocsis L, & Eke A (2001) *Fractal branching pattern in the pial vasculature in the cat.* Journal of Cerebral Blood Flow & Metabolism 21, pp.741–753, *Fig.6: Generations of fractals— von Koch, Mandelbrot tree, & tree model,* doi:10.1097/00004647-200106000-00012 (cited on p.96).

[341] (Herz-) Fischler R (1979) *The early relationship of Le Corbusier to the 'golden number'.* Environment and Planning B, 6:1, *traceés régulateurs ... très difficile de retrouver,* last para. p.102, doi:10.1068/b060095 (cited on pp.28, 124).

[342] —Ibid. *Le Corbusier and 1927 Matila Ghyka book [260]*, p.95, (cited on p.151).

[343] Herz-Fischler R (1987) *A mathematical history of Division in the Mean and Extreme Ratio*, (DEMR). Ontario Canada: Wilfred Laurier University Press, New York: Dover Publications, republished by Dover 1998 [346], (cited on pp.34, 518).

[344] ———. (1993) *A 'Very pleasant theorem'.* The College Maths Jnl. 24:4, p.320, doi:10.2307/2686347 (cited on p.128).

[345] —Ibid. *thankfulness to [Magirus]*, p.318, (cited on p.518).

[346] ———. (1998) *A mathematical history of the Golden Number.* Unabridged republication of *A mathematical history of Division in the Mean and Extreme Ratio.* [343] with new material by Herz-Fischler. New York: Dover Publications, (cited on p.518).

[347] ———. (2009) *The shape of the Great Pyramid.* Waterloo, Canada: Wilfrid Laurier Univ. Press, *theories,* Chapters 5–15, pp.30–123, (cited on p.6).

[348] —Ibid. *Kepler ... learned about the triangle from a music professor ... Magirus,* § 3, p.229, (see also Herz-Fischler [345]) (cited on pp.7, 128).

[349] —Ibid. *Cole survey 1925,* p.11, (cited on p.498).

[350] —Ibid. *Cole did not give a value for the height, ... most of the summit and outer casing stones ... are missing,* p.11; also (in personal correspondence with the author), RH-F noted that: *'Cole did not determine the height because as a surveyor he could not. His interest was strictly measuring what he could measure',* (cited on p.7).

[351] —Ibid. *missing not only virtually all of its casing stones, but also approximately 9m of the top portion of the core masonry,* note 9, p.192, (cited on p.7).

[352] —Ibid. *[RH-F estimates]original height ... approximately 146.6m,* p.11, (cited on p.7).

[353] —Ibid. *Kepler triangle theory,* pp.80–91, (cited on p.7).

[354] —Ibid. *design principle? ... [conclusion]... not possible,* p.165–168, (cited on p.7).

[355] ———. (1997) *Le nombre d'or en France de 1896 à 1927*. La Revue d l'art, 118:4, pp.9–16, ... *certains cercles artisitiques allemands*, p.9, (cited on p.67).

[356] —Ibid. *visite à Prague en 1896 de Paul Sérusier à Jan Verkade*, p.9, (cited on p.71).

[357] —Ibid. *all intrigued by ... golden rule or section*, p.10, (cited on p.71).

[358] —Ibid. *autobiographie Severini ... étudier ... après ... Raoul Bricard prof. de mathématiques*, p.11, (cited on p.71).

[359] —Ibid. *lettre ... Gris parlait du* [ϕ] *à Ozenfant*, p.14, (cited on p.71).

[360] —Ibid. *Gabrielle Buffet: name chosen by Metzinger & Duchamp —'philosophic & scientific research'*, p.14, (cited on p.70).

[361] —Ibid. *suggère que ... il a utilisé* [ϕ], p.14, (cited on p.71).

[362] —Ibid. *redessiné ... pour faire croire ...* , p.16, (also see Frings p.23 [229]), (cited on pp.71, 72).

[363] ———. (1983) *An examination of claims concerning Seurat and 'The Golden Number'*. Gazette des Beaux-Arts 101, 1370, Mar83, pp.109–112. Lhote—ref.2; RH-F also notes Golden decompositions of *La Parade* by Gonse, Dorra, & Marcou— ref.3, (cited on pp.28, 167, 517, 536).

[364] Heyrovská R (2005) *The Golden Ratio, ionic and atomic radii and bond lengths*. Mol. Phys., 103:6–8, 20Mar–20Apr2005, pp.877–882, doi:10.1080/00268970412331333591 (cited on p.308).

[365] ———. (2006) *Dependence of ion-water distances on covalent radii, ionic radii in water and distances of oxygen and hydrogen of water from ion/water boundaries*. Chem. Phys. Lett. 429, pp.600–605, cited by Chaplin [105], doi:10.1016/j.cplett.2006.08.073 (cited on p.308).

[366] ———. (2007) *Dependences of molar volumes in solids, partial molal and hydrated ionic volumes of alkali halides on covalent and ionic radii and the golden ratio*. Chem. Phys. Lett. 436, pp.287–293, cited by Chaplin [105], doi:10.1016/j.cplett.2007.01.042 (cited on p.308).

[367] Hiley BJ (2013) *The early history of the Aharonov-Bohm effect.* TPRU, Birkbeck, University of London, ArXiv article, http://arxiv. org/pdf/1304.4736.pdf accessed 30Nov2013, (cited on pp.362, 505).

[368] Hill A (1968) *Art and Mathesis: Mondrian's structures.* Leonardo, 1:3, Jul1968, Cambridge, MA: MIT Press, p.234, (cited on p.91).

[369] Hines TG (2003–) *Epidaurus plan.* On website *The Ancient Theatre Archive,* http://www.whitman.edu/theatre/theatretour/epidaurus/ commentary/epidaurus.commentary.htm accessed 25Apr2014, (cited on pp.17, 577).

[370] Hodges A (1983) *Alan Turing: the enigma.* (Bio.), Univ. California, Bennet Books. Cited by Hodge [380], (cited on p.58).

[371] Hoggatt VE Jr. (1969) *Fibonacci and Lucas numbers.* Houghton Mifflin Company, cited by Knott [480], available online in full from the Fibonacci Association [220]: http://www.fq.math.ca/ fibonacci-lucas.html accessed 13Jul2013, (cited on pp.529, 530, 530).

[372] —Ibid. *Pascal's triangle,* p.50, (cited on p.13).

[373] Hokenson J (2004) *Japan, France, and East-West Aesthetics.* New Jersey: Fairleigh Dickinson Univ Press, *Charles Henry, in his Sorbonne lectures of 1884,* p.180, (cited on p.149).

[374] Holty C (1957) *Mondrian in New York: a memoir.* Arts, 31:10, *(when Mondrian was asked about losing good pictures—by changing them into others as he worked),* p.21, cited by Galenson [231], (cited on pp.90, 508).

[375] Holtzman H & James MS (1987) *The new art—the new life: the collected writings of Piet Mondrian.* London: Thames & Hudson, p.15, (cited on p.492).

[376] Homola J, Vaisocherová H, Dostálek J, & Piliarik M (2005) *Multi-analyte surface plasmon resonance biosensing.* Methods 37, pp.26–36, doi:10.1016/j.ymeth.2005.05.003 (cited on p.378).

[377] Horadam AF (1988) *Elliptic functions and Lambert series in the summation of reciprocals in certain recurrence-generated sequences.* The Fibonacci Quarterly 26:2, *§ 1. Introduction* p.98, http://www.fq.math.ca/Issues/26-2.pdf accessed 27May2014, (cited on p.252).

[378] —Ibid. *special cases of p and q,* p.99, (cited on p.265).

[379] Horn PL (1991) *Handbook of French popular culture.* p.65, (cited on pp.71, 152).

[380] ————. (1995) *Alan Turing: the enigma.* Website, *Part 7—Turing at Manchester,* http://www.turing.org.uk/bio/part7.html accessed 27Nov2013, (cited on pp.58, 520).

[381] —Ibid. *Work on the Enigma code,* http://www.turing.org.uk/bio/part4. html accessed 27Nov2013, (cited on p.58).

[382] Howat R (1983) *Debussy in proportion, a musical analysis.* London: Cambridge University Press, … *halves … Golden Section (proportions 1, 2, & ϕ),* p.1, (cited on p.65).

[383] —Ibid. *La mer: Dialogue du vent et de la mer,* spiral analysis, Fig.8.1, p.97, (cited on pp.142, 151, 155).

[384] —Ibid. *Golden Section in Spleen. Analysis* pp.34–36, *comprehensiveness of* [ϕ] *in Spleen,* p.38 & *after 1885 … organized way … in Spleen,* p.166, (cited on pp.146, 150).

[385] —Ibid. *Charles Henry, mathematician to the Symbolists,* pp.164–167, (cited on pp.71, 75, 147, 150, 517).

[386] —Ibid. *esoteric involvement: neo-Rosicrucians,* p.167, (cited on pp.71, 147, 150).

[387] —Ibid. *Debussy letters from Rome,* p.167, (cited on p.149).

[388] —Ibid. *Paris arrondissements spiral,* p.173, (cited on p.143).

[389] —Ibid. *Ravel Miroirs: Fibonacci examples,* pp.189–192, (cited on p.150).

[390] —Ibid. *Ravel* $\sqrt{5}$, p.191, (cited on p.149).

[391] —Ibid. *Symbolists,* p.164, (cited on p.146).

[392] ————. (1983) *Bartók, Lendvai and the principles of proportional analysis.* Review article, Music Analysis, 2:1 pp.69–95, http://www.jstor.org/stable/853953 accessed 3Jan2014, (cited on p.490).

[393] —Ibid. *symmetric tonal system, halfway point (1st movement of MFSPC),* p.79, (cited on pp.65, 140).

[394] —Ibid. *3rd movement of Music for Strings, Percussion and Celeste,* (a) Lendvai & (c) Lowman: 21 first theme, 13 second theme, Fig.6, p.82, (cited on p.146).

[395] —Ibid. *spiral-based proportional analysis of 3rd movement MFSPC,* Fig.7, p.83, (cited on pp.142, 144, 151, 155, 582).

[396] —Ibid. *what many critics already regarded as over-cerebral music,* p.84, (cited on pp.113, 155).

[397] —Ibid. *... quietly disposed of sheets of calculation connected with his compositions,* p.87, (cited on p.65).

[398] Hunt BJ (2012) *Oliver Heaviside: a first-rate oddity.* Phys. Today 65:11, p.48, doi:10.1063/PT.3.1788 (cited on p.362).

[399] Huntley HE (1974) *The Golden Ellipse.* The Fibonacci Quarterly 12:1, Feb1974, p.38, http://www.fq.math.ca/Scanned/12-1/huntley1.pdf accessed 5Jul2013, (cited on p.4).

[400] Huszár V (1917) *Cover of the first issue of De Stijl Magazine.* Printed paper, Oct1917. The Hague: Gemeentemuseum Collection, *Mondrian in writing (Room 11),* http://mediation. centrepompidou.fr/education/ressources/ENS-mondrian/ENS-mondrian-en.html accessed 13Apr2014, (cited on p.93).

[401] IMU: International Mathematical Union (1902) *Compte Rendu du Deuxième Congrès International des Mathématiciens, tenu à Paris du 6 au 12 août 1900. Procès-Verbaux et Communications.* International Congress of Mathematicians (ICM), Hilbert D *Problèmes futurs des mathématiques.* Hilbert's Tenth Problem: 'De la possibilité de résoudre une équation de Diophante: On donne une équation de Diophante à un nombre quelconque d'inconnues et à coefficients entiers rationnels: On demande de trouver une méthode par laquelle, au moyen d'un nombre fini d'opérations, on pourra distinguer si l'équation est résoluble en nombres entiers rationnels.' Paris: Gauthiers-Villars, p.87, http://www.mathunion.org/ICM/ICM1900/ICM1900. ocr.pdf accessed 2Sep2014, (cited on p.450).

[402] Inductiveload (2009) *A Penrose tiling (P1).* http://en.wikipedia.org/ wiki/File:Penrose_Tiling_(P1).svg accessed 11Sep2013, (cited on pp.312, 585).

[403] ———. (2009) *A Penrose tiling (P3) using thick and thin rhombi.* http://en.wikipedia.org/wiki/File:Penrose_Tiling_(Rhombi).svg accessed 15May2013, (cited on pp.317, 586).

[404] ———. (2009) *Penrose Tiling (P1 over P3).* http://en.wikipedia.org/ wiki/File:Penrose_Tiling_(P1_over_P3).svg accessed 1Jun2014, (cited on pp.319, 586).

[405] ISO 16 (1975) *Acoustics—Standard tuning frequency (Standard musical pitch)*. International Standards Organization. Reviewed & confirmed in 2011, A=440Hz ±0.5 Hz. http://www.iso.org/iso/catalogue_detail.htm?csnumber=3601 accessed 22Oct2013, (cited on p.224).

[406] Jacobson ET (1990) *Distribution of the Fibonacci numbers mod* 2^k. Fib. Qtrly., 30:3, 1992, pp.211–215, http://www.fq.math.ca/Scanned/30-3/jacobson.pdf accessed 23Apr2013, (cited on pp.416, 419, 420).

[407] —Ibid. *Case 4, residue twos, mod 32,* p.214, (cited on p.416).

[408] Jacoby M (1999) *Quasicrystals: a new kind of order.* American Chemical Society, Chem. & Engineering News 77:11, 15Mar1999, pp.44–77, doi:10.1021/cen-v077n011.p044 (cited on p.339).

[409] Janson HW (1962) *History of art.* 3rd edn. 1986 revised & expanded by Janson AF, London: Thames and Hudson, *Mondrian's Composition with Red, Blue, and Yellow,* [658] colour plate 140, p.679, (cited on p.545).

[410] Janssen H (1997) *Bringing about light—the function of colour in Mondrian's painting.* Essay and chronology in Riley B (1997) *Mondrian, nature to abstraction,* (to accompany the exhibition 30Jul–30Nov), London: Tate Gallery Publishing. *Consistent sabotage of his own theoretical assumptions ... destruction of former positions,* p.22, (cited on pp.110, 113).

[411] —Ibid. *Dusk and dawn are the moments of transition on which he concentrates,* p.33, (cited on p.110).

[412] Java (1995–) *Java programming language.* Oracle Corporation. Java is a registered trademark of Oracle in USA & certain other countries. http://www.oracle.com/ accessed 21Nov2013, (cited on pp.134, 197).

[413] Jazbec S (2009) *The properties and applications of quasicrystals, Seminar II.* Univ. Ljubljana, Maths & Physics. *Aero and rocket,* p.15, http://mafija.fmf.uni-lj.si/seminar/files/2009_2010/Quasicrystals.pdf accessed 18Sep2013, (cited on p.341).

[414] —Ibid. H_2 *storage,* Fig.13, & Table 1, pp.15–16, (cited on p.340).

[415] Jean RV (1994) *Phyllotaxis: a systemic study in plant morphogenesis.* New York: Cambridge Univ. Press, p.209, (cited on p.52).

[416] —Ibid. *sunflower parastichies from (13, 21) through (89, 144)*, p.16, (cited on p.154).

[417] —Ibid. *Lucas angle*, p.24, (cited on pp.58, 292, 293).

[418] —Ibid. *Pine cones... a common pair of spiral numbers is (5, 8), other cones may show (2, 3) or (3, 5)*, p.16, (cited on pp.56, 154).

[419] —Ibid. *Sunflower parastichy pair (55, 89)*, p.26, and *scanning electron micrograph*, Fig. 1.7, p.27, (cited on p.154).

[420] —Ibid. *Landmark in molecular biology*, p.214, (cited on p.295).

[421] —Ibid. *Crick's protein structure experiments*, p.218, (cited on p.295).

[422] Jingyun Huang, Xudong Wang, & Zhong Lin Wang (2006) *Controlled replication of butterfly wings for achieving tunable photonic properties.* Nano Letters 6:10, pp.2325–2331, doi:10.1021/nl061851t (cited on p.365).

[423] Joannopoulos JD, Johnson SG, Joshua NW, & Meade RD (2008) *Photonic crystals, molding the flow of light.* 2nd edn. Princeton: Princeton Univ. Press (pdf available online), http://ab-initio.mit.edu/book/ accessed 19Jan2015, (cited on pp.397, 397, 589).

[424] —Ibid. *inverse opals*, p.103, (cited on p.365).

[425] —Ibid. *waveguide-cavity-waveguide filter*, p.209, (cited on p.398).

[426] Joosten JM (1998) *Piet Mondrian: Volume 2: Catalogue Raisonné of the work 1911–1944.* New York: Harry N Abrams, (Vol.1 by Welsh [941]), (cited on p.571).

[427] —Ibid. *Studies, Mondrian B352: Study of a composition, 1938–40 (?)*, p.428, *Mondrian B355: Study of a composition, 1938–40 (?)* p.429, *Mondrian B358: Study of a composition, 1938–40 (?)* p.430, (cited on p.98).

[428] —Ibid. *Unfinished works, Mondrian B249: Composition with red, (unfinished) 1934*, p.374, *Mondrian B290: Composition, (unfinished) 1938*, p.399, *Mondrian B291: Composition, (unfinished) 1938 or 1939*, p.399, *Mondrian B298: Composition with red, blue and yellow, (unfinished) 1940*, p.403, (cited on p.98).

[429] —Ibid. *Mondrian's letter to HP Bremmer, Paris, 29Jan1914*, p.105, (cited on p.492).

[430] —Ibid. *Mondrian B3: Study for the Gray Tree, 1911,* p.200, (cited on p.95).

[431] —Ibid. *Mondrian B4: The Gray Tree, 1911,* p.201, (cited on p.95).

[432] —Ibid. *Mondrian B10: Study of Trees 1: Study for 'Tableau No. 2 / Composition No. VII', 1912,* p.204, (cited on p.95).

[433] —Ibid. *Mondrian B11: Forest: Study for 'The Trees', 1912,* p.205, (cited on p.95).

[434] —Ibid. *Mondrian B15: Tree, 1912,* p.207, (cited on p.95).

[435] —Ibid. *Mondrian B19: Flowering Appletree, 1912,* p.210, (cited on p.95).

[436] —Ibid. *Mondrian B20: Flowering Appletree, 1912,* p.211, (cited on p.95).

[437] —Ibid. *Mondrian B21: Flowering Appletree, 1912,* p.212, (cited on p.95).

[438] —Ibid. *Mondrian B23: Composition Trees 1 (unfinished), 1912,* p.213, (cited on p.95).

[439] —Ibid. *Mondrian B24: Composition Trees 2, 1912–1913,* p.214, (cited on p.95).

[440] —Ibid. *Mondrian B30: The Tree A, 1913,* p.218, (cited on p.95).

[441] —Ibid. *Mondrian B34: Study of Trees 2: Study for Tableau No. 2 / Composition No. VII, 1913,* p.221, (cited on p.95).

[442] —Ibid. *Mondrian B207: Composition, with Yellow, Blue, black and Light Blue, 1929,* Oil on canvas 50.5 × 50.5 cm, p.343, (cited on pp.106, 151).

[443] —Ibid. *Mondrian B212: Composition II, with Red, Blue, black and Yellow, 1929,* Oil on canvas 45 × 45 cm, p.347, (cited on p.106).

[444] —Ibid. *Mondrian B214: Composition No. 1, with Red and Black, 1929,* Oil on canvas 52 × 52 cm, p.348, (cited on p.106).

[445] —Ibid. *Mondrian B216: Composition No. IV, with Red, Blue and Yellow, 1929,* Oil on canvas 52 × 51.5 cm, p.350, (cited on p.106).

[446] —Ibid. *Mondrian B217: Composition with Red, Blue and Yellow, 1930,* Oil on canvas 46 × 46 cm, p.351, (cited on pp.106, 151, 544, 580).

[447] —Ibid. *Mondrian B219: Composition No. II / Composition I / Composition en rouge bleu et jaune, 1930,* Oil on canvas 51 × 51 cm, p.352, (cited on pp.104, 106, 545).

[448] —Ibid. *Mondrian B220: Composition No. I, with Yellow and Light Gray, 1930,* Oil on canvas 50.5 × 50.5 cm, p.353, (cited on p.106).

[449] —Ibid. *Mondrian B224: Composition en Blanc et Noir II, with black lines, 1930,* Oil on canvas 50.5 × 50.5 cm, p.356, (cited on p.106).

[450] —Ibid. *Mondrian B235: Composition with Blue and Yellow, 1932,* Oil on canvas 45.4 × 45.4 cm, p.364, (cited on p.106).

[451] —Ibid. *Mondrian B236: Composition with Yellow and Blue, 1932,* Oil on canvas 41.3 × 33 cm, p.365, (cited on p.106).

[452] Joseph GG (1991) *The crest of the peacock, non-European roots of mathematics.* London: Tauris, p.5 et seq., Fig. 1.2 p.9, Fig. 1.3 p.10, (cited on p.6).

[453] —Ibid. *Cultural dependencies,* pp.9–10, (cited on p.6).

[454] —Ibid. *ya mā tā rā ja bhā na sa la gām,* p.55, (also see Knuth [501]), (cited on pp.13, 531).

[455] —Ibid. *Dating Piṅgala Nāga,* p.254, (cited on p.12).

[456] ———. (1968) *Psychology & alchemy.* Collected Works of CG Jung, Vol.12, translators Adler G & Hull RFC, Princeton NJ: Princeton University Press. Jung quotes Dorn: 'Transform yourselves into living philosophical stones'. 2nd Edn. (1980), p.148, (cited on p.154).

[457] Kalia S (2011) *The generalizations of the Golden Ratio: their powers, continued fractions, and convergents.* Dept. Maths, Massachusetts Institute of Technology, http://web.mit.edu/primes/materials/2011/Kalia-Generalizations.pdf accessed 2May2012, (cited on pp.426, 483).

[458] Kaplan CS (2006) *A meditation on Kepler's Aa.* Website article. David R Cheriton School of Computer Science, University of Waterloo, http://www.cgl.uwaterloo.ca/~csk/papers/kaplan_bridges2006.pdf accessed 16Jan2105, (cited on p.342).

[459] Kappraff J (1992) *The relationship between mathematics and mysticism of the Golden Mean through history.* In (1992) *Fivefold symmetry.* Ed. Hargittai I, Singapore: World Scientific, *Full set of Scottish Neolithic 'Platonic solids'* (carved in stone), Fig.1, p.35, from Critchlow K (1982) *Time stands still.* New York: St. Martin's Press, (cited on p.14).

[460] —Ibid. *full moon on a given day... will not repeat... until 19 years have elapsed,* p.57, (cited on p.116).

[461] Kelton KF (2011) *The influence of icosahedral ordering in metallic liquids on phase transitions.* Technion lecture: video 0h:37m:48s, http://www.s.com/watch?v=5m4cyP0YTk8 accessed 21Sep2013, (cited on pp.307, 567).

[462] ———. (2013) *Crystal nucleation in supercooled liquid metals.* (Review) Int. J. Microgravity Sci. Appl. 30:1, p.16, http://www.jasma. info/wp/wp-content/uploads/2013/02/2013_p011.pdf accessed 5Dec2013, (cited on pp.307, 308).

[463] Kelton KF, Lee GW, et al. (2003) *First X-ray scattering studies on electrostatically levitated metallic liquids: demonstrated influence of local icosahedral order on the nucleation barrier.* Phys. Rev. Lett. 90:19 16May2003, 195504/1–4, http://www.nasa. gov/centers/marshall/news/news/releases/2003/03-104.html accessed 21Sep2013, doi:10.1103/physrevlett.90.195504 (cited on pp.308, 508).

[464] Kenkō Y (1330–1332) *Essays in idleness—the Tzurezure Gusa of Yoshida Kenkō.* Orig. publ. 1911: Kenkō [465]. Transl. from Japanese by Keene D (1967), Columbia Univ. Press. 2nd edn. (1998) pp.xxii–xxiii, (cited on pp.154, 527).

[465] Kenkō Y & Sansom GB (1911) *The Tsuredzure Gusa of Yoshida No Kaneyoshi: being the meditations of a recluse in the 14th century.* Transactions of the Asiatic Society of Japan, 39, Shanghai: Kelly & Walsh, (and see Kenko [464]), (cited on pp.154, 527).

[466] Kennedy DP & Squire LR (2007) *An analysis of calendar performance in two autistic calendar savants.* Learning & Memory (Cold Spring Harbor Laboratory Press), 14, ('10%' is reported in intro.) p.533, doi:10.1101/lm.653607 (cited on p.320).

[467] Kepler J (1611) *Strena Seu De Nive Sexangula.* Frankfort-on-Main: Godfrey Tampach, Transl. from Latin by Hardie C (1906) as *The six cornered snowflake.* Oxford: Clarendon 1966, p.21, (cited on pp.22, 22, 51, 52, 303, 423).

[468] ———. (1619) *Harmonices Mundi*. Frankfurt. English transl. Aiton EJ, Duncan AM & Field JV, *The Harmony of the World*. Am. Phil. Soc. (1997) Book II, p.106. (cited on pp.22, 578).

[469] Khan L (2013) *The 216 letter hidden Name of God—revealed*. CreateSpace, Amazon Group, *The Fibonacci 60 digit repeat cycle*, pp.14–23, https://www.youtube.com/watch?v=4u_a-VTmqSU accessed 20Jun2015, (cited on p.248).

[470] Khan SA (2007) *Arab origins of the discovery of the refraction of light*. OSA OPN Oct2007. Washington DC: The Optical Society of America, p.22, http://www.osa-opn.org/Content/ViewFile.aspx?id=10890 accessed 28Aug2014, (cited on p.362).

[471] Kim JY (2008) *Magnetic properties of Ti-Zr-Ni quasicrystals*. Jnl. Korean Phys. Soc. 53:3, Sep2008, p.1593, doi:10.3938/jkps.53.1593 (cited on p.340).

[472] Kim S-H & Das MP (2013) *Artificial seismic shadow zone by acoustic metamaterials*. Mod. Phys. Lett. B 27:20, 1350140, doi:10.1142/S0217984913501406 (cited on p.382).

[473] King M (2009) *The American cinema of excess*. Jefferson NC: McFarland, [but] *possible that term* Asperger's *has gained a popular currency beyond its clinical definition*, p.181, (cited on p.106).

[474] Kircher A (1665) *Arithmologia, sive De Abditis Numerorum Mysteriis*. Frontispiece (detail), the Bavarian State Library, http://web.stanford.edu/group/kircher/cgi-bin/site/?attachment_id=619 accessed 2Aug2013, (cited on pp.184, 583).

[475] Kiš Žuvela S (2011) *The Golden Section as a source of consistency in 20th century music*. Summary of MSc thesis, Academy of Music, Univ. Zagreb, Sažeci Magistarskih I Doktorskih Radova, ARMUD6 42/2, p.276, http://www.academia.edu/3569720/The_Golden_Section_as_a_Source_of_Consistency_in_20th_Century_Music accessed 27Apr2014, (cited on p.130).

[476] —Ibid. *dynamic symmetry: ABC and C'AB described as 'primary' and 'secondary' Golden Sections*, p.277, (cited on p.141).

[477] —Ibid. *[Golden Section] appearing in the works of [composers]*, pp.276–277, (cited on p.62).

[478] Klein F (1884) *Vorlesungen über das Ikosaeder und die Auflösung der Gleichungen vom fünften Grade.* Leipzig: Teubner, (see Klein [479]), (cited on pp.28, 529).

[479] ———. (1884) *Lectures on the icosahedron.* Transl. from the German: Klein [478]. Phoenix 2nd revised edition (2003). New York: Dover Publications, (cited on pp.28, 529).

[480] Knott R (since 1996) *Fibonacci numbers and the Golden Section.* Dedicated website hosted by Maths Dept., Univ. Surrey, Guildford, UK, http://www.maths.surrey.ac.uk/hosted-sites/R.Knott/Fibonacci/fib.html accessed 22Apr2013, (cited on pp.28, 34, 491, 520, 531, 539, 555).

[481] —Ibid. *Binet's formula for non-integer values of n?* http://www.maths.surrey.ac.uk/hosted-sites/R.Knott/Fibonacci/fibFormula.html#binetReal accessed 20Nov2013, (cited on pp.133, 495, 574).

[482] —Ibid. $F_{-n} = (-1)^{n+1}F_n$, *Definitions & notation, extending Fibonacci series 'backwards',* Knott cites: Vajda-2 [911], Dunlap-5 [188] http://www.maths.surrey.ac.uk/hosted-sites/R.Knott/Fibonacci/fibFormulae.html#defs accessed 26May2013, (cited on pp.29, 195, 482).

[483] —Ibid. $L_{-n} = (-1)^n L_n$, *Definitions & notation, extending Lucas series 'backwards',* Knott cites: Vajda-4 [912], Dunlap-6 [188] http://www.maths.surrey.ac.uk/hosted-sites/R.Knott/Fibonacci/fibFormulae.html#defs accessed 26May2013, (cited on pp.29, 481, 482).

[484] —Ibid. *Petals on flowers,* e.g. 3 petals: lily; 8 petals: delphiniums; 55 & 89 petals: Michaelmas daisies; http://www.maths.surrey.ac.uk/hosted-sites/R.Knott/Fibonacci/fibnat.html#petals accessed 15Aug2014, (cited on p.51).

[485] —Ibid. $F_{n+2} + F_{n-2} = 3F_n$, *Linear sums of Fibonacci numbers,* Knott cites: B&Q 2003-Identity 7 [50], http://www.maths.surrey.ac.uk/hosted-sites/R.Knott/Fibonacci/fibFormulae.html#linear2fib accessed 26Feb2014, (cited on pp.173, 482).

[486] —Ibid. $L_{n+1} + L_{n-1} = 5F_n$, *Linear sums of Lucas numbers,* Knott cites: Vajda-5 [914], Dunlap-13 [188], Koshy-5.16 [506], B&Q 2003-Identity 34 [50], Hoggatt-I9 [371], http://www.maths.surrey.ac.uk/hosted-sites/R.Knott/Fibonacci/fibFormulae.html#linear2luc accessed 26Feb2014, (cited on pp.173, 482).

[487] —Ibid. $F_n + 2F_{n-1} = L_n$, *Linear Sums of Fibonacci numbers,* Knott cites: Dunlap-32 [188], http://www.maths.surrey.ac.uk/hosted-sites/ R.Knott/Fibonacci/fibFormulae.html#linear2fib accessed 27Jan2014, (cited on pp.119, 482).

[488] —Ibid. $F_{n+2} + F_n = L_{n+1}$, *Linear sums of Fibonacci numbers,* Knott cites: Vajda-6 [915], Hoggatt-I8 [371], Dunlap-14 [188], Koshy-5.14 [506], http://www.maths.surrey.ac.uk/hosted-sites/R.Knott/ Fibonacci/fibFormulae.html#linear2fib accessed 26May2013, (cited on pp.147, 173, 173, 292, 351, 437, 476, 482).

[489] —Ibid. $F_{n+1}F_{n-1} - F_n^2 = (-1)^n$, *Cassini's Formula,* [Kepler 1619], Cassini 1680, Simson 1753, Knott cites: Vajda-29, Hoggatt-I13, Dunlap-9, B&Q-Id.8, http://www.maths.surrey.ac.uk/hosted-sites/R. Knott/Fibonacci/fibFormulae.html#order2Fib accessed 31Jul2014, (cited on pp.24, 481).

[490] —Ibid. $\Sigma F_n = F_{n+2} - 1$, *Fibonacci and Lucas summations,* Knott cites: Hoggatt-I1 [371], Lucas(1878) [595], B&Q 2003-Identity 1 [50], http://www.maths.surrey.ac.uk/hosted-sites/R.Knott/Fibonacci/ fibFormulae.html#FLsummations accessed 26May2013, (cited on pp.171, 414, 416, 433, 435, 482).

[491] —Ibid. $\phi^n = F_n\phi + F_{n-1}$, *Golden Ratio with Fibonacci and Lucas, Phi power Fibonacci linear form,* Knott cites: Rabinowitz-28 [771], B&Q(2003)-Corollary 33 [50], http://www.maths.surrey.ac. uk/hosted-sites/R.Knott/Fibonacci/fibFormulae.html#gsFibLuc accessed 26May2013, (cited on pp.181, 195, 196, 197, 254, 353, 433, 434, 435, 477, 482).

[492] —Ibid. $F_{n+m} = F_mF_{n+1} + F_{m-1}F_n$, *Order 2 Formulae, Fibonacci,* Knott notes: an alternative to Dunlap-10, B&Q(2003)-Identity 3—due to Hansen [323], http://www.maths.surrey.ac.uk/hosted-sites/R. Knott/Fibonacci/fibFormulae.html#order2Fib accessed 15Jul2014, (cited on p.516).

[493] —Ibid. *The Silver Means,* (numbers that are each a whole number greater than their reciprocal), http://www.maths.surrey.ac.uk/ hosted-sites/R.Knott/Fibonacci/cfINTRO.html#silver accessed 16Sep2014, (cited on pp.266, 394, 424, 470, 470, 472, 479).

[494] —Ibid. *Phi's fascinating figures: numerical relationships between Phi and its powers,* http://www.maths.surrey.ac.uk/hosted-sites/ R.Knott/Fibonacci/propsOfPhi.html#numprops accessed 6Jun2013, (cited on p.425).

[495] —Ibid. $y = \phi x$, *Phi line graph.* http://www.maths.surrey.ac.uk/ hosted-sites/R.Knott/Fibonacci/phi.html#philine accessed 30Jul2013, (cited on pp.423, 590).

[496] —Ibid. *Pine cone illustrations,* right-hand side shows (5,8) parastichies, http://www.maths.surrey.ac.uk/hosted-sites/R.Knott/Fibonacci/fibnat.html#pinecones accessed 5Jan2014, (cited on p.154).

[497] —Ibid. *Bartók, Debussy, Schubert, Bach and Satie,* http://www.maths.surrey.ac.uk/hosted-sites/R.Knott/Fibonacci/fibInArt.html#bartokDebussy accessed 22Dec2013, (cited on pp.28, 62, 532, 538).

[498] Knuth DE (1968) *The art of computer programming. Volume 1.* London: Addison Wesley. Fundamental algorithms, p.100 and § 1.2.8. (cited on pp.439, 483).

[499] —Ibid. *Lucas proving $2^{127} - 1$ is prime,* p.79, (cited on p.28).

[500] ———. (2006) *The art of computer programming. Volume 4.* London: Addison Wesley. (Fascicle 4), *Indian prosody,* § 7.2.1.7 p.49, (cited on p.13).

[501] —Ibid. *ya mā tā rā ja bhā na sa la gām—earliest known example of a ... de Bruijn cycle,* § 7.2.1.7, p.51, (also see Joseph [454]), (cited on pp.13, 526).

[502] ———. (1968) *The art of computer programming. Vol. 2.* 2nd Edn. Reading MA: Addison Wesley. Seminumerical algorithms, *Lamé's theorem,* § 4.5.3, theorem F, p.343, (cited on p.26).

[503] Kohmoto M, Sutherland B, & Iguchi K (1987) *Localization in optics: quasiperiodic media.* Phys. Rev. Lett. 58, pp.2436–2438, doi:10.1103/PhysRevLett.58.2436 (cited on pp.389, 390, 589).

[504] Kohmoto M, Sutherland W, & Tang C (1987) *Critical wave functions and a Cantor-set spectrum of a one-dimensional quasicrystal model.* Phys. Rev. B, 35, p.1020, doi:10.1103/PhysRevB.35.1020 (cited on pp.561, 562).

[505] Koppens F (2012) *Graphene nano-electronics.* Slide presentation. ICFO, The Institute of Photonic Sciences, Barcelona, p.9 http://www.graphene-flagship.eu/GF/Files/Workshop_Graphene_Presentation_Koppens.pdf accessed 4Oct2013, (cited on p.379).

[506] Koshy T (2001) *Fibonacci and Lucas numbers with applications.* Wiley-Interscience. Cited by Knott [480], (cited on pp.529, 530).

[507] ———. (2002) *Elementary Number Theory with Applications.* San Diego, CA: Harcourt/Academic Press, *Binet matrix multiplication rule & Lamé's rediscovery,* p.129, (cited on pp.26, 485).

[508] Kothalkar A, Sharma AS, Tripathi G, Basu B, & Biswas K (2012) *HDPE-quasicrystal composite: fabrication and wear resistance.* Trans. Indian Inst. Met. Feb2012, 65:1, pp.13–20, doi:10.1007/s12666-012-0120-2 (cited on pp.339, 339, 339, 339, 340, 340).

[509] Kramer J (1973) *The Fibonacci series in twentieth-century music.* Journal of Music Theory, 17:1, pp.110–148, cited by Knott [497] doi:10.2307/843120 (cited on p.62).

[510] —Ibid. *'avoids periodicity … yet it is well ordered'.* Theoretical Aspects, p.114, (cited on p.65).

[511] Kramer P & Neri R (1984) *On periodic and non-periodic space fillings of E^m obtained by projection.* Acta Cryst. Sep1984 A40, pp.580–587, doi:10.1107/S0108767384001203 (cited on p.387).

[512] Kremsmünster Observatory (1758–) (owner of) *Portrait of Johannes Kepler.* by unknown artist. Image reproduced with kind permission of Sternwarte Kremsmünster, Austria. http://www.specula.at/english.htm accessed 7Nov2013, (cited on pp.22, 578).

[513] Krishan M (2007) *Review of Symmetry, Causality, Mind.* Cognitive Systems Research 8, p.131, refers to Leyton [550], (cited on p.108).

[514] Kroto HW, Heath JR, O'Brien SC, Curl RF, & Smalley RE (1985) C_{60}: *Buckminsterfullerene.* Nature, Nov1985 318, pp.162–163, doi:10.1038/318162a0 (cited on p.305).

[515] Kruk S, Helgert C, Decker M, Staude I, Menzel C, Etrich C, Rockstuhl C, Jagadish C, Pertsch T, Neshev DN, & Kivshar Y (2012) *Quasicrystal metamaterials: a route to optical isotropy.* Asia Communications and Photonics Conference, Guangzhou China, 7–10Nov2013, (cited on pp.383, 392).

[516] ———. (2013) *Optical metamaterials with quasicrystalline symmetry: Symmetry-induced optical isotropy.* Phys. Rev. B, 88, 201404(R), doi:10.1103/PhysRevB.88.201404 (cited on pp.383, 392).

[517] Kubrick S (1968) *2001 A space odyssey.* Film with screenplay by Arthur C Clark and Stanley Kubrick. Kubrick directed and produced. Metro-Goldwyn-Mayer, (cited on pp.94, 124).

[518] Kuo KH (2002) *Mackay, anti-Mackay, double-Mackay, and related icosahedral shell clusters.* Structural Chemistry, Aug2002 13:3–4, pp.221–230, doi:10.1023/A:1015847520094 (cited on pp.300, 540).

[519] Kushwaha MS, Halevi P, Martinez G, Dobrzynski L, & Djafari-Rouhani B (1993) *Acoustic band structure of periodic elastic composites.* Phys. Rev. Lett. 71:13, p.2022. Cited by Chen et al. [112], doi:10.1103/PhysRevLett.71.2022 (cited on p.380).

[520] Kymeta (2014) *Kymeta's flat antennas*: Metamaterials-based antennas for satellite comms. Redmond WA: Kymeta Corp., http://www.kymetacorp.com accessed 30Jan2015, (cited on p.373).

[521] Lagarias JC (1985) *The set of primes dividing the Lucas numbers has density 2/3.* Pacific Jnl. Maths 118:2, pp.449–461, (cited by Schiffman [804] pp.19–20), doi:10.2140/pjm.1985.118.449 (cited on p.174).

[522] Lagrange JL (1877) *Oeuvres de LaGrange.* Paris: Gauthiers Villars, pp.5–182, (cited on p.248).

[523] Larsen K (2005) *Stephen Hawking: a biography.* Westport CT: Greenwod Press, p.43, (cited on p.xi).

[524] Lauwerier H (1991) *Fractals: endlessly repeated geometric figures.* Transl. from the Dutch by Sophia Gill-Hoffstädt. Princeton, NJ: Princeton University Press, *Julia Sets,* p.124, (cited on p.95).

[525] Lawlor R (1982) *Sacred geometry, philosophy and practice.* London: Thames and Hudson, *Series C^1 = Series C ×2,* p.57, (cited on pp.187, 445).

[526] —Ibid. *Gnomonic expansion,* p.65, (cited on p.103).

[527] Ledermann A, Wegener M, & Freymann G von (2010) *Rhombicuboctahedral three-dimensional photonic quasi-crystals.* Advanced Materials 22:21, p.2363, 4Jun2010, cited by Wegener [939], doi:10.1002/adma.200903885 (cited on pp.334, 388).

[528] Lefaix H, Prima F, Zanna S, Vermaut P, Dubot P, Marcus P, Janičkovič J, & Švec P (2007) *Surface properties of a nano-quasicrystalline forming Ti based system.* Materials Transactions, 48:3, p.285, http://www.jim.or.jp/journal/e/pdf3/48/03/278.pdf accessed 5Dec2013, (cited on p.340).

[529] Lehmer DH (1930) *An extended theory of Lucas' functions.* Ann. Math., 31:3, pp.419–448. Reprinted in (1981) *Selected Papers.* Ed. McCarthy D, v.1, Ch. Babbage Res. Center, St. Pierre, Manitoba Canada, pp.11–48, cited by Caldwell [96], http://www.jstor.org/stable/1968235 accessed 3Sep2014, (cited on p.448).

[530] ———. (1935) *On Lucas's test for the primality of Mersenne's numbers.* J. London Math. Soc., 10, pp.162–165, cited by Caldwell [96], (cited on p.448).

[531] Lehrer J (2009) *Unlocking the mysteries of the artistic mind.* Psychology Today 1Jul2009, sub-heading: *Giving the mind a break / Semir Zeki,* http://www.psychologytoday.com/articles/200907/unlocking-the-mysteries-the-artistic-mind accessed 21May2014, (cited on p.108).

[532] Lei Wang & Baowen Li (2008) *Phononics gets hot.* Physicsworld.com, March2008, p.27 et seq., http://physicsworldarchive.iop.org/full/pwa-pdf/21/03/phwv21i03a31.pdf accessed 5Dec2013, (cited on pp.392, 566).

[533] Leighton T (2010) *Linear Recurrences.* Lecture 15, MIT 6.042J Mathematics for Computer Science, MIT OpenCourseWare video 1h:18m:19s. Massachusetts Institute of Technology, *any linear combination,* at 0h:17m:10s, https://www.youtube.com/watch?v=TWBB-JlmYUc accessed 23Aug2014, (cited on p.445).

[534] Lendvai E (1955) *Bartók stílusa.* Transl. from the Hungarian by Paul Merrick & Judit Pokoly (1999) as *Bartók's style.* Budapest: Akkord Music Publishers, *'struck by spiritual lightning'* in section *About the author,* (cited on p.63).

[535] ———. (1971) *Béla Bartók: an analysis of his music.* Stanmore Press, London: Kahn & Averill, *sunflowers & fir-cones,* p.29, (cited on pp.62, 490).

[536] —Ibid. *Fibonacci numbers chapter,* pp.27–34, (cited on p.141).

[537] —Ibid. *Fibonacci structure (1st movement MFSPC),* Fig. 22, p.28, (cited on pp.63, 64, 579).

[538] —Ibid. *positive and negative—types of Golden Section,* pp.20–21, (cited on p.141).

[539] —Ibid. *('Root Golden') spiral,* Fig. 25, p.31, (cited on pp.140, 141).

[540] ———. (1976) *Duality and synthesis in the music of Béla Bartók.* In *Bartók Studies.* Reprinted from The New Hungarian quarterly, 1961–1973, ed. Todd Crow, Detroit: Information Coordinators, p.39, cited by Simons HA in *Béla Bartók's sonata for two pianos and percussion.* University of Alberta doctoral essay (2000), front matter quote, p.iii & p.101, http://www.nlc-bnc.ca/obj/s4/f2/dsk2/ftp02/NQ59918.pdf accessed 6Jan2014, (cited on pp.64, 138).

[541] Leonardo da Vinci (1503) *The notebooks of Leonardo Da Vinci, complete.* The Project Gutenberg EBook #3, para.415, ftp://ftp.ibiblio.org/pub/docs/books/gutenberg/etext04/7ldvc10.txt accessed 2Sep2013, (cited on p.52).

[542] ———. (c.1512) *(Presumed) self-portrait.* Red chalk on paper, 33.3 × 21.3 cm. Turin: Biblioteca Reale, (cited on pp.21, 578).

[543] Leopold AC (1955) *Auxins and Plant Growth.* Oakland CA: University of California Press, p.202, (cited on p.289).

[544] Lesnie M (2012) *Five things you didn't know about Erik Satie.* Article, Limelight Magazine—classical music & arts website, 13Jun2012, http://www.limelightmagazine.com.au/Article/304597,five-things-you-didnt-know-about-erik-satie.aspx/0 accessed 27Feb2014, (cited on p.152).

[545] Lesure F (1977) *Debussy on music: the critical writings of the great French composer Claude Debussy.* Transl. from the French by Langham-Smith R, New York: AA Knopf; reprinted (1988) New York: Cornell University Press, pp.277–278, (cited on p.502).

[546] Lesure F & Nichols R (1987) *Debussy letters.* English edn. transl. Nichols R, London: Faber and Faber. Letter to his publisher (Jacques Durand) in 1903 about proofs for *Estampes III,* page 8 of *Jardins sous la pluie*: *'bar missing ... necessary [regarding] ... the divine number',* § 108, p.137, (cited on p.150).

[547] Lethbridge PG & Stace AJ (1989) *An investigation of the properties of large krypton cluster ions (development of the P=3 and P=4 Mackay icosahedral shells).* Journal of Chemical Physics, 91:12, found 'intense ion peaks for Kr_{147}^+ and Kr_{309}^+,' p.7685, cited by Baggott [24], doi:10.1063/1.457237 (cited on p.303).

[548] Levine D & Steinhardt PJ (1984) *Quasicrystals: a new class of ordered structures.* Phys. Rev. Lett. 53:26, pp.2477–2480, doi:10.1103/physrevlett.53.2477 (cited on pp.333, 333, 383).

[549] Levitov LS (1991) *Phyllotaxis of flux lattices in layered superconductors.* Phys. Rev. Lett. 66:2, 14Jan1991, pp.224–227, doi:10.1103/PhysRevLett.66.224 (cited on pp.293, 295).

[550] Leyton M (1992) *Symmetry, causality, mind.* Cambridge MA: MIT Press, *Platonic solids—no history, 'timeless',* p.35, (cited on pp.109, 532).

[551] —Ibid. *shape is time,* p.73, (cited on p.109).

[552] —Ibid. *energy transfer as cause,* p.74, (cited on pp.109, 111).

[553] —Ibid. *purpose of perception,* p.91, (cited on p.109).

[554] —Ibid. *representation is explanation,* p.157, (cited on p.108).

[555] —Ibid. *cognitive system ... creates and manipulates causal explanations,* p.157, (cited on p.109).

[556] Lewotsky K (2007) *The promise of plasmonics.* SPIE Professional 30July2007, *Surface-enhanced Raman spectroscopy (SERS),* doi:10.1117/2.4200707.07 (cited on p.378).

[557] Lhote A (1935) *Composition du tableau.* Paris: Encyclopédie Française, p.16.30-6-30-12, Fig.2, pp.6–7, cited by Herz-Fischler [363], (cited on p.84).

[558] Life Magazine (1943) *Life presents R Buckminster Fuller's Dymaxion World.* Photographic essay, 1Mar1943, pp.41–55, (cited on p.303).

[559] Lifshitz R (2009) *Nanotechnology and quasicrystals: from self-assembly to photonic applications.* In *Silicon versus carbon.* NATO Science for Peace and Security Series B: Physics and Biophysics 2009, p.121, (§2.) doi:10.1007/978-90-481-2523-4_10 (cited on pp.342, 514, 576).

[560] —Ibid. *new applications based on self-assembled nano-materials,* p.123, (cited on p.341).

[561] Lisi AG & Weatherall JO (2014) *A geometric Theory of Everything.* Scientific American, Aug2014, 23, pp.96–103, doi:10.1038/scientificamericanuniverse0814-96 (cited on p.298).

[562] Liverpool University (2014) *International Year of Crystallography 2014 Symposium.* Chair: Prof. Ronan McGrath, 17Jul2014, Victoria Gallery & Museum, http://www.liv.ac.uk/events/science-and-society/international-year-of-crystallography.php accessed 17Jul2014, (cited on p.331).

[563] Livio M (2003) *The Golden Ratio: the story of Phi, the world's most astonishing number.* New York: Broadway Books, *Naqadah, Egyptian pentagrams excavated,* p.44, (cited on pp.14, 28).

[564] —Ibid. *(Jeanneret/Le Corbusier's attitude to Golden Ratio—'originally ... skeptical, even negative')*, p.172, (cited on pp.71, 72, 151).

[565] —Ibid. *Kepler: two great treasures of geometry*, p.62, (cited on p.23).

[566] —Ibid. *[any] sum of 10 consecutive Fibonacci numbers ... evenly divisible by 11*, p.104, (cited on p.174).

[567] —Ibid. *Debussy associated with the group ... friend Verkade's 'measures' included Golden Ratio*, pp.168–169, (cited on p.71).

[568] —Ibid. *musicologists Tatlow & Griffiths*, p.190, (cited on p.64).

[569] —Ibid. *Parthenon*, p.72,

[570] —Ibid. *Kepler's independent discovery of Fibonacci numbers*, p.101, (cited on p.22).

[571] —Ibid. *Theœtetus first to construct Platonic solids*, p.67, (cited on p.14, 14).

[572] —Ibid. *Uruk, Sumerian pentagrams excavated*, p.43, (cited on pp.14, 28).

[573] Lokhorst G-J (2013) *Descartes and the pineal gland.* The Stanford Encyclopedia of Philosophy, Edward N Zalta (ed.), http://plato.stanford.edu/archives/fall2013/entries/pineal-gland/ accessed 7Jan2014, (cited on p.154).

[574] Lokki T, Southern A, Siltanen S, & Savioja L (2013) *Acoustics of Epidaurus—studies with room acoustics modelling methods.* Acta Acustica united with Acustica, 99, pp.40–47, doi:10.3813/AAA.918586 (cited on p.16).

[575] Lomas D (2006) *'Painting is dead—long live painting': Notes on Dalí and Leonardo.* Papers of Surrealism, Issue 4, Spring 2006, p.12, http://www.surrealismcentre.ac.uk/papersofsurrealism/journal4/acrobat%20files/Lomaspdf.pdf accessed 23Apr2014, (cited on pp.71, 72).

[576] Lord EA & Rangatham S (2001) *The Gummelt decagon as a 'quasi unit cell'.* Acta Cryst. A, *Foundations of Crystallography*, 57:5, pp.531–539, doi:10.1107/S0108767301007504 (cited on p.337).

[577] Lowman EA (1971) *An example of Fibonacci numbers used to generate rhythmic values in modern music.* Fib. Qtrly. 9:4, p.436, http://www.fq.math.ca/Scanned/9-4/lowman-a.pdf accessed 19Dec2013, http://www.fq.math.ca/Scanned/9-4/lowman-b.pdf accessed 19Dec2013, (cited on p.62).

[578] ———. (1971) *Some striking proportions in the music of Béla Bartók.* Fibonacci Qtrly. 9:5, p.527–528, p.536–537, cited by Knott [497], http://www.fq.math.ca/Scanned/9-5/lowman-a.pdf accessed 19Dec2013, http://www.fq.math.ca/Scanned/9-5/lowman-b.pdf accessed 19Dec2013, (cited on pp.62, 64, 579).

[579] Lu P (2007) *Quasicrystals in medieval Islamic architecture.* Harvard University Physics Dept. colloquium 3Dec2007, video 1h:05m:20s, http://www.youtube.com/watch?v=rldnu9rNpH8 accessed 25May2014, (cited on pp.20, 346).

[580] —Ibid. *girih tiles subdivision rules,* at 0h:27m:20s, (cited on p.347).

[581] —Ibid. *mapping girih tiles to Penrose patterns,* at 0h:39m:30s, (cited on p.20).

[582] —Ibid. *linear algebra approach to determining quasi-crystallinity,* at 0h:43m:40s, (cited on pp.347, 348).

[583] —Ibid. *counting tiles ... hexagon subdivision rule,* at 0h:47m:00s, (cited on pp.348, 348, 350).

[584] —Ibid. *subdivision transformation matrix* 3 × 3 ... *calculate eigenvalues and eigenvectors,* at 0h:48m:15s, (cited on p.350).

[585] —Ibid. *subdivision transformation matrix* 2 × 2 ... *with Fibonacci elements,* at 0h:50m:45s, (cited on p.351, 351).

[586] —Ibid. *ratio of bowties to hexagons found to be* ϕ *in the limit,* at 0h:51m:35s, (cited on p.353).

[587] Lu PJ & Steinhardt, PJ (2007) *Decagonal and quasi-crystalline tilings in medieval Islamic architecture.* Science New Series, American Association for the Advancement of Science, 315:5815, 23Feb2007, pp.1106–1110, Būzjānī note 7. doi:10.1126/science.1135491, see also supplement [590], http://www.jstor.org/stable/20039057 accessed 1Jun2013, (cited on pp.20, 346, 346, 578).

[588] —Ibid. *signature of quasiperiodicity,* p.1108, (cited on p.453).

[589] —Ibid. *girih tile subdivision,* Fig.3 (D & E), p.1109, (cited on pp.347, 349, 588).

[590] ———. (2007) *Supporting online material for decagonal and quasi-crystalline tilings in medieval Islamic architecture.*

Supplementary Fig. S4B, *Panel 50 from the Topkapı scroll,* http: //www.sciencemag.org/content/suppl/2007/02/20/315.5815.1106.DC1/Lu.SOM.pdf accessed 19Jun2014, (cited on pp.346, 346, 538, 588).

[591] Luca F (2000) *'Perfect Fibonacci and Lucas numbers' (not)—* an article showing their non-existence. Rendiconti del Circolo Matematico di Palermo, May2000, 49:2, pp.313–318, doi:10.1007/BF02904236 (cited on p.280).

[592] Lucas E (1876) *Comptes Rendus.* Paris, 82, pp.1303–1305, cited by Dickson [175] p.396, note 19, (cited on pp.69, 149, 252, 255, 256, 259, 272, 282, 439, 483, 484, 484).

[593] —Ibid. *Primality of Mersenne M_{127},* Th. XI, p.167, (cited on p.448).

[594] ———. (1877) *Bollettino di Bibliogr. e Storia dei Sci. Matem. e Fis.* x, p.129, (cited by Thompson [893] p.923), (cited on p.27).

[595] ———. (1877) *Sur la théory des fonctions numériques simplement périodiques.* Nouv. Corresp. Math., **3**, pp.369–376, pp.401–7, cited by Dickson [175] p.399, note 32, and cited by Knott [480], see also AJM [596] & English translation [598], (cited on pp.149, 252, 530, 557).

[596] ———. (1877) *Sur la théory des fonctions numériques simplement périodiques.* (1878) Reprint in *American Journal of Mathematics.* Johns Hopkins University, 1, pp.184–240, pp.289–321, *Girard isosceles,* p.186, http://www.forgottenbooks.org/books/ American_Journal_of_Mathematics_1878_v1_1000081216 accessed 13Jul2013, (cited on pp.23, 539).

[597] —Ibid. *Lamé's theorem of 1844,* p.187, (cited on p.26).

[598] ———. (1877) *The theory of simply periodic functions.* English language reprint (1969) transl. by Kravitz S, The Fibonacci Association, available in full online: http://www.fq.math.ca/Books/ Complete/simply-periodic.pdf accessed 13Jul2013, (cited on p.539).

[599] ———. (1882–1894) *Récréations mathématiques (4 volumes).* Paris: Gauthier-Villars, (cited on p.28).

[600] Lück R (2000) *Dürer—Kepler—Penrose, the development of pentagon tilings.* Materials Science and Engineering: A 294–296, 15Dec2000, pp.263–267, doi:10.1016/S0921-5093(00)01302-2 (cited on pp.23, 504).

[601] Lundeen JS & Steinberg AM (2009) *Experimental joint weak measurement on a photon pair as a probe of Hardy's paradox.* Phys. Rev. Lett. 16Jan2009 102:020404, doi:10.1103/PhysRevLett.102.020404 (cited on p.298).

[602] Lynn J (2013) *Flux lattice in superconductors and melting.* Web article, NIST Center for Neutron Research, http://www.ncnr.nist.gov/staff/jeff/flux_lattice_in_superconductors.html accessed 8Nov2013, (cited on p.293).

[603] MacDonald AH (1988) *Fibonacci Superlattices.* Interfaces, Quantum Wells, and Superlattices. NATO ASI Series 179, pp.347–378, Note 1: *'For a review of work to 1982 see Simon, Adv. Appl. Math. 3, 463 (1982).',* doi:10.1007/978-1-4613-1045-7_19 (cited on p.383).

[604] —Ibid. *Merlin et al. realized ... [Levine & Steinhardt model] could be fabricated in semiconductor multi-layer systems,* p.347, (cited on p.383).

[605] Maciel D (2010) *Puerta de la Luna—Tiahuanaco (Bolivia).* Photo of the Gate of the Moon at Tiwanaku, http://en.wikipedia.org/wiki/File:Puerta_de_la_Luna_-_Tiahuanaco_(Bolivia).jpg accessed 7May2014, (cited on pp.9, 11, 119, 273, 577).

[606] Mackay AL (1962) *A dense non-crystallographic packing of equal spheres.* Acta Cryst. 15, pp.916–918, cited by Kuo [518], doi:10.1107/S0365110X6200239X (cited on p.303).

[607] ———. (1976) *Crystal Symmetry.* Phys. Bull., Nov1976, p.495 & 496, (cited on p.302, 302).

[608] ———. (1981) *De Nive Quinquangula.* Krystallografiya, (Russian), Sep–Oct1981, 26:5, pp.910-919, (cited on pp.303, 304, 324, 541).

[609] ———. (1981) *De Nive Quinquangula: on the pentagonal snowflake.* Soviet Physics Crystallography, (English text ©1982 American Inst. Physics), 26, pp.517–522, (cited on pp.303, 304, 324, 541).

[610] —Ibid. *semiregular quasilattice,* p.519, (cited on p.324).

[611] —Ibid. *might go unrecognized,* p.522, (cited on pp.324, 330).

[612] ———. (1982) *Crystallography and the Penrose pattern.* Physica A, Jan1982, 114:1–3, pp.609–613, doi:10.1016/0378-4371(82)90359-4 (cited on pp.324, 325, 586).

[613] ———. (1984) *Quaternion transformation of molecular orientation.* Acta Cryst. A40, pp.165–166, (cited on p.361)., doi:10.1107/S0108767384000362 (cited on p.361).

[614] ———. (1986) *Science of form: proceedings of the first international symposium for science on form.* Gen. Ed. Ishizaka S, Eds. Kato Y, Takaki R, & Toriwaki J, Tokyo: KTK Scientific Publishers, *'quasi-lattice' defined in Mackay 1981, Russian [608] & English [609],* p.617, http://www.scipress.org/e-library/sof/pdf/0615.PDF accessed 23May2014, (cited on p.324).

[615] Maiman TH (1960) *Stimulated optical radiation in ruby.* Nature 187, pp.493–494, doi:10.1038/187493a0 (cited on p.364).

[616] Maini L (2010) *Metamaterials based on photonic quasicrystals: from superlensing to new photonic devices.* Tesi di dottorato, Università degli Studi di Milano-Bicocca, http://boa.unimib.it/bitstream/10281/18116/3/phd_unimib_033574.pdf accessed 24Nov2013, (cited on p.392).

[617] Makovicky E (1992) *800 year old pentagonal tiling from Marāgha, Iran, and the new varieties of aperiodic tiling it inspired.* In (1992) *Fivefold symmetry.* Ed. Hargittai I, Singapore: World Scientific, pp.67–86. *Conclusions,* p.85, (cited on p.344).

[618] —Ibid. *principal tiles and tile patches (examples of what are now called 'girih'),* Fig.3, p.72, (cited on p.346).

[619] Maldovan M (2013) *Sound and heat revolutions in phononics.* Nature 14Nov2013, 503, p.210, doi:10.1038/nature12608 (cited on p.381).

[620] —Ibid. *Thermal metamaterials,* p.214, (cited on p.381).

[621] Mancinelli C, Jenks CJ, Thiel PA, & Gellman JA (2003) *Tribological properties of a B2-type Al-Pd-Mn quasicrystal approximant.* J. Mater. Res. 18:6, Jun2003, p.1447, doi:10.1557/JMR.2003.0199 (cited on p.340).

[622] Mandelbrot BB (1967) *How long is the coast of Britain? Statistical self-similarity and fractional dimension.* Science, New Series, Vol.156:3775, 5May1967, pp.636–638, doi:10.1126/science.156.3775.636 (cited on pp.96, 230, 584).

[623] ———. (1982) *The fractal geometry of nature.* New York: WH Freeman, p.15, (cited on pp.96, 230, 304, 390, 556).

[624] —Ibid. *Koch fractal,* p.36, p.42, (cited on pp.233, 234).

[625] —Ibid. *Sierpiński fractal,* p.142, (cited on p.236).

[626] ———. (1990) *Negative fractal dimensions and multifractals.* North-Holland, Physica A 163, pp.306–315, doi:10.1016/0378-4371(90)90339-T http://users.math.yale.edu/mandelbrot/web_pdfs/123negativeFractalDimensions.pdf accessed 6Oct2013, (cited on p.242).

[627] Mandich ML (2006) *Noble gas clusters.* In *Springer handbook of atomic, molecular, and optical physics.* Ed. Drake GWF, New York: Springer Science+Business Media Inc., § 39.5, p.599, (cited on p.304).

[628] Marcia E (2012) *Quasicrystals and the quest for next generation thermoelectric materials.* Critical Reviews in Solid State and Materials Sciences, 37:4, p.215, doi:10.1080/10408436.2012.703978 (cited on p.340).

[629] Markowsky G (1992) *Misconceptions about the Golden Ratio.* The College Mathematics Journal, 23:1, Jan1992, p.8, doi:http://dx.doi.org/10.2307%2F2686193 http://www.math.nus.edu.sg/aslaksen/teaching/maa/markowsky.pdf accessed 8May2013,

[630] Marohnić L & Strmečki T (2012) *Plastic number: construction and applications.* Advanced Research in Scientific Areas (arsa-conf.com), pp.1523–1528, https://bib.irb.hr/datoteka/628836.Plastic_Number_-_Construct.pdf accessed 13Mar2015, (cited on p.94).

[631] Marshall DC, Odell E & Starbird M (2007) *Number theory through inquiry.* MAA Textbooks, Washington DC: Mathematical Association of America, p.40, (cited on p.282).

[632] Materialscientist (2010) *Electron diffraction pattern of an icosahedral Zn-Mg-Ho quasicrystal.* Photograph. http://en.wikipedia.org/wiki/File:Zn-Mg-HoDiffraction.JPG accessed 1Jun2013, (cited on pp.iv, 330, 355, 586, 587, 587, 588).

[633] Matile P (1990) *Albert Frey-Wyssling. 8 November 1900–30 August 1988.* Biogr. Mems Fell. R. Soc. 35:114–116, 1Mar1990, pp.1748–8494. Cited by Wikipedia, doi:10.1098/rsbm.1990.0005 (cited on p.290).

[634] Matiyasevich YV (1993) *Hilbert's 10th Problem.* Cambridge, Mass.; London: MIT Press, *Fibonacci numbers,* p.38, http://mitpress.mit.edu/books/hilberts-10th-problem accessed 2Sep2014, (cited on p.450).

[635] Matossian N (1986) *Xenakis.* London: Kahn & Averill, p.17, (cited on p.71).

[636] Matsui T, Agrawal A, Nahata A, & Vardeny ZV (2007) *Transmission resonances through aperiodic arrays of subwavelength apertures.* Nature 446 29Mar2007, pp.517–521, cited by Vardeny et al. [919], doi:10.1038/nature05620 (cited on p.391).

[637] Matsumoto M & Nishimura T (1998) *Mersenne Twister: a 623-dimensionally equidistributed uniform pseudo-random number generator.* ACM Transactions on Modeling and Computer Simulation (TOMACS), Jan1998, 8:1, pp.3–30, doi:10.1145/272991.272995 (cited on p.450).

[638] Maurice D (2004) *Bartók's viola concerto: the remarkable story of his swansong (studies in musical genesis and structure).* New York: Oxford University Press, p.88, (cited on p.62).

[639] Maxwell J Clerk (1865) *A dynamical theory of the electromagnetic field.* Phil. Trans. R. Soc. Lond. 155, Part III: *20 equations,* pp.480–488; *20 unknowns,* p.486, doi:10.1098/rstl.1865.0008 (cited on p.361).

[640] McManus IC (1980) *The aesthetics of simple figures.* British Journal of Psychology 71:4, Nov1980, pp.505–524, doi:10.1111/j.2044-8295.1980.tb01763.x (cited on pp.506, 575).

[641] Merlin R, Bajema K, Clarke R, Juany FY, & Battacharya PK (1985) *Quasiperiodic GaAs-AlAs heterostructures.* Phys. Rev. Lett. 55, 1768, doi:10.1103/PhysRevLett.55.1768 (cited on p.383).

[642] Mermin ND (1992) *Copernican crystallography.* Phys. Rev. Lett. 68, pp.1172–1175, (cited by Schewe & Stein, AIP Physics news update 67, 12Feb1992), doi:10.1103/PhysRevLett.68.1172 (cited on p.387).

[643] ———. (2004) *Could Feynman have said this?* Physics Today 57:5, doi:10.1063/1.1768652 (cited on p.297).

[644] Mhwater (2006) *The Gateway of the Sun from the Tiwanaku civilization in Bolivia.* http://en.wikipedia.org/wiki/File:Zonnepoort_tiwanaku.jpg accessed 7May2014, (cited on pp.9, 577).

[645] Miller AI (2001) *Einstein, Picasso: space, time and the beauty that causes havoc.* New York: Basic Books, p.4, (cited on p.71).

[646] —Ibid. *Picasso's admission and denial of use of mathematics,* p.87, & p.103, (cited on p.92).

[647] —Ibid. *Bateau-Lavoir Cubists,* p.100, (cited on p.70).

[648] —Ibid. *Princet/Derain,* p.102, (cited on p.70).

[649] Miller M (1973) *Gelöste und ungelöste mathematische Probleme.* Leipzig: Teubner Verlgsgesellschaft. *TL Heath proof of even perfect number > 6 is sum of cubes,* p.14f, cited by Jürgen Köller, http://www.mathematische-basteleien.de/cubenumber.htm accessed 15Sep2014, (cited on p.280).

[650] Milner J (1992) *Mondrian.* London: Phaidon, *[and] ... energy ... in action,* p.7, (cited on p.110).

[651] —Ibid. *Red ... earthbound and sensual ... blue associated with divinity,* p.85, (cited on p.112).

[652] Minamino R, Tateno M (2014) *Tree branching: Leonardo da Vinci's Rule versus biomechanical models.* PLoS ONE 9:4, e93535, doi:10.1371/journal.pone.0093535 (cited on p.95).

[653] Mizera K (2008) *Rayonnant north rose window of the Cathédrale Notre-Dame de Paris.* Image courtesy of Wikipedia, http:// en.wikipedia.org/wiki/File:Rozeta_Pary%C5%BC_notre-dame_chalger.jpg accessed 23Sep2013, (cited on pp.377, 589).

[654] Moivre, A de (1722) *De Fractionibus Algebraicis Radicalitate immunibus ad Fractiones Simpliciores reducendis, deque summandis Terminis quarumdam Serierum æquali Intervallo a se distantibus.* Philos. Trans., 32, Exemplum I, generating function $1/(1 - x - x^2)$, p.167, doi:10.1098/rstl.1722.0029 (cited on pp.25, 481).

[655] Mondrian PC (1918) *De nieuwe beelding.* In *De Stijl* 1:4, pp.42–43, (cited by Padovan [720] p.6 note 9), (cited on p.94).

[656] ——. (1926) *Tableau I: Lozenge with four lines and gray.* Oil on canvas, 113.7 × 111.8 cm (diagonal measurements), New York: MoMA, http://www.moma.org/collection/object.php?object_id=79059 accessed 27Apr2014, (cited on pp.92, 151).

[657] ——. (1930) *Composition with Red, Blue, and Yellow.* B217 in the *Catalogue Raisonné* [446], oil on canvas 46cm × 46cm. Zürich: Kunsthaus, (gift of Alfred Roth), also see Pompidou

Centre exhibition guide [758], http://www.snap-dragon.com/PM1930.html accessed 13Apr2014, (cited on pp.545, 554).

[658] ———. (1930) *Composition No.II / Composition I / Composition en Rouge, Bleu et Jaune.* B219 in the *Catalogue Raisonné* [447], very similar to B217 [657] also shown in Janson [409]. Oil on canvas 51cm × 51cm, http://www.snap-dragon.com/PM1930.html accessed 13Apr2014, (cited on p.523).

[659] ———. (1942) *A new realism.* Lecture given at Nierendorf Gallery (January 1942), New York, sponsored by American Abstract Artists, extract published in (1946) *Eleven Europeans in America.* Bulletin of MoMA, XII:4 & 5, New York: Museum of Modern Art, *Mondrian: 'The great struggle... ',* p.35, quoted by Padovan [726]. For the revised (1943) version see Mondrian [660], (cited on pp.108, 545, 551).

[660] ———. (1943) *A new realism.* Essay in (1945) *Plastic art and pure plastic art ... and other essays.* Ed. Holtzman H. Collected essays in English for memorial exhibition MoMA NY 20Mar–13May 1945. New York: Wittenborn, pp.16-26, revised version of lecture [659]: 'In plastic art, the static balance has to be transformed into the dynamic equilibrium which the universe reveals. It must be emphasized that it is important to discern two sorts of equilibrium: (1) a static balance and (2) a dynamic equilibrium. The first maintains the individual unity of particular forms, single or in plurality. The second is the unification of forms or elements of forms through continuous opposition. The first is limitation, the second is extension. Inevitably dynamic equilibrium destroys static balance.', p.25, (cited on p.545).

[661] Monnerot-Dumaine A (2009) *The Fibonacci word fractal.* (aka 'AMD fractal')—preprint paper. *Construction,* p.2, http://hal.archives-ouvertes.fr/hal-00367972/en accessed 7Sep2014, (cited on pp.238, 584).

[662] —Ibid. *derivation of Hausdorff dimension,* § 5.1, Eqn.17, p.9, (cited on p.239).

[663] —Ibid. *Koch-like variant, Hausdorff dimension expressed in continued fractions,* § 8.1, Eqn.41, p.22, and compare with § 5.1, Eqn.18, p.10, (cited on p.239).

[664] Morlock F (2007) *The very picture of a primal scene: Une leçon clinique à la Salpêtrière.* Visual Resources, 23:1–2, pp.129–146, doi:10.1080/01973760701219594 (cited on pp.82, 83).

[665] Mottron L, Dawson M, & Soulières I (2009) *Enhanced perception in savant syndrome: patterns, structure and creativity.* Phil. Trans. R. Soc. B, 27May2009, 364:1522, pp.1385–1391, doi:10.1098/rstb.2008.0333 (cited on p.320).

[666] Namavar F (2001) *Nanograin, quasicrystalline, multiphase coatings for reduced friction and wear.* Report for US Air Force Office of Scientific Research, Arlington, FR-60423, http://www.dtic. mil/dtic/tr/fulltext/u2/a395399.pdf accessed 16Sep2013, (cited on pp.339, 340).

[667] Narayana S & Sato Y (2012) *DC Magnetic Cloak.* Adv. Mater. 24, pp.71–74, doi:10.1002/adma.201104012 (cited on p.375).

[668] Nasibulin AG, Pikhitsa PV, Hua Jiang, Brown DP, Krasheninnikov AV, Anisimov AS, Queipo P, Moisala A, Gonzalez D, Lientschnig G, Hassanien A, Shandakov SD, Lolli G, Resasco DE, Mansoo Choi, Tománek D, & Kauppinen EI (2007) *A novel hybrid carbon material.* Nature Nanotechnology 2, pp.156–161, doi:10.1038/nnano.2007.37 (cited on p.305).

[669] NASA Jet Propulsion Laboratory (2014) *Cassini concept image.* Cassini-Huygens mission (arrived 2004). California Institute of Science and Technology, http://saturn.jpl.nasa.gov/multimedia/images/ artwork/images/IMG004628.jpg accessed 25Jul2015, (cited on p.24).

[670] National Bureau of Standards USA (1959) *Research highlights of the NBS.* US Dept. Commerce, Misc. Publication 229, p.13, (cited on p.486). (cited on p.486).

[671] National Gallery of Australia (undated) *Juan Gris, biographical notes.* Website article regarding *Checkerboard with playing cards,* (1915), http://nga.gov.au/International/Catalogue/Detail.cfm?IRN= 91632 accessed 1Sep2014, (cited on p.71).

[672] Nature Jnl. Editorial (2012) *Surface plasmon resurrection.* Nature Photonics 6 Nov2012, p.707, doi:10.1038/nphoton.2012.296 (cited on p.376).

[673] Necipoğlu G (1995) *The Topkapı scroll—geometry and ornament in Islamic architecture.* Los Angeles, CA: Getty Center Publication, cited by Cromwell [146], (cited on p.346).

[674] Neufert E (1936) *Bauentwurfslehre,* 'Architects' data', in many editions to date, Berlin SW 68: Bauvelt-Verlag, (cited on p.547).

[675] Neufert E & P (2000) *Architects' Data, Third Edition.* Eds. Baiche B & Walliman N (First edn. 1936 [674].) Oxford: Blackwell Science, *Zeising/Moessel,* p.15, (cited on p.132).

[676] —Ibid. *Le Corbusier/The Modulor,* p.30, (cited on p.132).

[677] —Ibid. *Theatre auditorium design,* p.479, (cited on p.132).

[678] Neufert E & Speer A (1943) *Bauordnungslehre.* Berlin: Volk und Reich Verlag, (cited on p.132).

[679] New World Encyclopedia contributors (2014) *Kairos.* New World Encyclopedia, http://www.newworldencyclopedia.org/p/index.php? title=Kairos&oldid=981846 accessed 21Aug2014, (cited on p.110).

[680] Nikolić H (2012) *EPR before EPR: a 1930 Einstein-Bohr thought experiment revisited.* Theoretical Physics Divn, Rudjer Bošković Institute, Zagreb, Croatia. http://arxiv.org/pdf/1203.1139v4.pdf accessed 7Jun2013, http://www.technologyreview.com/view/427174/ einsteins-spooky-action-at-a-distance-paradox-older-than-thought/ accessed 7Jun2013, (cited on p.297).

[681] Nishiyama Y (2012) *Five petals: the mysterious number '5' hidden in nature.* IJPAM (Internat. Jnl. of Pure & Applied Maths), Sofia: Academic Publications, 78:3, Table 1, p.351, (cited on pp.51, 51, 579).

[682] Nisoli C, Gabor NM, Lammert PE, Maynard JD, & Crespi VH (2009) *Static and dynamical phyllotaxis in a magnetic cactus.* Los Alamos National Laboratory. Phys. Rev. Lett. 102, p.186103, doi:10.1103/PhysRevLett.102.186103 (cited on pp.294, 585).

[683] ———. (2010) *Annealing a magnetic cactus into phyllotaxis.* Los Alamos National Laboratory, Phys. Rev. E, 81:4, 15Apr2010, doi:10.1103/PhysRevE.81.046107 http://arxiv.org/pdf/1002.0622.pdf accessed 3Jun2013, (cited on p.294).

[684] Noorden RV (2011) *Persistence pays off for crystal chemist.* Nature 478, pp.165-166, doi:10.1038/478165a http://www.nature.com/ news/2011/111010/full/478165a.html accessed 6Jun2013, (cited on p.333).

[685] Norman G (1977) *Nineteenth-century painters and painting: a dictionary.* Oakland: Univ. of California, p.195, (cited on p.71).

[686] Noro JJ (1990) *Mestres da música.* Vol.5, Barcelona: Planeta De Agostini, image *Claude Debussy au piano l'été 1893 dans la maison de Luzancy (chez son ami Ernest Chausson),* (cited on pp.142, 582).

[687] Notomi M, Suzuki H, Tamamura T, & Edagawa K (2004) *Lasing action due to the two-dimensional quasiperiodicity of photonic quasicrystals with a Penrose lattice.* Phys. Rev. Lett. 92, p.123906, cited by Vardeny [919], doi:10.1103/PhysRevLett.92.123906 (cited on p.391).

[688] O'Connor JJ & Robertson EF (1997) *The MacTutor history of mathematics archive.* University of St. Andrews, School of Maths. and Stats., http://www-history.mcs.st-andrews.ac.uk/Curves/Equiangular.html accessed 30May2013, (cited on p.39).

[689] ———. (1999) *Abu Kamil,* http://www-history.mcs.st-andrews.ac.uk/Biographies/Abu_Kamil.html accessed 2Jun2013, (cited on p.18).

[690] ———. (1999) *Al'Khwarizmi,* http://www-history.mcs.st-and.ac.uk/Mathematicians/Al-Khwarizmi.html accessed 2Nov2013, (cited on p.18, 18).

[691] ———. (2003) *Giovanni Domenico Cassini,* http://www-history.mcs.st-andrews.ac.uk/Biographies/Cassini.html accessed 31Jul2014, (cited on p.24).

[692] ———. (1999) *Conon of Samos,* http://www-history.mcs.st-andrews.ac.uk/Biographies/Conon.html accessed 12Aug2013, (cited on p.36).

[693] ———. (2000) *MC Escher,* http://www-history.mcs.st-and.ac.uk/Biographies/Escher.html accessed 27Jul2014, (cited on p.343).

[694] ———. (2014) *Pierre Fatou,* http://www-history.mcs.st-andrews.ac.uk/Biographies/Fatou.html accessed 27Nov2014, (cited on p.95).

[695] ———. (1998) *Leonardo Pisano Fibonacci,* http://www-history.mcs.st-andrews.ac.uk/Biographies/Fibonacci.html accessed 29Nov2014, (cited on p.19).

[696] ———. (2010) *Albert Girard biog,* http://www-history.mcs.st-and.ac.uk/Biographies/Girard_Albert.html accessed 14Feb2014, (cited on p.23, 23).

[697] ———. (2001) *The Golden Ratio,* http://www-history.mcs.st-andrews.ac.uk/HistTopics/Golden_ratio.html accessed 2Jun2013, (cited on pp.18, 21, 23, 23).

[698] ———. (2000) *Gaston Julia,* http://www-history.mcs.st-and.ac.uk/Biographies/Julia.html accessed 27Nov2014, (cited on p.95).

[699] ———. (2008) *Ernst Jacobsthal biography,* http://www-history.mcs.st-and.ac.uk/Biographies/Jacobsthal.html accessed 30May2013, (cited on p.275).

[700] ———. (2006) *Kepler's planetary laws,* with note of Kepler's use of the term 'orbit' in *Astronomia Nova* ch.1, http://www-history.mcs.st-and.ac.uk/HistTopics/Keplers_laws.html accessed 9Jul2013, (cited on pp.23, 170).

[701] ———. (2006) *Gabriel Lamé,* http://www-history.mcs.st-and.ac.uk/ Mathematicians/Lame.html accessed 13Aug2014, (cited on p.26).

[702] ———. (1996) *François Édouard Anatole Lucas biography,* http://www-history.mcs.st-andrews.ac.uk/Biographies/Lucas.html accessed 3Feb2014, (cited on pp.28, 448).

[703] ———. (2005) *Marin Mersenne biography,* http://www-history.mcs. st-and.ac.uk/Biographies/Mersenne.html accessed 13Jun2013, (cited on pp.37, 447).

[704] ———. (1999) *Luca Pacioli biography,* http://www-groups.dcs. st-andrews.ac.uk/~history/Biographies/Pacioli.html accessed 7Nov2013, (cited on pp.21, 23, 69).

[705] ———. (2005) *John Pell biography,* http://www-history.mcs.st-and.ac. uk/Biographies/Pell.html accessed 10Mar2015, (cited on p.268).

[706] ———. (2009) *Perfect numbers,* http://www-history.mcs.st-and.ac.uk/ HistTopics/Perfect_numbers.html accessed 6Feb2014, (cited on pp.280, 282, 490).

[707] ———. (2008) *Julia Robinson and Hilbert's 10th Problem.* http://www-history.mcs.st-and.ac.uk/Extras/Robinson_Hilbert_10th.html accessed 2Sep2014, (cited on pp.450, 488).

[708] ———. (2000) *Caspar Wessel biography.* http://www-history.mcs. st-and.ac.uk/Biographies/Wessel.html accessed 5Jan2014, (cited on p.134).

[709] Odd Perfect Organization (undated) *Odd Perfect Number Search.* Dedicated website, cited by Greathouse & Weisstein [270], http://oddperfect.org/against.html accessed 7Feb2014, (cited on p.284).

[710] Odom G (1983) *Golden Section arising from equilateral triangle inscribed in a circle – Problem E 3007.* Am. Math. Mthly 90, p.482, (cited on p.2).

[711] OEIS Foundation (2011) *The OEIS® reaches 200,000 sequences.* Press release, regarding Sloane website [839], http://oeisf.org/press3. pdf accessed 4Apr2014, (cited on p.34).

[712] OEIS (2014) *Aliquot divisors of n,* 'Divisors of n less than n.', http://oeis.org/wiki/Divisors accessed 26Dec2014, (cited on p.280).

[713] Omegatron (2006) *Simple sine wave.* Wikimedia Commons diagram http://commons.wikimedia.org/wiki/File:Simple_sine_wave.svg accessed 23Sep2009, (cited on pp.379, 589).

[714]   Onoda GY, Steinhardt PJ, DiVincenzo DP, & Socolar JES (1988) *Growing perfect quasicrystals.* Phys. Rev. Lett. 60:25, 20Jun1988, pp.2653–2656,   doi:10.1103/PhysRevLett.60.2653 cited by Peterson I, in *Tiling to infinity: the surprising solution to a tiling problem provides new insights into unusual forms of crystal growth.* Science News, 16Jul1988   http://www.thefreelibrary.com/Tiling+to+infinity%3A+the+surprising+solution+to+a+tiling+problem...-a06542300   accessed 12Sep2013, (cited on p.334).

[715]   Onstott S (2011) *The snowflake and the flower.* Website *Secrets in plain sight: mathematics, relationship between Phi and pi,*   http://www.secretsinplainsight.com/2011/07/19/the-snowflake-and-the-flower/   accessed 28Apr2013, (cited on p.184).

[716]   Pacioli L, illustr. by Leonardo da Vinci (1509) *Divina Proportione Opera a tutti glingegni perspicaci e curiosi necessaria ove ciascun studioso di Philosophia: Prospectiva Pictura Sculptura: Architectura: Musica: e altre Mathematice suavissima: sottile: e admirabile doctrina consequira: e delectarassi: cõ varie questione de secretissima scientia* (underline added). Book, from LP & LDV's three manuscript work of 1497, Venice: Paganinus de Paganinus,   (cited on p.21).

[717]   —Ibid. *Comme lapicidi ...* And: *ve ricordo ...* I,18, f.32v, cited by Frings [230],   (cited on p.67).

[718]   Padovan R (1999) *Proportion: science, philosophy, architecture.* London: Taylor & Francis, *underlying secret,* p.307,   (cited on p.67).

[719]   —Ibid. *RIBA, London, meeting 18Jun1957,* p.1,   (cited on p.156).

[720]   ——. (2002) *Towards universality, Le Corbusier, Mies & De Stijl.* London: Routledge, p.4,   (cited on pp.97, 544).

[721]   —Ibid. *coloured rectangles not bounded,* p.5,   (cited on p.94).

[722]   —Ibid. *black lines ... lie in front of the white ground,* (Padovan quoting Schapiro [825]), p.33,   (cited on p.94).

[723]   —Ibid. *extend in that virtual space outside,* (Padovan quoting Schapiro), p.33,   (cited on p.102).

[724]   —Ibid. *destroy the external and internal boundaries of the painting,* p.33,   (cited on p.102).

[725] —Ibid. *Boogie Woogie paintings (New York 1942–1944)*, p.34, (cited on p.106).

[726] —Ibid. *Padovan quoting Mondrian from 'Eleven Europeans in America',* see MoMA [659]. Quotation: *'The great struggle... '*, p.34, (cited on p.545).

[727] —Ibid. *Theosophical Society, Schoenmaekers [joined] in 1905,* p.134, (cited on p.106).

[728] Palais R (2001) *π is wrong!* The Mathematical Intelligencer, New York: Springer-Verlag, 23:3, pp.7–8, doi:10.1007/BF03026846 http://www.math.utah.edu/~palais/pi.html accessed 3Jun2013, (cited on p.33).

[729] Patañjali (1200 BC) *The Yoga Sutras of Patanjali.* Translated and introduced by Alistair Shearer (1982). New York: Bell Tower, 2002, Ch.3, p.124, (cited on p.361).

[730] Palin T (2011) *Photographs of author.* Palin Images, http://www.palinimages.com/ accessed 4May2013, (cited on pp.166, 576, 583).

[731] Patrickringgenberg (2008) *Samarkand, Shah-i Zinda, Uzbekistan, decoration of the Tuman Aqa complex.* Photograph. http://en.wikipedia.org/wiki/File:Samarkand_Shah-i_Zinda_Tuman_Aqa_complex_cropped2.jpg accessed 1Jun2013, (cited on pp.iv, 355, 588).

[732] Pauling L & Corey RB (1953) *Compound helical configurations of polypeptide chains: structure of proteins of the α-keratin type.* Nature, 171, 10Jan1953, p.59, cited by Frey-Wyssling [228], doi:10.1038/171059a0 (cited on p.290).

[733] Pendry JB, Holden AJ, Robbins DJ, & Stewart WJ (1999) *Magnetism from conductors and enhanced nonlinear phenomena.* Microwave Theory and Techniques, IEEE Transactions on, Nov1999, 47:11, pp.2075–2084, doi:10.1109/22.798002 (cited on p.370).

[734] Pendry JB (2000) *Negative refraction makes a perfect lens.* Phys. Rev. Lett. 85:18, 30Oct2000, pp.3966–3969, doi:10.1103/PhysRevLett.85.3966 (cited on p.372).

[735] ———. (2011) *Metamaterials: Going beyond nature.* New Scientist, 8Jan2011, http://www.newscientist.com/article/mg20927940.800-metamaterials-going-beyond-nature.html accessed 7Jan2015, (cited on p.373).

[736] Pendry JB, Schurig D, & Smith DR (2006) *Controlling electromagnetic fields.* Science 312:5781, 23Jun2006, pp.1780–1782, doi:10.1126/science.1125907 (cited on pp.370, 371).

[737] Penrose R (1974) *The role of aesthetics in pure and applied mathematical research.* Bull. Inst. Math. Appl. 10:7/8, pp.266–271, cited by Au-Yang & Perk [21], (cited on p.312).

[738] ———. (1978) *Pentaplexity: a class of non-periodic tilings of the plane.* The Mathematical Intelligencer 2, pp.32–37, [reprinted from (1978) Eureka 39, pp.16–22], cited by Au-Yang & Perk [21], doi:10.1007/BF03024384 (cited on p.319).

[739] —Ibid. *kites and darts, names suggested by John Conway,* p.33, (cited on p.313).

[740] ———. (1979) *Patent US 4133152 A. Set of tiles for covering a surface.* United States Patent and Trademark Office, https://www.google.com/patents/US4133152 accessed 1Jun2014, (cited on p.311).

[741] ———. (1989) *Tilings and quasicrystals: a nonlocal growth problem?* In *Introduction to the mathematics of quasicrystals.* Ed. Jaric M, Academic Press, pp.53–80, cited by Austin D in web essay (2013) *Penrose tiles talk across miles.* American Mathematical Society, http://www.ams.org/samplings/feature-column/fcarc-penrose accessed 21Jun2013, (cited on p.334).

[742] ———. (2013) *Forbidden crystal symmetry in mathematics and architecture.* Video 0h:58m:13s, London: Royal Institution, *architectural examples* 34m:08s onwards, http://www.youtube.com/watch?v=th3YMEamzmw accessed 18Mar2014, (cited on p.311).

[743] Penzel F (1978) *Theatre lighting before electricity.* Middletown, Conn: Wesleyan University Press, p.69, (cited on p.84).

[744] Pérez-Gómez R (1986) *The four regular mosaics missing in the Alhambra.* Computers & Mathematics with Applications, 14:2, *Four groups are: p2, pg, pgg and p3m1,* pp.133–137, doi:10.1016/0898-1221(87)90143-X (cited on p.344).

[745] Peshkin M & Tonomura A (1989) *The Aharonov-Bohm effect.* University of Michigan. New York, Berlin, Heidelberg: Springer-Verlag, cited by Wikipedia, (cited on p.362).

[746] Petrie WM Flinders (1883) *The pyramids and temples of Gizeh.* 1st edn. London: Field & Tuer, New York: Scribner & Welford, and see Petrie (online) [747], *The mean base being 9068.8 ± .5 inches, this yields a height of 5776.0 ± 7.0 inches,* § 25, p.15, (cited on pp.6, 498, 553).

[747] ———. (1883) *The pyramids and temples of Gizeh.* [746] Edited & republ. online by Ronald Birdsall (2003) Rev. 5Aug2012, http:// ronaldbirdsall.com/gizeh/ accessed 3Jul2014, http://ronaldbirdsall.com/gizeh/petrie/ c6.html#25 accessed 3Jul2014, (cited on pp.6, 553).

[748] ———. (1925) *Surveys of the Great Pyramids.* Nature, 26Dec1925, 116:2930, *Petrie restates the two mean base side measurements: Petrie(1881)—9068.8, & Cole(1925)—9069.4 inches,* p.942, doi:10.1038/116942a0 (cited on p.7).

[749] Pfender F & Ziegler GM (2004) *Kissing numbers, sphere packings, and some unexpected proofs.* Notices of the AMS, Sep2004, 51:8, p.875, http://www.ams.org/notices/200408/fea-pfender.pdf accessed 3Apr2014, (cited on p.300).

[750] Pielou EC (2011) *The world of northern evergreens, second edition.* (1st edn. 1988.) Comstock Publishing Associates, Ithaca & London: Cornell University Press, p.67, (cited on p.56).

[751] Piṅgala Nāga (c.450 BC or 200 BC) *Work on the Mountain of Cadence, now called Pascal's triangle.* The Wolfram Functions Site, Fibonacci number, http://functions.wolfram.com/IntegerFunctions/ Fibonacci/introductions/FibonacciLucasNumbers/ShowAll.html accessed 9Apr2013, (cited on pp.12, 13).

[752] Plato (380 BC) *The Republic of Plato.* Transl. by Cornford FM, Oxford: Clarendon, Book VI, 509D–511E, p.216, (cited on p.16).

[753] ———. (c. 360 BC) *The dialogues of Plato: Timæus.* Transl. Jowett B, impr. 1931, from 3rd edn. 1892, Vol.III, London: Lowe & Brydone, p.476, (cited on p.306).

[754] Poe EA (1843) *The Black Cat.* In *The Saturday Evening Post* magazine. French transl. Baudelaire mid-1850s, (cited on p.152).

[755] Poirier A (2014) *Watching boxing with Picasso and a ménage-à-trois at home: my life with the surrealist élite.* Newspaper article: recollections of Cécile Élouard. The Observer 13Apr2014, London: Guardian News and Media, (cited on p.112).

[756] Pollit JJ (1972) *Art and experience in classical Greece.* Cambridge: Cambridge University Press, p.200, (cited on p.555).

[757] Pólya, George (1924) *Über die Analogie der Kristallsymmetrie in der Ebene.* Zeitschrift für Kristallographie 60, pp.278–282, cited by Grünbaum, Grünbaum, & Shephard [278], (cited on p.343).

[758] Pompidou Centre (2011) *Mondrian/De Stijl—Album de l'exposition* (exhibition guide), Paris: Éditions du Centre Pompidou. *Mondrian Composition with Red, Blue, and Yellow (1930)* [657], front cover & p.20, http://boutique.centrepompidou.fr/ fr/albums-dexposition/mondrian-de-stijl-album-de-lexposition/170.html accessed 16Dec2014, (cited on pp.93, 100, 545).

[759] —Ibid. *1909, 25 May: Mondrian becomes a member of the Netherlands Theosophical Society,* p.55, (cited on p.89).

[760] —Ibid. *1919, 22 Jun: Mondrian leaves Holland and moves to Paris in June 1919,* p.55, (cited on p.89).

[761] —Ibid. *1932, Jan: Last issue of De Stijl, dedicated to van Doesburg,* p.55, (cited on p.89).

[762] Posamentier AS & Lehmann I (2011) *The glorious Golden Ratio.* New York: Prometheus Books, *16 examples of Golden Ratio construction,* pp.13–38, (cited on p.2).

[763] —Ibid. *Golden Ratio and fractals,* pp.269–276, (cited on p.237).

[764] Posnansky A (1903–1904) *Centro de la puerta* (Photograph of the central figure depicted on the Gate of the Sun, Tiwanaku, Bolivia), Vienna: Kunsthistorisches Museum, Inv.-Nr. VF_13245, (cited on pp.116, 581, 581).

[765] ——. (1945) *Tihuanacu the cradle of American Man.* (2 vols.) Transl. Shearer JF, New York: JJ Augustin, (cited on p.116).

[766] Prange SR (2009) *The tiles of infinity, Peter J Lu in Uzbekistan.* Saudi Aramco World, 60:5, Sep/Oct 2009, http://www.saudiaramcoworld.com/issue/200905/the.tiles.of.infinity.htm accessed 7Jun2013, (cited on p.20).

[767] Promenader, The (2007) *Plan of Paris.* Map, http://commons. wikimedia.org/wiki/File:Paris_arr_jms.gif accessed 18Feb2014, (cited on pp.143, 582).

[768] Protzen J-P & Nair SE (2000) *On reconstructing Tiwanaku architecture.* Jnl. Soc. Architectural Historians, 59:3, p.364, http://www.jstor.org/stable/991648 accessed 8May2014, (cited on p.8).

[769] Prusinkiewicz P & Lindenmayer A (1990) *The algorithmic beauty of plants.* Springer-Verlag. pp.101–107, http://algorithmicbotany.org/papers/#abop accessed 7May2013, (cited on p.56).

[770] —Ibid. *(Lindenmayer) L-Systems,* chap.1, (cited on pp.96, 304).

[771] Rabinowitz S (1996) *Algorithmic Manipulation of Fibonacci Identities.* In *Applications of Fibonacci numbers: proceedings of the sixth international research conference on Fibonacci numbers and their applications.* Cited by Knott [480]. Dordrecht: Kluwer Academic, pp.389–408, doi:10.1007/978-94-009-0223-7 (cited on p.530).

[772] Rabson DA (2012) *Toward theories of friction and adhesion on quasicrystals.* Progress in Surface Science, 87:9, pp.253–271, doi:10.1016/j.progsurf.2012.10.001 (cited on p.340).

[773] Rama (2007) *Photograph: Benoît Mandelbrot at the École Polytechnique Fédérale de Lausanne (EPFL), on 14 March 2007.* http://en.wikiquote.org/wiki/File:Benoit_Mandelbrot_mg_1845.jpg accessed 5Nov2013, (cited on pp.229, 584).

[774] Ravel M (1905) *Miroirs.* Suite in 5 movements for solo piano. No.3 and No.4 later orchestr. by Ravel, (No.3 in 1906 rev. 1926, No.4 in 1918). Publ. 1906 Paris: E Demets, (cited on pp.149, 150).

[775] Raven JE (1951) *Polyclitus and Pythagoreanism.* Classical Quarterly 45, pp.147–152, cited by Pollit [756], (cited on p.69).

[776] Rayleigh, Lord (1878) *The theory of sound / by John William Strutt, Baron Rayleigh.* Vol.2. London: Macmillan, *Whispering-gallery waves, St. Paul's Cathedral,* (cited on p.398).

[777] ———. (1887) *On the maintenance of vibrations by forces of double frequency, and on the propagation of waves through a medium endowed with a periodic structure.* Philosophical Magazine & Journal of Science, 5th Series 24:147, Aug1887, pp.145–159, cited by Yablonovitch [994]. doi:10.1080/14786448708628074 (cited on p.367, 367).

[778] Ricciardi A, Crescitelli A, Consales M, Galdi V, Esposito E, & Cusano A (2012) *Plasmonic-photonic resonances in nanostructured metallo-dielectric quasi-crystals: tuning and*

sensitivity analysis. Proc. SPIE 8351, Third Asia Pacific Optical Sensors Conference, 83511Q 31Jan2012, doi:10.1117/12.913430 (cited on p.393).

[779] Richards J (2005) *Stonehenge (guidebook).* London: English Heritage, p.10, (cited on p.120).

[780] Richardson LF (1961) *The problem of contiguity: an appendix to statistics of deadly quarrels.* General Systems Yearbook 6, pp.139–187, cited by Mandelbrot [623] p.27, (cited on pp.230, 584).

[781] Riley B (1997) *Mondrian, nature to abstraction.* (To accompany the exhibition 30Jul–30Nov), London: Tate Gallery Publishing, *equilibrium of the universal & particular,* p.18, (cited on p.90).

[782] Roberts GE (2012) *Béla Bartók and the Golden Section.* Lecture slides, Dept. Maths & Computer Science, College of the Holy Cross, Worcester, MA, *Studies at ... Catholic Gymnasium ... Pozsony ... excels in math and physics in addition to music* p.2, http://mathcs.holycross.edu/~groberts/Courses/Mont2/Handouts/ Lectures/Bartok-web.pdf accessed 20Jun2014, (cited on p.62).

[783] —Ibid. *too 'cerebral',* p.29, (cited on pp.113, 155).

[784] Robinson RM (1954) *Mersenne and Fermat Numbers.* Proc. Amer. Math. Soc. 5, *then 5 largest Mersenne primes, indices 521, 607, 1279, 2203, & 2281,* pp.842–846, http://www.ams.org/journals/ proc/1954-005-05/S0002-9939-1954-0064787-4/S0002-9939-1954-0064787-4.pdf accessed 6Jun2014, (cited on p.322).

[785] Robster1983 (2008) *The Giza-pyramids and Giza Necropolis, Egypt, seen from above.* Photograph, public domain, http://en. wikipedia.org/wiki/File:Giza-pyramids.JPG accessed 3Jul2014, (cited on pp.7, 577).

[786] Rocchini C (2012) *Example of Ammann-Beenker tiling.* Illustration, http://en.wikipedia.org/wiki/File:Ammann-Beenker_tiling1.svg accessed 6Dec2013, (cited on pp.337, 587).

[787] Roosendaal T et al. (c.1998–) *Blender: open-source 3D computer graphics software.* Amsterdam: Blender Foundation, http://www. blender.org/blenderorg/ accessed 13Jun2013, (cited on pp.15, 301, 389, 577, 584, 584, 584, 585).

[788] Rosaline SI & Raghavan S (2013) *Survey on metamaterials in bio-medicine.* IEEE International Conference on Computational Intelligence and Computing Research (ICCIC), doi:10.1109/ICCIC.2013.6724184 (cited on p.382).

[789] Roper C & The Ocean Portal Team (2013) *Giant Squid, Architeuthis dux.* Website article. Smithsonian National Museum of Natural History, http://ocean.si.edu/giant-squid accessed 16Dec2014, (cited on p.82).

[790] —Ibid. *snatching prey ... more than 10 meters (30 feet) from the animal's eye!* http://www.mnh.si.edu/natural_partners/squid4/ArchiteuthisFeeding.html accessed 16Dec2014, (cited on p.82).

[791] Rorem N (1988) *Bluebeard and Erwartung: a notebook.* In (2013) *Other Entertainment: Collected Pieces.* Kindle edn., New York: Open Road Media, *the palindrome is clear ...*, (cited on p.140).

[792] Rouse Ball WW & Coxeter HSM (1987) *Mathematical recreations and essays.* 13th edition, eds. Harold Scott & Macdonald Coxeter, New York: Dover Publications, p.161 (note text), cited by Gardner [252], (cited on pp.322, 342).

[793] Royal Swedish Academy of Sciences (2011) *The Nobel Prize in chemistry 2011, crystals of Golden proportions.* Web article about Dan Shechtman [829], http://www.nobelprize.org/nobel_prizes/chemistry/laureates/2011/popular-chemistryprize2011.pdf accessed 10May2013, (cited on p.330).

[794] Ruggles D (1963) *Some Fibonacci results using Fibonacci-type sequences.* The Fibonacci Quarterly 1:2, p.80, Note: Ruggles (p.75) swaps α and β as compared with Lucas [595], hence: $\beta^n = (L_n + F_n\sqrt{5})/2$, http://www.fq.math.ca/Scanned/1-2/ruggles.pdf accessed 2May2013, (cited on pp.181, 427, 478, 482).

[795] Sachs J (1882) *Textbook of botany: morphological and physiological.* Oxford: Clarendon Press, cited by Adler et al. [7], p.236 (cited on p.52).

[796] ——. (1989) *Icosahedral order, curved space and quasicrystals.* In *Extended Icosahedral Structures.* Editors: Jarić MV & Denis Gratias, London: Academic Press, p.167, (cited on p.308).

[797] Sammells CA (2012) *Ancient calendars and Bolivian modernity: Tiwanaku's Gateway of the Sun.* Jnl. of Latin American & Caribbean Anthropology, 17:2, p.302, doi:10.1111/j.1935-4940.2012.01221.x (cited on p.8).

[798] Sándor J & Crstici B (2004) *Handbook of Number Theory II, Volume 2.* Dordrecht: Kluwer Academic Publishers, *Catalan 1888 OPN restriction* p.25, (cited on p.284).

[799] Satie E (1892) *Sonneries de la Rose+Croix.* In (1974) *Erik Satie Piano music. Vol.1, Selected works 1887-1894.* 'With definitive corrections by Erik Satie himself', Paris: Éditions Salabert, pp.37–46, (cited on p.150).

[800] Sautoy M du (2011) *Penrose tiles at Wadham College, Oxford.* Maths in the City webpage, http://www.mathsinthecity.com/snapshots/penrose-tiles-wadham-college-oxford accessed 12Dec2013, (cited on p.311).

[801] ———. (2009) *Symmetry, reality's riddle.* TED video talk 0h:18m:19s, *incompleteness* at 0h:3m:56s, http://www.youtube.com/watch?v=415VX3QX4cU accessed 30Apr2014, with transcript: http://www.ted.com/talks/marcus_du_sautoy_symmetry_reality_s_riddle/transcript accessed 30Apr2014, (cited on p.154).

[802] Schaefer RJ, Bendersky LA, Schectman D, Boettinger WJ, & Biancaniello FS (1986) *Icosahedral and decagonal phase formations in Al-Mn alloys.* Metallurg. Trans. A, Dec1986 17:12, Fig.11: *photograph of conspicuously 5-fold symmetric crystals,* p.2122, doi:10.1007/BF02645910 (cited on pp.325, 586).

[803] Schattschneider D (2010) *The mathematical side of MC Escher.* Notices of the AMS, Jun2010, 57:6, pp.707 & 712, (cited on p.69).

[804] Schiffman JL (2012) *Divisibility and periodicity ideas in Fibonacci-like sequences.* 62nd Ohio Council of Teachers of Mathematics Conf., 18Oct2012, p.9, (cited on pp.174, 533).

[805] Schimper KF (1830) *Beschreibung des Symphytum Zeyheri.* (cited by Adler [3]), Geiger's Mag. Für Pharm. 29, pp.1–92, (cited on p.52).

[806] Schreiber P (1995) *A supplement to J Shallit's paper 'Origins of the analysis of the Euclidean algorithm'.* Historia Mathematica, 22:4, pp.422–424, doi:10.1006/hmat.1995.1033 (cited on p.23).

[807] Schurig D, Mock JJ, Justice BJ, Cummer SA, Pendry JB, Star AF, & Smith DR (2006) *Metamaterial electromagnetic cloak at microwave frequencies.* Science 10Nov2006 314, p.977, doi:10.1126/science.1133628 (cited on pp.374, 375).

[808] Schurig D, Pendry JB, & Smith DR (2006) *Calculation of material properties and ray tracing in transformation media.* OSA, Optics Express 14:21, 16Oct2006, Eqn. 12, p.9797, doi:10.1364/OE.14.009794 (cited on pp.371, 375, 588).

[809] ———. (2007) *Transformation-designed optical elements.* Opt. Express 15:22, 14772–14782, doi:10.1364/OE.15.014772 (cited on p.373).

[810] Schütte & Waerden BL van der (1953) *Das Problem der dreizehn Kugeln.* (The problem of 13 spheres.) Math. Ann. 125, pp.325–334, (cited on p.300).

[811] Scott TC & Marketos P (2014) *On the origin of the Fibonacci sequence.* MacTutor History of Mathematics, pp.27–29, http://www-groups.dcs.st-and.ac.uk/~history/Publications/fibonacci.pdf accessed 22Apr2014, (cited on pp.16, 511).

[812] —Ibid. *Bee family tree,* Figure 3, p.11, from Basin SL [42] (cited on pp.284, 491).

[813] —Ibid. *Golden Triangles near centre of the Theatre of Epidaurus,* Fig. 8, p.29, from Gerkan & Müller-Wiener [259], (cited on p.16).

[814] —Ibid. *Suggestion of $\sqrt{\phi}$ at theatre of Dodona, Epirus, Northern Greece,* (As ratio 19/15, in a seating rows/Fibonacci context.) p.30. Quoting work of Tsimpourakhs D (1985) *H Gewmetr'ia kai oi erg'ates ths sthn Arqa'ia Ell'ada,* Athens (self-publication in Greek), (cited on p.128).

[815] Secret Museum of Mankind (undated) *'One of the most ancient monuments in the Western hemisphere'—the Gate of the Sun.* Ian Macky website for (1935) book *The Secret Museum.* New York: Manhattan House, http://ian.macky.net/secretmuseum/page_1.35.html accessed 1May2014, (cited on pp.11, 115, 118, 577, 581, 581, 581).

[816] Senechal M (2004) *Mathematical communities: the mysterious Mr. Ammann.* Springer, The Mathematical Intelligencer, 26:4, *MC Escher,* p.10, doi:10.1007/BF02985414 (cited on p.316).

[817] —Ibid. *Ammann bars (his own overview),* p.14, (cited on p.321).

[818] —Ibid. *equally spaced—1st letter to Gardner,* p.14, (cited on p.387).

[819] —Ibid. *photographs in newspapers,* p.15, (cited on p.320).

[820] —Ibid. *autism/Asperger's?* p.16, (cited on p.320).

[821] ———. (2011) *Prof. Marjorie Senechal—quasicrystals gifts to mathematics.* Technion Video 0h:30m:56s, 12Jan2011, http://www.youtube.com/watch?v=pjao3H4z7-g accessed 6Dec2013, *Gold mine* at 0h:3m:30s, (cited on p.387).

[822] Seurat GP (1884) *Bathers at Asnières.* Oil on canvas, 201 × 300 cm, London: National Gallery, (cited on pp.77, 580).

[823] ———. (1886) *A Sunday afternoon on the Island of la Grande Jatte.* Oil on canvas, 207.5 × 308.1 cm, Chicago: Art Institute, 1926.224, (cited on pp.iv, 75, 77, 580, 580).

[824] ———. (1888) *La Parade du Cirque / Circus Sideshow.* Oil on canvas, 99.7 × 149.9 cm, New York: Metropolitan Museum of Art, Accession 61.101.17, Gallery 826. For note—*Circus Corvi, place de la Nation*—see online Gallery label: http://www.metmuseum.org/ collection/the-collection-online/search/437654 accessed 13Dec2014, (cited on pp.79, 79, 80, 580).

[825] Schapiro M (1982) *Modern Art.* New York: George Braziller, p.195, (cited by Galenson [231] p.3), (cited on p.550).

[826] Shanks D (2001) *Solved and Unsolved Problems in Number Theory.* Providence, Rhode Island: American Mathematical Soc., *Cataldi-Fermat* theorem 4, p.3, (cited on p.447).

[827] Shaw JEH (2001) *Some quotable quotes for statistics.* Online article. Quote: *A first-rate theory predicts...* by Aleksander Isaakovich Kitaigorodskii (1914–1985), lecture, IUC Amsterdam, August 1975, p.78, http://web.warwick.ac.uk/statsdept/staff/JEHS/data/ jehsquot.pdf accessed 13Nov2014, (cited on p.100).

[828] Shechtman D, Blech I (1985) *The Microstructure of rapidly solidified* Al_6Mn. Metallurgical Transactions A Jul1985 16:6, pp.1005–1012, doi:10.1007/BF02811670 (cited on p.328).

[829] Shechtman D, Blech I, Gratias D, & Cahn JW (1984) *Metallic phase with long-range orientational order and no translational symmetry.* Phys. Rev. Lett. 53:20, pp.1951–1954, doi:10.1103/PhysRevLett.53.1951 (cited on pp.324, 328, 330, 383, 557).

[830] Shtull-Trauring A (2011) *Clear as crystal.* Haaretz newspaper magazine article 1Apr2011, Shechtman quoting Pauling: '... *no such thing as quasicrystals, only quasi-scientists.*', http:// www.haaretz.com/weekend/magazine/clear-as-crystal-1.353504 accessed 2Jun2014, (cited on p.330).

[831] Siegel PH (2013) *Introduction to special issue on THz metamaterials and applications.* IEEE Trans. Terahertz Science & Tech., 3:6, doi:10.1109/TTHZ.2013.2285790 (cited on p.375).

[832] Siegel R (2013) *Ancient Greek theatre at Epidaurus, Saronic Gulf.* Photograph, http://www.ploync.de/reisen/5041-epidauros-peloponnes. html accessed 22Apr2014, (cited on pp.16, 577).

[833] Sigalas M & Economou EN (1992) *Elastic and acoustic wave band structure.* Journal of Sound and Vibrations, 158:2, p.377–382, cited by Steurer & Sutter-Widmer [871], doi:10.1016/0022-460X(92)90059-7 (cited on p.380).

[834] ———. (1993) *Band structure of elastic waves in two dimensional systems.* Solid State Communications, 86:3, Apr1993, pp.141–143, doi:10.1016/0038-1098(93)90888-T (cited on p.380).

[835] Simon B (1982) *Almost periodic Schrödinger operators: a review.* Adv. in Appl. Math. 3, p.463, cited by Kohmoto & S. [504] note 26, doi:10.1016/S0196-8858(82)80018-3 (cited on p.383).

[836] Singh P (1985) *The so-called Fibonacci numbers in ancient and medieval India.* Historia Mathematica 12, pp.229–244, doi:10.1016/0315-0860(85)90021-7 (cited on pp.13, 17, 569).

[837] —Ibid. *short and long syllables,* Table 1, p.231, (cited on pp.13, 13, 577).

[838] —Ibid. *dating Piṅgala & brother or uncle or...* p.232, (cited on pp.12, 12, 487, 569).

[839] Sloane NJA (1964–) *The online encyclopedia of integer sequences.* OEIS Foundation Inc., http://oeis.org accessed 12Mar2013, (cited on pp.34, 187, 268, 277, 277, 277, 277, 280, 280, 280, 281, 281, 390, 458, 549, 567).

[840] Smeardon J (2006) *The formation and characterisation of aperiodic ultra-thin films on the surfaces of quasicrystals.* Doctoral thesis, Univ. Liverpool. http://www.scribd.com/doc/34555414 accessed 4Oct2013, (cited on pp.316, 328).

[841] Smith DR (2013) *David Smith—Metamaterials Talk 2013.* Video 1h:08m:19s, *epsilon & mu: 4 regions* at 0h:32m:32s, https://www.youtube.com/watch?v=uNUsmezUhYg accessed 30Nov2013, (cited on pp.368, 369, 588).

[842] —Ibid. *Initial reaction to 'left-handed materials',* at 0h:44m:04s, (cited on p.373).

[843] —Ibid. *Gradient metamaterials,* at 0h:47m:30s, (cited on pp.373, 588).

[844] Smith MB (1937) *Plan of Gunbad-i-Quābūd, Marāgha, Iran.* Annotated field drawing—ink, pencil and colour on paper. Myron Bement Smith Collection, Freer Gallery of Art & Arthur M Sackler Gallery Archives, ref. A1169. Washington DC: Smithsonian Institution, (cited on pp.344, 587).

[845] Socolar JES & Taylor JM (2013) *Forcing nonperiodicity with a single tile.* The Mathematical Intelligencer, Mar2012, 34:1, pp.18–28, doi:10.1007/s00283-011-9255-y (cited on p.337).

[846] Sokoloff JB (1985) *Unusual band structure, wave functions and electrical conductance in crystals with incommensurate periodic potentials.* Phys. Rep. 126:4, p.189, cited by Kohmoto & S. [504] note 27, doi:10.1016/0370-1573(85)90088-2 (cited on p.383).

[847] Solarflare100 (2010) *Roger Penrose in the foyer of the Mitchell Institute Building at Texas A&M University.* Photograph: 14Mar2010, http://en.wikipedia.org/wiki/File:RogerPenroseTileTAMU2010.jpg accessed 3Nov2013, (cited on pp.iv, 311, 585).

[848] Somfai L (1996) *Béla Bartók: composition, concepts, and autograph sources.* The Ernest Bloch lectures. London: University of California Press, *'solid evidence'*, p.81, (cited on p.138).

[849] —Ibid. *arch form—symmetric—palindrome*, p.100, (cited on p.140).

[850] Soukoulis C (2013) *Photonic metamaterials: review, challenges and opportunities.* Masterclass, Physics@FOM Veldhoven 2013, 2h:23m:26s. *Secret garden ...* at 43m:20s, https://www.youtube.com/watch?v=EFVBZI2JNJQ accessed 13Nov2013, (cited on p.370).

[851] —Ibid. *Four regions epsilon and mu, plus and minus each,* at 0h:4m:37s, (cited on pp.369, 588).

[852] Spalding F (2012) *Mondrian and Nicholson: an artistic journey along parallel lines.* In The Guardian newspaper, Art & Design 3Feb2012, *Mondrian's train journey with Winifred Nicholson,* last para. (cited on p.94).

[853] Spinadel VW de (1998) *The metallic means family and multifractal spectra.* Nonlinear Analysis, 36:6, § 2, Eqn.2.2, p.724, doi:10.1016/S0362-546X(98)00123-0 (cited on p.429).

[854] ——. (1999) *The family of metallic means.* Visual Mathematics 1:3. Note that de Spinadel uses non-negated q, so her $q = 4$ is Lucas' $Q = -4$. http://www.mi.sanu.ac.rs/vismath/ spinadel/index.html accessed 27Apr2014, http://vixra.org/pdf/1403.0507v1.pdf accessed 27May2014, (cited on pp.252, 266, 274, 394, 429, 470, 470, 472, 479).

[855] —Ibid. *Bronze mean CF,* $p = 3$, $q = 1$, (Lucas: $P = 3$, $Q = -1$), §3, (cited on p.429).

[856] —Ibid. *Pisot numbers—natural candidates for coordinating quasicrystalline nodes (and Bragg peaks in related diffraction patterns),* §7, see Spinadel's references 24–27. Correspondences: $\phi = (1 + \sqrt{5})/2$ for 5- and 10-fold; $\delta = (1 + \sqrt{2})$ for 8-fold; and $(2 + \sqrt{3})$ for 12-fold; (cited on p.337).

[857] ——. (2011) *Excess continued fraction expansions.* Web article, *CF's for* ϕ^n, §III, p.4; $\phi^{2n} =$, §III, Eqn. 19, p.5, https: //archive.org/details/ExcessContinuedFractionExpansions accessed 26jan2015, (cited on pp.425, 426).

[858] Squibbs R (2011) *Aspects of compositional realization in Xenakis's pre-stochastic and early stochastic music.* Proceedings of the Xenakis Intl. Symposium Southbank Centre, London, 1– 3Apr2011, http://www.gold.ac.uk/media/03.1%20Ron%20Squibbs.pdf accessed 27Aug2014, (cited on p.130).

[859] Squire JS (2003) *Complex.* A Java® class for complex number arithmetic and trigonometric functions. http://www.csee.umbc.edu/ ~squire/math/Complex.java accessed 20Nov2013, (cited on p.134).

[860] Staffaroni M, Conway J, Vedantam S, Tang J, & Yablonovitch E (2012) *Circuit analysis in metal-optics.* Photonics and Nanostructures, Fundamentals and Applications 10, pp.166– 176, doi:10.1016/j.photonics.2011.12.002 (cited on p.397).

[861] Stakhov AP (2009) *Mathematics of harmony: from Euclid to contemporary mathematics and computer science.* Singapore: World Scientific, p.162, (cited on p.28).

[862] Stampfli P (1986) *A dodecagonal quasiperiodic lattice in two dimensions.* Helv. Phys. Acta 159, pp.1260–1263, (cited on p.390).

[863] Steinhardt PJ (2006) *Paul Steinhardt on impossible crystals.* Lecture at The Perimeter Institute 6Sep2006, video 0h:54m:16s, *had already studied 3d theoretical quasi-periodic structure,*

at 00h:28m:01s, http://www.youtube.com/watch?v=a0wo_yAh0Ps accessed 11Mar2014, (cited on p.333).

[864] —Ibid. *non-local interactions apparently required,* at 00h:31m:51s, (cited on p.334).

[865] —Ibid. *matching rules on vertices rather than on edges,* at 00h:39m:20s, (cited on p.334).

[866] Steinhardt PJ & Jeong HC (1996) *A simpler approach to Penrose tiling with implications for quasicrystal formation.* Nature 382, 1Aug1996, pp.431–433, doi:10.1038/382431a0 (cited on p.337).

[867] Steinhardt PJ, Jeong HC, Saitoh K, Tanaka M, Abe E, & Tsai AP (1998) *Experimental verification of the quasi-unit-cell model of quasicrystal structure.* Nature, 396:6706, 5Nov1998, pp.55–57, doi:10.1038/23902 (cited on p.337).

[868] Steinhardt PJ & Bindi L (2012) *In search of natural quasicrystals.* Reports on Progress in Physics 75:9, p.092601 doi:10.1088/0034-4885/75/9/092601, (cited on p.341).

[869] Steinlen TA (1896) *Tournée du Chat Noir avec Rodolphe Salis.* French poster: colour lithograph on paper. Height: 140.8 cm, width: 100 cm. Image source: BNF (Bibliothèque nationale de France), http://gallica.bnf.fr/ark:/12148/btv1b90140922/f1.highres accessed 26Oct2014, (cited on pp.153, 582).

[870] Sterckx P (2010) *Mondrian—Pompidou 2010.* Short video in French, 0h:07m:32s, (with jazz: Duke Ellington, 1929, *Black and Tan Fantasy,*) *'... un accord ...' (musical chord)* at 0h:01m:19s, http://www.youtube.com/watch?v=IF-anl6rz2c accessed 18Apr2014, (cited on p.106).

[871] Steurer W & Sutter-Widmer D (2007) *Photonic and phononic quasicrystals.* J. Phys. D: Appl. Phys. 40:13, p.R229, doi:10.1088/0022-3727/40/13/R01 (cited on pp.379, 383, 561).

[872] Stillwell JC (2006) *Yearning for the impossible: the surprising truths of mathematics.* Wellesley MA: AK Peters/CRC Press, p.4, (cited on p.182).

[873] Strang, WG (2005) *Lecture 21: Eigenvalues and Eigenvectors.* Course 18.06 Linear Algebra, MIT OpenCourseWare video 0h:51m:22s, Massachusetts Institute of Technology, https://www.youtube.com/watch?v=XM4GU8hPoZs accessed 30Jul2014, (cited on p.348).

[874]   —Ibid. *trace of a square matrix,* at 0h:14m:40s,   (cited on p.350).

[875]   —Ibid.   *how to find eigenvalues and eigenvectors,*   at 0h:16m:20s,

[876]   —Ibid. *'x = 0 is a useless eigenvector... if x ≠ 0, then matrix is singular,'* at 0h:17m:20s,   (cited on p.352).

[877]   Styer D (2000) *Hardy's test of quantum mechanics.* Oberlin College Physics Department, *φ-based result (for Hardy's probability),* Eqn.10 et seq., p.8,   http://www.oberlin.edu/physics/ dstyer/StrangeQM/Hardy.pdf   accessed 18Jun2013,   (cited on p.298).

[878]   Sun Y, Edwards B, Alù A, & Engheta N (2012) *Experimental realization of optical lumped nanocircuits at infrared wavelengths.* Nature Materials 11, pp.208–212, doi:10.1038/nmat3230   (cited on p.397).

[879]   Sunagawa I (1987) *Morphology of Crystals.* Tokyo: Terra Scientific Publishing, p.x,   (cited on p.342).

[880]   Szabolcsi B, with collaborators: Kodály Z, Lendvai E, Szöllösy A, & Demény J (1956) *Bartok sa vie et son oeuvre—1ère édition.* Budapest: Corvina,   (cited on pp.63, 140).

[881]   ———. (1968) *Bartok sa vie et son oeuvre—2ème édition.* Paris: Boosey & Hawkes, *nautilus ... Jules Verne,* p.135, & Ex.51 *spiral curve,* p.136,   (cited on pp.140, 141, 582).

[882]   Szabolcsi B (1957) *Das Leben Béla Bartóks* in: *Béla Bartók. Weg und Werk, Schriften und Briefe.* Comp. Bence Szabolcsi, Budapest: Corvina, p.45, cited by Ferenc Bónis in *Comments on Béla Bartók's working method in dealing with proofs for his Violin Concerto (1937–1938),* p.1,   http://memory.loc.gov/ammem/ collections/moldenhauer/2428109.pdf   accessed 6Jan2014,   (cited on p.140).

[883]   Tallián T (1995) *Bartók's reception in America, 1940–1945.* Transl. Laki P, essay in *Bartók and his world.* Ed. Laki P, Princeton: Princeton University Press, p.107,   (cited on pp.113, 155).

[884]   Tanabe T, Kameoka S, & Tsai AP (2010) *Microstructure of leached Al-Cu-Fe quasicrystal with high catalytic performance for steam reforming of methanol.* Appl. Catalysis A: Gen. 384:1– 2, 20Aug2010, pp.241–251,   doi:10.1016/j.apcata.2010.06.045   (cited on p.341).

[885] Tattersall JJ (1999) *Elementary Number Theory in Nine Chapters.* Cambridge: University Press, *Wantzel ... digital root ... unity,* p.129, (cited on pp.268, 280).

[886] —Ibid. *units digits of the Fibonacci numbers ... sequence with period 60,* p.26, (cited on p.248).

[887] Tavalaee AA, Hon PWC et al. (2010) *Zero-index terahertz quantum-cascade metamaterial lasers.* Quantum Electronics, IEEE Jnl. of, 46:7, doi:10.1109/JQE.2010.2043642 (cited on p.375).

[888] Taylor MC (1992) *Disfiguring: Art, Architecture, Religion.* Chicago: University of Chicago Press, *(deep red) anger, sensuality ... (deep blue) devotion, and (pale azure) union with the divine,* p.67, (cited on pp.112, 492).

[889] Tenniel J (1866) *Illustrations for Alice's adventures in wonderland* by Lewis Carroll [100], and for *Through the looking-glass and what Alice found there,* (cited on pp.395, 589).

[890] Tepapa Tongarewa (undated) *The eye of the colossal squid—the largest animal eye known.* Web article by the Museum of New Zealand Te Papa Tongarewa, http://squid.tepapa.govt.nz/anatomy/article/the-eye-of-the-colossal-squid accessed 16Dec2014, (cited on p.81).

[891] Terraneo M, Peyrard M, & Casati G (2002) *Controlling the energy flow in nonlinear lattices: a model for a thermal rectifier.* Phys. Rev. Lett. 88:9, 094302, 4Mar2002, cited by [532], doi:10.1103/PhysRevLett.88.094302 (cited on p.392).

[892] Thomas KD (2013) *From prime numbers to nuclear physics and beyond.* Inst. for Adv. Study Letter, Spring 2013, www.ias.edu accessed 29Mar2015, (cited on p.334).

[893] Thompson DW (1942) *On growth and form.* Cambridge: University Press, pp.922–925, (cited on pp.19, 58, 492, 539).

[894] Thompson N (undated) *Hawaiian voyaging traditions: on wayfinding.* Polynesian Voyaging Society, 32-direction navigation, http://pvs.kcc.hawaii.edu/ike/hookele/on_wayfinding.html accessed 11Dec2013, (cited on p.162).

[895] Thomson R (1985) *Seurat.* Oxford: Phaidon Press. (Referring to Seurat's sheet of diagrams and notes for *Parade*)—*... planned its geometrical structure with reference to Henry's ideas ... one of the diagrams relates to the Golden Section,* p.152, (cited on p.150).

[896] —Ibid. *Superville's theories, as discussed and developed by Blanc and Henry*, p.152, (cited on p.82).

[897] Thue A (1912) *Über eine Eigenshaft, die keine transcendente Grösse haben kann.* Skrifter utg. av Videnskapsselskapet i Kristiania., 1. Mat.-naturv. Kl., 20, (cited on p.72).

[898] Tobias R (2006) *Musical keyboard with 3 octaves, 21 white and 15 black keys, 36 keys total.* Diagram, http://commons.wikimedia.org/wiki/ File:Klaviatur-3-en.svg accessed 13May2014, (cited on pp.107, 581).

[899] Tognetti K (c.1999) *Fibonacci—his rabbits and his numbers and Kepler.* School of Mathematics and Applied Statistics University of Wollongong, Australia. Australian Mathematical Society Web Site, p.11, cited by Sloane in A000045 [839], http://www.austms.org. au/Modules/Fib/fib.pdf accessed 9Jul2013, (cited on p.444).

[900] —Ibid. *conics*, p.4, (cited on p.23).

[901] Torres-Silva H & Cabezas DT (2103) *Chiral seismic attenuation with acoustic metamaterials.* Journal of Electromagnetic Analysis and Applications, 5, pp.10–15, doi:10.4236/jemaa.2013.51003 (cited on p.381).

[902] Tovstra (2006) *half kite and half dart, initial and first generation deflation.* (Four .svg image files), http://en.wikipedia.org/wiki/Penrose_ tiling#CITEREFGardner1997 accessed 30May2014, (cited on pp.315, 315, 512, 586, 586).

[903] Trebin H-R Ed., Feuerbacher M, et al. (2006) *Quasicrystals: structure and physical properties.* Wiley online library, © 2003, Wiley-VCH Verlag GmbH & Co. KGaA, p.437, doi:10.1002/3527606572.ch5 (cited on p.339).

[904] Trevett S (2012) *Polypeptide alpha helix diagram.* © Institute of Physics, IOP Publishing http://stevetrevett.co.uk/?works= biological-physics-iop-publishing accessed 21Oct2013, (cited on pp.291, 585).

[905] Turing Digital Archive (2013) *Turing papers collection.* Archive Centre, King's College, Cambridge. *Fibonacci & phyllotaxis* in Folder AMT/C/25, http://www.turingarchive.org/browse.php/C/25 accessed 27Nov2013, (cited on p.58).

[906] Turnbull D (1949) *The subcooling of liquid metals.* J. Appl. Phys. 20, p.817, cited by [461], doi:10.1063/1.1698534 (cited on p.307).

[907] Turner JC (2003) *The Goldpoint snowflake.* In *Some fractals in Goldpoint geometry.* Fib. Qtrly. 41:1, p.65, (cited on p.235).

[908] Unknown artist (1877) *Engraving of a giant squid, beached in Trinity Bay, Newfoundland.* Source: Wikimedia, (cited on pp.81, 580).

[909] Urban KW (1998) *Quasicrystals: from tilings to coverings.* Nature 396, 5Nov1998, p.14, doi:10.1038/23806 (cited on p.335).

[910] Vaidyanathan G & Garvey JF (1996) *Magic numbers, reactivity, and ionization mechanisms in $Ar_n X_m$ heteroclusters.* In *Chemical reactions in clusters.* Ed. Bernstein ER, New York: Oxford University Press Inc., p.231, (cited on p.300).

[911] Vajda S (1989) *Fibonacci and Lucas numbers, and the Golden Section: theory and applications.* Chichester: Ellis Horwood. Republished 2008, New York: Dover, *Eqn. 2,* $F_{-n} = (-1)^{n+1} F_n$, p.10, (cited on pp.446, 529).

[912] —Ibid. *Eqn. 4,* $L_{-n} = (-1)^n L_n$, p.10, (cited on p.529).

[913] —Ibid. *Study of Mersenne numbers,* case (iii), 'another example', p.20; and note Vajda does not use 'negated Q', so his coefficients are $a_1 = 4$ and $a_2 = -1$, (cited on p.284). (cited on pp.284, 448).

[914] —Ibid. *Eqn. 5,* $L_{n-1} + L_{n+1} = 5F_n$, p.24, (cited on p.529).

[915] —Ibid. *Eqn. 17c,* $L_{2n} + 2(-1)^n = L_n^2$, p.27, (cited on pp.427, 530).

[916] —Ibid. *Eqn. 23,* $L_{2n} - 2(-1)^n = 5F_n^2$, p.29, (cited on p.427).

[917] Vakil R (1996) *A mathematical mosaic—patterns and problem solving.* Burlington, Ontario: Brendan Kelly, *The Ailles Rectangle,* p.87, (cited on p.405).

[918] Van Hout N (2009) *The unfinished, the uncompleted and the birth of Non-Finito.* In Vervoordt (ed.) *In-Finitum.* Wijnegem Belgium: Vervoordt Foundation, p.35, cited by Paite L in *A wounded surface.* Thesis, University of Cape Town, http:// uctscholar.uct.ac.za/PDF/43038_Palte,%20L.pdf accessed 20May2014, (cited on p.90).

[919] Vardeny ZV, Nahata A, & Agrawal A (2013) *Optics of photonic quasi-crystals.* Nature Photonics 7, 27Feb2013, Fig.1, *Non-resonant 1D Fibonacci quasicrystalline structure (a),* p.178, doi:10.1038/nphoton.2012.343 http://www.nature.com/nphoton/journal/v7/n3/pdf/ nphoton.2012.343.pdf accessed 3Sep2013, (cited on pp.389, 389, 393, 543, 548, 589).

[920] —Ibid. *not been widely recognized,* p.177, (cited on p.391).

[921] Veda Vyāsa (c.500–c.200 BC) *Bhagavad Gītā.* (Being part of the Bhīṣma-Parva section of the Mahābhārata.) Translation and commentary Chapters 1–6 by Maharishi Mahesh Yogi (1967). London: Penguin, Ch.2, v.48, p.135. The version quoted in this book's *Dedication* is MMY's own paraphrase, (cited on p.v).

[922] Vega FC de la (1999) *La enseñanza de la historia : Bolivia.* Bogota. Columbia: Convenio Andrés Bello, p.44, (cited on p.8).

[923] Velankar HD (1962) *Vṛttajatisamuchchaya of Kavi Virahāṅka.* Jodhpur: Rajasthan Oriental Research Institute, Intro. xxi–xxv & p.101, cited by Singh [836], (cited on pp.17, 18).

[924] Velter A (1996) *Les Poètes du Chat Noir.* Paris: Gallimard, p.14, (cited on pp.149, 152).

[925] Verba E (2012) *The Golden Ratio in time-based media.* Jnl. of Arts & Humanities, 1:1, p.59, http://www.theartsjournal.org/index.php/site/article/view/6/6 accessed 7Apr2014, (cited on p.94).

[926] Verne J (1870) *Twenty thousand leagues under the sea: an underwater tour of the world.* Series: *Voyages Extraordinaires.* Paris: Hetzel PJ *Nautilus submarine attacked by giant squid,* Ch.XVII, (cited on p.82).

[927] Veselago V (1968) *The electrodynamics of substances with simultaneously negative values of ϵ and μ.* (Russian text 1967). Sov. Phys. Usp. 10:4, pp.509–514, http://www.turpion.org/php/paper.phtml?journal_id=pu&paper_id=3699 accessed 18Sep2013, doi:10.1070/PU1968v010n04ABEH003699 (cited on p.368).

[928] Vijayaraghavan T (1941) *On the fractional parts of the powers of a number. II.* Proc. Cambridge Philos. Soc. 37, pp.349–357, (cited on p.72).

[929] Vinayasagar M (1965) *Vṛtta Mauktika of Chandrasekhara Bhatta.* Jodhpur: Rajasthan Oriental Research Institute, cited by Singh [838], (cited on p.12).

[930] Vogel H (1979) *A better way to construct the sunflower head.* Mathematical Biosciences, 44, pp.179–189, doi:10.1016/0025-5564(79)90080-4 (cited on pp.56, 482, 585).

[931] Wall DD (1960) *Fibonacci series modulo m*. American Mathematical Monthly, (MAA), 67:6, pp.525–532, doi:10.2307/2309169 (cited on p.416).

[932] Walser H (1996) *Der Goldene Schnitt*. Stuttgart: BG Teubner, (MAA English transl. [934]), also Walser [933], (cited on p.570).

[933] ———. (2013) *Der Goldene Schnitt*. 6th edition, Leipzig: Edition am Gutenbergplatz, (cited on p.570).

[934] ———. (2001) *The Golden Section*. Transl. from original German, Walser [932] by Hilton P & Pederson J, Washington DC: Mathematical Association of America, *constructions*, §3.1, p.23; *from sequence*, §5.5, p.81; *from quadratic*, §5.4, p.78; note that Walser uses non-negated *q*, (cited on pp.2, 253, 570).

[935] —Ibid. *round trip: linearizing powers of quadratic roots produces GFS; limit of quotient series of GFS leads back to quadratic*, §5.5, p.82, (cited on p.255).

[936] Wang H (1961) *Proving theorems by pattern recognition, II*. Bell Systems Tech. J. 40, pp.1–41, doi:10.1002/j.1538-7305.1961.tb03975.x (cited on p.310).

[937] Wang N, Chen H, & Kuo KH (1987) *Two-dimensional quasicrystal with eightfold rotational symmetry*. Phys. Rev. Lett. 59, pp.1010–1013, doi:10.1103/PhysRevLett.59.1010 cited by Suck JB et al. in (2002) *Quasicrystals: an introduction to structure, physical properties and applications*. Berlin-Heidelberg: Springer-Verlag, (cited on p.337).

[938] Ward AJ & Pendry JB (1996) *Refraction and geometry in Maxwell's equations*. J. Mod. Opt. 43:4, pp.773–793, doi:10.1080/09500349608232782 (cited on p.371).

[939] Wegener M (2013) *Plenary talk PW13: Three-dimensional metamaterials made by direct laser writing*. SPIE Photonics West, video 44m:20s. Rhombicuboctahedral QC at 5m:53s, https://www.youtube.com/watch?v=CQE9VnjQF-c accessed 24Nov2013, (cited on p.533).

[940] Wells D (1986) *The Penguin dictionary of curious and interesting numbers*. Middx., England: Penguin Books, cited by Weisstein [968] p.62, revised edn. (1997) p.43, (cited on p.572).

[941] Welsh RP (1998) *Piet Mondrian: Volume 1: Catalogue Raisonné of the naturalistic works (until early 1911)*. New York: Harry N Abrams, (Vol.2 by Joosten [426]), (cited on p.524).

[942] Wenk A (1975) *Claude Debussy and the poets*. Berkeley: Univ. California Press, p.277, quoting Debussy: 'Soutenons que la beauté d'une œuvre d'art restera toujours mystérieuse, c'est à-dire qu'on ne pourra jamais exactement vérifier «comment cela est fait.» Conservons, à tout prix, cette magie particulière à la musique. Par son essence, elle est plus susceptible d'en contenir que tout autre art.' from *Monsieur Croche et autres écrits*, c.1901–c.1915, ed. François Lesure, Paris: Gallimard, 1971, p.224, (cited on pp.65, 130, 151).

[943] Weisstein EW (1999–2014) *Binet's Fibonacci number formula— Binet JPM (1843)*. From MathWorld—a Wolfram Web Resource, http://mathworld.wolfram.com/BinetsFibonacciNumberFormula.html accessed 9Apr2013, (cited on p.25).

[944] —Ibid. *CF for Silver Ratio,* identities (1) & (2), http://mathworld.wolfram.com/SilverRatio.html accessed 12Aug2014, (cited on p.429).

[945] —Ibid. *Cosecant*$(\pi/10) = 2\phi$, identity (11), http://mathworld.wolfram.com/GoldenRatio.html accessed 4Jun2014, (cited on pp.16, 483).

[946] —Ibid. 2×2 *determinant defined,* Eqn.6, http://mathworld.wolfram.com/Determinant.html accessed 10Aug2014, (cited on p.485).

[947] —Ibid. *Determinant—expand with 'minors',* Eqn.7, http://mathworld.wolfram.com/Determinant.html accessed 10Aug2014, (cited on p.486).

[948] —Ibid. *Double angle formulæ*: $sin(2x) = 2sin(x)cos(x)$ and $tan(2x) = 2tan(x)/(1 - tan^2(x))$, http://mathworld.wolfram.com/Double-AngleFormulas.html accessed 4Jun2014, (cited on pp.432, 483, 483).

[949] —Ibid. *'Eadem mutata resurgo',* (Bernoulli), http://mathworld.wolfram.com/LogarithmicSpiral.html accessed 30May2013, (cited on p.39).

[950] —Ibid. *Fermat number,* http://mathworld.wolfram.com/FermatNumber.html accessed 1Jun2013, (cited on p.277).

[951] —Ibid. *Gnomon,* http://mathworld.wolfram.com/Gnomon.html accessed 28Nov2014, (cited on p.104).

[952] —Ibid. *Golden Ratio,* τ notation/tome, http://mathworld.wolfram.com/GoldenRatio.html accessed 12Aug2014, (cited on p.33).

[953] —Ibid. *Half angle formula,* http://mathworld.wolfram.com/Half-AngleFormulas.html accessed 28Sep2013, (cited on pp.404, 483).

[954] —Ibid. *Jacobsthal number,* http://mathworld.wolfram.com/ JacobsthalNumber.html accessed 1Jun2013, (cited on p.275).

[955] —Ibid. *Lamé's theorem,* http://mathworld.wolfram.com/LamesTheorem. html accessed 13Aug2014, (cited on p.26).

[956] —Ibid. *Lucas number, conjugation relation* $L_n = F_{n-1} + F_{n+1}$ Eqn.(11), http://mathworld.wolfram.com/LucasNumber.html accessed 23Apr2013, (cited on pp.199, 444).

[957] —Ibid. *Linear transformation represented as matrix equation,* Eqn.5, http://mathworld.wolfram.com/Matrix.html accessed 10Aug2014, (cited on p.485).

[958] —Ibid. *Lucas cannonball pyramid stacking,* http://mathworld. wolfram.com/CannonballProblem.html accessed 20Apr2014, (cited on p.299).

[959] —Ibid. *Pell number,* http://mathworld.wolfram.com/PellNumber.html accessed 28Feb2014, (cited on p.268).

[960] —Ibid. *Penrose triangle,* http://mathworld.wolfram.com/PenroseTriangle. html accessed 27Jul2014, (cited on p.310).

[961] —Ibid. *Pentagonal number,* http://mathworld.wolfram.com/ PentagonalNumber.html accessed 6Apr2014, (cited on p.34).

[962] —Ibid. *Perfect number,* http://mathworld.wolfram.com/PerfectNumber.html accessed 18aug2014, (cited on p.282).

[963] —Ibid. *Perfect number, sum of reciprocals of all the divisors,* Eqns. 22–24, http://mathworld.wolfram.com/PerfectNumber.html accessed 18aug2014, (cited on p.280).

[964] —Ibid. *Pisot number,* http://mathworld.wolfram.com/PisotNumber.html accessed 24Nov2014, (cited on pp.72, 457).

[965] —Ibid. *Platonic solid,* http://mathworld.wolfram.com/PlatonicSolid.html accessed 29Apr2013, (cited on p.14).

[966] —Ibid. *Quaternion,* http://mathworld.wolfram.com/Quaternion.html accessed 8Jan2015, (cited on p.361).

[967] —Ibid. *Rabbit sequence,* http://mathworld.wolfram.com/RabbitSequence. html accessed 8Sep2014, (cited on p.238).

[968] —Ibid. *Simson's proof 1753,* Weisstein cites Wells [940], http: //mathworld.wolfram.com/FibonacciNumber.html accessed 3Feb2014, (cited on pp.24, 570).

[969] —Ibid. *Trigonometric addition formulas, sine difference,* Identity (2), http://mathworld.wolfram.com/TrigonometricAdditionFormulas.html accessed 28Sep2013, (cited on pp.404, 483).

[970] —Ibid. *Trigonometry angles—pi/16,* with chained footnote links to *pi/8, pi/4 and pi/2,* http://mathworld.wolfram.com/ TrigonometryAnglesPi16.html accessed 22May2013, (cited on pp.402, 403).

[971] —Ibid. *Twin Primes,* http://mathworld.wolfram.com/TwinPrimes.html accessed 4May2014, (cited on p.116).

[972] Wersin W von (1956) *Das Buch vom Rechteck Gesetz und Gestik des Räumlichen.* (The book of rectangles, spatial law and gestures of the orthogons described), Ravensburg: Otto Maier Verlag Publishers, *Die Orthogon-Scheibe.* The (1558) 7 rectangles are 1:1, 1:2, 2:3, 3:4, 3:5, 4:5, & 1:$\sqrt{2}$, p.34, (cited on p.69).

[973] —Ibid. *Orthogon names and ratios:* quadrat 1, hemidiagon $\sqrt{5}/2$=1.118, trion $2/\sqrt{3}$=1.1547, quadriagon $(1+\sqrt{2})/2$=$\delta/2$=1.2071, biauron $4/(1+\sqrt{5})$=$2/\phi$=1.236, penton $\sqrt{6+2\sqrt{5}}$ / $\sqrt{10-2\sqrt{5}}$=1.3764, diagon $\sqrt{2}$=1.414, bipenton $2\sqrt{10-2\sqrt{5}}/\sqrt{6+2\sqrt{5}}$=1.453, hemiolion $3/2$=1.5, auron ϕ=1.618, sixton $\sqrt{3}$=1.732, doppelquadrat $\sqrt{4}$=2, p.39, p.41, (cited on p.69).

[974] West TG (1991) *In the mind's eye: visual thinkers, gifted people with dyslexia and other learning difficulties, computer images and the ironies of creativity.* (Updated edn. 1997), Amherst NY: Prometheus Books, Ch.1, (cited on p.361).

[975] White M (2003) *De Stijl and Dutch Modernism.* Manchester University Press, *influence (on Mondrian) of Mathieu Schoenmaekers,* p.24, (cited on pp.71, 106).

[976] —Ibid. *Van Doesburg... wrote to Kok,* p.24, (cited on p.106).

[977] —Ibid. *vocabulary of Neo-plasticism,* p.24, (cited on p.106).

[978] Whitten J & Goodeve M (2012) *Teachers' resource—Mondrian, Nicholson: in parallel.* Illustrated collection of essays. The Courtauld Gallery, p.21, http://www.courtauld.ac.uk/publicprogrammes/ documents/MondrianNicholsonteachersweb.pdf accessed 15Apr2014, (cited on p.144).

[979] Williams R (2013) *Sir John Pendry, pioneer of the 'invisibility cloak', wins top Institute of Physics honour.* The Independent newspaper article 1Jul2013, (cited on p.374).

[980] 'Willka' (Álvaro Rodrigo Zarate Huayta PhD.) (2013) *The lost calendar of the Andes: decoding the Tiwanaku Calendar and of the Muisca culture.* Harvard Anthropology and Archaeology School, Boston MA: Harvard Press, *calendar wall,* p.23, (cited on p.117).

[981] Williamson JH (2014) *Number spirals.* Dept. Computing Science, Univ. Glasgow, website article, section on *Vogel spiral,* http://www.dcs.gla.ac.uk/~jhw/spirals/index.html accessed 18Jan2014, (cited on p.58).

[982] —Ibid. *Eleven Chasqui icons,* p25, (cited on p.117).

[983] Wilson S (1991) *Tate Gallery: an illustrated companion.* Revised edition. London: Tate Gallery, *new plastic idea cannot ... take the form of ... natural representation,* p.144, http://www.tate.org. uk/art/artworks/mondrian-no-vi--composition-noii-t00915/text-illustrated-companion accessed 5Apr2014, (cited on p.90).

[984] Wiscombe G (2007) *Stonehenge.* Photograph, http://en.wikipedia. org/wiki/File:Stonehenge2007_07_30.jpg accessed 10May2014, (cited on pp.121, 581).

[985] Wolfram S (2003) *A new kind of science.* H Paul Rockwood Memorial Lecture, Institute for Neural Computation, video 01h:26m:42s, California: UCSD and UCSD TV, http://www.youtube. com/watch?v=_eC14GonZnU accessed 28May2014, (cited on p.304).

[986] Woltman G & Kurowski S (1996–) *Great Internet Mersenne Prime Search—GIMPS.* 'Web content, data & software downloads at Mersenne.org are owned and operated by Mersenne Research, Inc., a non-profit corporation.' http://www.mersenne.org accessed 6Feb2014, (cited on p.450).

[987] —Ibid. *Curtis Cooper discovers 48th Mersenne prime,* http://www. mersenne.org/primes/?press=M57885161 accessed 14Sep2014, (cited on pp.34, 450).

[988] Wood H (2014) *World Octopus Day. (Biography of William Evans Hoyle).* Website article. Cardiff: National Museum of Wales, http://www.museumwales.ac.uk/blog/2014-10-08/World-Octopus-Day accessed 26Dec2014, (cited on pp.82, 150).

[989] Wood JB (1995–) *The Cephalopod Page.* Dedicated website, http://www.thecephalopodpage.org/ accessed 8Jan2015, (cited on p.374).

[990] Wunderlich FJ, Shaw DE, & Hones MJ (1974) *Argand diagrams of extended Fibonacci and Lucas numbers.* Fibonacci Quarterly 12:3, pp.233–234, cited by Knott [481], http://www.fq.math.ca/Scanned/ 12-3/wunderlich.pdf accessed 21Nov2013, (cited on p.136).

[991] Xavier J, Probst J, et al. (2014) *Quasicrystalline-structured light harvesting nanophotonic silicon films on nanoimprinted glass for ultra-thin photovoltaics.* Optical Materials Express, 4:11, pp.2290–2299,   doi:10.1364/OME.4.002290   (cited on p.394).

[992] Xenakis I (1953–1954) *Metastaseis (B)* for orchestra of 61 performers—each playing a different part, 0h:07m:00s. London, New York: Boosey & Hawkes,   (cited on pp.130, 132, 151).

[993] ———. (1971) *Formalized music—thought and mathematics in composition.* Bloomington, London: Indiana Univ. Press. Rev'd edn. (1992) Stuyvesant NY: Pendragon Press, *String glissandi, bars 309–314 of Metastaseis,* Fig.1.2, p.5,   (cited on pp.131, 132, 582).

[994] Yablonovitch E (2013) *Photonic crystals in science, engineering and nature.* Technion lecture video 0h:20m:56s, at 0h:04m:54s http://www.youtube.com/watch?v=VlyQOns__cw   accessed 6Oct2013,   (cited on pp.367, 501, 555).

[995] Yan Y & Pennycook SJ (2000) *Structural model for the $Al_{72}Ni_{20}Co_8$ decagonal quasicrystals.* Phys. Rev. B, 61:21, pp.14291–14294,   doi:10.1103/PhysRevB.61.14291   (cited on p.337).

[996] Yokota K, Yamamoto T, Koashi M, & Imoto N (2009) *Direct observation of Hardy's paradox by joint weak measurement with an entangled photon pair.* New Journal of Physics, 4Mar2009, 11:033011 (9pp),   doi:10.1088/1367-2630/11/3/033011   (cited on p.298).

[997] Youmans EL & WJ (1880) *The Popular Science Monthly.* XVII May–Oct1880, NY: D Appleton. *Frontispiece,*   (cited on pp.361, 588).

[998] Yu D, Xue D, & Ratajczak H (2006) *Golden Ratio and bond-valence parameters of hydrogen bonds of hydrated borates.* Jnl. Mol. Struct. 783:1–3, 6Feb2006, pp.210–214, doi:10.1016/j.molstruc.2005.08.022 (cited on p.308).

[999] Yves Saint Laurent (1965) *'Mondrian' day dress, autumn 1965.* Wool jersey in colour blocks of white, red, blue, black, and yellow, (C.I.69.23). New York: The Metropolitan Museum of Art, (cited on p.97).

[1000] Zeising A (1854) *Neue Lehre von den Proportionen des menschlichen Körpers.* Leipzig, Germany: R Weigel, cited by McManus [640],   (cited on p.67).

[1001] Zeng X, Ungar G, Liu Y, Percec V, et al. (2004) *Supramolecular dendritic liquid quasicrystals.* 428, pp.157–160, cited by Lifshitz [559].   doi:10.1038/nature02368 (cited on p.341).

[1002] Zimmer B (2012) *Tracking down the roots of a 'Super' word.* Website article,   http://www.visualthesaurus.com/cm/wordroutes/tracking-down-the-roots-of-a-super-word/ accessed 26Feb2014, (cited on p.139).

[1003] Zolotoyabko E (2011) *Basic concepts of crystallography.* Weinheim: Wiley-VCH, *1784, René Haüy,* p.5, (cited on p.323).

[1004] Zong C (2013) *What is the Leech lattice?* Notices of the AMS Oct2013, 60:9, p.1169,   http://www.ams.org/notices/201309/rnoti-p1168.pdf accessed 19Mar2014, (cited on p.304).

About the author

© 2011 Trevor Palin [730].

Clive N. Menhinick is a computer scientist, geometer, and research photographer, based in Cheshire, England. In 2005, he recognized that 'state-of-the-art' developments in DSLR camera technology—particularly the vast increase in colour resolution ('bit depth')—opened up a new frontier for research in natural science. Over a ten-year period he has conducted his own pioneering project, developing digital-photographic techniques and custom software. In his analysis he found that many results appeared to depend on a simple angle-based geometry that he named 'Ori32'. It was the study of this geometry that provided a clear path to the discovery of the Fibonacci Resonance.

Clive read physics and mathematics at Liverpool University. He learnt Transcendental Meditation® while at university, and has practised TM for the past 40 years.

List of Figures

Index

INDEX

Lightning Source UK Ltd.
Milton Keynes UK
UKOW06n1939110216

268095UK00010B/39/P